Frontiers in Mathematics

Frontiers in Elliptic and Parabolic Problems

Series Editor

Michel Chipot, Institute of Mathematics, Zürich, Switzerland

The goal of this series is to reflect the impressive and ongoing evolution of dealing with initial and boundary value problems in elliptic and parabolic PDEs. Recent developments include fully nonlinear elliptic equations, viscosity solutions, maximal regularity, and applications in finance, fluid mechanics, and biology, to name a few. Many very classical notions have been revisited, such as degree theory or Sobolev spaces. Books in this series present the state of the art keeping applications in mind wherever possible.

The series is curated by the Series Editor.

Henrique Fernandes de Lima •
Giovanni Molica Bisci •
Marco Antonio Lázaro Velásquez

Immersions in Warped Product Spaces

Henrique Fernandes de Lima
Department of Mathematics
Federal University of Campina Grande
Campina Grande, Brazil

Marco Antonio Lázaro Velásquez
Department of Mathematics
Federal University of Campina Grande
Campina Grande, Brazil

Giovanni Molica Bisci
Department of Human Sciences and Quality of Life Promotion
San Raffaele University of Rome
Rome, Italy

ISSN 1660-8046   ISSN 1660-8054 (electronic)
Frontiers in Mathematics
ISSN 2730-549X   ISSN 2730-5503 (electronic)
Frontiers in Elliptic and Parabolic Problems
ISBN 978-3-031-78041-7   ISBN 978-3-031-78042-4 (eBook)
https://doi.org/10.1007/978-3-031-78042-4

Mathematics Subject Classification: 53B30, 53E10, 58J90, 53A10, 53C50, 58C40, 58J05, 53C20, 53C42

© The Editor(s) (if applicable) and The Author(s), under exclusive license to Springer Nature Switzerland AG 2025

This work is subject to copyright. All rights are solely and exclusively licensed by the Publisher, whether the whole or part of the material is concerned, specifically the rights of translation, reprinting, reuse of illustrations, recitation, broadcasting, reproduction on microfilms or in any other physical way, and transmission or information storage and retrieval, electronic adaptation, computer software, or by similar or dissimilar methodology now known or hereafter developed.
The use of general descriptive names, registered names, trademarks, service marks, etc. in this publication does not imply, even in the absence of a specific statement, that such names are exempt from the relevant protective laws and regulations and therefore free for general use.
The publisher, the authors and the editors are safe to assume that the advice and information in this book are believed to be true and accurate at the date of publication. Neither the publisher nor the authors or the editors give a warranty, expressed or implied, with respect to the material contained herein or for any errors or omissions that may have been made. The publisher remains neutral with regard to jurisdictional claims in published maps and institutional affiliations.

This book is published under the imprint Birkhäuser, www.birkhauser-science.com by the registered company Springer Nature Switzerland AG
The registered company address is: Gewerbestrasse 11, 6330 Cham, Switzerland

If disposing of this product, please recycle the paper.

"One, remember to look up at the stars and not down at your feet.

Two, never give up work. Work gives you meaning and purpose and life is empty without it.

Three, if you are lucky enough to find love, remember it is there and don't throw it away."

Stephen William Hawking
Interview with ABC's Diane Sawyer,
June 2010.

Dedicated to our Families
The Authors

Preface

This book is an interplay between Mathematical Analysis and Global Differential Geometry. More precisely, the main purpose is to investigate intrinsic geometrical properties of Warped Product Spaces through powerful analytic tools that are typically established in order to study the qualitative behavior of solutions to certain classes of elliptic problems arising in a Semi-Riemannian scenario.

Remarkable advances have been made in the last years in this field also in connection with General Relativity and there are now many results that are well established. Moreover, several surprising phenomena appear when studying Semi-Riemannian Warped Products emphasizing once again that the naive idea that what is valid in the Euclidean setting must be valid for manifolds is in general false.

The presentation is intended to be as self-contained as possible in order to make the book accessible to a large audience, including postgraduate students and specialists of other fields. To this reason, in many cases we prefer to state results under slightly stronger assumptions, when this makes the exposition much more clear and avoids some unnecessary technicalities.

The monograph is divided into five parts in which we present sharp refinements of some results obtained by ourselves or through direct cooperation with other mathematicians and already present in the current literature. After a first part containing Uniqueness Results, Height Estimates, and Half-Space Theorems, the book deals with Riemannian Immersions, geometrical properties of Submanifolds, Local Rigidity and Bifurcation Analysis in Semi-Riemannian Warped Products. More precisely:

Part I—from Chaps. 1 to 4—Uniqueness Results, Height Estimates and Half-Space Theorems.

Part II—from Chaps. 5 to 8—Riemannian Immersions in Weighted Semi-Riemannian Warped Products

Part III—from Chaps. 9 to 12—Submanifolds Immersed in Semi-Riemannian Warped Products

Part IV—from Chaps. 13 to 16—Stability of Riemannian Immersions in Semi-Riemannian Warped Products

Part V—from Chaps. 17 to 20—Local Rigidity and Bifurcation of Riemannian Immersions in Semi-Riemannian Warped Products

Each chapter contains a brief introduction, as well as the necessary background that will be used along it. Moreover, in order to give a direct presentation to the treated subject, along any chapter some repetitions, concerning definitions and notations, will be admitted. This choice sacrifices elegance in reason of clearness.

Finally, as customary, the bibliography does not escape the usual rule, being incomplete. Indeed, we have listed only papers that are closer to the topics discussed in this book. However, we are afraid that even for these arguments, the references are far from being exhaustive. We apologize for possible omissions.

This monograph would never have been written without the support of Professor Michel Chipot. With great pleasure we thank him and the other friends and colleagues who encourage us during the preparation of this volume.

A special thanks go to Dr. Frida Trotter from Birkhäuser for her constant and kindest help at all stages of the editing process.

Campina Grande, Brazil	Henrique Fernandes de Lima
Rome, Italy	Giovanni Molica Bisci
Campina Grande, Brazil	Marco Antonio Lázaro Velásquez
October 2024	

Contents

Part I Uniqueness Results, Height Estimates and Half-Space Theorems for Hypersurfaces in Semi-Riemannian Warped Products

1 Riemannian Immersions 3
   1.1   Introduction 3
   1.2   Higher Order Mean Curvatures of a Riemannian Immersion 3
   1.3   Newton Transformations and L_r Operators 5
   1.4   Semi-Riemannian Warped Products 8
   1.5   Height and Angle Functions of a Riemannian Immersion 9
   1.6   On the Omori-Yau's Generalized Maximum Principle 11

2 Uniqueness of Complete Hypersurfaces 15
   2.1   Introduction 15
   2.2   Uniqueness of Riemannian Immersions in Semi-Riemannian Warped Products 17
   2.3   A Further Rigidity Result in Spatially Closed GRW Spacetimes 25
   2.4   Rigidity of Hypersurfaces in Riemannian Warped Products and Pseudo-Hyperbolic Spaces 30
   2.5   Rigidity Results via Higher Order Mean Curvatures 39
   2.6   Further Uniqueness Results via Higher Order Mean Curvatures 42
       2.6.1   Rigidity of Complete Spacelike Hypersurfaces 42
       2.6.2   Rigidity of Complete Two-Sided Hypersurfaces 49

3 Height Estimates and Half-Space Theorems 53
   3.1   Introduction 53
   3.2   Auxiliary Results and Setup 55
   3.3   Height Estimates in Riemannian Warped Products 60
   3.4   Half-Space Theorems and Topology at Infinity in Riemannian Warped Products 75
   3.5   Height Estimates in GRW Spacetimes 92
   3.6   Half-Space Theorems and Topology at Infinity in GRW Spacetimes 104

4 Spacelike Hypersurfaces in Standard Static Spacetimes ... 113
- 4.1 Introduction ... 113
- 4.2 Standard Static Spacetimes ... 114
- 4.3 Some Auxiliary Results ... 117
- 4.4 Rigidity and Parabolicity Results ... 121
- 4.5 A Parabolicity Criterion ... 129
- 4.6 Further Rigidity Results ... 132
- 4.7 Entire Killing Graphs ... 134

Part II Riemannian Immersions in Weighted Semi-Riemannian Warped Products

5 Basic Facts Concerning Weighted Manifolds ... 145
- 5.1 Introduction ... 145
- 5.2 The Bakry-Émery-Ricci Tensor and the Drift Laplacian ... 145
- 5.3 Weighted Riemannian Warped Products ... 147
- 5.4 Weighted GRW Spacetimes ... 151

6 Two-Sided Hypersurfaces in Weighted Riemannian Warped Products ... 159
- 6.1 Introduction ... 159
- 6.2 Studying Two-Sided Hypersurfaces in Weighted Riemannian Warped Products ... 160
  - 6.2.1 Via Weak Omori-Yau Maximum Principle ... 160
  - 6.2.2 Via Integrability Constraints ... 164
  - 6.2.3 Via a φ-Parabolicity Criterion ... 168
- 6.3 Moser-Bernstein Type Results in Weighted Warped Products ... 173
- 6.4 Entire Graphs in Weighted Riemannian Product Spaces ... 179
- 6.5 Uniqueness and Nonexistence of Entire Translating Graphs ... 182
- 6.6 A Height Estimate and a Half-Space Type Result in Weighted Riemannian Product Spaces ... 187

7 Spacelike Hypersurfaces in Weighted GRW Spacetimes ... 191
- 7.1 Introduction ... 191
- 7.2 Uniqueness Results for Spacelike Hypersurfaces in Spatially Weighted GRW Spacetimes ... 192
  - 7.2.1 Comparison Inequalities Between Curvatures in the Presence of a Omori-Yau Maximum Principle ... 192
  - 7.2.2 Uniqueness Under Certain Integrability Conditions ... 196
  - 7.2.3 Uniqueness Under φ-Parabolicity Criteria ... 199
  - 7.2.4 Constant φ-Mean Curvature Spacelike Hypersurfaces ... 207
- 7.3 On the Existence of φ-Maximal Spacelike Hypersurfaces ... 209
  - 7.3.1 Additional Key Lemmas ... 210
  - 7.3.2 Proof of Theorem 7.14 ... 212
- 7.4 Calabi-Bernstein Type Results ... 215

	7.5	Spacelike Translating Solitons of the Mean Curvature Flow in Weighted Lorentzian Product Spaces	222
	7.6	A Height Estimate and a Half-Space Type Result in Weighted Lorentzian Product Spaces	226
8	**Spacelike Hypersurfaces in Weighted Standard Static Spacetimes**		**229**
	8.1	Introduction	229
	8.2	Weighted Standard Static Spacetimes	229
	8.3	A φ-Parabolicity Criterion for Spacelike Hypersurfaces in $(M^n \times_\rho \mathbb{R}_1)_\varphi$	232
	8.4	Rigidity Results for Spacelike Hypersurfaces in $M_\varphi^n \times_\rho \mathbb{R}_1$	235
	8.5	Further Rigidity Results in a Weighted Static Spacetime $M_\varphi^n \times \mathbb{R}_1$	243
	8.6	Entire Killing Graphs	247
		8.6.1 The φ-Mean Curvature Equation in $M_\varphi^n \times_\rho \mathbb{R}_1$	248

Part III Submanifolds Immersed in Semi-Riemannian Warped Products

9	**Submanifolds in a Riemannian Warped Product**		**259**
	9.1	Introduction	259
	9.2	Recalling Some Basic Facts	260
	9.3	Rigidity Results for n-Dimensional Submanifolds in $I \times_f M^{n+p-1}$	264
		9.3.1 Rigidity Via a Suitable Integrability Condition	264
		9.3.2 Rigidity Via Omori-Yau Maximum Principle	268
		9.3.3 Rigidity Via Stochastic Completeness	271
	9.4	A Duo-Graph in $\mathbb{R} \times \mathbb{H}^2 \times \mathbb{R}$	272
	9.5	Further Rigidity and Nonexistence Results	275
	9.6	Submanifolds in a Weighted Riemannian Warped Product	282
10	**Submanifolds Immersed in a Killing Warped Product**		**291**
	10.1	Introduction	291
	10.2	Recalling Some Basic Aspects of Submanifolds in a Killing Warped Product	292
	10.3	Rigidity Results for Submanifolds in a Killing Warped Product	294
		10.3.1 The Compact Case	295
		10.3.2 The Complete Case	297
	10.4	Further Results for Submanifolds in a Killing Warped Product	301
	10.5	Hypersurfaces in Killing Warped Products	305
	10.6	A Parabolicity Criterion for Two-Sided Hypersurfaces in a Killing Warped Product	306
	10.7	Rigidity Results for Two-Sided Hypersurfaces in a Killing Warped Product	308
	10.8	Entire Killing Graphs in $M^n \times_\rho \mathbb{R}$	310

	10.9	Mean Curvature Flow Solitons in Killing Warped Products	313
		10.9.1 Uniqueness of MCFS in Killing Warped Products	314
		10.9.2 Moser-Bernstein Type Results for MCFS	320

11 Weakly Trapped Submanifolds Immersed in a Generalized Robertson-Walker Spacetime ... 325
    11.1 Introduction ... 325
    11.2 Spacelike Submanifolds in a GRW Spacetime ... 326
    11.3 Weakly Trapped Submanifolds in GRW Spacetimes ... 328
    11.4 Codimension Reduction Results in GRW Spacetimes ... 329
        11.4.1 Codimension Reduction Via Certain Integrability Constraints ... 329
        11.4.2 Codimension Reduction Via Omori-Yau's Generalized Maximum Principle ... 333
    11.5 Constructing a Nontrivial Weakly Trapped Submanifold ... 338

12 Studying the Geometry of Weakly Trapped Submanifolds in Standard Static Spacetimes ... 343
    12.1 Introduction ... 343
    12.2 Spacelike Submanifolds in Standard Static Spacetimes ... 343
    12.3 Some Auxiliary Results ... 345
    12.4 Weakly Trapped Submanifolds in Standard Static Spacetimes ... 349
    12.5 Einstein Standard Static Spacetimes ... 351
    12.6 Further Results via a Suitable Density Function ... 352

Part IV Stability of Riemannian Immersions in Semi-Riemannian Warped Products

13 A Notion of Stability to Closed Hypersurfaces in the Hyperbolic Space ... 359
    13.1 Introduction ... 359
    13.2 A Warped Product Model for the Hyperbolic Space ... 360
    13.3 The Notion of Strong (r, k, a, b)-Stability ... 362
    13.4 Strongly (r, k, a, b)-Stable Hypersurfaces in $\mathbb{H}^{n+1}$ ... 368

14 Stable Closed Spacelike Hypersurfaces in the de Sitter Space ... 377
    14.1 Introduction ... 377
    14.2 A Warped Product Model for the de Sitter Space ... 378
    14.3 Description of the Variational Problem ... 379
    14.4 Strongly (r, s, a, b)-Stable Hypersurfaces in $\mathbb{S}_1^{n+1}$ ... 384

15 Stability in Certain Semi-Riemannian Manifolds with Density ... 391
    15.1 Introduction ... 391
    15.2 Stability of Conformal Killing Graphs in Foliated Riemannian Spaces with Density ... 392
        15.2.1 Conformal Killing Graphs ... 392

Contents

		15.2.2	An Auxiliary Lemma	396
		15.2.3	Stability of φ-Minimal Conformal Killing Graphs	400
		15.2.4	Stability of Constant φ-Mean Curvature Conformal Killing Graphs	404
	15.3	Stability of Spacelike Hypersurfaces in Weighted Spacetimes	407
		15.3.1	Background	407
		15.3.2	Description of the Variational Problem	408
		15.3.3	The φ-Laplacian of a Support Function	413
		15.3.4	Stability Results in Weighted Spacetimes	418
	15.4	A Notion of φ-Stability in Weighted Standard Static Spacetimes	421

16 L_φ-Stability of Zero φ-Mean Curvature Hypersurfaces ... 431
 16.1 Introduction ... 431
 16.2 The Weighted Jacobi Operator L_φ ... 432
 16.3 L_φ-Stability of φ-Minimal Hypersurfaces ... 434
 16.4 L_φ-Stability of φ-Maximal Hypersurfaces ... 438

Part V Local Rigidity and Bifurcation of Riemannian Immersions in Semi-Riemannian Warped Products

17 Bifurcation of H_2-Hypersurfaces in Riemannian Warped Products ... 445
 17.1 Introduction ... 445
 17.2 Some Preliminaries ... 448
 17.3 The Variational Problem and the Notion of Bifurcation Instants ... 450
  17.3.1 Description of the Variational Problem ... 450
  17.3.2 The Notion of Bifurcation Instants for H_2-Hypersurfaces ... 459
 17.4 Bifurcation and Locally Rigidity Results for H_2-Hypersurfaces ... 462

18 Bifurcation of Spacelike Hypersurfaces with Constant Mean Curvature in Spacetimes ... 473
 18.1 Introduction ... 473
 18.2 The Variational Problems (VP-1) and (VP-2) ... 474
 18.3 Notions of Bifurcation Instants for Spacelike H-Hypersurfaces ... 480
 18.4 Bifurcation and Locally Rigidity Results for (VP-1) ... 483
 18.5 Bifurcation and Locally Rigidity Results for (VP-2) ... 485

19 Bifurcation of φ-Minimal Hypersurfaces in a Weighted Killing Warped Product ... 495
 19.1 Introduction ... 495
 19.2 The Setup of Ambient Space ... 497
 19.3 The Second Variation of the Weighted Area Functional ... 498
 19.4 Bifurcation Instants for φ-Minimal Hypersurfaces in $M_\varphi^n \times_\rho \mathbb{R}$ ... 501
 19.5 Main Results of Bifurcation for φ-Minimal Hypersurfaces ... 503

20 Bifurcation of Hypersurfaces with Constant φ-Mean Curvature in $M_\varphi^n \times_\rho \mathbb{R}$ 507
   20.1 Introduction 507
   20.2 Description of the Variational Problem 508
   20.3 The Notion of Bifurcation for H_φ-Hypersurfaces in $M_\varphi^n \times_\rho \mathbb{R}$ 514
   20.4 Main Results of Bifurcation for H_φ-Hypersurfaces 517

References 521

Part I

Uniqueness Results, Height Estimates and Half-Space Theorems for Hypersurfaces in Semi-Riemannian Warped Products

Riemannian Immersions

1.1 Introduction

This chapter is devoted to establish the background concerning Riemannian immersions in a semi-Riemannian warped product space, whose notion was introduced in the classical paper of Bishop and O'Neill [86] and which widely generalizes the usual product of Riemannian manifolds.

To this aim, we assume a basic knowledge concerning semi-Riemannian manifolds, in particular some notions related to the fundamental equations of an isometric immersion in these ambient spaces; see [33, Chapter 1], [109, Chapters 1 and 2], and [133, Chapter 1], as well as [251, Chapter 3].

1.2 Higher Order Mean Curvatures of a Riemannian Immersion

Let $\overline{M}^{n+1}$ be a (connected) $(n+1)$-dimensional semi-Riemannian manifold with metric $\overline{g} = \langle \cdot, \cdot \rangle$ of index $\nu \leq 1$, and semi-Riemannian connection $\overline{\nabla}$. We let $C^\infty(\overline{M}^{n+1})$ denote the ring of smooth functions $F : \overline{M}^{n+1} \to \mathbb{R}$ and $\mathfrak{X}(\overline{M}^{n+1})$ the algebra of smooth vector fields on $\overline{M}^{n+1}$. For a vector field $X \in \mathfrak{X}(\overline{M}^{n+1})$, let $\epsilon_X = \langle X, X \rangle$; X is a *unit* vector field if $\epsilon_X = \pm 1$, and *timelike* if $\epsilon_X = -1$.

Here and in the sequel, we will consider *Riemannian immersions* $\psi : \Sigma^n \to \overline{M}^{n+1}$, namely, immersions from a connected, n-dimensional orientable differentiable manifold Σ^n into $\overline{M}^{n+1}$, such that the induced metric $g = \psi^*(\overline{g})$ turns Σ^n into a Riemannian manifold (in the Lorentz case $\nu = 1$ and we will refer to (Σ^n, g) as a *spacelike hypersurface* of $\overline{M}^{n+1}$), with Levi-Civita connection ∇. Furthermore, Σ^n is oriented by the choice of a unit normal vector field N on it.

Let us denote by $A : \mathfrak{X}(\Sigma^n) \to \mathfrak{X}(\Sigma^n)$ the shape operator (or Weingarten operator) of the Riemannian immersion Σ^n in $\overline{M}^{n+1}$ with respect to N, which is given by $AX = -\overline{\nabla}_X N$, where as already mentioned above $\overline{\nabla}$ stands for the Levi-Civita connection of $\overline{M}^{n+1}$. It is well-known that the curvature tensor R of the hypersurface Σ^n can be described in terms of the second fundamental form A and of the curvature tensor $\overline{R}$ of the ambient space $\overline{M}^{n+1}$ by the Gauss equation as follows

$$R(X, Y)Z = (\overline{R}(X, Y)Z)^\top + \epsilon (\langle AX, Z \rangle AY - \langle AY, Z \rangle AX) \qquad (1.1)$$

for every $X, Y, Z \in \mathfrak{X}(\Sigma^n)$, where $(\cdot)^\top$ denotes the tangential component of a vector field in $\mathfrak{X}(\overline{M})$ along Σ^n.

Associated with the shape operator A there are n algebraic invariants, which are the elementary symmetric functions S_r, given by

$$S_0 = 1 \quad \text{and} \quad S_r = S_r(\kappa_1, \ldots, \kappa_n) = \sum_{i_1 < \ldots < i_r} \kappa_{i_1} \ldots \kappa_{i_r}, \quad 1 \leq r \leq n,$$

where $\kappa_1, \ldots, \kappa_n$ are the principal curvatures of Σ^n.

For each $0 \leq r \leq n$, the r-mean curvature H_r of the hypersurface Σ^n is defined as follows

$$\binom{n}{r} H_r = \epsilon^r S_r(\kappa_1, \ldots, \kappa_n).$$

In particular, when $r = 1$, one has

$$H_1 = \epsilon \frac{1}{n} \sum_i \kappa_i = \epsilon \frac{1}{n} \mathrm{tr}(A) = H$$

is just the mean curvature of Σ^n, which is the main extrinsic curvature of the hypersurface Σ^n. When $r = 2$, H_2 defines a geometric quantity which is related to the (intrinsic) scalar curvature S of the hypersurface Σ^n.

Moreover, when the ambient space has constant sectional curvature c, the Gauss equation ensures that

$$S = n(n-1)(c + \epsilon H_2).$$

In general, the Gauss equation yields that when r is odd H_r is extrinsic (and its sign depends on the orientation of Σ^n), while when r is even, H_r is an intrinsic geometric quantity. We also observe that the characteristic polynomial of A can be written in terms of the r-mean curvature H_r. More precisely, one has

$$\det(xI - A) = \sum_{j=0}^{n} \binom{n}{j} (-\epsilon)^j H_j x^{n-j}. \tag{1.2}$$

It is usual to refer to the r-mean curvatures as the higher order mean curvatures of the hypersurface.

Finally, the Hilbert-Schmidt norm of the shape operator A of Σ^n has the form

$$|A|^2 = n^2 H^2 - n(n-1) H_2. \tag{1.3}$$

Regarding the higher order mean curvatures, it is a classical fact that it satisfy a very useful set of inequalities (Newton's inequalities) that we collect shortly here.

Lemma 1.1 *Let* $\psi : \Sigma^n \to \overline{M}^{n+1}$ *be a Riemannian immersion into a semi-Riemannian manifold* $\overline{M}^{n+1}$. *Suppose that there exists an elliptic point in* Σ^n. *If* H_{r+1} *is positive on* Σ^n, *we have that the same holds for* H_k, $k = 1, \ldots, r$. *Moreover*,

(a) $H_k H_{k+2} \leq H_{k+1}^2$ *for every* $k = 1, \ldots, r$;
(b) $H_1 \geq H_2^{1/2} \geq \ldots \geq H_r^{1/r}$,

and the equality holds only at umbilical points.

A detailed proof of Lemma 1.1 can be found in [198, Theorems 51 and 52] and [111, Proposition 3.2]; see also [96, Proposition 2.3].

We end this section withe the next definition that will be useful in the sequel.

By an elliptic point in a Riemannian immersion $\psi : \Sigma^n \to \overline{M}^{n+1}$ we mean a point $p_0 \in \Sigma^n$ where the principal curvatures $\kappa_i(p_0)$ are either positive or negative, with respect to an appropriate choice of the orientation N of Σ^n, according to the fact that $\overline{M}^{n+1}$ is either a Riemannian or a Lorentzian manifold.

1.3 Newton Transformations and L_r Operators

Let $r \in \mathbb{N}$ with $0 \leq r \leq n$. The r-th Newton transformation P_r on Σ^n is defined recursively by setting $P_0 = I$ (the identity operator) and

$$P_r = \epsilon_N^r S_r I - \epsilon_N A P_{r-1} \tag{1.4}$$

for every $r \in \mathbb{N}$ with $0 < r \leq n$, according to the notations given in Sect. 1.2.

By induction, Eq. (1.4) assume the form

$$P_r = \epsilon_N^r (S_r I - S_{r-1} A + S_{r-2} A^2 - \cdots + (-1)^r A^r). \tag{1.5}$$

Clearly, the classical Cayley-Hamilton theorem gives $P_n = 0$. Moreover, P_r can be viewed as self-adjoint operator that commutes with A, since P_r is a formal polynomial expression in A. Therefore, the bases of $T_p \Sigma^n$ diagonalizing A at $p \in \Sigma^n$ also diagonalize P_r at $p \in \Sigma^n$ for every $1 \le r \le n$.

Now, let $\{e_1, \ldots, e_n\}$ be an orthonormal frame on $T_p \Sigma^n$ which diagonalizes A_p, with $A_p(e_i) = \kappa_i(p) e_i$. Then, by (1.5) it follows that

$$(P_r)_p e_i = \epsilon_N^r \sum_{i_1 < \ldots < i_r, i_j \ne i} \kappa_{i_1}(p) \ldots \kappa_{i_r}(p) e_i. \tag{1.6}$$

Taking into account (1.6), we obtain the following preparatory result.

Lemma 1.2 *With the above notations, the following formulas hold:*

(a) $S_r(A_i) = S_r - \kappa_i S_{r-1}(A_i)$;

(b) $\operatorname{tr}(P_r) = \epsilon_N^r \sum_{i=1}^{n} S_r(A_i) = \epsilon_N^r (n-r) S_r = c_r H_r$;

(c) $\operatorname{tr}(A P_r) = \epsilon_N^r \sum_{i=1}^{n} \kappa_i S_r(A_i) = \epsilon_N^r (r+1) S_{r+1} = \epsilon_N c_r H_{r+1}$,

where $c_r = (n-r) \binom{n}{r}$.

See [60, Lemma 2.1] for additional comments and details.

Associated to each Newton transformation P_r one has the second order linear differential operator $L_r : C^\infty \to C^\infty(\Sigma^n)$, given by

$$L_r(u) = \operatorname{tr}(P_r \circ \operatorname{Hess} u) \tag{1.7}$$

for every $u \in C^\infty(\Sigma^n)$.

For a smooth $\varphi : \mathbb{R} \to \mathbb{R}$ and $u \in C^\infty(\Sigma^n)$, one has

$$L_r(\varphi \circ u) = \varphi'(u) L_r(u) + \varphi''(u) \langle P_r \nabla u, \nabla u \rangle, \tag{1.8}$$

thanks to the classical properties of the Hessian of a smooth function.

1.3 Newton Transformations and L_r Operators

Furthermore, we observe that

$$L_r(u) = \mathrm{tr}(P_r \circ \mathrm{Hess}\, u) = \sum_{i=1}^{n} \langle P_r(\nabla_{e_i} \nabla u), e_i \rangle$$

$$= \sum_{i=1}^{n} \langle \nabla_{e_i} \nabla u, P_r(e_i) \rangle = \sum_{i=1}^{n} \langle \nabla_{P_r(e_i)} \nabla u, e_i \rangle = \mathrm{tr}(\mathrm{Hess}\, u \circ P_r),$$

where $\{e_1, \ldots, e_n\}$ is a local orthonormal frame on Σ^n. Moreover, according to [266], we have

$$\mathrm{div}_{\Sigma^n}(P_r(\nabla u)) = \sum_{i=1}^{n} \langle (\nabla_{e_i} P_r)(\nabla u), e_i \rangle + \sum_{i=1}^{n} \langle P_r(\nabla_{e_i} \nabla u), e_i \rangle \quad (1.9)$$

$$= \langle \mathrm{div}_{\Sigma^n} P_r, \nabla u \rangle + L_r(u),$$

where the divergence of P_r on Σ^n is given by

$$\mathrm{div}_{\Sigma^n} P_r = \mathrm{tr}(\nabla P_r) = \sum_{i=1}^{n} (\nabla_{e_i} P_r)(e_i).$$

By Eq. (1.9), we conclude that the operator L_r is elliptic if and only if P_r is positive definite.

We notice that $L_0 = \Delta$ is always elliptic. The next lemma, which corresponds to [19, Lemma 3.2], gives a geometric condition which guarantees the ellipticity of the differential operator L_1.

Lemma 1.3 *Let $\psi : \Sigma^n \to \overline{M}^{n+1}$ be a Riemannian immersion in a semi-Riemannian manifold $\overline{M}^{n+1}$. If $H_2 > 0$ on Σ^n, then L_1 is elliptic or, equivalently, P_1 is positive definite (for an appropriate choice of the Gauss map N).*

When $r \geq 2$, the following lemma establishes sufficient conditions to guarantee the ellipticity of the differential operator L_r.

Lemma 1.4 *Let $\psi : \Sigma^n \to \overline{M}^{n+1}$ be a Riemannian immersion in a semi-Riemannian manifold $\overline{M}^{n+1}$. If there exists an elliptic point of Σ^n, with respect to an appropriate choice of the Gauss map N, and $H_{r+1} > 0$ on Σ^n, for $2 \leq r \leq n-1$, then for all $1 \leq k \leq r$ the operator L_k is elliptic or, equivalently, P_k is positive definite (for an appropriate choice of the Gauss map N, if k is odd).*

See [60, Proposition 3.2] for a detailed proof.

Remark 1.1 In the Riemannian case, as showed in [111, Proposition 3.2], Theorem 1.4 still holds under the weaker assumption that there exists a point $p \in \Sigma^n$ such that the principal curvatures of Σ^n at p are nonnegative.

1.4 Semi-Riemannian Warped Products

In order to study semi-Riemannian warped products, we define the notion of conformal vector fields. A vector field V on $\overline{M}^{n+1}$ is said to be *conformal* if

$$\mathcal{L}_V \langle \cdot, \cdot \rangle = 2\phi \langle \cdot, \cdot \rangle$$

for some function $\phi \in C^\infty(\overline{M}^{n+1})$, where $\mathcal{L}$ stands for the Lie derivative of the metric of $\overline{M}^{n+1}$. The function ϕ is called the *conformal factor* of V.

Since $\mathcal{L}_V(X) = [V, X]$ for every $X \in \mathfrak{X}(\overline{M}^{n+1})$, the tensorial character of $\mathcal{L}_V$ ensures that $V \in \mathfrak{X}(\overline{M}^{n+1})$ is conformal if and only if

$$\langle \overline{\nabla}_X V, Y \rangle + \langle X, \overline{\nabla}_Y V \rangle = 2\phi \langle X, Y \rangle,$$

for every $X, Y \in \mathfrak{X}(\overline{M}^{n+1})$. In particular, V is a *Killing vector field* relatively to $\overline{g}$ if $\phi \equiv 0$.

Let M^n be a connected, n-dimensional oriented Riemannian manifold, $I \subseteq \mathbb{R}$ be an open interval and $f : I \to \mathbb{R}$ be a positive smooth function. In the product differentiable manifold $\overline{M}^{n+1} = \epsilon I \times_f M^n$ let us denote by π_I and π_M the projections onto I and M^n, respectively.

A particular class of semi-Riemannian manifolds having conformal fields is given by the manifolds of the form $\overline{M}^{n+1} = \epsilon I \times_f M^n$ endowed with the metric locally defined by

$$\langle v, w \rangle_p = \epsilon \langle (\pi_I)_* v, (\pi_I)_* w \rangle + f(p)^2 \langle (\pi_M)_* v, (\pi_M)_* w \rangle, \qquad (1.10)$$

for every $p \in \overline{M}^{n+1}$ and any $v, w \in T_p \overline{M}^{n+1}$, where $\epsilon = \epsilon_{\partial_t}$ and ∂_t denotes the standard unit vector field tangent to I.

Moreover (cf. [242] and [243]), the vector field defined by

$$V = (f \circ \pi_I) \partial_t$$

is conformal and closed (in the sense that its dual 1-form is closed), with conformal factor $\phi = f' \circ \pi_I$, where, as usual f' denotes the first derivative of f with respect to $t \in I$.

The space defined above is a meaningful prototype of a semi-Riemannian *warped product*, and, from now on, when no confusion arises, we shall write $\overline{M}^{n+1}$ instead of $\epsilon I \times_f M^n$.

By using the previous notations, if $\psi : \Sigma^n \to \epsilon I \times_f M^n$ is a Riemannian immersion, with Σ^n oriented by the unit vector field N, we have that $\epsilon = \epsilon_{\partial_t} = \epsilon_N$. Moreover, when $\epsilon = 1$ Eq. (1.10) gives a Riemannian metric on $\overline{M}^{n+1} = I \times_f M^n$ and the corresponding Riemannian manifold $\overline{M}^{n+1}$, will be called a Riemannian warped product. On the other hand, if $\epsilon = -1$ then (1.10) is a Lorentzian metric on $\overline{M}^{n+1}$. In this case, the corresponding Lorentzian manifold, $\overline{M}^{n+1} = -\epsilon I \times_f M^n$, is called a Lorentzian warped product.

According to the mathematical literature on the subject, if M^n has constant sectional curvature, we refer to the warped product $\overline{M}^{n+1} = -I \times_f M^n$ as a Robertson-Walker (RW) spacetime. This nomenclature is strongly motivated by the fact that, for $n = 3$, the warped product $\overline{M}^4 = -I \times_f M^3$ yields an exact solution of the Einstein's field equations; see [251, Chapter 12].

Similarly, after [24], the warped product $\overline{M}^{n+1} = -I \times_f M^n$ has usually been referred to as a generalized Robertson-Walker (GRW) spacetime, and we will stick to this usage along this part.

Remark 1.2 For $t_0 \in \mathbb{R}$, we orient the *slice* $M_{t_0}^n = \{t_0\} \times M^n$ by using the unit normal vector field ∂_t. According to [24], $M_{t_0}^n$ has constant r-th mean curvature

$$H_r = -\epsilon \left(\frac{f'(t_0)}{f(t_0)} \right)^r$$

with respect to ∂_t. For additional comments and remarks, see also [242] and [243].

1.5 Height and Angle Functions of a Riemannian Immersion

Given a Riemannian immersion $\psi : \Sigma^n \to \epsilon I \times_f M^n$, we will consider two particular functions naturally attached to it, namely, the (vertical) *height function* $h = (\pi_I)|_{\Sigma^n}$ and the *angle function* $\Theta = \langle N, \partial_t \rangle$. Let us denote by $\overline{\nabla}$ and ∇ the gradients with respect to the metrics of $\epsilon I \times_f M^n$ and Σ^n, respectively. A simple computation shows that the gradient of π_I on $\epsilon I \times_f M^n$ is given by

$$\overline{\nabla} \pi_I = \epsilon \langle \overline{\nabla} \pi_I, \partial_t \rangle = \epsilon \partial_t,$$

so that the gradient of h on Σ^n is

$$\nabla h = (\overline{\nabla} \pi_I)^\top = \epsilon \partial_t^\top = \epsilon \partial_t - \Theta N. \tag{1.11}$$

In particular, we get

$$|\nabla h|^2 = \epsilon \left(1 - \Theta^2\right), \qquad (1.12)$$

where $|\cdot|$ denotes the norm of a vector field on Σ^n. So, the angle function also satisfies $|\Theta| \leq 1$ when the ambient space is Riemannian and $|\Theta| \geq 1$ if the ambient space is Lorentzian.

Next, we recall some key formulas concerning height and angle functions, which were obtained in [31] and [19], respectively, in the Riemannian case and in the Lorentzian setting.

More precisely, the following result holds.

Proposition 1.1 *Let $\psi : \Sigma^n \to \overline{M}^{n+1}$ be an orientable Riemannian immersion immersed into a semi-Riemannian warped product $\overline{M}^{n+1} = \epsilon I \times_f M^n$. For every $r = 0, \ldots, n-1$, the following formulas hold:*

(a) The height function h satisfies

$$\operatorname{Hess} h(X, Y) = \epsilon (\log f)'(h)(\langle X, Y \rangle - \epsilon \langle \nabla h, X \rangle \langle \nabla h, Y \rangle) + \Theta \langle AX, Y \rangle$$

and

$$L_r h = \epsilon (\log f)'(h) \left(c_r H_r - \epsilon \langle P_r \nabla h, \nabla h \rangle \right) + \epsilon c_r \Theta H_{r+1}, \qquad (1.13)$$

where $c_r = (n-r)\binom{n}{r} = (r+1)\binom{n}{r+1}$.

(b) Let σ be a primitive of f. Then, it follows that

$$L_r \sigma(h) = \epsilon c_r \left(f'(h) H_r + f(h) \Theta H_{r+1} \right) \qquad (1.14)$$

(c) Set $\tilde{\Theta} = f(h)\Theta$. Then, one has

$$L_r \tilde{\Theta} = -\epsilon \frac{c_r f(h)}{r+1} \langle \nabla H_{r+1}, \nabla h \rangle - \epsilon c_r f'(h) H_{r+1}$$

$$-\epsilon \frac{c_r f(h) \Theta}{r+1} (n H_1 H_{r+1} - (n-r-1) H_{r+2})$$

$$-\epsilon \frac{\tilde{\Theta}}{f^2(h)} \sum_{i=1}^{n} \mu_{i,r} K_M(N^*, E_i^*) |N^* \wedge E_i^*|^2$$

$$-\tilde{\Theta} (\log f)''(h) \left(c_r |\nabla h|^2 H_r - \langle P_r \nabla h, \nabla h \rangle \right),$$

where $\{E_1, \ldots, E_n\}$ is an orthonormal frame on Σ^n diagonalizing A, K_M denotes the sectional curvature of the fiber M^n, $\mu_{i,r}$ stands for the eigenvalues of P_r, for every vector field $X \in \mathfrak{X}(\overline{M}^{n+1})$, X^* is the orthogonal projection on TM^n and

$$|X^* \wedge Y^*|^2 = |X^*|^2|Y^*|^2 - \langle X^*, Y^* \rangle^2.$$

For more details see [31, Proposition 6 and Lemma 27], as well as [19, Lemma 4.1 and Corollary 8.5].

We close this section quoting a lemma which gives a sufficient condition to guarantee the existence of an elliptic point in Riemannian immersions into a semi-Riemannian warped product.

Lemma 1.5 *Let $\overline{M}^{n+1} = \epsilon I \times_f M^n$ be a semi-Riemannian warped product, and let $\psi : \Sigma^n \to \overline{M}^{n+1}$ be a Riemannian immersion. If $-\epsilon f(h)$ attains a local minimum at some $p \in \Sigma^n$, such that $f'(h(p)) \neq 0$, then p is an elliptic point for Σ^n.*

See [26, Lemma 5.4] for an exaustive proof.

1.6 On the Omori-Yau's Generalized Maximum Principle

In this section, we briefly recall a generalized version of the Omori-Yau's maximum principle for trace type differential operators, proved in [33], as well as the well known Omori-Yau's maximum principle for the Laplacian operator.

Let Σ^n be a Riemannian manifold and let $\mathcal{L} = \text{tr}(\mathcal{P} \circ \text{Hess})$ be a semi-elliptic operator, where $\mathcal{P} : \mathfrak{X}(\Sigma^n) \to \mathfrak{X}(\Sigma^n)$ is a positive semi-definite symmetric tensor. Following the terminology introduced in [256], we say that the Omori-Yau maximum's principle holds on Σ^n for the operator $\mathcal{L}$ if, for any function $u \in C^2(\Sigma^n)$ with $u^* = \sup u < +\infty$, there exists a sequence of points $\{p_j\} \subset \Sigma^n$ satisfying

$$u(p_j) > u^* - \frac{1}{j}, \quad |\nabla u(p_j)| < \frac{1}{j} \quad \text{and} \quad \mathcal{L}u(p_j) < \frac{1}{j}$$

for every $j \in \mathbb{N} \setminus \{0\}$.

Equivalently, for any smooth function $u \in C^2(\Sigma^n)$ with $u_* = \inf u > -\infty$ there exists a sequence of points $\{p_j\} \subset \Sigma^n$ satisfying

$$u(p_j) < u_* + \frac{1}{j}, \quad |\nabla u(p_j)| < \frac{1}{j} \quad \text{and} \quad \mathcal{L}u(p_j) > -\frac{1}{j}$$

for every $j \in \mathbb{N} \setminus \{0\}$.

In this sense, the classical result due to Omori [250] and Yau [297] states that the Omori-Yau's maximum principle holds for the Laplacian on every complete Riemannian manifold with Ricci curvature bounded from below.

More precisely, the following result holds.

Lemma 1.6 *Let Σ^n be a complete Riemannian manifold whose Ricci curvature is bounded from below and $u \in C^2(\Sigma^n)$ satisfying $u^* < +\infty$. Then, there exists a sequence of points $\{p_j\} \subset \Sigma^n$ such that*

$$u(p_j) > u^* - \frac{1}{j}, \quad |\nabla u(p_j)| < \frac{1}{j} \quad \text{and} \quad \Delta u(p_j) < \frac{1}{j}.$$

As observed in [256], the validity of Omori-Yau's maximum principle on Σ^n does not depend on curvatures bounds as much as one would expect.

For instance, the Omori-Yau's maximum principle holds on every Riemannian manifolds which is properly immersed into a Riemannian space form with controlled mean curvature; see [256, Example 1.14]. In particular, it holds for every constant mean curvature hypersurface properly immersed into a Riemannian space form.

More generally, and following again the terminology introduced in [256], the weak Omori-Yau's maximum principle is said to hold on an (not necessarily complete) n-dimensional Riemannian manifold Σ^n if, for any smooth function $u \in C^2(\Sigma^n)$ with $u^* < +\infty$ there exists a sequence of points $\{p_j\} \subset \Sigma^n$ with the properties

$$u(p_j) > u^* - \frac{1}{j} \quad \text{and} \quad \Delta u(p_j) < \frac{1}{j}.$$

As proved in [254, 256], the fact that the weak Omori-Yau's maximum principle holds on Σ^n is equivalent to the stochastic completeness of the manifold.

Indeed, the following result holds.

Lemma 1.7 *A Riemannian manifold Σ^n is stochastically complete if and only if for every $u \in C^2(\Sigma^n)$ satisfying $u^* < +\infty$, there exists a sequence of points $\{p_j\} \subset \Sigma^n$ such that*

$$u(p_j) > u^* - \frac{1}{j} \quad \text{and} \quad \Delta u(p_j) < \frac{1}{j}.$$

In particular, the weak Omori-Yau's maximum principle holds on every parabolic Riemannian manifold; see [194, Corollary 6.4].

Furthermore, on a complete noncompact Riemannian manifold a suitable version of the Omori-Yau's maximum principle for trace type differential operators can be found in literature.

The result, whose proof is given in [33, Theorem 6.13], is stated below.

1.6 On the Omori-Yau's Generalized Maximum Principle

Lemma 1.8 *Let Σ^n be a complete noncompact Riemannian manifold; let $o \in \Sigma^n$ be a reference point and denote by r_o the Riemannian distance function from o. Assume that the sectional curvature of Σ^n satisfies*

$$K_{\Sigma^n} \geq -G^2(r_o), \tag{1.15}$$

with $G \in C^1([0, +\infty))$ is such that

$$G(0) > 0, \quad G'(t) \geq 0 \quad \text{and} \quad \frac{1}{G(t)} \notin \mathcal{L}^1(+\infty). \tag{1.16}$$

Let $\mathcal{P}$ be a positive semi-definite symmetric tensor on Σ^n. If $\sup \operatorname{tr}(\mathcal{P}) < +\infty$, then the Omori-Yau's maximum principle holds on Σ^n for the semi-elliptic operator $\mathcal{L} = \operatorname{tr}(\mathcal{P} \circ \operatorname{Hess})$.

We emphasize that Lemma 1.8 remains valid if we replace the condition given in (1.15) by a stronger geometrical assumption on Σ^n. More precisely, the above result continue to holds requiring that:

Σ^n is an hypersurface with sectional curvature bounded from below by a constant.

Remark 1.3 As it is well-known, examples of functions G satisfying condition (1.16) in Lemma 1.8 are given by

$$G(t) = t \prod_{j=1}^{N} \log^j(t), \quad t \gg 1,$$

where $\log^j$ stands for the j-th iterated logarithm; see [33, 256].

Finally, if the ambient space is a semi-Riemannian warped product, we mention here the following version of the Omori-Yau generalized maximum principle for the differential operator L_r.

Lemma 1.9 *Let $\psi : \Sigma^n \to \epsilon I \times_f M^n$ be a complete Riemannian immersion with nonnegative sectional curvature, and $F \in C^\infty(\Sigma^n)$ a smooth function bounded from above. If, for some choice of the Gauss map, the r-th Newton transformation P_r of Σ^n is positive semi-definite and the r-th mean curvature H_r is bounded from above, then there exists a sequence $\{p_k\}$ of points of Σ^n, such that*

$$\lim_k F(p_k) = \sup_{\Sigma^n} F, \quad \lim_k |\nabla F(p_k)| = 0 \quad \text{and} \quad \limsup_k L_r(F)(p_k) \leq 0.$$

A detailed proof of Lemma 1.9 can be found in [99, Corollary 3.3].

Remark 1.4 In the last years, several authors have studied new forms of the Omori-Yau's maximum principle in order to extend the investigation to a much more general class of differential operators which generalize the classical Laplacian operator. We refer to the interested reader the comprehensive book [33] for a complete background about this topic.

Finally, for the sake of completeness, we refer the interested reader to [109] as useful general reference for some of the notions briefly defined and considered in this introductory chapter.

Uniqueness of Complete Hypersurfaces

2.1 Introduction

An important thematic into the theory of isometric immersions is the study of geometric properties concerning submanifolds immersed in a semi-Riemannian warped product space of the form $\epsilon I \times_f M^n$, where M^n is an n-dimensional connected oriented Riemannian manifold, $I \subset \mathbb{R}$ is an open interval, $f : I \to \mathbb{R}$ is a positive smooth function and $\epsilon = \pm 1$, being $\epsilon = 1$ when $\overline{M}^{n+1}$ is a Riemannian warped product and $\epsilon = -1$ when $\overline{M}^{n+1}$ is, according to the terminology established in [24], a generalized Robertson-Walker (GRW) spacetime.

In this context, an interesting question to pay attention is the uniqueness of complete hypersurfaces immersed in $\epsilon I \times_f M^n$, under reasonable restrictions on their higher order mean curvatures.

In recent years, several authors have obtained a considerable amount of results in this context. For instance, Alías and Dajczer [20] studied complete surfaces immersed in a warped product $\mathbb{R} \times_f M^2$ such that the fiber M^2 is a complete surface with nonnegative Gaussian curvature. Under certain restrictions on the constant mean curvature, they showed nonexistence results of surfaces contained between two slices M_{t_1}, M_{t_2} with $t_1 < t_2$ of the foliation $M_t = \{t\} \times M^2$, as well as results in which these immersions are slices of the trivial totally umbilical foliation.

The same authors in [21] generalized the results of Montiel [242] for compact hypersurfaces of constant mean curvature immersed into $\mathbb{R} \times_f M^n$, treating the case of complete hypersurfaces via Omori-Yau maximum principle.

Successively, Romero et al. [260] presented results for complete noncompact maximal hypersurfaces in spatially parabolic GRW spacetimes, that is, the fiber is a complete noncompact Riemannian manifold such that the only superharmonic functions on it which are bounded from below are the constants.

Moreover, in [263], under curvature assumptions on the Riemannian fiber of the ambient space and some conditions on the warping function, these authors also studied complete maximal hypersurfaces in spatially open GRW spacetimes via different maximum principles. Assuming that the ambient spacetime satisfies the Null Convergence Condition (NCC), which means that the Ricci curvature is nonnegative along null directions, Pelegrín and Rigoli [252] also obtained uniqueness and nonexistence results for complete spacelike hypersurfaces of constant mean curvature immersed in a GRW spacetime.

Along this direction, Alías et al. [29, 31] investigated compact and complete noncompact hypersurfaces with constant higher order mean curvatures H_r, $2 \leq r \leq n$, immersed into semi-Riemannian warped products $\overline{M}^{n+1}$ via a generalized version of the Omori-Yau maximum principle for a divergence-type operator $\mathcal{L}_r$ associated to each globally defined Newton transformation P_r, $0 \leq r \leq n$, which can be regarded as a natural extension of the standard Laplacian operator.

Motivated by the work of Alías and Dajczer [21], García-Martínez et al. [189] proved height estimates for compact hypersurfaces of constant positive higher order mean curvature in Riemannian warped product spaces with boundary contained in a slice and applied such estimates in the study of properly immersed complete hypersurfaces in pseudo-hyperbolic spaces $\mathbb{R} \times_{e^t} M^n$ or $\mathbb{R} \times_{\cosh t} M^n$ contained in certain half-spaces. Subsequently, under appropriated constraints on the higher order mean curvatures, Aquino et al. [49] established suitable sufficient conditions to guarantee the rigidity of hypersurfaces immersed in $\epsilon I \times M^n$ via generalized version of the Omori-Yau maximum principle.

Furthermore, many works have approached problems in the context of entire graphs in semi-Riemannian warped products. For instance, Caminha and de Lima [98] obtained necessary conditions for the existence of complete vertical graphs with constant mean curvature in the hyperbolic and steady state spaces $\epsilon \mathbb{R} \times_{e^t} \mathbb{R}^n$.

To this aim, they deduced suitable formulas for the Laplacian of the height function and of a support-like function naturally attached to the graph. Later on, Alías et al. [32] considered restrictions on the mean curvature H_r to obtain uniqueness results for entire graphs in a warped product satisfying a standard curvature condition. As application, they obtained rigidity results for minimal and radial graphs over the Euclidean sphere.

More recently, Antonia et al. [44] investigated complete hypersurfaces with some positive higher order mean curvature in a semi-Riemannian warped product space. Under standard curvature conditions on the ambient space and appropriate constraints on the higher order mean curvatures, they established rigidity and nonexistence results via Liouville type results and suitable maximum principles related to the divergence of smooth vector fields on a complete noncompact Riemannian manifold. They also presented applications to standard warped product models, like the Schwarzschild, Reissner-Nordström and pseudo-hyperbolic spaces, as well as steady state type spacetimes.

When the ambient space is a pp-wave spacetime (that is, a connected Lorentzian manifold admitting a parallel and lightlike vector field), Molica Bisci et al. [84] studied some geometric aspects of the higher order mean curvatures of a spacelike hypersurface. Initially, they developed general Minkowski-type integral formulas for compact (without boundary) spacelike hypersurfaces and they applied these formulas to get uniqueness and nonexistence results for compact spacelike hypersurfaces in terms of their higher order mean curvatures. For the noncompact case, they used suitable forms of maximum principles on complete noncompact Riemannian manifolds due to Caminha [97] and Alías et al. [37] to infer the nonexistence of these hypersurfaces via some geometric conditions involving some higher order mean curvature.

Into this setting, our purpose in this chapter is to apply appropriated generalized maximum principles in order to study the uniqueness of complete hypersurfaces immersed in a semi-Riemannian warped product, which is supposed to obey a suitable convergence condition. By assuming a natural comparison inequality between the r-th mean curvatures of the hypersurface and that ones of the slices of the slab where the hypersurface is contained, we establish sufficient conditions to guarantee that such a hypersurface must be a slice.

The results presented in this chapter are based mainly on the papers [47,49,50,55,137].

2.2 Uniqueness of Riemannian Immersions in Semi-Riemannian Warped Products

In what follows, motivated by Remark 1.2, we will assume that the orientation N of the spacelike hypersurface $\psi : \Sigma^n \to -I \times_f M^n$ is such that its angle function $\Theta = \langle N, \partial_t \rangle$ satisfies

$$\Theta \leq -1.$$

The mean curvature H with respect to this orientation N is called the *future mean curvature* of Σ^n.

Theorem 2.1 *Let $\overline{M}^{n+1} = -I \times_f M^n$ be a Lorentzian warped product whose fiber M^n has sectional curvature K_M satisfying the following convergence condition*

$$K_M \geq \sup_I (ff'' - f'^2). \tag{2.1}$$

Moreover, let $\psi : \Sigma^n \to \overline{M}^{n+1}$ be a complete spacelike hypersurface contained in a slab $[t_1, t_2] \times M^n$ of $\overline{M}^{n+1}$, with $f'(t) > 0$ for $t_1 \le t \le t_2$. Suppose that the future mean curvature H is bounded, satisfying

$$H \ge \frac{f'}{f}(h). \tag{2.2}$$

If

$$|\nabla h| \le \inf_{\Sigma^n}\left(H - \frac{f'}{f}(h)\right), \tag{2.3}$$

then Σ^n is a slice.

Proof By Eq. (1.8), we have that

$$\Delta f(h) = f''(h)|\nabla h|^2 + f'(h)\Delta h. \tag{2.4}$$

Thus, by Proposition 1.1, we obtain

$$\Delta f(h) = \left(\frac{f''f - f'^2}{f}(h)\right)|\nabla h|^2 - nf'(h)\left(\frac{f'}{f}(h) + H\Theta\right). \tag{2.5}$$

On the other hand, since the convergence condition (2.1) holds, and taking into account that Σ^n is contained in a slab of $-I \times_f M^n$, we claim that the Ricci curvature of Σ^n is bounded from below.

In order to prove our claim, fix $X \in \mathfrak{X}(\Sigma^n)$ and let $\{E_1, \ldots, E_n\}$ be a local orthonormal frame on an open set $U \subset \Sigma^n$. By the Gauss equation it follows that the Ricci curvature Ric of Σ^n satisfies

$$\operatorname{Ric}(X, X) \ge \sum_{i=1}^{n}\langle \overline{R}(X, E_i)X, E_i\rangle - \frac{n^2 H^2}{4}|X|^2. \tag{2.6}$$

Thus, if the mean curvature H of Σ^n is supposed to be bounded, then $\operatorname{Ric}(X, X)$ is bounded from below if and only if

$$\sum_{i=1}^{n}\langle \overline{R}(X, E_i)X, E_i\rangle$$

is also bounded from below.

To give an estimate for the first sum at the right hand side of (2.6), let (by changing U with a smaller open set, if necessary) $E_i^* = (\pi_M)_*(E_i)$ and $X^* = (\pi_M)_*(X)$.

2.2 Uniqueness of Riemannian Immersions in Semi-Riemannian Warped...

By [251, Proposition 7.42] and Eq. (1.11), we have

$$\sum_{i=1}^{n} \langle \overline{R}(X, E_i)X, E_i \rangle = \sum_{i=1}^{n} \langle R_M(X^*, E_i^*)X^*, E_i^* \rangle + (n-1)((\log f)')^2 |X|^2$$
$$- (n-2)(\log f)'' \langle X, \nabla h \rangle^2 - (\log f)'' |\nabla h|^2 |X|^2, \quad (2.7)$$

where R_M denotes the curvature tensor of M^n. By writing $X^* = X + \langle X, \partial_t \rangle \partial_t$, we can easily estimate the first summand at the right hand side of (2.7) to get

$$\sum_{i=1}^{n} \langle R_M(X^*, E_i^*)X^*, E_i^* \rangle = f^2(|X^*|_M^2 |E_i^*|_M^2 - \langle X^*, E_i^* \rangle_M^2) K_{M^n}(X^*, E_i^*)$$

$$\geq \frac{1}{f^2}((n-1)|X|^2 + |\nabla h|^2 |X|^2$$

$$+ (n-2)\langle X, \nabla h \rangle^2) \min_i K_M(X^*, E_i^*).$$

On the other hand, since our ambient space obeys the convergence condition (2.1), we have that

$$\sum_{i=1}^{n} \langle R_M(X^*, E_i^*)X^*, E_i^* \rangle \geq ((n-1)|X|^2 + |\nabla h|^2 |X|^2$$
$$+ (n-2)\langle X, \nabla h \rangle^2)(\log f)''. \quad (2.8)$$

Substituting (2.8) into 2.7, we get

$$\sum_{i=1}^{n} \langle \overline{R}(X, E_i)X, E_i \rangle \geq ((n-1)|X|^2 + (n-2)\langle X, \nabla h \rangle^2 + |\nabla h|^2 |X|^2)(\log f)''$$
$$+ (n-1)((\log f)')^2 |X|^2 - (n-2)\langle X, \nabla h \rangle^2 (\log f)''$$
$$- |\nabla h|^2 |X|^2 (\log f)''$$
$$= (n-1)\frac{f''}{f}|X|^2. \quad (2.9)$$

Then, by (2.6) and 2.9, we finally have

$$\mathrm{Ric}(X, X) \geq \left((n-1)\frac{f''}{f} - \frac{n^2 H^2}{4}\right)|X|^2. \quad (2.10)$$

Consequently, since Σ^n is contained in a slab of $-I \times_f M^n$ and H is supposed to be bounded, by (2.10) we conclude that the Ricci curvature of Σ^n is bounded from below.

Hence, we can apply Lemma 1.6 to the height function h obtaining a sequence $\{p_k\} \subset \Sigma^n$ such that

$$\lim_k h(p_k) = \sup_{\Sigma^n} h, \quad \lim_k |\nabla h(p_k)| = 0 \text{ and } \limsup_k \Delta(h)(p_k) \leq 0.$$

Consequently, since $f' > 0$ in a slab where Σ^n is contained, by (2.4) we have that

$$\limsup_k \Delta f(h)(p_k) \leq 0.$$

Moreover, by (1.12) and taking into account the orientation N of Σ^n, it follows that

$$\lim_k \Theta(p_k) = -1.$$

Thus, by (2.2) and (2.5) we get

$$0 \geq \limsup_k \Delta f(h)(p_k) = nf'(\sup_{\Sigma^n} h) \limsup_k \left(H - \frac{f'}{f}(h) \right)(p_k) \geq 0. \tag{2.11}$$

Then, by (2.11) we have that

$$\limsup_k \left(H - \frac{f'}{f}(h) \right)(p_k) = 0.$$

Hence

$$\inf_{\Sigma^n} \left(H - \frac{f'}{f}(h) \right) = 0.$$

Therefore, by (2.3), we conclude that Σ^n is a slice of $-I \times_f M^n$. □

We recall that a hypersurface $\psi : \Sigma^n \to \overline{M}^{n+1}$ is said to be *two-sided* if its normal bundle is trivial, that is, there is on it a globally defined unit normal vector field N. Moreover, taking into account once more Remark 1.2, we will assume that the orientation N of the hypersurface ψ is such that its angle function satisfies

$$-1 \leq \Theta \leq 0. \tag{2.12}$$

Theorem 2.2 *Let $\overline{M}^{n+1} = I \times_f M^n$ be a Riemannian warped product whose fiber M^n has sectional curvature K_M satisfying the following convergence condition*

$$K_M \geq \sup_I (f'^2 - ff''). \tag{2.13}$$

2.2 Uniqueness of Riemannian Immersions in Semi-Riemannian Warped...

Moreover, let $\psi : \Sigma^n \to \overline{M}^{n+1}$ be a complete two-sided hypersurface contained in a slab $[t_1, t_2] \times M^n$ of $\overline{M}^{n+1}$, with $f'(t) > 0$ for $t_1 \leq t \leq t_2$. Suppose that the angle function Θ is negative on Σ^n, the mean curvature H satisfies

$$0 < H \leq \frac{f'}{f}(h), \tag{2.14}$$

and H_2 is bounded from below. If

$$|\nabla h| \leq \inf_{\Sigma^n} \left(\frac{f'}{f}(h) - H \right), \tag{2.15}$$

then Σ^n is a slice.

Proof By Proposition 1.1, we obtain

$$\Delta f(h) = \left(\frac{f''f - f'^2}{f}(h) \right) |\nabla h|^2 + nf'(h) \left(\frac{f'}{f}(h) + H\Theta \right). \tag{2.16}$$

Now, we observe that, since H is bounded and H_2 is bounded from below, by Eq. (1.3) we get the boundedness of the second fundamental form of Σ^n.

In order to show that the Ricci curvature Ric of Σ^n is bounded from below, we will proceed as in the proof of Theorem 2.1. So, let us consider $X \in \mathfrak{X}(\Sigma^n)$ and take a local orthonormal frame $\{E_1, \ldots, E_n\}$ of $\mathfrak{X}(\Sigma^n)$.

Then, the Gauss equation yields

$$\mathrm{Ric}(X, X) \geq \sum_{i=1}^{n} \langle \overline{R}(X, E_i)X, E_i \rangle - (n|H||A| + |A|^2)|X|^2. \tag{2.17}$$

Thus, if the mean curvature H and the second fundamental form A are supposed to be bounded, then $\mathrm{Ric}(X, X)$ is bounded from below if and only if

$$\sum_{i=1}^{n} \langle \overline{R}(X, E_i)X, E_i \rangle$$

is bounded from below.

On the other hand, a straightforward computation ensures that

$$\overline{R}(X, E_i)X = \overline{R}(X^*, E_i^*)X^* + \langle X, \partial_t \rangle \overline{R}(X^*, E_i^*)\partial_t + \langle X, \partial_t \rangle \langle E_i, \partial_t \rangle \overline{R}(X^*, \partial_t)\partial_t$$
$$+ \langle E_i, \partial_t \rangle \overline{R}(X^*, \partial_t)X^* + \langle X, \partial_t \rangle \overline{R}(\partial_t, E_i^*)X^*$$
$$+ \langle X, \partial_t \rangle^2 \overline{R}(\partial_t, E_i^*)\partial_t,$$

where $X^* = X - \langle X, \partial_t \rangle \partial_t$ and $E_i^* = E_i - \langle E_i, \partial_t \rangle \partial_t$ denotes the projections of the tangent vector fields X and E_i onto the fiber M^n, respectively.

Now, by using [251, Proposition 7.42] and by Eq. (1.11), we get

$$\sum_{i=1}^n \langle \overline{R}(X, E_i)X, E_i \rangle = \sum_{i=1}^n \langle R_M(X^*, E_i^*)X^*, E_i^* \rangle - \frac{ff''}{f^2}|X|^2$$

$$+ \frac{f'^2}{f^2}\left(|\nabla h|^2 - (n-1)\right)|X|^2$$

$$+ (n-2)\left(\frac{f'^2 - ff''}{f^2}\right)\langle X, \nabla h \rangle^2,$$

where R_M denotes the curvature tensor of the fiber M^n.

Furthermore, is not difficult to verify that

$$\sum_{i=1}^n \langle R_M(X^*, E_i^*)X^*, E_i^* \rangle = \frac{1}{f^2}\sum_{i=1}^n K_M(X^*, E_i^*)(|X|^2 - \langle \nabla h, E_i \rangle^2 |X|^2$$

$$- \langle X, \nabla h \rangle^2 - \langle X, E_i \rangle^2 + 2\langle X, \nabla h \rangle \langle X, E_i \rangle \langle \nabla h, E_i \rangle).$$

Thus, by using the convergence condition (2.13), a direct computation ensures that the following inequality holds

$$\sum_{i=1}^n \langle \overline{R}(X, E_i)X, E_i \rangle \geq -\frac{f''}{f}\left(n - |\nabla h|^2\right)|X|^2 \geq -n\frac{|f''|}{f}|X|^2. \tag{2.18}$$

Consequently, since Σ^n is contained in a slab of $I \times_f M^n$, by (2.17) and (2.18) we conclude that the Ricci curvature of Σ^n is bounded from below.

Hence, we can apply Lemma 1.6 to the height function h obtaining a sequence $\{p_k\} \subset \Sigma^n$ such that

$$\lim_k h(p_k) = \sup_{\Sigma^n} h, \quad \lim_k |\nabla h(p_k)| = 0 \text{ and } \limsup_k \Delta(h)(p_k) \leq 0.$$

As in the proof of Theorem 2.1, by 1.12 and taking into account the orientation N of Σ^n, it follows that

$$\lim_k \Theta(p_k) = -1.$$

2.2 Uniqueness of Riemannian Immersions in Semi-Riemannian Warped... 23

Then, by (2.14) and (2.16) we get

$$0 \geq \limsup_k \Delta f(h)(p_k) = nf'(\sup_{\Sigma^n} h) \limsup_k \left(\frac{f'}{f}(h) - H\right)(p_k) \geq 0. \qquad (2.19)$$

Thus, by (2.19) we have that

$$\limsup_k \left(\frac{f'}{f}(h) - H\right)(p_k) = 0$$

and, hence

$$\inf_{\Sigma^n} \left(\frac{f'}{f}(h) - H\right) = 0.$$

Therefore, by (2.15), we conclude that Σ^n is a slice of $I \times_f M^n$. $\square$

Remark 2.1 If the ambient space is the Lorentz product $-\mathbb{R} \times \mathbb{H}^n$, a result similar to Theorem 2.1 was obtained in [8, Theorems 3.1 and 3.2]. Moreover, in [141, Theorem 3.1] it was treated the corresponding case of Theorem 2.2 to the context of the Riemannian product $\mathbb{R} \times \mathbb{H}^n$. We note that the product spaces $-\mathbb{R} \times \mathbb{H}^n$ and $\mathbb{R} \times \mathbb{H}^n$ are standard examples of semi-Riemannian warped products that do not obey the convergence conditions (2.1) and (2.13), respectively.

Now, we apply a parabolicity result, showed by Huber in [208], in order to obtain rigidity results related to complete surfaces of nonnegative Gaussian curvature in 3-dimensional warped products spaces.

Theorem 2.3 *Let $\overline{M}^3 = -I \times_f M^2$ be a Lorentzian warped product such that $\log f$ is convex and whose fiber M^2 has nonnegative Gaussian curvature. Moreover, let $\psi : \Sigma^2 \to \overline{M}^3$ be a complete spacelike surface with nonnegative Gaussian curvature contained below a slice $\{t_1\} \times M^2$ of $\overline{M}^3$, with $f'(t) > 0$ for every $t \leq t_1$. If the future mean curvature H satisfies*

$$H \geq \frac{f'}{f}(h), \qquad (2.20)$$

then Σ^2 is a slice.

Proof By Proposition 1.1, we obtain that

$$\Delta f(h) = \left(\frac{f''f - f'^2}{f}(h)\right)|\nabla h|^2 - 2f'(h)\left(\frac{f'}{f}(h) + H\Theta\right).$$

Thus, since $\log f$ is convex and $f'(t) > 0$ for every $t \leq t_1$, by (2.20) and taking into account that $\Theta \leq -1$, we get

$$\Delta(f(t_1) - f(h)) \leq 0.$$

Consequently, the function $f(t_1) - f(h)$ is a superharmonic nonnegative function on Σ^2. However, the quoted result obtained by Huber in [208] assures that complete surfaces of nonnegative Gaussian curvature must be parabolic.

Therefore, by using again that $f'(h) > 0$, we conclude that h is constant and, hence, the hypersufaces Σ^2 is a slice. □

We observe that, by (2.10), the Gaussian curvature K_{Σ^2} of a spacelike surface Σ^2 with mean curvature H and immersed in a Lorentz warped product $-I \times_f M^2$, which obeys the convergence condition (2.1), also satisfies the inequality below

$$K_{\Sigma^2} \geq \frac{f''}{f} - H^2.$$

Consequently, we can apply Theorem 2.3 to get the following result.

Corollary 2.1 *Let $\overline{M}^3 = -I \times_f M^2$ be a Lorentzian warped product which obeys the convergence condition (2.1) and such that $\log f$ is convex. Let $\psi : \Sigma^2 \to \overline{M}^3$ be a complete spacelike surface contained below a slice $\{t_1\} \times M^2$ of $\overline{M}^3$, with $f'(t) > 0$ for every $t \leq t_1$. If the future mean curvature H is positive and satisfies*

$$\frac{f''}{f}(h) \geq H^2 \geq \left(\frac{f'}{f}(h)\right)^2,$$

then Σ^2 is a slice.

Arguing as in the proof of Theorem 2.3 the following result holds.

Theorem 2.4 *Let $\overline{M}^3 = I \times_f M^2$ be a Riemannian warped product such that $\log f$ is convex and whose fiber M^2 has nonnegative Gaussian curvature. Let $\psi : \Sigma^2 \to \overline{M}^3$ be a complete surface with nonnegative Gaussian curvature contained below a slice $\{t_1\} \times M^2$ of $\overline{M}^3$, with $f'(t) > 0$ for every $t \leq t_1$. If the angle function Θ is negative on Σ^2 and the* *mean curvature H satisfies*

$$0 < H \leq \frac{f'}{f}(h),$$

then Σ^2 is a slice.

2.3 A Further Rigidity Result in Spatially Closed GRW Spacetimes

We start this section establishing some key preparatory results that will be useful in the sequel.

The first lemma gives a suitable lower bound for the Ricci curvature of a spacelike hypersurface $\psi : \Sigma^n \to \overline{M}^{n+1}$ immersed in a *spatially closed* GRW spacetime $\overline{M}^{n+1} = -\mathbb{R} \times_f M^n$, which means that the Riemannian fiber M^n is compact (without boundary).

The main result reads as follows.

Lemma 2.1 *Let $\psi : \Sigma^n \to \overline{M}^{n+1}$ be a spacelike hypersurface immersed in a spatially closed GRW spacetime $\overline{M}^{n+1} = -\mathbb{R} \times_f M^n$. Then, for every $X \in \mathfrak{X}(\Sigma^n)$, the Ricci curvature of Σ^n satisfies the following inequality*

$$\mathrm{Ric}_{\Sigma^n}(X, X) \geq \alpha(n-1)|X|^2 + (\alpha - (\log f)'')(|X|^2|\nabla h|^2 + (n-2)\langle X, \nabla h\rangle^2)$$
$$+ \left(\frac{f'}{f}\right)^2 (n-1)|X|^2 + nH\langle AX, X\rangle + |AX|^2,$$

where h and H denote, respectively, the height function and the future mean curvature of Σ^n and $\alpha = \min_M K_M$, with K_M being the sectional curvature of the Riemannian fiber M^n.

Proof Let us consider $X \in \mathfrak{X}(\Sigma^n)$ and a local orthonormal frame $\{E_1, \ldots, E_n\}$ of $\mathfrak{X}(\Sigma^n)$. The Gauss equation yields

$$\mathrm{Ric}_{\Sigma^n}(X, X) = \sum_{i=1}^{n} \langle \overline{R}(X, E_i)X, E_i\rangle + nH\langle AX, X\rangle + |AX|^2. \qquad (2.21)$$

On the other hand, with a straightforward computation, we deduce that

$$\overline{R}(X, E_i)X = \overline{R}(X^*, E_i^*)X^* - \langle X, \partial_t\rangle \overline{R}(X^*, E_i^*)\partial_t - \langle E_i, \partial_t\rangle \overline{R}(X^*, \partial_t)X^*$$
$$+\langle E_i, \partial_t\rangle\langle X, \partial_t\rangle \overline{R}(X^*, \partial_t)\partial_t - \langle X, \partial_t\rangle \overline{R}(\partial_t, E_i^*)X^*$$
$$+\langle X, \partial_t\rangle^2 \overline{R}(\partial_t, E_i^*)\partial_t,$$

where $X^* = X + \langle X, \partial_t \rangle \partial_t$ and $E_i^* = E_i + \langle E_i, \partial_t \rangle \partial_t$ are the projections on the tangent vector fields X and E_i onto the fiber M^n, respectively.

Now, by using [251, Proposition 7.42] and (1.11), we obtain

$$\langle \overline{R}(X, E_i)X, E_i \rangle = \langle \overline{R}(X^*, E_i^*)X^*, E_i^* \rangle - \frac{f''}{f}(|X|^2 + \langle X, \nabla h \rangle^2)\langle E_i, \nabla h \rangle^2$$

$$+ 2\frac{f''}{f}(\langle X, E_i \rangle + \langle X, \nabla h \rangle\langle E_i, \nabla h \rangle)\langle E_i, \nabla h \rangle\langle X, \nabla h \rangle$$

$$- \frac{f''}{f}\langle X, \nabla h \rangle^2(1 + \langle E_i, \nabla h \rangle^2).$$

The above formula and (2.23) yield

$$\mathrm{Ric}_{\Sigma^n}(X, X) = \sum_{i=1}^n \langle \overline{R}(X^*, E_i^*)X^*, E_i^* \rangle - \frac{f''}{f}(|X|^2 + \langle X, \nabla h \rangle^2)|\nabla h|^2$$

$$+ 2\frac{f''}{f}(1 + |\nabla h|^2)\langle X, \nabla h \rangle^2 - \frac{f''}{f}(n + |\nabla h|^2)\langle X, \nabla h \rangle^2$$

$$+ nH\langle AX, X \rangle + |AX|^2.$$

Hence, direct computations ensures that

$$\mathrm{Ric}_{\Sigma^n}(X, X) = \sum_{i=1}^n \langle \overline{R}(X^*, E_i^*)X^*, E_i^* \rangle - \frac{f''}{f}(|X|^2|\nabla h|^2$$

$$+ (n-2)\langle X, \nabla h \rangle^2) + nH\langle AX, X \rangle + |AX|^2. \tag{2.22}$$

By using again the Gauss equation, we immediately have

$$\sum_{i=1}^n \langle \overline{R}(X^*, E_i^*)X^*, E_i^* \rangle = \sum_{i=1}^n K_M(X^*, E_i^*)|X^* \wedge E_i^*|^2 + \left(\frac{f'}{f}\right)^2 \sum_{i=1}^n |X^* \wedge E_i^*|^2. \tag{2.23}$$

On the other hand, it is not difficult to check that

$$\langle X^*, X^* \rangle_M \langle E_i^*, E_i^* \rangle_M = (1 + \langle E_i, \nabla h \rangle^2)(|X|^2 + \langle X, \nabla h \rangle^2)$$

and

$$\langle X^*, E_i^* \rangle_M^2 = \langle X, E_i \rangle^2 + 2\langle X, \nabla h \rangle\langle E_i, \nabla h \rangle\langle X, E_i \rangle + \langle X, \nabla h \rangle^2\langle E_i, \nabla h \rangle^2.$$

2.3 A Further Rigidity Result in Spatially Closed GRW Spacetimes

Hence, combining the above expressions we deduce

$$\sum_{i=1}^{n} |X^* \wedge E_i^*|^2 = (n-1)|X|^2 + |X|^2|\nabla h|^2 + (n-2)\langle X, \nabla h\rangle^2. \tag{2.24}$$

Now, by (2.23) and (2.22) and by using (2.24), we infer

$$\mathrm{Ric}_{\Sigma^n}(X,X) = \sum_{i=1}^{n} K_M(X^*, E_i^*)|X^* \wedge E_i^*|^2 - (\log f)''(|X|^2|\nabla h|^2$$

$$+ (n-2)\langle X, \nabla h\rangle^2) + nH\langle AX, X\rangle + |AX|^2$$

$$+ (n-1)\left(\frac{f'}{f}\right)^2 |X|^2. \tag{2.25}$$

Therefore, the hypothesis on the sectional curvature K_M of the Riemannian fiber M^n in addition to (2.25) and (2.24) allow us to conclude that the Ricci curvature of Σ^n satisfies the inequality below

$$\mathrm{Ric}_{\Sigma^n}(X,X) \geq \alpha(n-1)|X|^2 + (\alpha - (\log f)'')(|X|^2|\nabla h|^2 + (n-2)\langle X, \nabla h\rangle^2)$$

$$+ \left(\frac{f'}{f}\right)^2 (n-1)|X|^2 + nH\langle AX, X\rangle + |AX|^2.$$

The proof is now complete. □

The next key lemma is a direct consequence of the Bochner's formula [87]. By using some ideas contained in [228, Proposition 3.1], the next result can be proved.

Lemma 2.2 *Let* $\psi : \Sigma^n \to \overline{M}^{n+1}$ *be a spacelike hypersurface immersed with constant future mean curvature H in a spatially closed GRW spacetime* $\overline{M}^{n+1} = -\mathbb{R} \times_f M^n$. *Then*

$$\frac{1}{2}\Delta|\nabla h|^2 \geq \left(\alpha(n-1) - (\log f)''n\right)|\nabla h|^2(1+|\nabla h|^2)$$

$$+ \left|A(\nabla h) - \frac{f'}{f}\Theta\nabla h\right|^2 + nH\frac{f'}{f}\Theta|\nabla h|^2$$

$$+ \left(\frac{f'}{f}\right)^2 (n+|\nabla h|^2)|\nabla h|^2 + |\mathrm{Hess}\, h|^2,$$

where h denotes the height function on Σ^n *and* $\alpha = \min_M K_M$, *with* K_M *being the sectional curvature of* M^n.

Proof Let us observe that, by item (a) of Proposition 1.1 we obtain that

$$\nabla \Delta h = \nabla \left(\frac{f'}{f}\right)(-n - |\nabla h|^2) - \frac{f'}{f}\nabla|\nabla h|^2 - nH\nabla\Theta,$$

which can be rewritten, by using (1.12), as follows

$$\nabla \Delta h = \left(\frac{f''f - (f')^2}{f^2}\right)(-n - |\nabla h|^2)\nabla h - 2\frac{f'}{f}\Theta\nabla\Theta - nH\nabla\Theta. \tag{2.26}$$

Next, by (1.11), in addition to [251, Proposition 7.35], we get that

$$X\Theta = \langle AX, \nabla h\rangle + \frac{f'}{f}\langle X, \partial_t\rangle\Theta = \left\langle X, A(\nabla h) - \frac{f'}{f}\Theta\nabla h\right\rangle,$$

for any $X \in \mathfrak{X}(\Sigma^n)$. Thus, one has

$$\nabla\Theta = A(\nabla h) - \frac{f'}{f}\Theta\nabla h. \tag{2.27}$$

Hence, we use (2.27) and (2.26) to deduce that

$$\nabla \Delta h = (\log f)''(-n - |\nabla h|^2)\nabla h - 2\frac{f'}{f}\Theta A\nabla h$$

$$+ 2\left(\frac{f'}{f}\right)^2 \Theta^2 \nabla h - nHA(\nabla h) + nH\frac{f'}{f}\Theta\nabla h. \tag{2.28}$$

Now, the Bochner's formula [87] and (2.28) yield

$$\frac{1}{2}\Delta|\nabla h|^2 = \langle \nabla\Delta h, \nabla h\rangle + \mathrm{Ric}_{\Sigma^n}(\nabla h, \nabla h) + |\mathrm{Hess}\, h|^2$$

$$= (\log f)''(-n - |\nabla h|^2)|\nabla h|^2 - 2\frac{f'}{f}\Theta\langle A\nabla h, \nabla h\rangle$$

$$+ 2\left(\frac{f'}{f}\right)^2 \Theta^2|\nabla h|^2 - nH\langle A(\nabla h), \nabla h\rangle$$

$$+ nH\frac{f'}{f}\Theta|\nabla h|^2 + \mathrm{Ric}_{\Sigma^n}(\nabla h, \nabla h) + |\mathrm{Hess}\, h|^2. \tag{2.29}$$

2.3 A Further Rigidity Result in Spatially Closed GRW Spacetimes

On the other hand, by applying Lemma 2.1 to the vector field ∇h, we have

$$\mathrm{Ric}_{\Sigma^n}(\nabla h, \nabla h) \geq \alpha(n-1)|\nabla h|^2 + (\alpha - (\log f)'')(n-1)|\nabla h|^2$$
$$+ \left(\frac{f'}{f}\right)^2 (n-1)|\nabla h|^2 + nH\langle A(\nabla h), \nabla h\rangle + |A(\nabla h)|^2.$$

Therefore, the last inequality, combined with (2.29), immediately yields

$$\frac{1}{2}\Delta|\nabla h|^2 = \left(-(\log f)''(n+|\nabla h|^2) + (\alpha - (\log f)''(n-1)|\nabla h|^2 + \alpha(n-1)\right)|\nabla h|^2$$
$$+ \left|A(\nabla h) - \frac{f'}{f}\Theta\nabla h\right|^2 + \left(\frac{f'}{f}\right)^2 \Theta|\nabla h|^2$$
$$+ nH\frac{f'}{f}\Theta|\nabla h|^2 + \left(\frac{f'}{f}\right)^2 (n-1)|\nabla h|^2 + |\mathrm{Hess}\, h|^2.$$

We get

$$\frac{1}{2}\Delta|\nabla h|^2 \geq \left(\alpha(n-1) - (\log f)''n\right)|\nabla h|^2(1+|\nabla h|^2)$$
$$+ \left|A(\nabla h) - \frac{f'}{f}\Theta\nabla h\right|^2 + nH\frac{f'}{f}\Theta|\nabla h|^2$$
$$+ \left(\frac{f'}{f}\right)^2 (n+|\nabla h|^2)|\nabla h|^2 + |\mathrm{Hess}\, h|^2.$$

The proof is now complete. □

The following consequence of the generalized maximum principle of Omori–Yau [250, 297] is due to Nishikawa [248].

Lemma 2.3 *Let Σ^n denote an n-dimensional complete Riemannian manifold having Ricci curvature bounded from below. If $\zeta \in C^2(\Sigma^n)$ is nonnegative and satisfies $\Delta\zeta \geq a\zeta^b$, for some real numbers $a > 0$ and $b > 1$, then ζ vanishes identically on Σ^n.*

We are in position now to state and prove the following rigidity result.

Theorem 2.5 *Let $\overline{M}^{n+1} = -I \times_f M^n$ be a spatially closed GRW spacetime, whose sectional curvature of the Riemannian fiber M^n is positive. Moreover, let $\psi : \Sigma^n \to \overline{M}^{n+1}$ be a complete spacelike hypersurface immersed in $\overline{M}^{n+1}$ such that $-\log f$ is convex on Σ^n. If the future mean curvature H of Σ^n is constant and satisfies $Hf' \leq 0$, then Σ^n must be a totally geodesic slice of $\overline{M}^{n+1}$.*

Proof By Lemma 2.2 we have that

$$\Delta |\nabla h|^2 \geq 2\alpha(n-1)|\nabla h|^4,$$

where $\alpha = \min_M K_M > 0$. Thus, taking into account Lemma 2.1, we can apply Lemma 2.3 to obtain that $|\nabla h|^2$ vanishes identically on Σ^n and, consequently, Σ^n is a slice of $\overline{M}^{n+1}$. Moreover, since $Hf' \leq 0$, it easily seen that the mean curvature H of Σ^n is zero and Σ^n must be a totally geodesic slice of $\overline{M}^{n+1}$. □

2.4 Rigidity of Hypersurfaces in Riemannian Warped Products and Pseudo-Hyperbolic Spaces

In this section, motivated by the results contained in [55], we study the rigidity of a complete two-sided hypersurface $\psi : \Sigma^n \to \overline{M}^{n+1}$ contained in a slab $[t_1, t_2] \times M^n$ of a Riemannian warped product $\overline{M}^{n+1} = I \times_f M^n$, via a suitable conformal change of metric.

Along this section, in order to simplify the notation, we will denote by g the induced metric $\langle \cdot, \cdot \rangle$ of Σ^n via ψ.

Lemma 2.4 *Let $\overline{M}^{n+1} = I \times_f M^n$ be a warped product which satisfies the convergence condition (2.13) and let $\psi : \Sigma^n \to \overline{M}^{n+1}$ be a two-sided hypersurface with bounded second fundamental form and lying in a slab $[t_1, t_2] \times M^n$ of $\overline{M}^{n+1}$. Then, the Ricci curvature $\widehat{\text{Ric}}$ of Σ^n with respect to the conformal metric*

$$\hat{g} = \frac{1}{f(h)^2} g$$

is bounded from below.

Proof By [80, Section 1J] we have the following equation

$$\widehat{\text{Ric}}(X, X) = \text{Ric}(X, X) + \frac{1}{f(h)^2} \{(n-2) f(h) \text{Hess} f(h)(X, X)$$

$$+ (f(h) \Delta f(h) - (n-1)|\nabla f(h)|^2)|X|^2\}, \quad (2.30)$$

where Ric stands for the Ricci tensor of Σ^n with respect to its induced metric g; see also [224, Section A], as well as [277, p. 168].

2.4 Rigidity of Hypersurfaces in Riemannian Warped Products and...

Now, by (2.30), we get

$$\widehat{\mathrm{Ric}}(X, X) = \mathrm{Ric}(X, X) + \frac{1}{f(h)^2}\{(n-2)f(h)(f''(h)g(\nabla h, X)^2$$
$$+ f'(h)\nabla^2 h(X, X)) + (f(h)(f''(h)|\nabla h|^2 + f'(h)\Delta h)$$
$$- (n-1)f'(h)^2|\nabla h|^2)|X|^2\}. \quad (2.31)$$

Hence, by using (1.12), item (*a*) of Proposition 1.1, (2.17), (2.18) and (2.31), we deduce the following lower estimate

$$\widehat{\mathrm{Ric}}(X, X) \geq -\frac{1}{f(h)}\{(2n-1)|f''(h)| + (n+\sqrt{n}-2)|f'(h)||A|$$
$$+ (\sqrt{n}+1)f(h)|A|^2\}|X|^2. \quad (2.32)$$

Therefore, taking into account once more that $|A|$ is bounded and that Σ^n lies in a slab of the ambient space, by (2.32) we conclude that $\widehat{\mathrm{Ric}}$ is bounded from below. □

From now on, we will orient the two-sided hypersurfaces $\psi : \Sigma^n \to \overline{M}^{n+1}$ in such a way that Θ satisfies (2.12).

In this setting, we obtain the following rigidity result which can be viewed as a complement of Theorem 2.2.

Theorem 2.6 *Let $\overline{M}^{n+1} = I \times_f M^n$ be a warped product whose fiber M^n is complete with sectional curvature obeying the convergence condition* (2.13). *Moreover, let $\psi : \Sigma^n \to \overline{M}^{n+1}$ be a complete two-sided hypersurface with bounded second fundamental form and lying in a slab $[t_1, t_2] \times M^n$, with $f'(t) > 0$ for every $t \in [t_1, t_2]$. If the height function h and the mean curvature function H satisfy*

$$0 < -\frac{f'(h)}{f(h)}\Theta \leq H \quad (2.33)$$

and

$$|\nabla h| \leq \inf_{\Sigma^n}\left|H - \frac{f'(h)}{f(h)}\right|, \quad (2.34)$$

then Σ^n is a slice.

Proof Let us consider on Σ^n the metric

$$\hat{g} = \frac{1}{f(h)^2} g,$$

which is conformal to its induced metric g. If we denote by $\hat{\Delta}$ the Laplacian with respect to the metric $\hat{g}$, by (1.12) and item (*a*) of Proposition 1.1, we get

$$\hat{\Delta} h = f(h)^2 \Delta h - (n-2) f(h) f'(h) |\nabla h|^2$$
$$= n f(h) f'(h) \Theta^2 + f(h) f'(h) |\nabla h|^2 + n H f(h)^2 \Theta. \qquad (2.35)$$

With a direct computation, by (2.35) we obtain

$$\hat{\Delta} f(h) = f''(h) \hat{g}(\widehat{\nabla} h, \widehat{\nabla} h) + f'(h) \hat{\Delta} h$$
$$= f''(h) f(h)^2 |\nabla h|^2 + f'(h) \Big(n f(h) f'(h) \Theta^2$$
$$+ f(h) f'(h) |\nabla h|^2 + n H f(h) \Theta \Big)$$
$$= n f(h) f'(h)^2 + n H f'(h) f(h)^2 \Theta + f(h)^3 \Big((\log f)''(h)$$
$$-(n-2) \frac{f'(h)^2}{f(h)^2} \Big) |\nabla h|^2. \qquad (2.36)$$

Given a positive real number α, we have that

$$\hat{\Delta} f(h)^{-\alpha} = \alpha(\alpha+1) f(h)^{-\alpha-2} \hat{g}(\widehat{\nabla} f(h), \widehat{\nabla} f(h)) - \alpha f(h)^{-\alpha-1} \hat{\Delta} f(h). \qquad (2.37)$$

By using (2.36) and (2.37), we get

$$\hat{\Delta} f(h)^{-\alpha} = -\alpha n f(h)^{-\alpha} f'(h)^2 - \alpha n H f'(h) f(h)^{-\alpha+1} \Theta$$
$$+ \alpha(\alpha+1) f(h)^{-\alpha} f'(h)^2 |\nabla h|^2$$
$$- \alpha f(h)^{-\alpha+2} \Big((\log f)''(h) - (n-2) \frac{f'(h)^2}{f(h)^2} \Big) |\nabla h|^2. \qquad (2.38)$$

Now, by (1.12), we have

$$-\alpha n f(h)^{-\alpha} f'(h)^2 = -\alpha n f(h)^{-\alpha} f'(h)^2 |\nabla h|^2 - \alpha n f(h)^{-\alpha} f'(h)^2 \Theta^2. \qquad (2.39)$$

2.4 Rigidity of Hypersurfaces in Riemannian Warped Products and...

Thus, by (2.38) and (2.39), we obtain

$$\hat{\Delta} f(h)^{-\alpha} = -n\alpha f(h)^{-\alpha} \{f'(h)^2 \Theta^2 + H f(h) f'(h) \Theta\}$$
$$-\alpha f(h)^{-\alpha+2} \left[(\log f)''(h) - (\alpha - 1) \frac{f'(h)^2}{f(h)^2} \right] |\nabla h|^2. \quad (2.40)$$

On the other hand, since $|A|$ is bounded and taking into account that Σ^n lies in a slab of the ambient space (which obeys the convergence condition (2.13)), Lemmas 1.6 and 2.4 guarantee the existence of a sequence of points $\{p_k\} \subset \Sigma^n$, such that

$$\lim_k f(h)^{-\alpha}(p_k) = \sup_{\Sigma^n} f(h)^{-\alpha}, \quad \lim_k |\widehat{\nabla} f(h)^{-\alpha}(p_k)| = 0$$

and

$$\limsup_k \hat{\Delta} f(h)^{-\alpha}(p_k) \leq 0.$$

Since $\widehat{\nabla} f(h)^{-\alpha} = -\alpha f(h)^{1-\alpha} f'(h) \nabla h$, taking into that Σ^n lies in a slab and that $f'(h) > 0$ in this slab, we get that

$$\lim_k |\nabla h(p_k)| = 0. \quad (2.41)$$

Moreover, by (1.12) we also have that

$$\lim_k \Theta(p_k) = -1. \quad (2.42)$$

So, by using (2.33), (2.40), (2.41) and (2.42), it is easily seen that

$$0 \geq \limsup_k \hat{\Delta} f(h)^{-\alpha}(p_k)$$
$$\geq n\alpha \sup_{\Sigma^n} f(h)^{-\alpha} \limsup_k -\{f'(h)^2 - H f(h) f'(h)\}(p_k) \geq 0. \quad (2.43)$$

Thus, by (2.43), we infer that

$$\lim_k \left(H - \frac{f'(h)}{f(h)} \right)(p_k) = 0.$$

Hence

$$\inf_{\Sigma^n}\left|H - \frac{f'(h)}{f(h)}\right| = 0.$$

Therefore, by (2.34) we conclude that Σ^n must be a slice of $I \times_f M^n$. □

According to the terminology introduced by Tashiro [282], when the warping function is exponential, the corresponding warped product $I \times_{e^t} M^n$ is referred to as a *pseudo-hyperbolic space*.

Tashiro's terminology is due to the fact that the $(n + 1)$-dimensional hyperbolic space $\mathbb{H}^{n+1}$ is isometric to the warped product $\mathbb{R} \times_{e^t} \mathbb{R}^n$, where the slices constitute a family of horospheres sharing a fixed point in the asymptotic boundary $\partial_\infty \mathbb{H}^{n+1}$ and giving a complete foliation of $\mathbb{H}^{n+1}$.

We observe that a pseudo-hyperbolic space $I \times_{e^t} M^n$, whose fiber M^n has nonnegative sectional curvature, satisfies condition (2.13).

Taking into account the above definitions and remarks, by Theorem 2.6, the following meaningful consequence holds.

Corollary 2.2 *Let* $\psi : \Sigma^n \to \overline{M}^{n+1}$ *be a complete two-sided hypersurface immersed in a slab of a pseudo-hyperbolic space* $\overline{M}^{n+1} = I \times_{e^t} M^n$ *whose fiber M^n is complete with nonnegative sectional curvature, and has bounded second fundamental form. If $H \geq 1$ and $|\nabla h| \leq \inf_{\Sigma^n}(H - 1)$, then Σ^n is a slice.*

Inspired by [15], in the next results we will assume that the warping function f of the ambient space $\overline{M}^{n+1} = I \times_f M^n$ satisfies the following inequality

$$(\log f)'' \leq \gamma[(\log f)']^2, \tag{2.44}$$

for some constant $\gamma > -1$.

We notice that pseudo-hyperbolic spaces satisfy (2.44) for $\gamma = 0$.

For our purposes, we also need to quote a suitable extension of the classical Hopf's Theorem on a complete Riemannian manifold (Σ^n, g) due to Yau in [298].

More precisely, let us consider the following set

$$\mathcal{L}^1_g(\Sigma^n) = \left\{\varsigma : \Sigma^n \to \mathbb{R} : \int_{\Sigma^n} |\varsigma| d\Sigma^n < +\infty\right\},$$

where $d\Sigma$ stands for the measure related to the Riemannian metric g.

2.4 Rigidity of Hypersurfaces in Riemannian Warped Products and...

The Yau's results reads as follows.

Lemma 2.5 *Let ζ be a smooth function defined on a complete Riemannian manifold (Σ^n, g), such that $\Delta \zeta$ does not change sign on Σ^n. If $|\nabla \zeta| \in \mathcal{L}^1_g(\Sigma^n)$, then $\Delta \zeta$ vanishes identically on Σ^n.*

By using Lemma 2.5, we get the following result.

Theorem 2.7 *Let $\overline{M}^{n+1} = I \times_f M^n$ be a warped product whose warping function satisfies (2.44), holding the equality only at isolated points of I, and with complete fiber M^n. Let $\psi : \Sigma^n \to \overline{M}^{n+1}$ be a complete two-sided hypersurface which lies in a slab of $\overline{M}^{n+1}$. If the height function h and the mean curvature function H satisfy (2.33) and $|\nabla h| \in \mathcal{L}^1_g(\Sigma^n)$, then Σ^n is a slice.*

Proof Let us consider the conformal metric

$$\hat{g} = \frac{1}{f(h)^2} g.$$

It is easily seen that

$$|\hat{\nabla} f(h)^{-\alpha}|_{\hat{g}} = \alpha f(h)^{-\alpha} |f'(h)| |\nabla h|, \qquad (2.45)$$

for any positive constant α. Consequently, since Σ^n lies in a slab of $\overline{M}^{n+1}$ and $|\nabla h| \in \mathcal{L}^1_g(\Sigma^n)$, by (2.45) we get that $|\hat{\nabla} f(h)^{-\alpha}|_{\hat{g}} \in \mathcal{L}^1_{\hat{g}}(\Sigma^n)$.

Moreover, taking $\alpha = 1 + \gamma$, by (2.40) we also have that $\hat{\Delta} f(h)^{-\alpha} \geq 0$. Thus, we can apply Lemma 2.5 to infer that $\hat{\Delta} f(h)^{-\alpha} = 0$ on Σ^n.

Hence, since we are assuming that equality occurs in (2.44) only at isolated points of I, thanks to (2.40), we conclude that $|\nabla h|$ must vanish identically on Σ^n.

Therefore, Σ^n is a slice. $\qquad \square$

A slightly different version of Theorem 2.7 is presented below.

Theorem 2.8 *Let $\overline{M}^{n+1} = I \times_f M^n$ be a warped product whose warping function satisfies (2.44), and has complete fiber M^n. Let $\psi : \Sigma^n \to \overline{M}^{n+1}$ be a complete two-sided hypersurface which lies in a slab $[t_1, t_2] \times M^n$, with $f'(t) > 0$ for every $t \in [t_1, t_2]$. If the height function h and the mean curvature function H satisfy (2.33) and $|\nabla h| \in \mathcal{L}^1_g(\Sigma^n)$, then Σ^n is a slice.*

Proof As in the proof of Theorem 2.7, taking $\alpha = 1 + \gamma$, we get that $\hat{\Delta} f(h)^{-\alpha} = 0$ on Σ^n. Moreover, since Σ^n lies in a slab of $I \times_f M^n$, we can prove that $|\hat{\nabla} f(h)^{-2\alpha}|_{\hat{g}} \in \mathcal{L}^1_{\hat{g}}(\Sigma^n)$.

Now, we notice that

$$\hat{\Delta} f(h)^{-2\alpha} = 2f(h)^{-\alpha}\hat{\Delta} f(h)^{-\alpha} + 2|\widehat{\nabla} f(h)^{-\alpha}|_{\hat{g}}^2 = 2|\widehat{\nabla} f(h)^{-\alpha}|_{\hat{g}}^2 \geq 0.$$

Thus, we can apply Lemma 2.5 again to obtain that $\hat{\Delta} f(h)^{-2\alpha} = 0$. Hence, since we assume that $f'(t) > 0$ for every $t \in [t_1, t_2]$, by (2.45) we obtain that $|\nabla h| = 0$ on the hyperfurface Σ^n.

Therefore, Σ^n must be a slice. □

When the ambient space is a pseudo-hyperbolic space, Theorem 2.8 reads as follows.

Corollary 2.3 *Let* $\psi : \Sigma^n \to \overline{M}^{n+1}$ *be a complete two-sided hypersurface immersed in a slab of a pseudo-hyperbolic space* $\overline{M}^{n+1} = I \times_{e^t} M^n$ *with complete fiber* M^n. *If* $H \geq 1$ *and* $|\nabla h| \in \mathcal{L}_g^1(\Sigma^n)$, *then* Σ^n *is a slice.*

Now, we will apply a technique developed in [260, 261].

To this aim, we recall that a noncompact Riemannian manifold is said to be *parabolic* if the only subharmonic functions on it that are bounded from above are the constants.

On the other hand, given two Riemannian manifolds (Σ, g) and (Σ', g'), a diffeomorphism ϕ from Σ onto Σ' is called a *quasi-isometry* if there exists a constant $c \geq 1$ such that

$$c^{-1}|v|_g \leq |d\phi(v)|_{g'} \leq c|v|_g,$$

for any $v \in T_p\Sigma$, $p \in \Sigma$.

By [216, Theorem 1], we have the following preparatory result.

Lemma 2.6 *Let* (Σ, g) *and* (Σ', g') *be two complete Riemannian manifolds. If* Σ *and* Σ' *are quasi-isometric, then* Σ *and* Σ' *are either both parabolic or neither is parabolic.*

See also [194, Corollary 5.3].
We can use Lemma 2.6 to prove the following parabolicity criterion.

Lemma 2.7 *Let* $\psi : \Sigma^n \to \overline{M}^{n+1}$ *be a complete noncompact hypersurface immersed in a warped product* $\overline{M}^{n+1} = I \times_f M^n$, *whose fiber* (M^n, g_M) *has parabolic universal covering. If* Θ *is bounded away from zero, then* $(\Sigma^n, \hat{g})$, *endowed with the conformal metric*

$$\hat{g} = \frac{1}{f(h)^2} g,$$

is parabolic.

2.4 Rigidity of Hypersurfaces in Riemannian Warped Products and...

Proof Given $p \in \Sigma^n$ and $v \in T_p\Sigma^n$, by (1.10) and (1.12) we have

$$g(v, v) = g(v, \nabla h)^2 + f(h)^2 g_M(d\pi(v), d\pi(v)). \tag{2.46}$$

Thus, by (2.46) we get

$$\hat{g}(v, v) = \frac{1}{f(h)^2} g(v, v) \geq g_M(d\pi(v), d\pi(v)). \tag{2.47}$$

On the other hand, by using (1.12) and the Cauchy-Schwarz inequality in (2.46), we also have

$$\Theta^2 g(v, v) \leq f(h)^2 g_M(d\pi(v), d\pi(v)). \tag{2.48}$$

Since Θ is bounded away from zero, there exists a positive constant β such that $\Theta^2 \geq \beta^2$. Consequently, by (2.48), we get

$$\beta^2 g(v, v) \leq \Theta^2 g(v, v) \leq f(h)^2 g_M(d\pi(v), d\pi(v)). \tag{2.49}$$

Thus, by (2.49) we have

$$\hat{g}(v, v) \leq \frac{1}{\beta^2} g_M(d\pi(v), d\pi(v)). \tag{2.50}$$

Hence, by using (2.47) and (2.50), it follows that

$$g_{M^n}(d\pi(v), d\pi(v)) \leq \hat{g}(v, v) \leq \frac{1}{\beta^2} g_M(d\pi(v), d\pi(v)). \tag{2.51}$$

So, taking $c = 1/\beta^2 \geq 1$, by (2.51), we have

$$\frac{1}{c} g_M(d\pi(v), d\pi(v)) \leq \hat{g}(v, v) \leq c g_M(d\pi(v), d\pi(v)), \tag{2.52}$$

which means that π is a quasi-isometry between Σ^n and M^n.

Let Σ' be the universal Riemannian covering of Σ^n with projection $\pi_\Sigma : \Sigma' \to \Sigma^n$. Then, the map $\pi_0 = \pi \circ \pi_{\Sigma^n} : \Sigma' \to M^n$ is a covering map.

If M' is the universal Riemannian covering of M^n with projection $\pi' : M' \to M^n$, then there exists a diffeomorphism $\phi : \Sigma' \to M'$ such that $\pi' \circ \phi = \pi_0$.

Moreover, by (2.52) it is not difficult to verify that ϕ is also a quasi-isometry. Therefore, since the universal Riemannian covering of M^n is parabolic, it follows by Lemma 2.6 that the universal Riemannian covering of Σ^n is parabolic.

Consequently, Σ^n must be also parabolic with respect to the metric $\hat{g}$. □

Now, we recall that a function $\zeta : I \to (0, +\infty)$ is said to be globally constant when $I = \mathbb{R}$ and ζ is constant.

Taking into account the above definition, we are ready to state and prove the next rigidity result.

Theorem 2.9 *Let $\overline{M}^{n+1} = I \times_f M^n$ be a warped product whose fiber M^n is complete with parabolic universal covering and such that its warping function f is not globally constant and satisfies (2.44). Moreover, let $\psi : \Sigma^n \to \overline{M}^{n+1}$ be a complete two-sided hypersurface with $\Theta \leq -\beta < 0$, for some positive constant β, and such that $\inf_{\Sigma^n} f(h) > 0$. If $f'(h)H \geq 0$ and*

$$\frac{f'(h)^2}{f(h)^2}\Theta^2 \leq H^2,$$

then Σ^n is a slice.

Proof We first notice that Lemma 2.7 guarantees that $(\Sigma^n, \hat{g})$ is parabolic. Moreover, it follows by (2.40) that $f^{-\alpha}(h)$ (where $\alpha = 1 + \gamma$) is subharmonic on Σ^n. Thus, since the hypothesis $\inf_{\Sigma^n} f(h) > 0$ implies that $f(h)^{-\alpha}$ is bounded from above, by the parabolicity of $(\Sigma^n, \hat{g})$ it, follows that $f(h)$ is constant on Σ^n.

Consequently, thanks to (2.40), we get

$$H^2 = \frac{f'(h)^2}{f(h)^2}\Theta^2. \tag{2.53}$$

Now, arguing by contradiction, let us suppose that h is not constant. So, $J = \mathrm{Im}\, h$ is a subinterval of I and $f|_J$ is constant, which implies that $f'(t) = 0$ for every $t \in J$. Hence, $f'(h)$ vanishes identically and, by (2.53), we conclude that Σ^n is minimal.

Thus, by (2.35) it follows that h is a harmonic function on the parabolic Riemannian manifold $(\Sigma^n, \hat{g})$. Finally, since $f(h)$ is constant and taking into account that f is not globally constant, it is easily seen that h must be bounded either from below or from above. Then, h is constant on Σ^n, leading us to a contradiction.

Therefore, Σ^n must be a slice. □

A direct application of Theorem 2.9 is the following

Corollary 2.4 *Let $\overline{M}^{n+1} = I \times_{e^t} M^n$ be a pseudo-hyperbolic space whose fiber M^n is complete with parabolic universal covering. Let $\psi : \Sigma^n \to \overline{M}^{n+1}$ be a complete two-sided hypersurface with $\Theta \leq -\beta < 0$, for some positive constant β, and such that $\inf_{\Sigma^n} h > -\infty$. If $H \geq 1$, then Σ^n is a slice.*

2.5 Rigidity Results via Higher Order Mean Curvatures

The next result can be viewed as an extension of Theorem 2.1 to the context of the r-th mean curvatures.

As in the previous section, we will assume that the orientation N of the spacelike hypersurface Σ^n is such that $\Theta \leq -1$.

Furthermore, for every $r \in \mathbb{N}$ with $1 \leq r \leq n$, the r-th mean curvatures H_r, with respect to this orientation N, are called the *future r-th mean curvatures* of Σ^n.

The following result holds.

Theorem 2.10 *Let $\overline{M}^{n+1} = -I \times_f M^n$ be a Lorentzian warped product whose fiber M^n has nonnegative sectional curvature. Moreover, let $\psi : \Sigma^n \to \overline{M}^{n+1}$ be a complete spacelike hypersurface of nonnegative sectional curvature which is contained in a slab $[t_1, t_2] \times M^n$ of $\overline{M}^{n+1}$, with $f'(t) > 0$ for every $t_1 \leq t \leq t_2$. Suppose that there exist positive constants α and β such that, for some $1 \leq r \leq n - 1$, the future r-th mean curvature satisfies $\alpha \leq H_r \leq \beta$, and*

$$\frac{H_{r+1}}{H_r} \geq \frac{f'}{f}(h). \tag{2.54}$$

If h has a local minimum on Σ^n and

$$|\nabla h| \leq \inf_{\Sigma^n} \left(\frac{H_{r+1}}{H_r} - \frac{f'}{f}(h) \right), \tag{2.55}$$

then Σ^n is a slice.

Proof We observe that, by Eq. (1.8), one has

$$L_r f(h) = f''(h) \langle P_r \nabla h, \nabla h \rangle + f'(h) L_r h. \tag{2.56}$$

Thus, with the aid of Lemma 1.2 and Proposition 1.1, we obtain

$$L_r f(h) = \left(\frac{f''f - f'^2}{f}(h) \right) \langle P_r \nabla h, \nabla h \rangle - b_r f'(h) H_r \left(\frac{f'}{f}(h) + \frac{H_{r+1}}{H_r} \Theta \right), \tag{2.57}$$

where $b_r = (n - r) \binom{n}{r}$.

On the other hand, since $f'(h) > 0$ and h has a local minimum at some $p \in \Sigma^n$, by Lemma 1.5 we have that p is an elliptic point for Σ^n.

Thus, since H_{r+1} is positive on Σ^n, Lemma 1.4 guarantees that P_r is positive definite. Hence, since Σ^n is contained in a slab of $-I \times_f M^n$, we can apply Lemma 1.9 to the height function h obtaining a sequence $\{p_k\} \subset \Sigma^n$ such that

$$\lim_k h(p_k) = \sup_{\Sigma^n} h, \quad \lim_k |\nabla h(p_k)| = 0 \text{ and } \limsup_k L_r(h)(p_k) \leq 0.$$

We also observe that

$$0 \leq \langle P_r \nabla h, \nabla h \rangle \leq \text{tr}(P_r)|\nabla h|^2 = b_r H_r |\nabla h|^2 \leq b_r \beta |\nabla h|^2. \tag{2.58}$$

Now, since $f'(h) > 0$ in a slab where Σ^n is contained, by (2.56) we have that

$$\limsup_k L_r f(h)(p_k) \leq 0. \tag{2.59}$$

Moreover, by (1.12) and taking into account the orientation N of Σ^n, we have that

$$\lim_k \Theta(p_k) = -1.$$

Consequently, by (2.54), (2.57), (2.58) and (2.59), we get

$$0 \geq \limsup_k L_r f(h)(p_k)$$

$$\geq b_r \alpha f'(\sup_{\Sigma^n} h) \limsup_k \left(\frac{H_{r+1}}{H_r} - \frac{f'}{f}(h) \right)(p_k) \geq 0. \tag{2.60}$$

Thus, by (2.60) we have that

$$\limsup_k \left(\frac{H_{r+1}}{H_r} - \frac{f'}{f}(h) \right)(p_k) = 0.$$

Hence

$$\inf_{\Sigma^n} \left(\frac{H_{r+1}}{H_r} - \frac{f'}{f}(h) \right) = 0.$$

Therefore, by (2.55), we conclude that Σ^n is a slice of $-I \times_f M^n$. □

Taking into account Lemma 1.3, Theorem 2.10 ensures the validity of the following result.

Corollary 2.5 *Let* $\overline{M}^{n+1} = -I \times_f M^n$ *be a Lorentzian warped product whose fiber* M^n *has nonnegative sectional curvature. Moreover, let* $\psi : \Sigma^n \to \overline{M}^{n+1}$ *be a complete spacelike hypersurface of nonnegative sectional curvature which is contained in a slab*

2.5 Rigidity Results via Higher Order Mean Curvatures

$[t_1, t_2] \times M^n$ of $\overline{M}^{n+1}$, with $f'(t) > 0$ for every $t_1 \leq t \leq t_2$. Suppose that the future mean curvature satisfies $\alpha \leq H \leq \beta$, for some positive constants α and β. If

$$\frac{H_2}{H} \geq \frac{f'}{f}(h)$$

and

$$|\nabla h| \leq \inf_{\Sigma^n} \left(\frac{H_2}{H} - \frac{f'}{f}(h) \right),$$

then Σ^n is a slice.

Similar arguments used along the proof of Theorem 2.10 ensures that the next result holds.

Theorem 2.11 Let $\overline{M}^{n+1} = I \times_f M^n$ be a Riemannian warped product whose fiber M^n has nonnegative sectional curvature. Moreover, let $\psi : \Sigma^n \to \overline{M}^{n+1}$ be a complete two-sided hypersurface of nonnegative sectional curvature which is contained in a slab $[t_1, t_2] \times M^n$ of $\overline{M}^{n+1}$, with $f'(t) > 0$ for every $t_1 \leq t \leq t_2$. Suppose that, for some $1 \leq r \leq n-1$, there exist positive constants α and β such that $\alpha \leq H_r \leq \beta$, H_{r+1} is positive and

$$\frac{H_{r+1}}{H_r} \leq \frac{f'}{f}(h).$$

If the angle function Θ is negative on Σ^n, h has a local maximum and

$$|\nabla h| \leq \inf_{\Sigma^n} \left(\frac{f'}{f}(h) - \frac{H_{r+1}}{H_r} \right),$$

then Σ^n is a slice.

The next corollary is a consequence of Theorem 2.11.

Corollary 2.6 Let $\overline{M}^{n+1} = I \times_f M^n$ be a Riemannian warped product whose fiber M^n has nonnegative sectional curvature. Moreover, let $\psi : \Sigma^n \to \overline{M}^{n+1}$ be a complete two-sided hypersurface of nonnegative sectional curvature which is contained in a slab $[t_1, t_2] \times M^n$ of $\overline{M}^{n+1}$, with $f'(t) > 0$ for every $t_1 \leq t \leq t_2$. Suppose that there exist positive constants α and β such that $\alpha \leq H \leq \beta$ and that H_2 is positive. If the angle function Θ is negative on Σ^n,

$$\frac{H_2}{H} \leq \frac{f'}{f}(h)$$

and

$$|\nabla h| \leq \inf_{\Sigma^n}\left(\frac{f'}{f}(h) - \frac{H_2}{H}\right),$$

then Σ^n is a slice.

Remark 2.2 We observe that in [29,31] the authors studied the problem of uniqueness for hypersurfaces immersed with some constant r-th mean curvature in a semi-Riemannian warped product. Their approach is based on the use of a suitable generalized version of the Omori-Yau maximum principle for certain trace-type differential operators.

2.6 Further Uniqueness Results via Higher Order Mean Curvatures

In this section, under appropriated constraints on the higher order mean curvatures and without the requirement of the existence of an elliptic point, we are able to obtain other sufficient conditions which guarantee that such a hypersurface must be a slice of the ambient space. Our approach is also based on a suitable generalized maximum principle applied to the linearized operators of the higher order mean curvatures.

2.6.1 Rigidity of Complete Spacelike Hypersurfaces

Taking into account the terminology established in [6], we say that a hypersurface $\psi : \Sigma^n \to \epsilon I \times_f M^n$ is *bounded away from the future infinity* of $\epsilon I \times_f M^n$ if there exists $\bar{t} \in I$ such that

$$\psi(\Sigma^n) \subset \{(t, x) \in \epsilon I \times_f M^n : t \leq \bar{t}\}.$$

Analogously, we say that Σ^n is *bounded away from the past infinity* of $\epsilon I \times_f M^n$ if there exists $\underline{t} \in I$ such that

$$\psi(\Sigma^n) \subset \{(t, x) \in \epsilon I \times_f M^n : t \geq \underline{t}\}.$$

2.6 Further Uniqueness Results via Higher Order Mean Curvatures

As in the previous section, we will assume that the orientation N of the spacelike hypersurface $\psi : \Sigma^n \to -I \times_f M^n$ is such that its angle function satisfies $\Theta \leq -1$.
The main result reads as follows.

Theorem 2.12 *Let $\overline{M}^{n+1} = -I \times_f M^n$ be a Lorentzian warped product whose fiber M^n has sectional curvature K_M satisfying the convergence condition*

$$K_M \geq \sup_I (ff'' - (f')^2). \tag{2.61}$$

Moreover, let $\psi : \Sigma^n \to \overline{M}^{n+1}$ be a complete spacelike hypersurface with sectional curvature K_{Σ^n} bounded from below and such that

$$K_{\Sigma^n} \leq \frac{f''}{f}(h). \tag{2.62}$$

Suppose that, for some $r \in \{0, 1, \ldots, n-1\}$, the future r-th mean curvature is such that $\alpha \leq H_r \leq \beta$, for some positive constants α and β, and that one of the following conditions is satisfied

(i) *Σ^n is bounded away from the future infinity of $\overline{M}^{n+1}$ and $\dfrac{f'}{f}(h) \leq \dfrac{H_{r+1}}{H_r}$.*

(ii) *Σ^n is bounded away from the past infinity of $\overline{M}^{n+1}$ and $\dfrac{f'}{f}(h) \geq \dfrac{H_{r+1}}{H_r}$.*

If $|\nabla h| \leq \inf_{\Sigma^n} \left| \dfrac{H_{r+1}}{H_r} - \dfrac{f'}{f}(h) \right|$, then Σ^n is a slice $\{t\} \times M^n$.

Proof To our aim, let us define the self-adjoint operator $\mathcal{P}_r : \mathfrak{X}(\Sigma^n) \to \mathfrak{X}(\Sigma^n)$ as follows

$$\mathcal{P}_r = H_r P_r.$$

For each $p \in \Sigma^n$, let us take a local orthonormal frame $\{e_1, \ldots, e_n\}$ such that $Ae_i = \kappa_i e_i$. By (1.6) we have that

$$P_r e_i = (-1)^r \sum_{i_1 < \ldots < i_r, i_j \neq i} \kappa_{i_1} \cdot \ldots \cdot \kappa_{i_r} e_i.$$

Thus, for any $i \in \{1, \ldots, n\}$, we get

$$\langle \mathcal{P}_r e_i, e_i \rangle = \binom{n}{r}^{-1} \sum_{i_1 < \ldots < i_r, i_j \neq i, j_1 < \ldots < j_r} (\kappa_{i_1} \kappa_{j_1}) \cdot \ldots \cdot (\kappa_{i_r} \kappa_{j_r}).$$

On the other hand, by Gauss' equation, we have that

$$K_{\Sigma^n}(e_i, e_j) = \overline{K}(e_i, e_j) - \kappa_i \kappa_j, \tag{2.63}$$

where K_{Σ^n} and $\overline{K}$ are the sectional curvatures of Σ^n and $\overline{M}^{n+1}$, respectively.

With a straightforward computation, by using a general relations involving the curvature tensor of a warped product as well as the curvature tensor of its base and its fiber (cf. [251, Proposition 7.42]; see also equation (6.6) of [19]), we obtain that

$$\overline{R}(U, V)W = R_M(U^*, V^*)W^* + ((\log f)'(h))^2(\langle U, W \rangle V - \langle V, W \rangle U)$$
$$- (\log f)''(h)\langle W, \partial_t \rangle(\langle U, \partial_t \rangle V - \langle V, \partial_t \rangle U)$$
$$- (\log f)''(h)(\langle U, W \rangle \langle V, \partial_t \rangle - \langle U, \partial_t \rangle \langle V, W \rangle)\partial_t, \tag{2.64}$$

for arbitrary vector fields U, V, W in $\overline{M}^{n+1}$, where

$$U^* = (\pi_M)_* U = U + \langle U, \partial_t \rangle \partial_t.$$

Then, for an orthonormal basis $\{X, Y\}$ of an arbitrary 2-plane tangent to Σ^n, Eq. (2.64) gives

$$\overline{K}(X, Y) = \frac{1}{f^2(h)} K_M(X^*, Y^*)|X^* \wedge Y^*|^2 +$$
$$+ ((\log f)'(h))^2(\langle X, X \rangle \langle Y, Y \rangle - \langle Y, X \rangle \langle X, Y \rangle)$$
$$- (\log f)''(h)\langle X, \partial_t \rangle(\langle X, \partial_t \rangle \langle Y, Y \rangle - \langle Y, \partial_t \rangle \langle X, Y \rangle)$$
$$- (\log f)''(h)(\langle X, X \rangle \langle Y, \partial_t \rangle - \langle X, \partial_t \rangle \langle Y, X \rangle)\langle \partial_t, Y \rangle$$
$$= \frac{1}{f^2(h)} K_M(X^*, Y^*)|X^* \wedge Y^*|^2 + ((\log f)'(h))^2$$
$$- (\log f)''(h)(\langle X, \partial_t \rangle^2 + \langle Y, \partial_t \rangle^2). \tag{2.65}$$

Since $\nabla h = -\partial_t^\top = -\partial_t - \Theta N$, we have that

$$\langle X, \partial_t \rangle^2 = \langle X, -\nabla h - \Theta N \rangle^2 = \langle X, \nabla h \rangle^2. \tag{2.66}$$

Moreover, one has

$$|X^* \wedge Y^*|^2 = |X^*|^2 |Y^*|^2 - \langle X^*, Y^* \rangle^2$$
$$= \langle X^*, X^* \rangle \langle Y^*, Y^* \rangle - \langle X^*, Y^* \rangle^2$$
$$= (1 + \langle X, \partial_t \rangle^2)(1 + \langle Y, \partial_t \rangle^2) - \langle X, \partial_t \rangle^2 \langle Y, \partial_t \rangle^2$$

2.6 Further Uniqueness Results via Higher Order Mean Curvatures

$$= 1 + \langle X, \partial_t \rangle^2 + \langle Y, \partial_t \rangle^2$$
$$= 1 + \langle X, \nabla h \rangle^2 + \langle Y, \nabla h \rangle^2. \tag{2.67}$$

Consequently, by (2.66), (2.67) and (2.65), we get

$$\overline{K}(X, Y) = \frac{1}{f^2(h)} K_M(X^*, Y^*)(1 + \langle X, \nabla h \rangle^2 + \langle Y, \nabla h \rangle^2) + ((\log f)'(h))^2$$
$$- (\log f)''(h)(\langle X, \nabla h \rangle^2 + \langle Y, \nabla h \rangle^2)$$
$$= \frac{1}{f^2(h)} K_M(X^*, Y^*) + ((\log f)'(h))^2 +$$
$$+ \left(\frac{1}{f^2(h)} K_M(X^*, Y^*) - (\log f)''(h) \right) (\langle X, \nabla h \rangle^2 + \langle Y, \nabla h \rangle^2)$$
$$= \frac{1}{f^2(h)} K_M(X^*, Y^*) + \left(\frac{f'}{f}(h) \right)^2 + \frac{1}{f^2(h)} (K_M(X^*, Y^*)$$
$$- ff'' + (f')^2)(\langle X, \nabla h \rangle^2 + \langle Y, \nabla h \rangle^2). \tag{2.68}$$

Since we are assuming the convergence condition (2.61), by (2.68) we deduce the following inequality

$$\overline{K}(X, Y) \geq \frac{f''}{f}(h). \tag{2.69}$$

Thus, by (2.63) and (2.69) we obtain

$$\kappa_i \kappa_j = \overline{K}(e_i, e_j) - K_{\Sigma^n}(e_i, e_j) \geq \frac{f''}{f}(h) - K_{\Sigma^n}(e_i, e_j). \tag{2.70}$$

Consequently, by (2.62) and (2.70) we have

$$\kappa_i \kappa_j \geq 0, \quad \forall i, j \in \{1, 2, \ldots, n\}, \ i \neq j.$$

Hence, it follows that

$$\langle \mathcal{P}_r e_i, e_i \rangle = \binom{n}{r}^{-1} \sum (\kappa_{j_1} \kappa_{i_1}) \cdot \ldots \cdot (\kappa_{j_r} \kappa_{i_r}) \geq 0. \tag{2.71}$$

Therefore, by (2.71) we conclude that $\mathcal{P}_r$ is positive semi-definite. In addition, since H_r is bounded on Σ^n, the same holds for $tr(\mathcal{P}_r) = c_r H_r^2$, where $c_r = (n-r)\binom{n}{r}$.

Now, by (1.14) we have that

$$L_r \sigma(h) = c_r f(h) \left(-\Theta H_{r+1} - \frac{f'}{f}(h) H_r \right). \qquad (2.72)$$

Thus, we consider the operator $\mathcal{L}_r : C^\infty(\Sigma^n) \to C^\infty(\Sigma^n)$ given by

$$\mathcal{L}_r u = \operatorname{tr}(\mathcal{P}_r \circ \operatorname{Hess} u). \qquad (2.73)$$

Hence, by (2.72) and (2.73), we get

$$\mathcal{L}_r(\sigma(h)) = c_r f(h) H_r^2 \left(-\Theta \frac{H_{r+1}}{H_r} - \frac{f'}{f}(h) \right). \qquad (2.74)$$

Let us suppose now that Σ^n is bounded away from the future infinity of $\overline{M}^{n+1}$. So, by Lemma 1.8 we obtain a sequence $\{p_k\} \subset \Sigma^n$ such that

$$\limsup_k \mathcal{L}_r(\sigma(h(p_k))) \leq 0. \qquad (2.75)$$

Thus, by (2.74) and (2.75) we have

$$\lim_k c_r f(h(p_k)) H_r^2(p_k) \left(-\Theta(p_k) \frac{H_{r+1}}{H_r}(p_k) - \frac{f'}{f}(h(p_k)) \right) \leq 0 \qquad (2.76)$$

and

$$\lim_k |\nabla \sigma(h(p_k))| = \lim_k (f(h(p_k))|\nabla h(p_k)|) = 0. \qquad (2.77)$$

We notice that

$$\lim_k |\nabla h(p_k)| = 0.$$

Indeed, arguing by contradiction, suppose that there exists a subsequence $\{p_{k_j}\}$ of $\{p_k\}$ such that

$$|\nabla h(p_{k_j})| \geq c > 0, \qquad (2.78)$$

for every $j \in \mathbb{N}$.

Then, for each $\varepsilon > 0$, there exists $j_0 \in \mathbb{N}$ such that for every $j > j_0$, one has

$$f(h(p_{k_j}))|\nabla h(p_{k_j})| < c\varepsilon \Rightarrow cf(h(p_{k_j})) < c\varepsilon \Rightarrow f(h(p_{k_j})) < \varepsilon,$$

2.6 Further Uniqueness Results via Higher Order Mean Curvatures

that is

$$\lim_j f(h(p_{k_j})) = 0. \tag{2.79}$$

Hence, up to a subsequence, we have

$$0 < f(\sup_j h(p_{k_j})) = f(\lim_j h(p_{k_j})) = \lim_j f(h(p_{k_j})). \tag{2.80}$$

So, by (2.79) and (2.80) we have a contradiction.

Thus

$$\lim_k |\nabla h(p_k)| = 0,$$

and, hence

$$\lim_k \Theta(p_k) = -1.$$

We also observe that

$$\frac{f'}{f}(\sup_j h(p_{k_j})) = \frac{f'}{f}(\lim_j h(p_{k_j})) = \lim_j \frac{f'}{f}(h(p_{k_j})).$$

Now, by (i) we get

$$\lim_j (-\Theta)(p_{k_j}) \frac{H_{r+1}}{H_r}(p_{k_j}) = \lim_j \frac{H_{r+1}}{H_r}(p_{k_j}) \geq \lim_j \frac{f'}{f}(h(p_{k_j})).$$

Moreover, by (2.76) we have that

$$c_r \lim_j \left(H_r^2(p_{k_j}) f(h(p_{k_j})) \right) \lim_j \left(\frac{H_{r+1}}{H_r}(p_{k_j}) - \frac{f'}{f}(h(p_{k_j})) \right) \leq 0. \tag{2.81}$$

On the other hand, since $\frac{H_{r+1}}{H_r} \geq \frac{f'}{f}(h)$ by assumption, we obtain

$$c_r \lim_j \left(H_r^2(p_{k_j}) f(h(p_{k_j})) \right) \lim_j \left(\frac{H_{r+1}}{H_r}(p_{k_j}) - \frac{f'}{f}(h(p_{k_j})) \right) \geq 0, \tag{2.82}$$

Now, since $0 < \alpha^2 \leq H_r^2 \leq \beta^2$, we have

$$\lim_j \left(H_r^2(p_{k_j}) f(h(p_{k_j})) \right) > 0.$$

Thus, by (2.81) and (2.82) we conclude that

$$\lim_j \left(\frac{H_{r+1}}{H_r}(p_{k_j}) - \frac{f'}{f}(h(p_{k_j})) \right) = 0$$

and, consequently

$$\inf_{\Sigma^n} \left(\frac{H_{r+1}}{H_r} - \frac{f'}{f}(h) \right) = 0.$$

Therefore

$$0 \leq |\nabla h| \leq \inf_{\Sigma^n} \left(\frac{H_{r+1}}{H_r} - \frac{f'}{f}(h) \right) = 0.$$

Hence, Σ^n is a slice of $\overline{M}^{n+1}$.

It remains to consider the case in which the hypersurface Σ^n is bounded away from the past infinity of $\overline{M}^{n+1}$.

Arguing as before, by using once more Lemma 1.8, we obtain a sequence $\{p_{k_j}\} \subset \Sigma^n$ such that

$$c_r \lim_j \left(H_r^2(p_{k_j}) f(h(p_{k_j})) \right) \lim_k \left(\frac{H_{r+1}}{H_r}(p_{k_j}) - \frac{f'}{f}(h(p_{k_j})) \right) \geq 0.$$

We also have

$$c_r \lim_j \left(H_r^2(p_{k_j}) f(h(p_{k_j})) \right) \lim_k \left(\frac{H_{r+1}}{H_r}(p_{k_j}) - \frac{f'}{f}(h(p_{k_j})) \right) \leq 0.$$

Since

$$\lim_j \left(H_r^2(p_{k_j}) f(h(p_{k_j})) \right) > 0,$$

then

$$\lim_j \left(\frac{H_{r+1}}{H_r}(p_{k_j}) - \frac{f'}{f}(h(p_{k_j})) \right) = 0.$$

Hence

$$\inf_{\Sigma^n} \left(\frac{f'}{f}(h) - \frac{H_{r+1}}{H_r} \right) = 0.$$

Therefore, as in the previous case, the hypersuface Σ^n must be a slice of $\overline{M}^{n+1}$. □

2.6 Further Uniqueness Results via Higher Order Mean Curvatures

2.6.2 Rigidity of Complete Two-Sided Hypersurfaces

In this section we present the Riemannian dual of Theorem 2.12.

The main result assume the following form.

Theorem 2.13 *Let $\overline{M}^{n+1} = I \times_f M^n$ be a Riemannian warped product whose fiber M^n has sectional curvature K_M satisfying the convergence condition*

$$K_M \geq \sup_I((f')^2 - ff''). \tag{2.83}$$

Moreover, let $\psi : \Sigma^n \to \overline{M}^{n+1}$ be a complete two-sided hypersurface with angle function $\Theta \leq 0$ and whose sectional curvature K_{Σ^n} is bounded from below and such that

$$K_{\Sigma^n} \geq \frac{1}{f^2(h)}(K_M - (f'(h))^2). \tag{2.84}$$

Suppose that, for some $r \in \{0, 1, \ldots, n-1\}$, there exist positive constants α and β such that $\alpha \leq H_r \leq \beta$ and that one of the following conditions is satisfied

(i) Σ^n is bounded away from the future infinity of $\overline{M}^{n+1}$ and $\dfrac{f'}{f}(h) \geq \dfrac{H_{r+1}}{H_r}$.

(ii) Σ^n is bounded away from the past infinity of $\overline{M}^{n+1}$ and $\dfrac{f'}{f}(h) \leq \dfrac{H_{r+1}}{H_r}$.

If $|\nabla h| \leq \inf_{\Sigma^n} \left| \dfrac{H_{r+1}}{H_r} - \dfrac{f'}{f}(h) \right|$, then Σ^n is a slice $\{t\} \times M^n$.

Proof As in the proof of Theorem 2.12, we define the self-adjoint operator $\mathcal{P}_r : \mathfrak{X}(\Sigma^n) \to \mathfrak{X}(\Sigma^n)$ as follows

$$\mathcal{P}_r = H_r P_r.$$

For each $p \in \Sigma^n$, we take a local orthonormal frame $\{e_1, \ldots, e_n\}$ such that $Ae_i = \kappa_i e_i$. By (1.6) we have that

$$P_r e_i = \sum_{i_1 < \ldots < i_r, i_j \neq i} \kappa_{i_1} \cdot \ldots \cdot \kappa_{i_r}.$$

Thus, for any $i \in \{1, \ldots, n\}$, we get

$$\langle \mathcal{P}_r e_i, e_i \rangle = \binom{n}{r}^{-1} \sum_{i_1 < \ldots < i_r, i_j \neq i, j_1 < \ldots < j_r} (\kappa_{i_1} \kappa_{j_1}) \cdot \ldots \cdot (\kappa_{i_r} \kappa_{j_r}).$$

Now, the Gauss equation yields

$$K_{\Sigma^n}(e_i, e_j) = \overline{K}(e_i, e_j) + \kappa_i \kappa_j.$$

Thus, by using [251, Proposition 7.42], it follows that

$$\overline{R}(U, V)W = R_M(U^*, V^*)W^* - ((\log f)'(h))^2(\langle U, W \rangle V - \langle V, W \rangle U)$$
$$- (\log f)''(h)\langle W, \partial_t \rangle(\langle U, \partial_t \rangle V - \langle V, \partial_t \rangle U)$$
$$- (\log f)''(h)(\langle U, W \rangle \langle V, \partial_t \rangle - \langle U, \partial_t \rangle \langle V, W \rangle)\partial_t,$$

for arbitrary vector fields U, V, W in $\overline{M}^{n+1}$, where

$$U^* = (\pi_M)_* U = U - \langle U, \partial_t \rangle \partial_t.$$

Then, for an orthonormal basis $\{X, Y\}$ we find that

$$\overline{K}(X, Y) = \frac{1}{f^2(h)} K_M(X^*, Y^*) |X^* \wedge Y^*|^2$$
$$- ((\log f)'(h))^2 - (\log f)''(h)(\langle X, \nabla h \rangle^2 + \langle Y, \nabla h \rangle^2).$$

Moreover, since

$$|X^* \wedge Y^*|^2 = 1 - \langle X, \nabla h \rangle^2 - \langle Y, \nabla h \rangle^2,$$

we easily have

$$\overline{K}(X, Y) = \frac{1}{f^2(h)} K_M(X^*, Y^*)(1 - (\langle X, \nabla h \rangle^2 + \langle Y, \nabla h \rangle^2))$$
$$- ((\log f)'(h))^2 - (\log f)''(h)(\langle X, \nabla h \rangle^2 + \langle Y, \nabla h \rangle^2)$$
$$= \frac{1}{f^2(h)} K_M(X^*, Y^*) - ((\log f)'(h))^2$$
$$- \left(\frac{1}{f^2(h)} K_M(X^*, Y^*) + (\log f)''(h) \right) (\langle X, \nabla h \rangle^2 + \langle Y, \nabla h \rangle^2)$$
$$= \frac{1}{f^2(h)} K_M(X^*, Y^*) - \left(\frac{f'}{f}(h) \right)^2$$
$$- \left(\frac{1}{f^2(h)} K_M(X^*, Y^*) + \frac{ff'' - (f')^2}{f^2}(h) \right) (\langle X, \nabla h \rangle^2 + \langle Y, \nabla h \rangle^2).$$

2.6 Further Uniqueness Results via Higher Order Mean Curvatures

We also notice that

$$\kappa_i \kappa_j = K_{\Sigma^n}(e_i, e_j) - \overline{K}(e_i, e_j)$$

$$= K_{\Sigma^n}(e_i, e_j) + \left(\frac{f'}{f}(h)\right)^2 - \frac{1}{f^2(h)} K_M(X^*, Y^*) +$$

$$+ \frac{1}{f^2(h)}(K_M(X^*, Y^*) + (ff'' - (f')^2)(h))(\langle X, \nabla h \rangle^2 + \langle Y, \nabla h \rangle^2).$$

Hence, taking into account the convergence condition (2.83) and (2.84), we get

$$\kappa_i \kappa_j \geq 0$$

for every $i, j \in \mathbb{N}, i \neq j$. Therefore, it follows that

$$\langle \mathcal{P}_r e_i, e_i \rangle \geq 0,$$

and $\mathcal{P}_r$ is positive semi-definite. In addition, the operator $\text{tr}(\mathcal{P}_r) = c_r H_r^2$ is bounded on Σ^n, since H_r is too.

Now, assume that Σ^n is bounded away from the future infinity of $\mathbb{R} \times_f M^n$, that is, $h^* = \sup_{\Sigma^n} h < +\infty$. By (1.14) we get

$$L_r \sigma(h) = c_r f(h) H_r \left(\frac{f'}{f}(h) + \Theta \frac{H_{r+1}}{H_r}\right).$$

Consequently, considering the operator $\mathcal{L}_r$ defined in (2.73), we have

$$\mathcal{L}_r(\sigma(h)) = c_r f(h) H_r^2 \left(\frac{f'}{f}(h) + \Theta \frac{H_{r+1}}{H_r}\right).$$

By Lemma 1.8 we obtain a sequence $\{p_k\} \subset \Sigma^n$ such that

$$\limsup_k \mathcal{L}_r \sigma(h(p_k)) \leq 0$$

and

$$\lim_k |\nabla \sigma(h(p_k))| = \lim_k (f(h(p_k))|\nabla h(p_k)|) = 0,$$

that is

$$\lim_k |\nabla h(p_k)| = 0.$$

Hence, since $\Theta \leq 0$, we get that $\lim_k \Theta(p_k) = -1$. Therefore, arguing as in the proof of Theorem 2.12, we obtain

$$\inf_{\Sigma^n} \left(\frac{f'}{f}(h) - \frac{H_{r+1}}{H_r} \right) = 0.$$

In conclusion, taking into account the main assumption on $|\nabla h|$, we conclude that Σ^n must be a slice.

Similarly, if Σ^n is bounded away from the past infinity of $\overline{M}^{n+1}$ and $\frac{f'}{f}(h) \leq \frac{H_{r+1}}{H_r}$, one has

$$\inf_{\Sigma^n} \left(\frac{H_{r+1}}{H_r} - \frac{f'}{f}(h) \right) = 0,$$

which means that Σ^n is a slice. □

Remark 2.3 We notice that in Theorem 2.13 if Σ^n is locally a graph over the fiber M^n, then its angle function Θ is either $\Theta < 0$ or $\Theta > 0$ along Σ^n. Hence, the assumption that Θ does not change sign is generally weaker than that of Σ^n being a local graph.

Furthermore, by Remark 1.2, one has that the differential inequalities requested in items (i) and (ii) of Theorems 2.12 and 2.13 are natural constrained involving the r-th mean curvatures, that, in general, cannot be constants.

Finally, we also observe that the convergence conditions (2.61) and (2.83) are already present in the current literature. They appear in the works [19, 29, 31]. In particular, according to the terminology established in [19], the convergence condition (2.61) corresponds to the *strong null convergence condition*.

Indeed, the strong null convergence condition is a special case of the so-called *null convergence condition*, which means that the Ricci curvature of the ambient spacetime is nonnegative on null or lightlike directions.

More details concerning the above convergence conditions can be found in [19,25,243].

3 Height Estimates and Half-Space Theorems

3.1 Introduction

The last decades have seen a steadily growing interest in the study of a priori estimates of the height function of constant mean curvature compact graphs or, more generally, compact hypersurfaces with boundary having some constant higher order mean curvature immersed into semi-Riemannian product spaces of the type $\mathbb{R} \times M^n$ or $-\mathbb{R} \times M^n$, where M^n is an arbitrary Riemannian manifold.

This problem has gained special attention, being considered by several authors probably motivated by the fact that these estimates turn out to be a very useful tool in order to investigate existence and uniqueness results of complete hypersurfaces with constant higher order mean curvature, as well as to obtain information on the topology at infinity of such hypersurfaces; see, for instance, [14, 21, 111, 140, 154, 175, 205] Riemannian setting, as well as [120, 136, 234] for the Lorentzian case.

In the context of the Riemannian geometry, in 1969 the first height estimate of compact graphs with positive constant mean curvature in the Euclidean space $\mathbb{R}^3$ and boundary on a plane was obtained by Heinz [202]. More specifically, denoting by H the mean curvature, Heinz proved that the height of such a graph can rise at most $1/H$. In [222], Korevaar, Kusner, Meeks, III, and Solomon obtained a sharp bound of compact embedded hypersurfaces in the hyperbolic space $\mathbb{H}^{n+1}$ with nonzero constant mean curvature and boundary contained into a totally geodesic hyperplane.

Next, Rosenberg [222] gave height estimates of compact hypersurfaces with some constant higher order mean curvature and with zero boundary values embedded either in the Euclidean space $\mathbb{R}^{n+1}$ or in the hyperbolic space $\mathbb{H}^{n+1}$, generalizing the previous estimates of [202] and [222].

Later on, Cheng and Rosenberg [111] were able to generalize these estimates for compact graphs with some constant higher order mean curvature in the product manifold

$\mathbb{R} \times M^n$, with boundary contained into a slice $\{t_0\} \times M^n$, for some $t_0 \in \mathbb{R}$. When the fiber M^n is compact, as application of their height estimates, they used the Alexandrov's reflection method in order to obtain some information on the topology at infinity of noncompact properly embedded hypersurface having constant higher order mean curvature, proving that such a hypersurface must have at least two ends or, equivalently, it cannot lie in a half-space.

Afterwards, Alías and Dajczer [21] and García-Martínez et al. [189] gave extensions of the aforementioned results to the case of hypersurfaces immersed into the so called pseudo-hyperbolic spaces, that is, following the terminology introduced by Tashiro [282], Riemannian warped products of the type $\mathbb{R} \times_{e^t} M^n$ or $\mathbb{R} \times_{\cosh t} M^n$.

More precisely, in [21] the authors studied constant mean curvature hypersurfaces, as well as, in [189], the case of hypersurfaces with some constant higher order mean curvature is investigated. Moreover, in [189], some topological results, for noncompact properly immersed hypersurfaces of constant mean and higher order mean curvature of these pseudo-hyperbolic spaces, have been proved in the same spirit of [111], by assuming that the fiber M^n is compact.

Towards the Lorentzian context, the first result in this direction is due to López [234], who proved that compact spacelike surfaces with constant mean curvature in the 3-dimensional Lorentz-Minkowski spacetime $\mathbb{R}_1^3$ with boundary on a plane can reach at most a height of $(|H|A)/2\pi$, where A is the area of the region of the surface above the plane containing its boundary.

Later on, Montiel [244] obtained height estimates of compact spacelike graphs in the steady state spacetime and he applied them to prove some existence and uniqueness theorems for complete spacelike hypersurfaces in the de Sitter spacetime with constant mean curvature $H > 1$ and prescribed asymptotic future boundary.

Also, de Lima studied height estimates and obtained a sharp estimate of compact spacelike hypersurfaces with some constant higher order mean curvature in the Lorentz-Minkowski spacetime $\mathbb{R}_1^{n+1}$ and with boundary contained into a spacelike hyperplane; see [136]. Successively, in a joint work with Colares, they were able to generalize these estimates to the case of compact spacelike hypersurfaces of positive constant higher order mean curvature in Lorentzian product spacetime $-\mathbb{R} \times M^n$, whose fiber has nonnegative sectional curvature, and with boundary contained into a slice; see [120].

Finally, García-Martínez and Impera [188] extended the height estimates proved by Colares and de Lima for spacelike hypersurfaces of constant higher order mean curvature in a Lorentzian warped product $-\mathbb{R} \times_f M^n$, the so called generalized Robertson-Walker (GRW) spacetimes, with boundary contained into a slice. As application they obtained information on the topology at infinity of constant higher order mean curvature complete spacelike hypersurfaces immersed into a spatially closed GRW spacetime. Moreover, using a version of the Omori-Yau's maximum principle for trace type differential operators, they also gave some half-space results concerning complete spacelike hypersurfaces of constant higher order mean curvature immersed into the non-spatially closed GRW spacetime.

In this chapter we present height estimates and study the topology at infinity, in form of half-space theorems, of generalized linear Weingarten hypersurfaces immersed into either a Riemannian warped product or a GRW spacetime.

The results are mainly based on the papers [122, 139, 140, 154].

3.2 Auxiliary Results and Setup

Given a semi-Riemannian warped product $\overline{M}^{n+1} = \epsilon I \times_f M^n$, let us denote by $\overline{R}$ and R_M the curvatures tensors of the ambient space $\overline{M}^{n+1}$ and of the fiber M^n, respectively. Here, by following [251], the curvature tensor of $\overline{M}^{n+1}$ is defined as follows: if $U, V, W \in \mathfrak{X}(\overline{M}^{n+1})$ then

$$\overline{R}(U, V)W = \overline{\nabla}_{[X,Y]} Z - [\overline{\nabla}_X, \overline{\nabla}_Y] Z,$$

where $\overline{\nabla}$ stands for the Levi-Civita connection of $\overline{M}^{n+1}$ and $[\cdot, \cdot]$ denotes the standard Lie bracket.

A proof of the fundamental properties collected in the next lemma can be found in [251, Lemma 7.35 and Proposition 7.42].

Lemma 3.1 *Let* $\overline{M}^{n+1} = \epsilon I \times_f M^n$ *be a semi-Riemannian warped product. Moreover, let* $f : I \to \mathbb{R}$ *be a smooth function on I and $U, V, W \in \mathfrak{X}(\overline{M}^{n+1})$. We have:*

(a) $\overline{\nabla}(f \circ \pi_I) = \epsilon f' \partial_t$;
(b) $\overline{R}(U, \partial_t) \partial_t = \dfrac{f''}{f} U$;
(c) $\overline{R}(\partial_t, \partial_t) U = \overline{R}(U, V) \partial_t = 0$;
(d) $\overline{R}(\partial_t, U) V = \dfrac{\langle U, V \rangle}{f} \epsilon f'' \partial_t$;
(e) $\overline{R}(U, V)W = R_M(U, V)W - \dfrac{\epsilon (f')^2}{f^2} (\langle U, W \rangle V - \langle V, W \rangle U)$.

In order to simplify the notation, when no confusion occurs, we will write f instead of $f \circ \pi_I$ and $\overline{\nabla} f$ instead of $\overline{\nabla}(f \circ \pi_I)$.

A meaningful consequence of Lemma 3.1 is given in Lemma 3.2 below. Here, we shall prove a very useful relation involving the curvatures tensors $\overline{R}$ and R_M, as well as the derivatives of the warping function f of a semi-Riemannian warped product $\epsilon I \times_f M^n$.

The aforementioned result is not easily available in the literature. For this reason we report below a self-contained proof of the next lemma.

Lemma 3.2 Let $\overline{M}^{n+1} = \epsilon I \times_f M^n$ be a semi-Riemannian warped product. For every $U, V, W \in \mathfrak{X}(\overline{M}^{n+1})$, the following relation holds:

$$\overline{R}(U,V)W = R_M(U^*, V^*)W^* - \epsilon[(\log f)']^2 (\langle U, W\rangle V - \langle V, W\rangle U)$$
$$- (\log f)'' \langle W, \partial_t\rangle (\langle U, \partial_t\rangle V - \langle V, \partial_t\rangle U)$$
$$- (\log f)'' (\langle U, W\rangle\langle V, \partial_t\rangle - \langle V, W\rangle\langle U, \partial_t\rangle) \partial_t,$$

where we are writing $U^* = U - \epsilon \langle U, \partial_t\rangle \partial_t$ to denote the orthogonal projection of U onto TM^n.

Proof To prove this result we will use as main tool Lemma 3.1. So, by using the C^∞-linearity of the tensor $\overline{R}$ we get

$$\overline{R}(U,V)W = \overline{R}(U^*, V)W + \epsilon\langle U, \partial_t\rangle \overline{R}(\partial_t, V)W$$
$$= \overline{R}(U^*, V^*)W + \epsilon\langle V, \partial_t\rangle \overline{R}(U^*, \partial_t)W$$
$$+ \epsilon\langle U, \partial_t\rangle \overline{R}(\partial_t, V^*)W + \langle U, \partial_t\rangle\langle V, \partial_t\rangle \overline{R}(\partial_t, \partial_t)W$$
$$= \overline{R}(U^*, V^*)W^* + \epsilon\langle W, \partial_t\rangle \overline{R}(U^*, V^*)\partial_t + \epsilon\langle V, \partial_t\rangle \overline{R}(U^*, \partial_t)W^*$$
$$+ \langle V, \partial_t\rangle\langle W, \partial_t\rangle \overline{R}(U^*, \partial_t)\partial_t + \epsilon\langle U, \partial_t\rangle \overline{R}(\partial_t, V^*)W^*$$
$$+ \langle U, \partial_t\rangle\langle W, \partial_t\rangle \overline{R}(\partial_t, V^*)\partial_t + \langle U, \partial_t\rangle\langle V, \partial_t\rangle \overline{R}(\partial_t, \partial_t)W.$$

By Lemma 3.1-item (c), we have $\overline{R}(U^*, V^*)\partial_t = \overline{R}(\partial_t, \partial_t)W = 0$. The previous equation becomes

$$\overline{R}(U,V)W = \overline{R}(U^*, V^*)W^* + \epsilon\langle V, \partial_t\rangle \overline{R}(U^*, \partial_t)W^*$$
$$+ \langle V, \partial_t\rangle\langle W, \partial_t\rangle \overline{R}(U^*, \partial_t)\partial_t + \epsilon\langle U, \partial_t\rangle \overline{R}(\partial_t, V^*)W^*$$
$$+ \langle U, \partial_t\rangle\langle W, \partial_t\rangle \overline{R}(\partial_t, V^*)\partial_t,$$

which can be rewritten as

$$\overline{R}(U,V)W = \overline{R}(U^*, V^*)W^* - \epsilon\langle V, \partial_t\rangle \overline{R}(\partial_t, U^*)W^*$$
$$+ \langle V, \partial_t\rangle\langle W, \partial_t\rangle \overline{R}(U^*, \partial_t)\partial_t + \epsilon\langle U, \partial_t\rangle \overline{R}(\partial_t, V^*)W^*$$
$$- \langle U, \partial_t\rangle\langle W, \partial_t\rangle \overline{R}(V^*, \partial_t)\partial_t, \quad (3.1)$$

where we use the symmetries properties of the curvature tensor $\overline{R}$.

3.2 Auxiliary Results and Setup

Let us compute each term of the right side of (3.1). To this aim, Lemma 3.1-item (e) implies that

$$\overline{R}(U^*, V^*)W^* = R_M(U^*, V^*)W^* - \frac{\langle \overline{\nabla} f, \overline{\nabla} f \rangle}{f^2} \left(\langle U^*, W^* \rangle V^* - \langle V^*, W^* \rangle U^* \right)$$

$$= R_M(U^*, V^*)W^* - \frac{\epsilon(f')^2}{f^2} \left(\langle U^*, W^* \rangle V^* - \langle V^*, W^* \rangle U^* \right). \quad (3.2)$$

Moreover, we observe that

$$\langle V^*, W^* \rangle = \langle V, W \rangle - \epsilon \langle V, \partial_t \rangle \langle W, \partial_t \rangle,$$

which gives

$$\langle U^*, W^* \rangle V^* = \langle U, W \rangle V - \epsilon \langle U, W \rangle \langle V, \partial_t \rangle \partial_t - \epsilon \langle U, \partial_t \rangle \langle W, \partial_t \rangle V$$
$$+ \langle U, \partial_t \rangle \langle V, \partial_t \rangle \langle W, \partial_t \rangle,$$

and

$$\langle V^*, W^* \rangle U^* = \langle V, W \rangle U - \epsilon \langle V, W \rangle \langle U, \partial_t \rangle \partial_t - \epsilon \langle V, \partial_t \rangle \langle W, \partial_t \rangle U$$
$$+ \langle U, \partial_t \rangle \langle V, \partial_t \rangle \langle W, \partial_t \rangle.$$

Hence, we find

$$\langle U^*, W^* \rangle V^* - \langle V^*, W^* \rangle U^* = \langle U, W \rangle V - \langle V, W \rangle U$$
$$+ \epsilon \left(\langle V, W \rangle \langle U, \partial_t \rangle - \langle U, W \rangle \langle V, \partial_t \rangle \right) \partial_t$$
$$- \epsilon \langle W, \partial_t \rangle \left(\langle U, \partial_t \rangle V - \langle V, \partial_t \rangle U \right).$$

The above equation and (3.2) yield

$$\overline{R}(U^*, V^*)W^* = R_M(U^*, V^*)W^* - \frac{\epsilon(f')^2}{f^2} \left(\langle U, W \rangle V - \langle V, W \rangle U \right)$$
$$- \frac{(f')^2}{f^2} \left(\langle V, W \rangle \langle U, \partial_t \rangle - \langle U, W \rangle \langle V, \partial_t \rangle \right) \partial_t$$
$$+ \frac{(f')^2}{f^2} \langle W, \partial_t \rangle \left(\langle U, \partial_t \rangle V - \langle V, \partial_t \rangle U \right). \quad (3.3)$$

Next, by Lemma 3.1-item (d) we have that

$$\overline{R}(\partial_t, U^*)W^* = \frac{\langle U^*, W^*\rangle}{f}\overline{\nabla}_{\partial_t}\overline{\nabla}f$$

$$= \left(\frac{\langle U, W\rangle - \epsilon\langle U, \partial_t\rangle\langle W, \partial_t\rangle}{f}\right)\epsilon f''\partial_t$$

$$= \frac{\epsilon f''}{f}\langle U, W\rangle\partial_t - \frac{f''}{f}\langle U, \partial_t\rangle\langle W, \partial_t\rangle\partial_t. \quad (3.4)$$

and

$$\overline{R}(\partial_t, V^*)W^* = \frac{\epsilon f''}{f}\langle V, W\rangle\partial_t - \frac{f''}{f}\langle V, \partial_t\rangle\langle W, \partial_t\rangle\partial_t. \quad (3.5)$$

By using again Lemma 3.1-item (b) we also have

$$\overline{R}(U^*, \partial_t)\partial_t = \frac{f''}{f}U^* = \frac{f''}{f}U - \epsilon\frac{f''}{f}\langle U, \partial_t\rangle\partial_t \quad (3.6)$$

and

$$\overline{R}(V^*, \partial_t)\partial_t = \frac{f''}{f}V^* = \frac{f''}{f}V - \epsilon\frac{f''}{f}\langle V, \partial_t\rangle\partial_t \quad (3.7)$$

Then, by (3.3), (3.4), (3.5), (3.6), as well as (3.7), in addition to (3.1), we get the following computation

$$\overline{R}(U, V)W = R_M(U^*, V^*)W^* - \frac{\epsilon(f')^2}{f^2}(\langle U, W\rangle V - \langle V, W\rangle U)$$

$$- \frac{(f')^2}{f^2}(\langle V, W\rangle\langle U, \partial_t\rangle - \langle U, W\rangle\langle V, \partial_t\rangle)\partial_t$$

$$+ \frac{(f')^2}{f^2}\langle W, \partial_t\rangle(\langle U, \partial_t\rangle V - \langle V, \partial_t\rangle U)$$

$$- \frac{f''}{f}\langle U, W\rangle\langle V, \partial_t\rangle\partial_t + \frac{f''}{f}\langle V, W\rangle\partial_t$$

$$+ \frac{f''}{f}\langle V, \partial_t\rangle\langle W, \partial_t\rangle U - \frac{f''}{f}\langle U, \partial_t\rangle\langle W, \partial_t\rangle V$$

$$= R_M(U^*, V^*)W^* - \epsilon[(\log f)']^2(\langle U, W\rangle V - \langle V, W\rangle U)$$

3.2 Auxiliary Results and Setup

$$-\left(\frac{f''}{f} - \frac{(f')^2}{f^2}\right) \langle W, \partial_t \rangle (\langle U, \partial_t \rangle V - \langle V, \partial_t \rangle U)$$

$$-\left(\frac{f''}{f} - \frac{(f')^2}{f^2}\right) (\langle U, W \rangle \langle V, \partial_t \rangle - \langle V, W \rangle \langle U, \partial_t \rangle) \partial_t.$$

Finally, we conclude that

$$\overline{R}(U, V)W = R_M(U^*, V^*)W^* - \epsilon [(\log f)']^2 (\langle U, W \rangle V - \langle V, W \rangle U)$$

$$-(\log f)'' \langle W, \partial_t \rangle (\langle U, \partial_t \rangle V - \langle V, \partial_t \rangle U)$$

$$-(\log f)'' (\langle U, W \rangle \langle V, \partial_t \rangle - \langle V, W \rangle \langle U, \partial_t \rangle) \partial_t$$

as desired. □

As in the previous chapter, we will deal here with (connected) Riemannian immersions $\psi : \Sigma^n \to \overline{M}^{n+1}$ into a semi-Riemannian warped product $\overline{M}^{n+1} = \epsilon I \times_f M^n$. If $\overline{M}^{n+1}$ is a Riemannian warped product, we will assume that Σ^n is a two-sided hypersurface, which means that its normal bundle is trivial. In other words, there exists an unitary normal vector field N globally defined on Σ^n.

Otherwise, if $\overline{M}^{n+1}$ is a GRW spacetime, Σ^n is assumed to be a spacelike hypersurface, i.e. the induced metric on Σ^n via the immersion ψ is a Riemannian metric. In the latter case, since ∂_t is a timelike vector field globally defined on $\overline{M}^{n+1}$, there exists a unique unitary timelike normal vector field (also denoted by) N globally defined on Σ^n which is either in the same time-orientation of ∂_t, so that $\Theta = \langle N, \partial_t \rangle \leq -1$, or in the opposite time-orientation of ∂_t, that is, $\Theta \geq 1$.

In this case, we will refer to the normal vector field N as been the future-pointing Gauss map of Σ^n when N is in the same time-orientation of ∂_t. Its opposite will be referred as been the past-pointing Gauss map of Σ^n. As usual, in both the cases we also denote by $\langle \cdot, \cdot \rangle$ the metric of Σ^n induced via ψ.

We conclude this section by introducing a wide class of Riemannian immersions, the so called generalized linear Weingarten hypersurfaces.

Let $\psi : \Sigma^n \to \overline{M}^{n+1}$ be a Riemannian immersion into a semi-Riemannian warped product $\overline{M}^{n+1} = \epsilon I \times_f M^n$.

We say that Σ^n is (r, s)-linear Weingarten, for some $0 \leq r \leq s \leq n - 1$, if there exist nonnegative real numbers $b_r, \ldots, b_s$ (at least one of them nonzero) such that the following linear relation holds on Σ^n:

$$\sum_{k=r}^{s} b_k H_{k+1} = d \in \mathbb{R}. \tag{3.8}$$

Thus, naturally attached to a (r, s)-linear Weingarten Riemannian immersion we have the constant d given by (3.8).

We notice that the (r, r)-linear Weingarten Riemannian immersions are exactly the Riemannian immersions having $d = H_{r+1}$ constant.

In particular, this class of hypersurfaces is more general than those having some constant higher order mean curvature.

On the other hand, if the ambient space has zero sectional curvature and taking into account that in this case $\overline{S} = \epsilon H_2$, where $\overline{S}$ stands for the normalized scalar curvature of Σ^n, we observe that the $(0, 1)$-linear Weingarten Riemannian immersions are called simply linear Weingarten hypersurfaces.

Throughout Part II, we will always denote by d the constant given by (3.8).

3.3 Height Estimates in Riemannian Warped Products

The main intention of this section is to provide height estimates of compact (r, s)-linear Weingarten hypersurfaces immersed into Riemannian warped product spaces of the type $\mathbb{R} \times_f M^n$. Moreover, some particular cases are also studied.

More precisely, we will focus our attention to the cases in which the ambient space is either a pseudo-hyperbolic space, that is, either $f(t) = e^t$ or $f(t) = \cosh t$, or a standard product $\mathbb{R} \times M^n$, i.e. the warping function is constant.

The starting point is to prove that, under suitable assumptions on a (not necessarily constant) linear combination involving some of the higher order mean curvatures, any compact two-sided hypersurface immersed into a Riemannian warped product $\mathbb{R} \times_f M^n$ with non-empty boundary and contained into a slice $\{t_0\} \times M^n$, for some $t_0 \in \mathbb{R}$, must lie entirely in one of the two regions of the ambient space bounded by the slice.

We point out that the above fact was proved in [189] considering only one of the higher order mean curvatures being constant; see [189, Proposition 3.3]. For that reason and for the sake of completeness we give here a detailed proof of this claim.

In order to do this we recall an interesting tangency principle due to Fontenele and Silva [181]. Let Σ_1 and Σ_2 be a two hypersurfaces in an arbitrary Riemannian manifold $\overline{N}^{n+1}$ that are tangent at a common point p_0, that is $T_{p_0}\Sigma_1 = T_{p_0}\Sigma_2$.

Fix a normal vector η_0 at p_0 and locally parametrize both hypersurfaces in a neighborhood U of zero in $T_{p_0}\Sigma_1 = T_{p_0}\Sigma_2$ making use of the exponential map of $\overline{N}^{n+1}$ as follows:

$$\zeta_i(x) = \exp_{p_0}(x + \delta_i(x)\eta_0), \quad i = 1, 2,$$

where $\delta_i : U \to \mathbb{R}$ are well-determined functions satisfying $\delta_i(0) = 0$.

Following [181], we say that Σ_1 remains above Σ_2 in a neighborhood of p_0 with respect to η_0 if $\delta_1(x) \geq \delta_2(x)$ in a neighborhood of zero.

3.3 Height Estimates in Riemannian Warped Products

This fact is equivalent to require that the geodesics of the ambient space $\overline{N}^{n+1}$ normal to the hypersurface $\exp_{p_0}(U)$ in a neighborhood of p_0 in the orientation determined by η_0 intercept Σ_2 before Σ_1.

Taking into account the above remarks, we can state the following result; for more details, see [181, Theorem 1.1].

Lemma 3.3 *Let Σ_1 and Σ_2 be hypersurfaces as above such that Σ_1 remains above Σ_2 in a neighborhood of p_0 with respect to η_0. Assume that, for some $0 \le r \le n-1$, we have*

$$H^{\Sigma_1}_{r+1}(x) \le H^{\Sigma_2}_{r+1}(x)$$

in a neighborhood of zero in $T_{p_0}\Sigma_1 = T_{p_0}\Sigma_2$ and, if $r \ge 1$, the principal curvature vector

$$\kappa^{\Sigma_2}(0) = \left(\kappa_1^{\Sigma_2}(0), \ldots, \kappa_n^{\Sigma_2}(0)\right)$$

of Σ_2 at zero belongs to connected component in $\mathbb{R}^{n+1}$, Γ_{r+1}, of the set $\{S_{r+1} > 0\}$ containing $(1, \ldots, 1)$. Then, Σ_1 and Σ_2 coincide in a neighborhood of p_0.

Remark 3.1 We notice that if $H^{\Sigma_2}_{r+1}$ is positive, the assumption $\kappa^{\Sigma_2}(0) \in \Gamma_{r+1}$ trivially holds.

The result below can be viewed as a generalization of [189, Proposition 3.3].

Proposition 3.1 *Let $\psi : \Sigma^n \to \mathbb{R} \times_f M^n$ be a compact two-sided hypersurface with positive $(s+1)$-mean curvature, for some $0 \le s \le n-1$, and boundary $\partial \Sigma$ contained into the slice $\{t_0\} \times M^n$ for some $t_0 \in \mathbb{R}$. The following holds:*

(a) If f' does not change sign on $(-\infty, t_0]$ and, for some $0 \le r \le s$, we have

$$\sum_{k=r}^{s} b_k H_{k+1} \ge \sum_{k=r}^{s} b_k \sup_{(-\infty, t_0]} [(\log f)']^{k+1}$$

for certain nonnegative constants b_k and, when $s \ge 2$, there exists an elliptic point in Σ^n, then $h \ge t_0$;

(b) If $f' > 0$ on $[t_0, +\infty)$ and, for some $0 \le r \le s$, we have

$$\sum_{k=r}^{s} b_k H_{k+1} \le \sum_{k=r}^{s} b_k \inf_{[t_0, +\infty)} [(\log f)']^{k+1}$$

for certain nonnegative constants b_k and, when $s \ge 2$, there exists an elliptic point in Σ^n, then $h \le t_0$.

Proof In order to prove (a) let us observe that there exists $i_0 \in \{r, \ldots, s\}$ satisfying

$$H_{i_0+1}(p) \geq \sup_{(-\infty, t_0]} [(\log f)']^{i_0+1}, \qquad (3.9)$$

for all $p \in \Sigma^n$.

Clearly we can suppose that $r < s$ and, by Lemma 1.1, we have that H_{k+1} is positive for every $k = r, \ldots, s$. Since f' does not change sign on $(-\infty, t_0]$, we have two possibilities: either $f' \geq 0$ on $(-\infty, t_0]$ or $f' \leq 0$ on $(-\infty, t_0]$.

In the first case, we will prove the claim arguing by contradiction. Thus, let us assume that (3.9) is false.

So, we can choose a point $p \in \Sigma^n$ such that

$$H_{r+1}(p)^{1/(r+1)} < \sup_{(-\infty, t_0]} (\log f)'. \qquad (3.10)$$

On the other hand, by our hypothesis, there exists $r < i \leq s$ such that

$$H_{i+1}(p)^{1/(i+1)} > \sup_{(-\infty, t_0]} (\log f)', \qquad (3.11)$$

which gives

$$H_{r+1}(p)^{1/(r+1)} < H_{i+1}(p)^{1/(i+1)}, \qquad (3.12)$$

leading to a contradiction thanks to Lemma 1.1.

Consequently, taking $i_0 = r$, inequality (3.9) holds. In the latter case, the claim follows since at least one of the numbers $r+1, \ldots, s+1$ is odd.

In order to conclude part (a), arguing again by contradiction, let us assume that $h \geq t_0$ is false. Then, there is an interior point $p_1 \in \Sigma^n$ such that

$$\min h = h(p_1) = t_1 < t_0.$$

Setting $\Sigma_1 = \{t_1\} \times M^n$ and $\Sigma_2 = \Sigma^n$, we see that Σ_1 and Σ_2 are tangent at the common point p_1 and that Σ_1 remains above Σ_2 in a neighborhood of p_1 with respect to $-\partial_t$. Moreover, the claim above yields

$$H_{i_0+1}^{\Sigma_2} = H_{i_0+1} \geq \sup_{(-\infty, t_0]} [(\log f)']^{i_0+1} \geq (\log f)'(t_1)^{i_0+1} = H_{i_0+1}^{\Sigma_1}.$$

Hence, we can apply the tangency principle (see Theorem 3.3) to conclude that h is constantly equal to t_1 in a neighborhood of p_1.

Therefore, the set $\{h = t_1\}$ is open and closed in Σ^n. Since Σ^n is a connected set, we get a contradiction.

3.3 Height Estimates in Riemannian Warped Products

Now we prove part (b). Arguing as above, one has the

$$H_{s+1}(p) \leq \inf_{[t_0,+\infty)} [(\log f)']^{s+1} \qquad (3.13)$$

for every $p \in \Sigma^n$.

Assuming again by contradiction that $h \leq t_0$ is false, there exists an interior point $p_2 \in \Sigma^n$ satisfying

$$\max h = h(p_2) = t_2 > t_0.$$

Taking $\Sigma_1 = \Sigma^n$ and $\Sigma_2 = \{t_2\} \times M^n$, we obtain that Σ_1 and Σ_2 are tangent at the common point p_2, Σ_1 remains above Σ_2 in a neighborhood of p_2 with respect to $-\partial_t$ and

$$H_{s+1}^{\Sigma_1} = H_{s+1} \leq \inf_{[t_0,+\infty)} [(\log f)']^{s+1} \leq [(\log f)'(t_2)]^{s+1} = H_{s+1}^{\Sigma_2}.$$

Now, since $f' > 0$, it follows that p_2 is an elliptic point of Σ_2. From now on, arguing as in the proof of part (a) we have a contradiction. $\square$

Remark 3.2 If $r = s$, it is easy to see that we do not need to assume in part (a) of Proposition 3.1 any sign condition on f'.

In our next results, we focus our attention on Riemannian warped product spaces of the type $\mathbb{R} \times_f M^n$ satisfying the convergence condition

$$K_M \geq \sup\{(f')^2 - ff''\}, \qquad (3.14)$$

where K_M stands for the sectional curvature of the fiber M^n.

Warped products satisfying this convergence condition have been studied, for instance, in [31, 189]. The case in which this condition is assumed on the Ricci curvature instead sectional curvature, is also studied; see, for instance, [21, 32, 242].

Now we are ready to state and prove our first main result. More precisely, we will establish an estimate for the height function h of compact (r, s)-linear Weingarten two-sided hypersurfaces in Riemannian warped product spaces of the type $\mathbb{R} \times_f M^n$.

Let us recall again that we will always denote by d the constant given in Eq. (3.8). The main result reads as follows.

Theorem 3.1 *Let $\mathbb{R} \times_f M^n$ be a Riemannian warped product satisfying the convergence condition in (3.14) and with non-decreasing warping function. Moreover, let $\psi : \Sigma^n \to \mathbb{R} \times_f M^n$ be a compact (r, s)-linear Weingarten two-sided hypersurface with positive $(s + 1)$-mean curvature, for some $0 \leq s \leq n - 1$, boundary $\partial \Sigma^n$ contained into the slice*

$\{t_0\} \times M^n$, for some $t_0 \in \mathbb{R}$, and

$$d \geq \sum_{k=r}^{s} b_k \sup_{(-\infty, t_0]} [(\log f)']^{k+1}. \tag{3.15}$$

Suppose that

$$H_1 \geq \max |H_{r+1}|^{1/(r+1)} \tag{3.16}$$

and, if $s \geq 2$, there exists an elliptic point in Σ^n. If the angle function Θ does not change sign on Σ^n, then

$$\Sigma^n \subset \left[t_0, t_0 + \frac{f(\max h)}{f(t_0) \min H_1} \right] \times M^n.$$

Proof We can assume without loss of generality that Σ^n is not a slice, otherwise there is nothing to prove. By Proposition 3.1 we have $h \geq t_0$. In particular, we can choose an interior point p_0 in Σ^n such that the height function attains its maximum.

By Proposition 1.1 we get, at p_0, that

$$0 \geq \Delta h(p_0) = n(\log f)'(h(p_0)) + n\Theta(p_0) H_1(p_0) \geq n\Theta(p_0) H_1(p_0).$$

Since, by Lemma 1.1, the mean curvature is positive, we obtain that $\Theta(p_0) \leq 0$ and Θ is a nonpositive function.

Let us consider on Σ^n the smooth function $\zeta = c\sigma(h) + \tilde{\Theta}$, where $c \in \mathbb{R}$ is a positive constant to be chosen in an appropriate way, σ is a primitive of the warping function f and $\tilde{\Theta} = f\Theta$. Then, Proposition 1.1 yields

$$L_k \zeta = -\frac{c_k f(h)}{k+1} \langle \nabla H_{k+1}, \nabla h \rangle + c_k f'(h) (c H_k - H_{k+1})$$

$$- f(h) \Theta \binom{n}{k+1} (n H_1 H_{k+1} - (n-k-1) H_{k+2} - (k+1) c H_{k+1})$$

$$- \frac{\tilde{\Theta}}{f^2(h)} \sum_{i=1}^{n} \mu_{i,k} K_{M^n}(N^*, E_i^*) |N^* \wedge E_i^*|^2$$

$$- \tilde{\Theta}(\log f)''(h) \left(c_k |\nabla h|^2 H_k - \langle P_k \nabla h, \nabla h \rangle \right), \tag{3.17}$$

where $\{E_1, \ldots, E_n\}$ is an orthonormal frame on Σ^n diagonalizing A with $P_k E_i = \mu_{i,k} E_i$, for every $i = 1, \ldots, n$ and $k = r, \ldots, s$, X^* denotes the orthogonal projection on TM^n and the constants c_k are defined in Lemma 1.2.

3.3 Height Estimates in Riemannian Warped Products

Since H_{s+1} is positive and Σ^n has an elliptic point, by Lemma 1.1, we get that

$$H_1 H_{k+1} - H_{k+2} \geq H_1 H_{k+1} - H_{k+1}^2 H_k^{-1} = \frac{H_{k+1}}{H_k}(H_1 H_k - H_{k+1}).$$

By using once more Lemma 1.1 it follows that

$$H_1 H_{k+1} - H_{k+2} \geq \frac{H_{k+1}}{H_k}\left(H_1 H_k - H_k^{(k+1)/k}\right) = H_{k+1}(H_1 - H_k^{1/k}) \geq 0.$$

Then the previous inequality gives

$$nH_1 H_{k+1} - (n-k-1)H_{k+2} - (k+1)cH_{k+1} = (k+1)H_{k+1}(H_1 - c)$$
$$+ (n-k-1)(H_1 H_{k+1} - H_{k+2})$$
$$\geq (k+1)H_{k+1}(H_1 - c) \geq 0, \quad (3.18)$$

provided that $c = \min H_1$.

The main assumption on H_1 and Lemma 1.1 yield

$$cH_k - H_{k+1} \geq H_{k+1}^{1/(k+1)}(H_k - H_{k+1}^{k/(k+1)}) \geq 0. \quad (3.19)$$

Hence, we can apply Lemma 1.4 (or Lemma 1.3 if $s = 1$) in order to guarantee the ellipticity of the operators L_k for every $k = r, \ldots, s$ (equivalently, every P_k is positive definite).

In particular, the correspondent eigenvalues $\mu_{i,k}$ are all positive on Σ^n, and by the convergence condition in (3.14) we have

$$\mu_{i,k} K_M(N^*, E_i^*)|N^* \wedge E_i^*|^2 \geq \mu_{i,k} C |N^* \wedge E_i^*|^2, \quad (3.20)$$

for every $i = 1, \ldots, n$ and $k = r, \ldots, s$, where we are writing $C = \sup\{(f')^2 - ff''\}$.

By a direct computation, one has that

$$|N^* \wedge E_i^*|^2 = |\nabla h|^2 - \langle E_i, \nabla h \rangle^2,$$

that, in addition to (3.20), imply

$$\sum_{i=1}^n \mu_{i,k} K_M(N^*, E_i^*)|N^* \wedge E_i^*|^2 \geq C\left(\text{tr}(P_k)|\nabla h|^2 - \sum_{i=1}^n \mu_{i,k}\langle E_i, \nabla h\rangle^2\right)$$
$$= C\left(\text{tr}(P_k)|\nabla h|^2 - \langle P_k \nabla h, \nabla h\rangle\right).$$

Now, since $\operatorname{tr}(P_k) = c_k H_k$ and $C/f^2(h) + (\log f)''(h) \geq 0$, we conclude that

$$\frac{\sum_{i=1}^{n} \mu_{i,k} K_M(N^*, E_i^*)|N^* \wedge E_i^*|^2}{f^2(h)} + (\log f)''(h)\Big(c_k|\nabla h|^2 H_k - \langle P_k \nabla h, \nabla h\rangle\Big) \geq 0,$$

where the last inequality is due to the fact that P_k is positive definite.

Hence, by using (3.18), (3.19) and (3.21), taking into account that $\Theta \leq 0$, since the warping function is non-decreasing, by (3.17) we infer that

$$L_k \zeta \geq -\frac{c_k f(h)}{k+1} \langle \nabla H_{k+1}, \nabla h\rangle. \tag{3.21}$$

Next, let us introduce the second order linear differential operator $L : C^\infty(\Sigma^n) \to C^\infty(\Sigma^n)$ defined by

$$L = \sum_{k=r}^{s} (k+1) c_k^{-1} b_k L_k = \operatorname{tr}(P \circ \operatorname{Hess}), \tag{3.22}$$

where the tensor $P : \mathfrak{X}(\Sigma^n) \to \mathfrak{X}(\Sigma^n)$ is given by

$$P = \sum_{k=r}^{s} (k+1) c_k^{-1} b_k P_k. \tag{3.23}$$

Since $(k+1) c_k^{-1} b_k > 0$ for every $k = r, \ldots, s$ and each operator L_k is elliptic (equivalently, each P_k is positive definite) one has that the tensor P is positive definite and, consequently, the operator L is elliptic.

So, by (3.21) and taking into account that Σ^n is (r, s)-linear Weingarten, it follows that

$$L\zeta \geq 0.$$

Since Σ^n is compact, we can apply the weak maximum principle to the elliptic operator L and, since Θ is a nonpositive function, we find

$$c\sigma(h) - f(h) \leq \zeta \leq \max_{\partial \Sigma^n} \zeta = c\sigma(t_0) + f(t_0) \max_{\partial \Sigma^n} \Theta \leq c\sigma(t_0), \tag{3.24}$$

which implies

$$c(\sigma(h) - \sigma(t_0)) \leq f(h) \leq f(\max h). \tag{3.25}$$

3.3 Height Estimates in Riemannian Warped Products

By using once again that f is non-decreasing and σ is increasing, it easily seen that, for any $t \geq t_0$ it holds

$$\sigma(t) - \sigma(t_0) \geq f(t_0)(t - t_0).$$

Since the height function satisfies $h \geq t_0$, the above inequality and (3.25) yield

$$cf(t_0)(h - t_0) \leq f(\max h).$$

Therefore, we conclude that

$$h \leq t_0 + \frac{f(\max h)}{f(t_0) \min H_1}.$$

The proof is now complete. □

We emphasize that if $r = s$, i.e. the hypersurface has constant positive $(s+1)$-mean curvature H_{s+1}, the main assumption $H_1 \geq \max |H_{s+1}|^{1/(s+1)}$ in Theorem 3.1 trivially holds on account of Lemma 1.1.

In particular, the next result holds.

Corollary 3.1 *Let $\mathbb{R} \times_f M^n$ be a Riemannian warped product satisfying the convergence condition in (3.14) and with non-decreasing warping function. Moreover, let $\psi : \Sigma^n \to \mathbb{R} \times_f M^n$ be a compact two-sided hypersurface with constant positive $(s+1)$-mean curvature, for some $0 \leq s \leq n-1$, boundary $\partial \Sigma^n$ contained into the slice $\{t_0\} \times M^n$, for some $t_0 \in \mathbb{R}$, and*

$$H_{s+1} \geq \sup_{(-\infty, t_0]} [(\log f)']^{s+1}.$$

Suppose that there exists an elliptic point in Σ^n. If the angle function Θ does not change sign on Σ^n, then

$$\Sigma^n \subset \left[t_0, t_0 + \frac{f(\max h)}{f(t_0) \min H_1} \right] \times M^n.$$

If the warping function f is either the exponential or the hyperbolic cosine function, following the terminology introduced in [282], the corresponding warped product $\mathbb{R} \times_{e^t} M^n$ or $\mathbb{R} \times_{\cosh t} M^n$ has been referred to as a pseudo-hyperbolic space.

The so-called Tashiro's terminology is due to the fact that, with suitable choices of the fiber M^n, we obtain warped products which are isometric to the hyperbolic space. More details about these spaces can be found in [20, 21, 189, 242].

If the warping function is given by $f(t) = e^t$, we are able to improve the estimate of Theorem 3.1 so that it does not depend on the height function h of the hypersurface Σ^n. Along this direction, the following result is achieved.

Theorem 3.2 *Let $\mathbb{R} \times_{e^t} M^n$ be a pseudo-hyperbolic space whose fiber has nonnegative sectional curvature. Let $\psi : \Sigma^n \to \mathbb{R} \times_{e^t} M^n$ be a compact (r, s)-linear Weingarten two-sided hypersurface with positive $(s + 1)$-mean curvature, for some $0 \leq s \leq n - 1$, boundary $\partial \Sigma^n$ contained into the slice $\{t_0\} \times M^n$, for some $t_0 \in \mathbb{R}$, and*

$$d > \sum_{k=r}^{s} b_k.$$

Suppose that

$$H_1 \geq \max |H_{r+1}|^{1/(r+1)},$$

and, if $s \geq 2$, there exists an elliptic point in Σ^n. If the angle function Θ does not change sign on Σ^n, then

$$\Sigma^n \subset \left[t_0, t_0 + \log\left(\frac{\min H_1}{\min H_1 - 1}\right)\right] \times M^n.$$

Proof By Proposition 3.1 it is clear that $\min H_1 > 1$ and the height function satisfies $h \geq t_0$. We also notice that, in this case, the function ζ in Theorem 3.1 is given by $\zeta = e^h(c + \Theta)$, where $c = \min H_1$ and Θ must be a nonpositive function.

So, by Eq. (3.24), it follows that

$$e^h(\min H_1 - 1) \leq e^{t_0} \min H_1,$$

proving the result. □

For hypersurfaces with constant $(s + 1)$-mean curvature H_{s+1}, Theorem 3.2 improves the estimate obtained in [189, Theorem 3.10].

Indeed, in the above cited result, the height function h satisfies

$$t_0 \leq h \leq t_0 + \log\left(\frac{H_{s+1}^{1/(s+1)}}{H_{s+1}^{1/(s+1)} - 1}\right).$$

3.3 Height Estimates in Riemannian Warped Products

Now, since the real function $f(x) = \dfrac{x}{x-1}$ is decreasing on $(1, +\infty)$, it is easily seen that the following inequality holds

$$\log\left(\frac{\min H_1}{\min H_1 - 1}\right) \leq \log\left(\frac{H_{s+1}^{1/(s+1)}}{H_{s+1}^{1/(s+1)} - 1}\right)$$

for every $s = 0, \ldots, n-1$, which proves the claim.

Next, we consider the case $f(t) = \cosh t$. In this framework, as in Theorem 3.2, the estimate of Theorem 3.1 can be improved as follows.

Theorem 3.3 *Let $\mathbb{R} \times_{\cosh t} M^n$ be a pseudo-hyperbolic space whose fiber has sectional curvature satisfying $K_M \geq -1$. Moreover, let $\psi : \Sigma^n \to \mathbb{R} \times_{\cosh t} M^n$ be a compact (r,s)-linear Weingarten two-sided hypersurface with positive $(s+1)$-mean curvature, for some $0 \leq s \leq n-1$, boundary $\partial \Sigma^n$ contained into the slice $\{0\} \times M^n$ and*

$$d > \sum_{k=r}^{s} b_k.$$

Suppose that

$$H_1 \geq \max |H_{r+1}|^{1/(r+1)}$$

and, if $s \geq 2$, there exists an elliptic point in Σ^n. If the angle function Θ does not change sign on Σ^n, then

$$\Sigma^n \subset \left[0, \tanh^{-1}\left(\frac{1}{\min H_1}\right)\right] \times M^n.$$

Proof By Proposition 3.1 we easily seen that $h \geq 0$. Moreover, in this case, the smooth function ζ defined in Theorem 3.1 is given by $\zeta = c \sinh h + \Theta \cosh h$, where $c = \min H_1$ and the angle function Θ is taken nonpositive.

Since $\sinh t \geq 0$ for every $t \geq 0$, arguing as in the proof of Theorem 3.1 we can prove that (3.24) holds. Then

$$c \sinh h - \cosh h \leq c \sinh 0 = 0.$$

Therefore, the previous inequality yields

$$h \leq \tanh^{-1}\left(\frac{1}{\min H_1}\right),$$

and the theorem is achieved. □

If some higher order mean curvature is constant, Theorem 3.3 improves the estimate given in [189, Theorem 3.11]. Indeed, by [189, Theorem 3.11] it follows that

$$0 \leq h \leq \tanh^{-1}\left(\frac{1}{H_{s+1}^{1/(s+1)}}\right).$$

On the other hand, it is also clear that the following inequality holds

$$\tanh^{-1}\left(\frac{1}{\min H_1}\right) \leq \tanh^{-1}\left(\frac{1}{H_{s+1}^{1/(s+1)}}\right)$$

for every $s = 0, \ldots, n-1$.

Now, let us consider the simple case in which the warping function is constant. In this setting, Theorem 3.1 assume the following form.

Theorem 3.4 *Let $\mathbb{R} \times M^n$ be a product space whose fiber has nonnegative sectional curvature K_{M^n}. Let $\psi : \Sigma^n \to \mathbb{R} \times M^n$ be a compact (r, s)-linear Weingarten two-sided hypersurface with $(s+1)$-mean curvature $H_{s+1} \neq 0$ on Σ^n, for some $0 \leq s \leq n-1$, and boundary $\partial \Sigma^n$ contained into the slice $\{t_0\} \times M^n$ for some $t_0 \in \mathbb{R}$. Suppose that the angle function Θ does not change sign on Σ^n. Then, either*

(a) $\max h \neq t_0$ and

$$\Sigma^n \subset \left[t_0, t_0 + \frac{1}{\min H_1}\right] \times M^n,$$

or

(b) $\min h \neq t_0$ and

$$\Sigma^n \subset \left[t_0 - \frac{1}{\min H_1}, t_0\right] \times M^n.$$

Proof First of all it is clear from our hypothesis on the $(s+1)$-mean curvature that either $\max h \neq t_0$ or $\min h \neq t_0$. So, we begin by assuming that $\max h \neq t_0$ and let us choose an interior point p_0 of Σ^n such that the height function reaches its maximum and the orientation so that $\Theta \leq 0$. Then, from Proposition 1.1 we get

$$0 \geq \text{Hess } h(p_0)(v, v) = \Theta(p_0)\langle Av, v\rangle(p_0),$$

for all $v \in T_{p_0}\Sigma^n$, that is, at p_0 the principal curvatures are nonnegative. Since we are assume that $H_{s+1} \neq 0$ on Σ^n, we must have $H_{s+1} > 0$ on Σ^n. In particular, we can apply Lemma 1.4 (or Lemma 1.3 if $s = 1$; see also Remark 1.1) to guarantee the ellipticity of

3.3 Height Estimates in Riemannian Warped Products

the operator L_k and that H_{k+1} is positive for every $k = 0, \ldots, s$. So, for instance, we have

$$L_s h = c_s \Theta H_{s+1} \leq 0$$

and, consequently, by the weak maximum principle, we obtain that $h \geq t_0$ on Σ^n. Hence, by applying Theorem 3.1, part (a) follows.

In the case $\min h \neq t_0$, we choose an interior point q_0 of Σ^n satisfying $\min h = h(q_0)$ and the orientation so that $\Theta \geq 0$. Then,

$$0 \leq \operatorname{Hess} h(q_0)(v, v) = \Theta(q_0)\langle Av, v\rangle(q_0),$$

for all $v \in T_{p_0}\Sigma^n$, that is, at q_0 the principal curvatures must be nonnegative. So, reasoning as previous case we see that each operator L_k is elliptic and H_{k+1} is positive for every $k = 0, \ldots, s$. Besides must be $h \leq t_0$ on Σ^n.

Moreover, keeping the notation of Theorem 3.1, by Eqs. (3.17) and (3.18) and our assumption on K_{M^n}, it follows that $\zeta = ch + \Theta$ satisfies

$$L_k \zeta \leq -\frac{c_k}{k+1}\langle \nabla H_{k+1}, \nabla h\rangle,$$

which implies that $L\zeta \leq 0$, where the operator L is given by (3.22). Therefore, the weak maximum principle ensures that

$$\zeta \geq \min_{\partial \Sigma^n} \zeta \geq ct_0,$$

that is,

$$h \geq t_0 - \frac{1}{\min H_1}.$$

The proof is now complete. □

Remark 3.3 We emphasize that, in the above result, the existence of an elliptic point on the hypersurface Σ^n is not request.

Remark 3.4 We notice that the estimate given in Theorem 3.4 is sharp in the sense that it is attained by the hemisphere $\Sigma_+ = \{x \in \mathbb{S}^n : x_1 \geq 0\}$ of the standard sphere $\mathbb{S}^n$ in $\mathbb{R}^{n+1}$. Indeed, it easily seen that Σ_+ is a totally umbilical hypersurface (in fact, it is a vertical graph) with $H_1 = 1$, boundary $\{0\} \times \mathbb{S}^{n-1} \subset \{0\} \times \mathbb{R}^n$ and has the maximum height equal to 1.

It is worth pointing out that for hypersurface having constant $(s+1)$-mean curvature H_{s+1}, Theorem 3.4 improves [111, Theorem 4.1 - Part (i)]. Indeed, in [111] the authors showed that

$$t_0 \leq h \leq t_0 + \frac{1}{H_{s+1}^{1/(s+1)}}.$$

Here, it is easily seen that the following inequality holds

$$\frac{1}{\min H_1} \leq \frac{1}{H_{s+1}^{1/(s+1)}}$$

for every $s = 0, \ldots, n-1$. Moreover, this result can be viewed as an extension of [21, Theorem 3.5] (case $\alpha = 0$), as well as of [205, Proposition 1] (case $\tau = 0$).

Now, we are able to relax the assumption on the sectional curvature K_{M^n} of the fiber M^n. More precisely, we require that this sectional curvature is bounded from below by a negative constant. To this goal, we assume that the mean curvature satisfies a certain condition, which automatically holds if the sectional curvature of the fiber M^n is nonnegative.

Let $c = \min H_1$, the following result holds.

Theorem 3.5 *Let* $\mathbb{R} \times M^n$ *be a product space whose fiber has sectional curvature satisfying* $K_{M^n} \geq -\alpha$, *for some positive constant* $\alpha \in \mathbb{R}$. *Moreover, let* $\psi : \Sigma^n \to \mathbb{R} \times M^n$ *be a compact* (r, s)-*linear Weingarten two-sided hypersurface with* $(s+1)$-*mean curvature* $H_{s+1} \neq 0$ *on* Σ^n, *for some* $0 \leq s \leq n-1$, *and boundary* $\partial \Sigma^n$ *contained into the slice* $\{t_0\} \times M^n$ *for some* $t_0 \in \mathbb{R}$. *Suppose that the angle function* Θ *does not change sign on* Σ^n *and*

$$c(r+1)\min H_{k+1} > \alpha(s+1)\max H_k$$

for every $k = r \ldots, s$. *Then, either*

(a) $\max h \neq t_0$ *and*

$$\Sigma^n \subset \left[t_0, t_0 + \frac{(r+1)d}{(r+1)dc - (s+1)\alpha\beta}\right] \times M^n,$$

where d is given by (3.8) and $\beta = \sum_{k=r}^{s} b_k \max H_k$,

3.3 Height Estimates in Riemannian Warped Products

or

(b) $\min h \neq t_0$ and

$$\Sigma^n \subset \left[t_0 - \frac{(r+1)d}{(r+1)dc - (s+1)\alpha\beta}, t_0 \right] \times M^n,$$

where d is given by (3.8) and $\beta = \sum_{k=r}^{s} b_k \max H_k$.

Proof In what follows, we keep the notations established in Theorem 3.1. Let us first suppose $\max h \neq t_0$. Then, as in Theorem 3.4, taking the angle function Θ nonpositive, it is easily seen that, for every $k = 0, \ldots, s$, the operator L_k is elliptic, the $(k+1)$-mean curvature H_{k+1} is positive and $h \geq t_0$.

Moreover, by (3.17) and (3.18) we get that the function ζ defined in Theorem 3.1 satisfies (note that in this case $\zeta = ch + \Theta$)

$$L_k \zeta \geq -\frac{c_k}{k+1} \langle \nabla H_{k+1}, \nabla h \rangle - \Theta \sum_{i=1}^n \mu_{i,k} K_{M^n}(N^*, E_i^*) |N^* \wedge E_i^*|^2. \tag{3.26}$$

Since the eigenvalues $\mu_{i,k}$ are all positive on Σ^n, by using the main assumption on K_{M^n}, we have

$$\mu_{i,k} K_M(N^* \wedge E_i^*) |N^* \wedge E_i^*|^2 \geq -\mu_{i,k} \alpha |N^* \wedge E_i^*|^2 \geq -\mu_{i,k}\alpha, \tag{3.27}$$

for every $i = 1, \ldots, n$ and $k = r, \ldots, s$, since $|N^* \wedge E_i^*|^2 = |\nabla h|^2 - \langle E_i, \nabla h \rangle^2 \leq 1$.
Then, Eq. (3.27) yields

$$\sum_{i=1}^n \mu_{i,k} K_M(N^* \wedge E_i^*) |N^* \wedge E_i^*|^2 \geq -\alpha \mathrm{tr}(P_k) = -\alpha c_k H_k \geq -\alpha c_k \max H_k.$$

The above inequality and (3.26) yields

$$L\zeta \geq \sum_{k=r}^{s} (k+1)\alpha\Theta b_k \max H_k \geq (s+1)\alpha\beta\Theta, \tag{3.28}$$

where $\beta = \sum_{k=r}^{s} b_k \max H_k$.

On the other hand, by using Proposition 1.1 we have that

$$Lh = \sum_{k=r}^{s}(k+1)c_k^{-1}b_k L_k h = \sum_{k=r}^{s}(k+1)\Theta b_k H_{k+1} \leq (r+1)d\Theta. \tag{3.29}$$

So, let us consider on Σ^n the smooth function given by

$$\tilde{\zeta} = \zeta - \frac{(s+1)\alpha\beta}{(r+1)d}h = \frac{(r+1)dc - (s+1)\alpha\beta}{(r+1)d}h + \Theta.$$

Then, Eqs. (3.28) and (3.29) yield

$$L\tilde{\zeta} = L\zeta - \frac{(s+1)\alpha\beta}{(r+1)d}Lh \geq (s+1)\alpha\beta\Theta - (s+1)\alpha\beta\Theta = 0.$$

Hence, the weak maximum principle ensures that

$$\tilde{\zeta} \leq \max_{\partial \Sigma^n} \tilde{\zeta} \leq \frac{(r+1)dc - (s+1)\alpha\beta}{(r+1)d}t_0,$$

that is

$$\frac{(r+1)dc - (s+1)\alpha\beta}{(r+1)d}(h - t_0) \leq 1. \tag{3.30}$$

Now, by using the assumption on c, we get

$$(r+1)dc - (s+1)\alpha\beta = (r+1)c\sum_{k=r}^{s}b_k H_{k+1} - (s+1)\alpha\sum_{k=r}^{s}b_k \max H_k$$

$$= \sum_{k=r}^{s}b_k\left((r+1)cH_{k+1} - (s+1)\alpha \max H_k\right) > 0.$$

Therefore, by (3.30), we have

$$h \leq t_0 + \frac{(r+1)d}{(r+1)dc - (s+1)\alpha\beta},$$

as desired.

Since the case $\min h \neq t_0$ is similar, we omit the details. The proof is now complete. $\square$

In the case of hypersurfaces with constant $(s+1)$-mean curvature H_{s+1}, the assumption on the constant c in Theorem 3.5 assume the form $cH_{s+1} > \alpha \max H_s$. We notice that this assumption is weaker with respect to condition (7.77) of [33, Theorem 7.19], namely

$$H_{s+1}^{(s+2)/(s+1)} > \alpha \max H_s.$$

Indeed, in this case, by Lemma 1.1, one has $c \geq H_{s+1}^{1/(s+1)}$. Moreover, the constant

$$\frac{(r+1)d}{(r+1)dc - (s+1)\alpha\beta}$$

is just given by

$$\frac{H_{s+1}}{cH_{s+1} - \alpha \max H_s}.$$

Furthermore

$$\frac{H_{s+1}}{cH_{s+1} - \alpha \max H_s} \leq \frac{H_{s+1}}{H_{s+1}^{(s+2)/(s+1)} - \alpha \max H_s}.$$

In this setting, the main estimate improves that given in [33, Theorem 7.19] for compact hypersurfaces.

3.4 Half-Space Theorems and Topology at Infinity in Riemannian Warped Products

The aim of this section is to obtain information on the topology at infinity, in the form of half-space theorems, regarding complete two-sided hypersurfaces in Riemannian warped product spaces of the type $\mathbb{R} \times_f M^n$.

We point out that the results presented here do not assume that the hypersurface has some constant higher order mean curvature. The main theorems generalize the results contained in [111, 189, 205].

In the first part of this section, we study the case of hypersufaces whose fiber M^n is compact. Then as consequence of the height estimates proved in the previous section, we get various half-space theorems.

Successively, we analyze hypersurfaces in Riemannian warped products whose fiber is not necessarily compact. By using a generalized version of the Omori-Yau's maximum principle for trace type differential operators, we prove new half-space theorems which seem to be of independent theoretical interest.

Let us start the main analysis with the definition below.

Definition 3.1 We say that a two-sided hypersurface in a Riemannian warped product space $\mathbb{R} \times_f M^n$ *lies in an upper or lower half-space* if it is, respectively, contained into a region of $\mathbb{R} \times M^n$ of the form

$$[a, +\infty) \times M^n \quad \text{or} \quad (-\infty, a] \times M^n,$$

for some real number $a \in \mathbb{R}$.

Dealing with a Riemannian warped product, the first half-space theorem given in Theorem 3.6 below can be viewed as an application of Proposition 3.1.

More precisely, the main result reads as follows.

Theorem 3.6 *Let $\mathbb{R} \times_f M^n$ be a Riemannian warped product whose fiber is compact. Moreover, let $\psi : \Sigma^n \to \mathbb{R} \times_f M^n$ be a noncompact two-sided properly immersed hypersurface with positive $(s+1)$-mean curvature, for some $0 \leq s \leq n-1$. The following holds:*

(a) *If f' does not change sign on $\mathbb{R}$ and, for some $0 \leq r \leq s$, we have*

$$\sum_{k=r}^{s} b_k H_{k+1} \geq \sum_{k=r}^{s} b_k \sup[(\log f)']^{k+1}$$

for certain nonnegative constants b_k and, if $s \geq 2$, there exists an elliptic point in Σ^n, then Σ^n cannot lie in an upper half-space. In particular, Σ^n must have at least one bottom end.

(b) *If $f' > 0$ on $\mathbb{R}$ and, for some $0 \leq r \leq s$, we have*

$$\sum_{k=r}^{s} b_k H_{k+1} \leq \sum_{k=r}^{s} b_k \inf[(\log f)']^{k+1}$$

for certain nonnegative constants b_k and, if $s \geq 2$, there exists an elliptic point in Σ^n, then Σ^n cannot lie in a lower half-space. In particular, Σ^n must have at least one top end.

Proof We first prove part (a). Let us assume by contradiction that Σ^n lies in an upper half-space, that is, $\Sigma^n \subset [a, +\infty) \times M^n$, for some $a \in \mathbb{R}$.

For any number $t_0 > a$, let Σ_{t_0} be the hypersurface

$$\Sigma_{t_0} = \{(t, p) \in \Sigma^n : t \leq t_0\}.$$

3.4 Half-Space Theorems and Topology at Infinity in Riemannian...

Now, clearly Σ_{t_0} is a compact two-sided hypersurface with boundary contained into the slice $\{t_0\} \times M$, since M^n is compact and the immersion is proper.

Moreover, the following inequality holds

$$\sum_{k=r}^{s} b_k H_{k+1} \geq \sum_{k=r}^{s} b_k \sup[(\log f)']^{k+1} \geq \sum_{k=r}^{s} b_k \sup_{(-\infty, t_0]} [(\log f)']^{k+1}.$$

Hence, by Proposition 3.1 we conclude that the height function of Σ_{t_0} satisfies $h \geq t_0$, leading to a contradiction since t_0 is arbitrary.

Let us prove now part (b). Arguing again by contradiction, let us assume that Σ^n is contained into a lower half-space of the form $(-\infty, a] \times M^n$, for some $a \in \mathbb{R}$.

We set

$$\Sigma_{t_0} = \{(t, p) \in \Sigma^n : t \geq t_0\},$$

where $t_0 < a$ is arbitrary.

It follows that Σ_{t_0} is a compact two-sided hypersurface with boundary contained into the slice $\{t_0\} \times M$.

Moreover, we also have

$$\sum_{k=r}^{s} b_k H_{k+1} \leq \sum_{k=r}^{s} b_k \inf[(\log f)']^{k+1} \leq \sum_{k=r}^{s} b_k \inf_{[t_0, +\infty)} [(\log f)']^{k+1}.$$

By using once more Proposition 3.1 we get $h \leq t_0$ obtaining a contradiction.

This completes the proof. □

Considering a pseudo-hyperbolic space of the type $\mathbb{R} \times_{e^t} M^n$, by using Theorem 3.2, a stronger result than Theorem 3.6 which gives a generalization of [189, Theorem 4.3] can be obtained as follows.

Theorem 3.7 *Let $\mathbb{R} \times_{e^t} M^n$ be a pseudo-hyperbolic space whose fiber is compact and has nonnegative sectional curvature. Moreover, let $\psi : \Sigma^n \to \mathbb{R} \times_{e^t} M^n$ be a noncompact (r, s)-linear Weingarten two-sided properly immersed hypersurface with positive $(s+1)$-mean curvature, for some $0 \leq s \leq n-1$, and*

$$d > \sum_{k=r}^{s} b_k.$$

Suppose that

$$H_1 \geq \sup |H_{r+1}|^{1/(r+1)}$$

and, if $s \geq 2$, there exists an elliptic point in Σ^n. If the angle function Θ does not change sign on Σ^n, then Σ^n cannot lie in a half-space. In particular, Σ^n must have at least one bottom and one top end.

Proof It is immediate by Theorem 3.6 that Σ^n cannot lie in an upper half-space. On the other hand, if $\Sigma^n \subset (-\infty, a] \times M^n$, for some $a \in \mathbb{R}$, then as above let us consider for any $t_0 < a$ the hypersurface

$$\Sigma_{t_0} = \{(t, p) \in \Sigma^n : t \geq t_0\}.$$

Thus Σ_{t_0} satisfies the assumptions of Theorem 3.2, which implies that

$$a - t_0 \leq \log\left(\frac{\min_{\Sigma_{t_0}} H_1}{\min_{\Sigma_{t_0}} H_1 - 1}\right). \tag{3.31}$$

Moreover, taking into account the assumptions on d and H_1, we must have $\inf H_1 > 1$. Hence, Eq. (3.31) yields

$$a - t_0 \leq \log\left(\frac{\min_{\Sigma_{t_0}} H_1}{\min_{\Sigma_{t_0}} H_1 - 1}\right) \leq \log\left(\frac{\inf H_1}{\inf H_1 - 1}\right).$$

If t_0 is small enough we reached a contradiction. $\square$

In the case of a pseudo-hyperbolic space of the form $\mathbb{R} \times_{\cosh t} M^n$, arguing as in the proof of Proposition 3.1, we obtain the following generalization of [189, Theorem 4.5].

Theorem 3.8 *Let $\mathbb{R} \times_{\cosh t} M^n$ be a pseudo-hyperbolic space whose fiber is compact. Moreover, let $\psi : \Sigma^n \to \mathbb{R} \times_{\cosh t} M^n$ be a noncompact two-sided properly immersed hypersurface with positive $(s + 1)$-mean curvature, for some $0 \leq s \leq n - 1$. Suppose that for some $0 \leq r \leq s$ we have*

$$\sum_{k=r}^{s} b_k H_{k+1} \geq \sum_{k=r}^{s} b_k$$

for certain nonnegative constants b_k. If $s \geq 2$, assume that there exists an elliptic point in Σ^n. Then, Σ^n cannot lie in an upper half-space. In particular, Σ^n must have at least one bottom end.

3.4 Half-Space Theorems and Topology at Infinity in Riemannian...

Proof We argue once more by contradiction, that is, let us assume that Σ^n is contained into an upper half-space $[a, +\infty) \times M^n$, for some $a \in \mathbb{R}$. As in the proof of Proposition 3.1 we get $H_{r+1} \geq 1$. Given $t_0 > a$, we set the hypersurface

$$\Sigma_{t_0} = \{(t, p) \in \Sigma^n : t \leq t_0\}.$$

It is clear that Σ_{t_0} is a compact two-sided hypersurface with boundary $\partial \Sigma_{t_0} \subset \{t_0\} \times M^n$. Furthermore, one has

$$\sup_{(-\infty, t_0]} (\tanh t)^{r+1} \leq 1 \leq H_{r+1}.$$

Finally, by Proposition 3.1 we conclude that $h \geq t_0$, which gives a contradiction; see also Remark 3.2. □

Next, we prove half-space theorems regarding two-sided hypersurfaces immersed into a standard product space of the form $\mathbb{R} \times M^n$. Along this direction, by Theorem 3.4, the following result holds.

Theorem 3.9 *Let $\mathbb{R} \times M^n$ be a product space whose fiber is compact and has nonnegative sectional curvature. Moreover, let $\psi : \Sigma^n \to \mathbb{R} \times M^n$ be a noncompact (r, s)-linear Weingarten two-sided properly immersed hypersurface with bounded away from zero $(s + 1)$-mean curvature, for some $0 \leq s \leq n - 1$, and such that its angle function Θ does not change sign. Then, Σ^n cannot lie in a half-space. In particular, Σ^n must have at least one bottom and one top end.*

Proof Let us assume by contradiction that Σ^n lies in an upper half-space, that is, $\Sigma^n \subset [a, +\infty) \times M^n$, for some $a \in \mathbb{R}$. As in the proof of Theorem 3.6, we denote by Σ_{t_0} the hypersurface

$$\Sigma_{t_0} = \{(t, p) \in \Sigma^n : t \leq t_0\},$$

where $t_0 > a$ is arbitrary. Then, Σ_{t_0} is a compact (r, s)-linear Weingarten two-sided hypersurface with boundary contained into the slice $\{t_0\} \times M$ and $\min h \neq t_0$. Hence, by Theorem 3.4 we must have $H_{s+1} > 0$ on Σ_{t_0} and $\Sigma_{t_0} \subset [t_0 - \frac{1}{c(t_0)}, t_0] \times M^n$, where $c(t_0) = \min_{\Sigma_{t_0}} H_1 > 0$, that is

$$t_0 - a \leq \frac{1}{c(t_0)}.$$

Now, since H_{s+1} is bounded away from zero, we get $\inf H_{s+1} > 0$, which implies $\inf H_1 > 0$. Thus

$$t_0 - a \leq \frac{1}{c(t_0)} \leq \frac{1}{\inf H_1}.$$

Then, by choosing t_0 large enough we reached a contradiction.

Similar arguments stated above can be applied if Σ^n is contained in a lower half-space. The conclusion is achieved. □

Arguing as in the proof of Theorem 3.9 the following result, where we keep the notation $c = \min H_1$, can be achieved as a biproduct of Theorem 3.5.

Theorem 3.10 *Let $\mathbb{R} \times M^n$ be a product space whose fiber is compact with sectional curvature satisfying $K_M \geq -\alpha$, for some positive constant $\alpha \in \mathbb{R}$. Moreover, let $\psi : \Sigma^n \to \mathbb{R} \times M^n$ be a noncompact (r, s)-linear Weingarten two-sided properly immersed hypersurface with bounded away from zero $(s+1)$-mean curvature, for some $0 \leq s \leq n-1$, and such that its angle function Θ does not change sign. Suppose that*

$$c(r+1) \min H_{k+1} > \alpha(s+1) \max H_k$$

for every $k = r, \ldots, s$. Then, Σ^n cannot lie in a half-space. In particular, Σ^n must have at least one bottom and one top end.

We observe that the results obtained here generalize those obtained in [111] and [205] in which the authors studied the case of hypersurfaces $\psi : \Sigma^n \to \mathbb{R} \times M^n$ with mean curvature (or some higher order mean curvature) is constant.

In order to treat the case in which the fiber is not compact, we will make use of a generalized version of the Omori-Yau's maximum principle for trace type differential operators proved in [33]; for more details, see Sect. 1.6 and Lemma 1.8.

Let us recall that, given a Riemannian manifold Σ^n and a semi-elliptic operator $\mathcal{L} = \mathrm{tr}(\mathcal{P} \circ \mathrm{Hess})$, where $\mathcal{P} : \mathfrak{X}(\Sigma^n) \to \mathfrak{X}(\Sigma^n)$ is a positive semi-definite symmetric tensor, we say that the Omori-Yau's maximum principle holds on Σ^n for the operator $\mathcal{L}$ if, for any function $u \in C^2(\Sigma^n)$ with $u^* = \sup u < +\infty$, there exists a sequence of points $\{p_j\} \subset \Sigma^n$ satisfying

$$u(p_j) > u^* - \frac{1}{j}, \quad |\nabla u(p_j)| < \frac{1}{j} \quad \text{and} \quad \mathcal{L}u(p_j) < \frac{1}{j}$$

for every $j \in \mathbb{N} \setminus \{0\}$.

3.4 Half-Space Theorems and Topology at Infinity in Riemannian...

Equivalently, for any smooth function $u \in C^2(\Sigma^n)$ with $u_* = \inf u > -\infty$ there exists a sequence of points $\{p_j\} \subset \Sigma^n$ satisfying

$$u(p_j) < u_* + \frac{1}{j}, \quad |\nabla u(p_j)| < \frac{1}{j} \quad \text{and} \quad \mathcal{L}u(p_j) > -\frac{1}{j}$$

for every $j \in \mathbb{N} \setminus \{0\}$.

Now, we are ready to state and prove the next half-space theorem.

Theorem 3.11 *Let $\mathbb{R} \times_f M^n$ be a Riemannian warped product satisfying the convergence condition in (3.14) and with non-decreasing warping function. Let $\psi : \Sigma^n \to \mathbb{R} \times_f M^n$ be a complete noncompact (r, s)-linear Weingarten two-sided hypersurface with positive $(s+1)$-mean curvature, for some $1 \leq r \leq s \leq n-1$, and*

$$d > \sum_{k=r}^{s} b_k \sup[(\log f)']^{k+1}.$$

Suppose that

$$\sup \frac{f''(h)}{f(h)} < +\infty,$$

and, if $s \geq 2$, there exists an elliptic point in Σ^n. Assume further that $\sup |H_r| < +\infty$ and the second fundamental form satisfies $|A| \leq G(r_o)$, where $G \in C^1([0, +\infty))$ is such that

$$G(0) > 0, \quad G'(t) \geq 0 \quad \text{and} \quad \frac{1}{G(t)} \notin L^1(+\infty)$$

and r_o is the distance function from a reference point in Σ^n. Then, either $\sup \Theta > 0$ or Σ^n cannot lie in an upper half-space.

Proof We begin by stating that the sectional curvature K_{Σ^n} of Σ^n satisfies the assumption (1.15) of Lemma 1.8. Indeed, denoting by $\overline{K}$ the sectional curvature of the ambient space, by the Gauss equation (1.1) that if $\{X, Y\}$ is an orthonormal basis for an arbitrary plane tangent to Σ^n, then

$$K_{\Sigma^n}(X, Y) = \overline{K}(X, Y) + \langle AX, X \rangle \langle AY, Y \rangle - \langle AX, Y \rangle^2$$

$$\geq \overline{K}(X, Y) - |AX||AY| - |AX|^2$$

$$\geq \overline{K}(X, Y) - 2|A|^2, \tag{3.32}$$

where the last inequality follows from the fact that

$$|AX|^2 \leq \operatorname{tr}(A^2)|X|^2 = |A|^2$$

for every unitary vector X tangent to Σ^n.

On the other hand, Lemma 3.2 gives

$$\overline{R}(U, V)W = R_M(U^*, V^*)W^* - [(\log f)']^2(\langle U, W\rangle V - \langle V, W\rangle U)$$
$$-(\log f)''\langle W, \partial_t\rangle(\langle U, \partial_t\rangle V - \langle V, \partial_t\rangle U)$$
$$-(\log f)''(\langle U, W\rangle\langle V, \partial_t\rangle - \langle V, W\rangle\langle U, \partial_t\rangle)\partial_t,$$

for every vector U, V, W tangent to $\mathbb{R} \times_f M^n$, where U^* denotes the orthogonal projection of U on TM^n. In particular, for the base $\{X, Y\}$ we get

$$\overline{K}(X, Y) = \langle \overline{R}(X, Y)X, Y\rangle$$
$$= \frac{1}{f^2(h)} K_M(X^*, Y^*)|X^* \wedge Y^*|^2 - (\log f)'(h)^2$$
$$-(\log f)''(h)(\langle X, \partial_t\rangle^2 + \langle Y, \partial_t\rangle^2).$$

By using the convergence condition (3.14) and taking into account that $|X^* \wedge Y^*|^2 = 1 - \langle X, \partial_t\rangle^2 - \langle Y, \partial_t\rangle^2$, we obtain

$$\overline{K}(X, Y) \geq -(\log f)''(h) - (\log f)'(h)^2 = -\frac{f''(h)}{f(h)}.$$

Hence, since the shape operator satisfies $|A| \leq G(r_o)$, by (3.32) we infer that

$$K_{\Sigma^n} \geq -\sup \frac{f''(h)}{f(h)} - 2G^2(r_o), \qquad (3.33)$$

which concludes the claim.

From now on, we assume that the angle function Θ is nonpositive. Arguing by contradiction, suppose that Σ^n lies in an upper half-space. Equivalently, the height function of Σ^n satisfies $h_* = \inf h > -\infty$.

By [31], for each $k = r, \ldots, s$, let $\mathcal{L}_k^+ : C^\infty(\Sigma^n) \to C^\infty(\Sigma^n)$ be the second order linear differential operator given by

$$\mathcal{L}_k^+ = \sum_{i=0}^k (-1)^i \frac{c_k}{c_i} (\log f)'(h)^{k-i} \Theta^i L_i$$

$$= \operatorname{tr}(\mathcal{P}_k^+ \circ \operatorname{Hess}), \qquad (3.34)$$

3.4 Half-Space Theorems and Topology at Infinity in Riemannian...

where $\mathcal{P}_k^+ : \mathfrak{X}(\Sigma^n) \to \mathfrak{X}(\Sigma^n)$ is defined by

$$\mathcal{P}_k^+ = \sum_{i=0}^{k} (-1)^i \frac{c_k}{c_i} (\log f)'(h)^{k-i} \Theta^i P_i. \tag{3.35}$$

In particular, as showed in [31, Section 6], the following equality holds

$$\mathcal{L}_k^+ \sigma(h) = c_k f(h) \left((\log f)'(h)^{k+1} + (-1)^k \Theta^{k+1} H_{k+1} \right), \tag{3.36}$$

where σ denotes a primitive of the warping function f. We also notice that, by Lemma 1.4, since the angle function was supposed to be nonpositive, it follows that $\mathcal{P}_k^+$ is a positive semi-definite symmetric tensor for every $k = r, \ldots, s$.

Besides, since $d > b_k[\sup(\log f)']^{k+1}$ for every $k = r, \ldots, s$ and taking into account that Lemma 1.2 assures that $\mathrm{tr}(P_i) = c_i H_i$, by Lemma 1.1 we get

$$\mathrm{tr}(\mathcal{P}_k^+) \le c_k \sum_{i=0}^{k} \left(\frac{d}{b_k} \right)^{(k-i)/(k+1)} H_r^{i/r}.$$

Consequently $\sup \mathrm{tr}(\mathcal{P}_k^+) < +\infty$ for every $k = r, \ldots, s$, bearing in mind the assumption on H_r.

Let us consider the following second order linear differential operator $\mathcal{L}^+ : C^\infty(\Sigma^n) \to C^\infty(\Sigma^n)$ defined by

$$\mathcal{L}^+ = \sum_{k=r}^{s} b_k c_k^{-1} \mathcal{L}_k^+ = \mathrm{tr}(\mathcal{P}^+ \circ \mathrm{Hess}),$$

where $\mathcal{P}^+ : \mathfrak{X}(\Sigma^n) \to \mathfrak{X}(\Sigma^n)$ is given by

$$\mathcal{P}^+ = \sum_{k=r}^{s} b_k c_k^{-1} \mathcal{P}_k^+.$$

Then $\mathcal{P}^+$ is a positive semi-definite symmetric tensor with $\sup \mathrm{tr}(\mathcal{P}^+) < \infty$. In particular, $\mathcal{L}^+$ is a semi-elliptic operator.

So, we are ready to apply Lemma 1.8 to guarantee that the Omori-Yau's maximum principle holds on Σ^n for the operator $\mathcal{L}^+$.

Now, it is clear that σ satisfies $\sigma(h) \ge \sigma(h_*) > -\infty$. Hence, there exists a sequence of points $\{p_j\} \subset \Sigma^n$ with the following properties

$$\lim_j \sigma(h)(p_j) = \sigma(h_*), \quad |\nabla \sigma(h)(p_j)| = f(h(p_j))|\nabla h(p_j)| < \frac{1}{j},$$

and

$$\mathcal{L}^+\sigma(h)(p_j) > -\frac{1}{j}.$$

In particular, Eq. (3.36) yields

$$-\frac{1}{j} < \mathcal{L}^+\sigma(h)(p_j) = \sum_{k=r}^{s} f(h(p_j))(b_k(\log f)'(h(p_j))^{k+1}$$

$$+ (-1)^k \Theta(p_j)^{k+1} b_k H_{k+1}(p_j)).$$

Letting $j \to +\infty$, we must have $\Theta(p_j) \to -1$, since $|\nabla h|^2 = 1 - \Theta^2$.
Consequently

$$d \leq \sum_{k=r}^{s} b_k(\log f)'(h_*)^{k+1} \leq \sum_{k=r}^{s} b_k \sup[(\log f)']^{k+1},$$

reaching a contradiction. The proof is now complete. □

Let us observe that the proof of Theorem 3.11 remains true requiring the stronger assumption that K_{Σ^n} is bounded from below by a constant, which implies the validity of the Omori-Yau's maximum principle. For instance, if we assume that

$$\sup \frac{f''(h)}{f(h)} < +\infty,$$

arguing as in the proof of Theorem 3.11, we easily see that K_{Σ^n} is bounded from below since $\sup |A|^2 < +\infty$.

On the other hand, in Theorem 3.11 the hypothesis $\sup |H_r| < +\infty$ can be replaced by $\sup H_1 < +\infty$, thanks to Lemma 1.1. In this case, taking into account the relation

$$|A|^2 = n^2 H_1^2 - n(n-1)H_2,$$

it follows that the condition $\sup |A|^2 < +\infty$ is equivalent to $\sup H_1 < +\infty$.

Owing to the above remarks, the following result holds.

Corollary 3.2 *Let $\mathbb{R} \times_f M^n$ be a Riemannian warped product satisfying the convergence condition in (3.14) and with non-decreasing warping function. Moreover, let $\psi : \Sigma^n \to \mathbb{R} \times_f M^n$ be a complete noncompact (r, s)-linear Weingarten two-sided hypersurface with* positive $(s+1)$-mean curvature, for some $0 \leq r \leq s \leq n-1$, and

$$d > \sum_{k=r}^{s} b_k \sup[(\log f)']^{k+1}.$$

Suppose that

$$\sup |H_1| < +\infty \quad \text{and} \quad \sup \frac{f''(h)}{f(h)} < +\infty,$$

and, if $s \geq 2$, there exists an elliptic point in Σ^n. Then, either $\sup \Theta > 0$ or Σ^n cannot lie in an upper half-space.

In the case of hypersurfaces having constant mean curvature the assumption that the warping function is non-decreasing can be dropped as follows.

Corollary 3.3 *Let $\mathbb{R} \times_f M^n$ be a Riemannian warped product satisfying the convergence condition in (3.14). Moreover, let $\psi : \Sigma^n \to \mathbb{R} \times_f M^n$ be a complete noncompact two-sided hypersurface with constant mean curvature satisfying*

$$H_1 > \sup(\log f)'.$$

Suppose that

$$\inf H_2 > -\infty \quad \text{and} \quad \sup \frac{f''(h)}{f(h)} < +\infty.$$

Then Σ^n cannot lie in an upper half-space.

Proof Arguing by contradiction, let us suppose that Σ^n lies in an upper half-space, that is, $\inf h = h_* > -\infty$.

As in the proof of Theorem 3.11 and by using the remark above, the Omori-Yau's maximum principle holds on Σ^n for the Laplacian.

Then, there is a sequence of points $\{p_j\} \subset \Sigma^n$ satisfying

$$\lim_j h(p_j) = h_*, \quad |\nabla h(p_j)| < \frac{1}{j} \quad \text{and} \quad \Delta h(p_j) > -\frac{1}{j}.$$

By Proposition 1.1, one has

$$-\frac{1}{j} < \Delta h(p_j) = (\log f)'(h(p_j)) \left(n - |\nabla h(p_j)|^2 \right) + n\Theta(p_j) H_1$$

Since the angle function is bounded, passing to the limit and by choosing the orientation so that $H_1 > 0$, we conclude that

$$H_1 \leq \sup(\log f)',$$

which gives a contradiction. □

If the warping function is $f(t) = e^t$ we get the next result.

Corollary 3.4 *Let $\mathbb{R} \times_{e^t} M^n$ be a pseudo-hyperbolic space whose fiber has nonnegative sectional curvature. Moreover, let $\psi : \Sigma^n \to \mathbb{R} \times_{e^t} M^n$ be a complete noncompact two-sided hypersurface with constant mean curvature satisfying $H_1 > 1$. Suppose that $\inf H_2 > -\infty$. Then Σ^n cannot lie in an upper half-space.*

For hypersurfaces having some constant higher order mean curvature, the following result holds.

Corollary 3.5 *Let $\mathbb{R} \times_f M^n$ be a Riemannian warped product satisfying the convergence condition in (3.14) and with non-decreasing warping function. Moreover, let $\psi : \Sigma^n \to \mathbb{R} \times_f M^n$ be a complete noncompact two-sided hypersurface with constant $(s+1)$-mean curvature satisfying*

$$H_{s+1} > \sup[(\log f)']^{s+1}$$

for some $1 \leq s \leq n-1$. Suppose that

$$\sup |H_1| < +\infty \quad \text{and} \quad \sup \frac{f''(h)}{f(h)} < +\infty,$$

and, if $s \geq 2$, there exists an elliptic point in Σ^n. Then, either $\sup \Theta > 0$ or Σ^n cannot lie in an upper half-space.

In particular, in the pseudo-hyperbolic space $\mathbb{R} \times_{e^t} M^n$ we have the following meaningful consequence of the main result.

Corollary 3.6 *Let $\mathbb{R} \times_{e^t} M^n$ be a pseudo-hyperbolic space whose fiber has nonnegative sectional curvature. Moreover, let $\psi : \Sigma^n \to \mathbb{R} \times_{e^t} M^n$ be a complete noncompact two-sided hypersurface with constant $(s+1)$-mean curvature satisfying $H_{s+1} > 1$, for some $1 \leq s \leq n-1$. Suppose that $\sup |H_1| < +\infty$ and, if $s \geq 2$, there exists an elliptic point in Σ^n. Then, either $\sup \Theta > 0$ or Σ^n cannot lie in an upper half-space.*

3.4 Half-Space Theorems and Topology at Infinity in Riemannian...

In the next result we present a suitable version of Theorem 3.11 in which the warping function is non-increasing.

Theorem 3.12 *Let $\mathbb{R} \times_f M^n$ be a Riemannian warped product satisfying the convergence condition in (3.14) and with non-increasing warping function. Moreover, let $\psi : \Sigma^n \to \mathbb{R} \times_f M^n$ be a complete noncompact (r, s)-linear Weingarten two-sided hypersurface with positive $(s+1)$-mean curvature, for some $1 \leq r \leq s \leq n-1$, and*

$$d > \sum_{k=r}^{s} b_k \sup[-(\log f)']^{k+1}.$$

Suppose that

$$\sup \frac{f''(h)}{f(h)} < +\infty,$$

and, if $s \geq 2$, there exists an elliptic point in Σ^n. Assume further that $\sup |H_r| < +\infty$ and the second fundamental form satisfies $|A| \leq G(r_o)$, where $G \in C^1([0, +\infty))$ is such that

$$G(0) > 0, \quad G'(t) \geq 0 \quad \text{and} \quad \frac{1}{G(t)} \notin L^1(+\infty)$$

and r_o is the distance function from a reference point in Σ^n. Then, either $\inf \Theta < 0$ or Σ^n cannot lie in a lower half-space.

Proof Arguing by contradiction, suppose that Θ is a nonnegative function and that Σ^n lies in a lower half-space, that is, the height function of Σ^n satisfies $h^* = \sup h < +\infty$.

We consider, for each $k = r, \ldots, s$, the second order linear differential operator $\mathcal{L}_k^- : C^\infty(\Sigma^n) \to C^\infty(\Sigma^n)$ defined by

$$\mathcal{L}_k^- = \sum_{i=0}^{k} (-1)^{k-i} \frac{c_k}{c_i} (\log f)'(h)^{k-i} \Theta^i L_i = \text{tr}\left(\mathcal{P}_k^- \circ \text{Hess}\right), \quad (3.37)$$

where $\mathcal{P}_k^- : \mathfrak{X}(\Sigma^n) \to \mathfrak{X}(\Sigma^n)$ is the positive semi-definite symmetric tensor given by

$$\mathcal{P}_k^- = \sum_{i=0}^{k} (-1)^{k-i} \frac{c_k}{c_i} (\log f)'(h)^{k-i} \Theta^i P_i. \quad (3.38)$$

As in Theorem 3.11, the sectional curvature K_{Σ^n} of Σ^n satisfies condition (3.33), as well as $\text{tr}(\mathcal{P}_k^-) < +\infty$.

Hence, by Lemma 1.8 the Omori-Yau's maximum principle holds on Σ^n for the semi-elliptic second order linear differential operator $\mathcal{L}^- : C^\infty(\Sigma^n) \to C^\infty(\Sigma^n)$ given by

$$\mathcal{L}^- = \sum_{k=r}^{s} b_k c_k^{-1} \mathcal{L}_k^- = \operatorname{tr}(\mathcal{P}^- \circ \operatorname{Hess}),$$

where $\mathcal{P}^- : \mathfrak{X}(\Sigma^n) \to \mathfrak{X}(\Sigma^n)$, defined by

$$\mathcal{P}^- = \sum_{k=r}^{s} b_k c_k^{-1} \mathcal{P}_k^-,$$

is a positive semi-definite symmetric tensor and satisfies $\operatorname{tr}(\mathcal{P}^-) < +\infty$.

Now let σ be a primitive of the warping function, which must satisfies $\sigma(h) \leq \sigma(h^*)$. Then, there exists a sequence of points $\{q_j\} \subset \Sigma^n$ with the following properties

$$\lim_j \sigma(h)(q_j) = \sigma(h^*), \quad |\nabla \sigma(h)(q_j)| < \frac{1}{j} \quad \text{and} \quad \mathcal{L}^- \sigma(h)(q_j) < \frac{1}{j}.$$

Taking into account that $\mathcal{L}_k^- = (-1)^k \mathcal{L}_k^+$, where $\mathcal{L}_k^+$ is defined in the proof of Theorem 3.11 and thanks to (3.36), we find

$$\frac{1}{j} > \mathcal{L}^- \sigma(h)(q_j) = \sum_{k=r}^{s} f(h(q_j)) \Big(-b_k [-(\log f)'(h(q_j))]^{k+1}$$
$$+ \Theta(q_j)^{k+1} b_k H_{k+1}(q_j) \Big).$$

Therefore, making $j \to +\infty$ it follows that $\Theta(q_j) \to 1$. Consequently

$$d \leq \sum_{k=r}^{s} b_k \sup[-(\log f)']^{k+1},$$

against our assumption. The proof is complete. □

As in Theorem 3.11, the main Theorem 3.12 remains valid if we replace the conditions $|A| \leq G(r_0)$ and $\sup|H_r| < +\infty$ by a stronger assumption on the mean curvature H_1. More precisely, Theorem 3.12 remains true requiring that $\sup H_1 < +\infty$.

The main result reads as follows.

Corollary 3.7 *Let $\mathbb{R} \times_f M^n$ be a Riemannian warped product satisfying the convergence condition in (3.14) and with non-increasing warping function. Moreover, let $\psi : \Sigma^n \to \mathbb{R} \times_f M^n$ be a complete noncompact (r, s)-linear Weingarten two-sided hypersurface with*

3.4 Half-Space Theorems and Topology at Infinity in Riemannian...

positive $(s+1)$-mean curvature, for some $1 \leq r \leq s \leq n-1$, and

$$d > \sum_{k=r}^{s} b_k \sup[-(\log f)']^{k+1}.$$

Suppose that

$$\sup |H_1| < \infty \quad \text{and} \quad \sup \frac{f''(h)}{f(h)} < +\infty,$$

and, if $s \geq 2$, there exists an elliptic point in Σ^n. Then, either $\inf \Theta < 0$ or Σ^n cannot lie in a lower half-space.

The following result, where we do not require that the warping function is non-increasing, can be applied to hypersurfaces with constant mean curvature.

Corollary 3.8 *Let $\mathbb{R} \times_f M^n$ be a Riemannian warped product satisfying the convergence condition in (3.14). Moreover, let $\psi : \Sigma^n \to \mathbb{R} \times_f M^n$ be a complete noncompact two-sided hypersurface with constant mean curvature satisfying*

$$H_1 > \sup -(\log f)'.$$

Suppose that

$$\inf H_2 > -\infty \quad \text{and} \quad \sup \frac{f''(h)}{f(h)} < +\infty.$$

Then Σ^n cannot lie in a lower half-space.

In the case of hypersurfaces having some constant higher order mean curvature, Theorem 3.12 has the following form.

Corollary 3.9 *Let $\mathbb{R} \times_f M^n$ be a Riemannian warped product satisfying the convergence condition in (3.14) and with non-increasing warping function. Moreover, let $\psi : \Sigma^n \to \mathbb{R} \times_f M^n$ be a complete noncompact two-sided hypersurface with constant $(s+1)$-mean curvature satisfying*

$$H_{s+1} > \sup[-(\log f)']^{s+1},$$

for some $1 \leq r \leq n-1$. Suppose that

$$\sup |H_1| < +\infty \quad \text{and} \quad \sup \frac{f''(h)}{f(h)} < +\infty,$$

and, if $s \geq 2$, there exists an elliptic point in Σ^n. Then, either $\inf \Theta < 0$ or Σ^n cannot lie in a lower half-space.

Finally, we close this section by stating the following result in the case of product spaces $\mathbb{R} \times M^n$, which is a consequence of Theorems 3.11 and 3.12.

It is worth pointing out that the conclusion in Theorem 3.13 below, as well as in its consequences, is stronger with respect to the conclusions attained in Theorems 3.11 and 3.12.

Theorem 3.13 Let $\mathbb{R} \times M^n$ be a product space whose fiber has sectional curvature satisfying $K_M \geq -\alpha$, for some positive constant $\alpha \in \mathbb{R}$. Moreover, let $\psi : \Sigma^n \to \mathbb{R} \times M^n$ be a complete noncompact (r, s)-linear Weingarten two-sided hypersurface with positive $(s+1)$-mean curvature, for some $1 \leq r \leq s \leq n-1$. Suppose that $\sup |H_r| < +\infty$ and, if $s \geq 2$, there exists an elliptic point in Σ^n. Assume further that the shape operator satisfies $|A| \leq G(r_o)$, where $G \in C^1([0, +\infty))$ is such that

$$G(0) > 0, \quad G'(t) \geq 0 \quad \text{and} \quad \frac{1}{G(t)} \notin L^1(+\infty)$$

and r_o is the distance function from a reference point in Σ^n. The following holds:

(a) Either $\sup \Theta > 0$ or Σ^n cannot lie in an upper half-space;
(b) Either $\inf \Theta < 0$ or Σ^n cannot lie in a lower half-space.

In particular, when the hypersurface $\psi : \Sigma^n \to \mathbb{R} \times M^n$ has constant mean curvature or some higher order mean curvature, Theorem 3.13 enables us to draw the following conclusion.

Corollary 3.10 Let $\mathbb{R} \times M^n$ be a product space whose fiber has sectional curvature satisfying $K_M \geq -\alpha$, for some positive constant $\alpha \in \mathbb{R}$. Moreover, let $\psi : \Sigma^n \to \mathbb{R} \times M^n$ be a complete noncompact two-sided hypersurface with positive constant mean curvature and such that $\inf H_2 > -\infty$. The following holds.

(a) Either $\sup \Theta > 0$ or Σ^n cannot lie in an upper half-space;
(b) Either $\inf \Theta < 0$ or Σ^n cannot lie in a lower half-space.

3.4 Half-Space Theorems and Topology at Infinity in Riemannian...

In other words, we have:

(a′) *There is no complete noncompact two-sided hypersurface having positive constant mean curvature, $\inf H_2 > -\infty$, angle function nonpositive and contained into an upper half-space;*

(b′) *There is no complete noncompact two-sided hypersurface having positive constant mean curvature, $\inf H_2 > -\infty$, angle function nonnegative and contained into a lower half-space.*

As well as, the next result holds.

Corollary 3.11 *Let $\mathbb{R} \times M^n$ be a product space whose fiber has sectional curvature satisfying $K_M \geq -\alpha$, for some positive constant $\alpha \in \mathbb{R}$. Let $\psi : \Sigma^n \to \mathbb{R} \times M^n$ be a complete noncompact two-sided hypersurface with positive constant $(s+1)$-mean curvature, for some $1 \leq s \leq n-1$. Suppose that $\sup |H_1| < +\infty$ and, if $s \geq 2$, there exists an elliptic point in Σ^n. The following holds.*

(a) *Either $\sup \Theta > 0$ or Σ^n cannot lie in an upper half-space;*
(b) *Either $\inf \Theta < 0$ or Σ^n cannot lie in a lower half-space.*

In other words, we have:

(a′) *There is no complete noncompact two-sided hypersurface having $H_{s+1} > 0$, an elliptic point, with $\sup |H_1| < +\infty$, angle function nonpositive and contained into an upper half-space;*

(b′) *There is no complete noncompact two-sided hypersurface having $H_{s+1} > 0$, an elliptic point, with $\sup |H_1| < +\infty$, angle function nonnegative and contained into a lower half-space.*

Finally, the last result of this section is given below.

Corollary 3.12 *Let $\mathbb{R} \times M^n$ be a product space whose fiber has sectional curvature satisfying $K_M \geq -\alpha$, for some positive constant $\alpha \in \mathbb{R}$. Moreover, let $\psi : \Sigma^n \to \mathbb{R} \times M^n$ be a complete noncompact two-sided hypersurface with positive constant $(s+1)$-mean curvature, for some $0 \leq s \leq n-1$. In addition, if $s = 0$ assume that $\inf H_2 > -\infty$. Suppose further that $\sup |H_1| < +\infty$ and, if $s \geq 2$, there exists an elliptic point in Σ^n. Then, either Θ does not vanish identically or Σ^n cannot lie in a half-space.*

3.5 Height Estimates in GRW Spacetimes

The goal of this section is to give height estimates of compact generalized linear Weingarten spacelike hypersurfaces immersed into a GRW spacetime $-\mathbb{R} \times_f M^n$. To do this, in general way, we follows the techniques used in the Riemannian setting in Sect. 3.3. However, as we shall see, in this case our estimates are considerably different of those obtained in the Riemannian case. For this reason, as well as, for the sake of completeness, we would like to present the details of the main proofs.

In this setting, under a suitable assumption on a linear combination involving some of the higher order mean curvatures (not necessarily constants), any compact spacelike hypersurface immersed into a GRW spacetime $-\mathbb{R} \times_f M^n$ with non-empty boundary contained into a slice must lie entirely in one of the two regions of the spacetime bounded by the slice.

We point out that, when the warping function is increasing and the Gauss map N is future-pointing, the above conclusion was obtained in [188] considering a condition that involves only one of the higher order mean curvatures; see the paper [188, Proposition 14].

We also observe that in [188] the authors consider the case in which the warping function is decreasing. Here, we also consider the case in which the warping function is decreasing requiring, in addition, that the Gauss map N is past-pointing. In such a case, we are able to prove some new estimates for the height function of these hypersurfaces; see Theorem 3.16 below.

Proposition 3.2 *Let* $\psi : \Sigma^n \to -\mathbb{R} \times_f M^n$ *be a compact spacelike hypersurface with positive $(s + 1)$-mean curvature, for some $0 \leq s \leq n - 1$, and boundary $\partial \Sigma^n$ contained into the slice $\{t_0\} \times M^n$ for some $t_0 \in \mathbb{R}$. The following holds:*

(a) *If $f'(h) > 0$ and, for some $0 \leq r \leq s$, we have*

$$\sum_{k=r}^{s} b_k H_{k+1} \geq \sum_{k=r}^{s} b_k \sup[(\log f)']^{k+1}$$

for certain nonnegative constants b_k and, when $s \geq 2$, there exists an elliptic point in Σ^n with respect the future-pointing Gauss map, then $h \leq t_0$;

(b) *If $f'(h) < 0$ and, for some $0 \leq r \leq s$, we have*

$$\sum_{k=r}^{s} b_k H_{k+1} \geq \sum_{k=r}^{s} b_k \sup[-(\log f)']^{k+1}$$

for certain nonnegative constants b_k and, when $s \geq 2$, there exists an elliptic point in Σ^n with respect the past-pointing Gauss map, then $h \geq t_0$.

3.5 Height Estimates in GRW Spacetimes

Proof Let us prove part (a). As in the proof of Proposition 3.1 we can show that

$$H_{r+1}^{1/(r+1)}(p) \geq \sup(\log f)',$$

for all $p \in \Sigma^n$.

Indeed, if $r = s$ there is nothing to prove. Otherwise, arguing by contradiction, assume that there is a point $p \in \Sigma^n$ such that

$$H_{r+1}(p)^{1/(r+1)} < \sup(\log f)'.$$

Hence, there exists $r < i \leq s$ with

$$H_{i+1}(p)^{1/(i+1)} > \sup(\log f)',$$

which gives the contradiction

$$H_{r+1}(p)^{1/(r+1)} < H_{i+1}(p)^{1/(i+1)}.$$

From now on, we follow the ideas contained in [188, Proposition 14]. Let $\mathcal{L}_r^+ : C^\infty(\Sigma^n) \to C^\infty(\Sigma^n)$ be the operator defined in (3.34),

$$\mathcal{L}_r^+ = \sum_{i=0}^r (-1)^i \frac{c_r}{c_i} (\log f)'(h)^{r-i} \Theta^i L_i = \mathrm{tr}\left(\mathcal{P}_r^+ \circ \mathrm{Hess}\right),$$

where the tensor $\mathcal{P}_r^+ : \mathfrak{X}(\Sigma^n) \to \mathfrak{X}(\Sigma^n)$ given by (3.35), has the form

$$\mathcal{P}_r^+ = \sum_{i=0}^r (-1)^i \frac{c_r}{c_i} (\log f)'(h)^{r-i} \Theta^i P_i.$$

In this case the angle function satisfies $\Theta \leq -1$, consequently by Lemmas 1.3 and 1.4, we infer that $\mathcal{L}_r^+$ is an elliptic operator.

Besides, equation (3.4) of [29] yields

$$\mathcal{L}_r^+ \sigma(h) = c_r f(h) \left(-[(\log f)'(h)]^{r+1} + (-\Theta)^{r+1} H_{r+1}\right), \tag{3.39}$$

where σ is a primitive of the warping function f. Hence, the claim proved above ensures that

$$\mathcal{L}_r^+ \sigma(h) \geq c_r f(h) H_{r+1} \left(-1 + (-\Theta)^{r+1}\right) \geq 0.$$

Then, by the weak maximum principle, $\sigma(h)$ must attain its maximum on $\partial\Sigma^n$, in others words, $\sigma(h) \leq \sigma(t_0)$. Since σ is an increasing function, this implies that $h \leq t_0$, which proves part (a).

Now, we prove part (b). To this end, on account of the above computations and remarks, it easily seen that

$$H_{r+1}^{1/(r+1)}(p) \geq -\sup(\log f)',$$

for all $p \in \Sigma^n$.

Next, let us consider the second order linear differential operator defined in (3.37) and given by

$$\mathcal{L}_r^- = \sum_{i=0}^{r}(-1)^{r-i}\frac{c_r}{c_i}(\log f)'(h)^{r-i}\Theta^i L_i = \mathrm{tr}\left(\mathcal{P}_r^- \circ \mathrm{Hess}\right),$$

where the tensor $\mathcal{P}_r^- : \mathfrak{X}(\Sigma^n) \to \mathfrak{X}(\Sigma^n)$ is defined as in (3.38) and given by

$$\mathcal{P}_r^- = \sum_{i=0}^{r}(-1)^{r-i}\frac{c_r}{c_i}(\log f)'(h)^{r-i}\Theta^i P_i.$$

Since $f' < 0$, $\Theta \geq 1$ and the operator L_i is elliptic for every $i = 0, \ldots, r$, one has that $\mathcal{L}_r^-$ must be elliptic. Moreover, as already observed, we have that $\mathcal{L}_r^- = (-1)^r \mathcal{L}_r^+$.

So, by equation (3.4) of [29] we find

$$\mathcal{L}_r^-\sigma(h) = c_r f(h)\left([-(\log f)'(h)]^{r+1} - \Theta^{r+1}H_{r+1}\right). \tag{3.40}$$

Consequently, it follows that

$$\mathcal{L}_r^-\sigma(h) \leq c_r f(h)H_{r+1}\left(1 - \Theta^{r+1}\right) \leq 0.$$

Finally, using again the weak maximum principle, we conclude that $h \geq t_0$. The proof is now complete. □

Remark 3.5 As for Proposition 3.2, the condition on Σ^n to have an elliptic point when the warping function is either increasing and the Gauss map is future-pointing, or decreasing and the Gauss map is past-pointing, is a natural condition. Indeed, for instance, in [19] the authors proved that if the GRW spacetime is spatially closed, then any compact spacelike hypersurface immersed in it admit an elliptic point under the same assumptions on the warping function and the Gauss map N of the hypersurface; see [19, Lemma 5.3].

3.5 Height Estimates in GRW Spacetimes

Following the terminology introduced in [19], we recall that a GRW spacetime $-I \times_f M^n$ satisfies the strong null convergence condition (strong NCC, for short) if the sectional curvature K_M of the fiber M^n satisfies

$$K_M \geq \sup\{ff'' - (f')^2\}.$$

We are ready to state and prove our next result regarding a suitable estimate of the height function of compact (r, s)-linear Weingarten spacelike hypersurfaces in a GRW spacetime.

Theorem 3.14 *Let $-\mathbb{R} \times_f M^n$ be a GRW spacetime satisfying the strong NCC and with increasing warping function. Moreover, let $\psi : \Sigma^n \to -\mathbb{R} \times_f M^n$ be a compact (r, s)-linear Weingarten spacelike hypersurface with positive $(s+1)$-mean curvature, boundary $\partial \Sigma^n$ contained into the slice $\{t_0\} \times M^n$, for some $t_0 \in \mathbb{R}$, and*

$$d \geq \sum_{k=r}^{s} b_k \sup[(\log f)']^{k+1}.$$

Suppose that

$$H_1 \geq \sup |H_{r+1}|^{1/(r+1)},$$

and, if $s \geq 2$, there exists an elliptic point in Σ^n with respect the future-pointing Gauss map. Then,

$$\Sigma^n \subset [t_0 - \alpha, t_0] \times M^n,$$

where

$$\alpha = \frac{\dfrac{f(t_0)}{f(\min h)} \max_{\partial \Sigma^n}(-\Theta) - 1}{\min H_1}.$$

Proof We argue as in the proof of Theorem 3.1. Let us consider on Σ^n the smooth function $\zeta = c\sigma(h) + \tilde{\Theta}$, where $c \in \mathbb{R}$ is a positive constant to be chosen in an appropriate way, σ is a primitive of f and $\tilde{\Theta} = f\Theta$.

By Proposition 1.1 we have

$$L_k \zeta = \frac{c_k f(h)}{k+1} \langle \nabla H_{k+1}, \nabla h \rangle + c_k f'(h)(H_{k+1} - cH_k)$$

$$+ f(h)\Theta \binom{n}{k+1}(nH_1 H_{k+1} - (n-k-1)H_{k+2} - (k+1)cH_{k+1})$$

$$+\frac{\tilde{\Theta}}{f^2(h)}\sum_{i=1}^{n}\mu_{i,k}K_{M^n}(N^*,E_i^*)|N^*\wedge E_i^*|^2$$
$$-\tilde{\Theta}(\log f)''(h)\left(c_k|\nabla h|^2 H_k - \langle P_k \nabla h, \nabla h\rangle\right), \tag{3.41}$$

where $P_k E_i = \mu_{i,k} E_i$, for every $i=1,\ldots,n$ and $k=r,\ldots,s$.
Since H_{s+1} is positive and Σ^n has an elliptic point, Lemma 1.1 gives that

$$H_1 H_{k+1} \geq H_{k+2}.$$

Hence, it follows that

$$nH_1 H_{k+1} - (n-k-1)H_{k+2} - (k+1)cH_{k+1} = (k+1)H_{k+1}(H_1 - c)$$
$$+ (n-k-1)(H_1 H_{k+1} - H_{k+2})$$
$$\geq (k+1)H_{k+1}(H_1 - c) \geq 0, \tag{3.42}$$

provided that $c = \min H_1$.

In particular, with this choice of c, by our hypothesis on H_1 and Lemma 1.1 we have that

$$H_{k+1} - cH_k \leq H_{k+1}^{1/(k+1)}(H_{k+1}^{k/(k+1)} - H_k) \leq 0. \tag{3.43}$$

On the other hand, by our assumptions we can apply Lemma 1.4 (or Lemma 1.3 if $s=1$) to obtain the ellipticity of the operator L_k for every $k=r,\ldots,s$.

In other words, P_k is positive definite. In particular, its eigenvalues $\mu_{i,k}$ are all positive on Σ^n, and by the strong NCC we get

$$\mu_{i,k} K_M(N^*, E_i^*)|N^* \wedge E_i^*|^2 \geq \mu_{i,k} C |N^* \wedge E_i^*|^2, \tag{3.44}$$

for every $i=1,\ldots,n$ and $k=r,\ldots,s$, where we are writing $C = \sup\{ff'' - (f')^2\}$.
A straightforward computation ensures that

$$|N^* \wedge E_i^*|^2 = |N^*|^2 |E_i^*|^2 - \langle N^*, E_i^*\rangle^2 = |\nabla h|^2 - \langle E_i, \nabla h\rangle^2.$$

Now, the above equation in addition to (3.44) imply

$$\sum_{i=1}^{n}\mu_{i,k}K_M(N^*,E_i^*)|N^* \wedge E_i^*|^2 \geq C\left(\text{tr}(P_k)|\nabla h|^2 - \sum_{i=1}^{n}\mu_{i,k}\langle E_i, \nabla h\rangle^2\right)$$
$$= C\left(\text{tr}(P_k)|\nabla h|^2 - \langle P_k \nabla h, \nabla h\rangle\right).$$

3.5 Height Estimates in GRW Spacetimes

Then, since $\operatorname{tr}(P_k) = c_k H_k$ and $C/f^2(h) - (\log f)''(h) \geq 0$, we obtain that

$$\frac{\sum_{i=1}^{n} \mu_{i,k} K_M(N^*, E_i^*) |N^* \wedge E_i^*|^2}{f^2(h)} - (\log f)''(h) \left(c_k |\nabla h|^2 H_k - \langle P_k \nabla h, \nabla h \rangle \right) \geq 0, \tag{3.45}$$

where the last inequality holds true since P_k is positive definite.

Hence, putting (3.42), (3.43) and (3.45) into (3.41) and taking into account that the warping function is increasing and $\Theta < 0$, we infer that

$$L_k \zeta \leq \frac{c_k f(h)}{k+1} \langle \nabla H_{k+1}, \nabla h \rangle. \tag{3.46}$$

Now, let us consider the operator $L : C^\infty(\Sigma^n) \to C^\infty(\Sigma^n)$ defined in Eq. (3.22) and given by

$$L = \sum_{k=r}^{s} (k+1) c_k^{-1} b_k L_k = \operatorname{tr}(P \circ \operatorname{Hess}),$$

where the tensor $P : \mathfrak{X}(\Sigma^n) \to \mathfrak{X}(\Sigma^n)$, given in Eq. (3.23), has the form

$$P = \sum_{k=r}^{s} (k+1) c_k^{-1} b_k P_k.$$

As in the proof of Theorem 3.1, we have that L is an elliptic operator, since $(k+1) c_k^{-1} b_k > 0$ and each operator L_k is elliptic, for every $k = r, \ldots, s$.

Consequently, taking into account Eq. (3.46), since Σ^n is (r,s)-linear Weingarten, one has that $L\zeta \leq 0$, i.e.

$$L(-\zeta) \geq 0.$$

Now, we observe that the weak maximum principle holds for the elliptic operator L since Σ^n is compact. Then, we must have

$$-c\sigma(h) - f(h)\Theta \leq \max_{\partial \Sigma^n}(-\zeta) = -c\sigma(t_0) + f(t_0) \max_{\partial \Sigma^n}(-\Theta),$$

which implies

$$c(\sigma(h) - \sigma(t_0)) \geq f(h) - f(t_0) \max_{\partial \Sigma^n}(-\Theta). \tag{3.47}$$

By using once more that f and σ are increasing functions, is not difficult to see that, for any $t \leq t_0$, the following inequality holds

$$\sigma(t_0) - \sigma(t) \geq f(t)(t_0 - t).$$

Since Proposition 3.2 says that $h \leq t_0$, we can apply Eq. (3.47) to get

$$cf(h)(h - t_0) \geq f(h) - f(t_0) \max_{\partial \Sigma^n}(-\Theta).$$

Therefore, we conclude that

$$c(h - t_0) \geq 1 - \frac{f(t_0)}{f(h)} \max_{\partial \Sigma^n}(-\Theta),$$

that is

$$h \geq t_0 - \frac{\frac{f(t_0)}{f(\min h)} \max_{\partial \Sigma^n}(-\Theta) - 1}{\min H_1}.$$

The proof is now complete. □

As in the Riemannian case, by Lemma 1.1 it turns out that for hypersurfaces with constant $(s+1)$-mean curvature H_{s+1} our assumption $H_1 \geq \sup |H_{s+1}|^{1/(s+1)}$ in Theorem 3.14 trivially holds. Moreover, we claim that, in this case, Theorem 3.14 improves the estimate obtained in [188, Theorem 16], which states that the hight function satisfies

$$t_0 - \frac{\frac{f(t_0)}{f(\min h)} \max_{\partial \Sigma^n}(-\Theta) - 1}{H_{s+1}^{1/(s+1)}} \leq h \leq t_0.$$

Indeed, since the inequality

$$\frac{\frac{f(t_0)}{f(\min h)} \max_{\partial \Sigma^n}(-\Theta) - 1}{\min H_1} \leq \frac{\frac{f(t_0)}{f(\min h)} \max_{\partial \Sigma^n}(-\Theta) - 1}{H_{s+1}^{1/(s+1)}}$$

holds for every $s = 0, \ldots, n-1$, our claim is proved.

We also observe that Theorem 3.14 does not contemplate the case in which the warping function is constant. However, a similar argument used in [120] allows us to obtain the next result settled on standard Lorentzian product spaces of the type $-\mathbb{R} \times M^n$, which improves [120, Theorem 3.3].

3.5 Height Estimates in GRW Spacetimes

Theorem 3.15 *Let $-\mathbb{R} \times M^n$ be a Lorentzian product whose fiber has nonnegative sectional curvature. Let $\psi : \Sigma^n \to -\mathbb{R} \times M^n$ be a compact (r,s)-linear Weingarten spacelike hypersurface with positive $(s+1)$-mean curvature and boundary $\partial \Sigma^n$ contained into the slice $\{t_0\} \times M^n$. Suppose that, if $s \geq 2$, there exists an elliptic point in Σ^n with respect the future-pointing Gauss map. Then, one has*

$$\Sigma^n \subset [t_0 - \alpha, t_0] \times M^n,$$

where

$$\alpha = \frac{\max_{\partial \Sigma^n}(-\Theta) - 1}{\min H_1}.$$

Proof The strategy of the proof is similar to that Theorem 3.4. We first observe that, we can apply Lemma 1.4 (or Lemma 1.3 if $s = 1$) in order to assures the ellipticity of the operators L_k for every $k = r, \ldots, s$. By Proposition 1.1, one has

$$L_r h = -c_r H_{r+1} \Theta \geq 0,$$

which gives $h \leq t_0$ on Σ^n.

Now, let $\zeta = ch + \Theta$ be the smooth function on Σ^n where $c = \min H_1$. By using once more Proposition 1.1, we have

$$L_k \zeta = \frac{c_k}{k+1} \langle \nabla H_{k+1}, \nabla h \rangle$$

$$+ \Theta \binom{n}{k+1}(nH_1 H_{k+1} - (n-k-1)H_{k+2} - (k+1)cH_{k+1})$$

$$+ \Theta \sum_{i=1}^{n} \mu_{i,k} K_M(N^*, E_i^*)|N^* \wedge E_i^*|^2,$$

where $P_k E_i = \mu_{i,k} E_i$, for every $i = 1, \ldots, n$ and $k = r, \ldots, s$.

Then, one has

$$L_k \zeta \leq \frac{c_k}{k+1} \langle \nabla H_{k+1}, \nabla h \rangle.$$

Arguing as in the proof of Theorem 3.14, we easily seen that $-\zeta \leq \max_{\partial \Sigma^n}(-\zeta)$, that is

$$-ch + 1 \leq -ct_0 + \max_{\partial \Sigma^n}(-\Theta).$$

The conclusion is achieved. □

As aforementioned before, we also consider the case in which the warping function is decreasing and the Gauss map is past-pointing, keeping positive $(s+1)$-mean curvature. This is the subject of the next theorem.

Theorem 3.16 *Let $-\mathbb{R} \times_f M^n$ be a GRW spacetime satisfying the strong NCC and with decreasing warping function. Moreover, let $\psi : \Sigma^n \to -\mathbb{R} \times_f M^n$ be a compact (r, s)-linear Weingarten spacelike hypersurface with positive $(s+1)$-mean curvature, boundary $\partial \Sigma^n$ contained into the slice $\{t_0\} \times M^n$, for some $t_0 \in \mathbb{R}$, and*

$$d \geq \sum_{k=r}^{s} b_k \sup[-(\log f)']^{k+1},$$

Suppose that

$$H_1 \geq \sup |H_{r+1}|^{1/(r+1)},$$

and, if $s \geq 2$, there exists an elliptic point in Σ^n with respect the past-pointing Gauss map. Then, one has

$$\Sigma^n \subset [t_0, t_0 + \beta] \times M^n,$$

where

$$\beta = \frac{\dfrac{f(t_0)}{f(\max h)} \max_{\partial \Sigma^n}(\Theta) - 1}{\min H_1}.$$

Proof For simplicity, we keep here the same notation of Theorem 3.14. Since $f' < 0$ and $\Theta \geq 1$, by (3.42), (3.43), (3.45) and (3.41) it follows that

$$L_k \zeta \geq \frac{c_k f(h)}{k+1} \langle \nabla H_{k+1}, \nabla h \rangle,$$

which gives

$$L\zeta \geq 0.$$

By the weak maximum principle we get

$$c\sigma(h) + f(h)\Theta \leq \max_{\partial \Sigma^n} \zeta = c\sigma(t_0) + f(t_0) \max_{\partial \Sigma^n} \Theta,$$

3.5 Height Estimates in GRW Spacetimes

that is,

$$c(\sigma(h) - \sigma(t_0)) \leq f(t_0) \max_{\partial \Sigma^n} \Theta - f(h). \tag{3.48}$$

On the other hand, it is easy to see that for every $t \geq t_0$ the inequality

$$\sigma(t) - \sigma(t_0) \geq f(t)(t - t_0) \tag{3.49}$$

holds. Now, Proposition 3.2 gives $h \geq t_0$. Hence, by (3.48) and (3.49) one has

$$cf(h)(h - t_0) \leq f(t_0) \max_{\partial \Sigma^n} \Theta - f(h).$$

Therefore

$$c(h - t_0) \leq \frac{f(t_0)}{f(h)} \max_{\partial \Sigma^n} \Theta - 1.$$

The proof is now complete. □

Similarly to Theorem 3.15, the following result holds.

Theorem 3.17 *Let* $-\mathbb{R} \times M^n$ *be a Lorentzian product whose fiber has nonnegative sectional curvature. Moreover, let* $\psi : \Sigma^n \to -\mathbb{R} \times M^n$ *be a compact (r, s)-linear Weingarten spacelike hypersurface with positive $(s + 1)$-mean curvature and boundary $\partial \Sigma^n$ contained into the slice $\{t_0\} \times M^n$. Suppose that, if $s \geq 2$, there exists an elliptic point in Σ^n with respect the past-pointing Gauss map. Then, one has*

$$\Sigma^n \subset [t_0, t_0 + \beta] \times M^n,$$

where

$$\beta = \frac{\max_{\partial \Sigma^n}(\Theta) - 1}{\min H_1}.$$

To conclude this section let us consider as ambient space the Lorentz-Minkowski spacetime $\mathbb{L}^{n+1}$. For convenience, we will adopt as model for the Lorentz-Minkowski spacetime the product manifold $-\mathbb{R} \times \mathbb{R}^n$ endowed with the Lorentzian metric

$$\langle \cdot, \cdot \rangle = -\pi_{\mathbb{R}}^*(dt^2) + \pi_{\mathbb{R}^n}^*(dx^2),$$

where $\pi_{\mathbb{R}}^*$ and $\pi_{\mathbb{R}^n}^*$ denote the canonical projections from $\mathbb{R} \times \mathbb{R}^n$ on each factor, $dx^2 = dx_1^2 + \cdots + dx_n^2$ is the canonical Riemannian metric on the n-dimensional Euclidean space $\mathbb{R}^n$ and $-\mathbb{R}$ stands for the line $\mathbb{R}$ furnished with the metric $-dt^2$.

We also note that the Gauss map $N \in \mathfrak{X}^\perp(\Sigma^n)$ of a spacelike hypersurface Σ^n immersed into the Lorentz-Minkowski spacetime can be regarded as a map $N : \Sigma^n \to \mathbb{H}^n$, where $\mathbb{H}^n$ denotes the n-dimensional hyperbolic space, that is

$$\mathbb{H}^n = \{p \in \mathbb{L}^{n+1} : \langle p, p \rangle = -1, \ p_1 \geq 0\}.$$

In this setting, the image $N(\Sigma^n)$ will be called the *hyperbolic image* of Σ^n. Furthermore, given a geodesic ball $B(a, \varrho) \subset \mathbb{H}^n$ centered at a point $a \in \mathbb{H}^n$ and radius $\varrho > 0$, it is well-known that $B(a, \varrho)$ is characterized as

$$B(a, \varrho) = \{p \in \mathbb{H}^n : -\cosh \varrho \leq \langle p, a \rangle \leq -1\}.$$

In particular, if the hyperbolic image of Σ^n is contained into some geodesic ball $B(a, \varrho)$, then

$$1 \leq |\langle N, a \rangle| \leq \cosh \varrho.$$

Hence if Σ^n is compact (necessarily with non-empty boundary; see, for instance [22, Section 2]), one has

$$\max_{\partial \Sigma^n} |\Theta| \leq \cosh \varrho,$$

where ϱ is the radius of a geodesic ball of center $\partial_t = e_1 = (1, 0, \ldots, 0)$.

With this preliminaries, we are ready to prove the following result, where the assumption on the existence of an elliptic point of the hypersurface is replaced by a boundedness condition on its hyperbolic image.

Theorem 3.18 *Let $\psi : \Sigma^n \to \mathbb{L}^{n+1}$ be a compact (r, s)-linear Weingarten spacelike hypersurface immersed into the Lorentz-Minkowski space such that H_{s+1} has strict sign on it and whose boundary $\partial \Sigma$ is contained into the hyperplane $\{0\} \times \mathbb{R}^n$. If the hyperbolic image of Σ^n is contained into a geodesic ball of center $e_1 \in \mathbb{H}^n$ and radius $\varrho > 0$, then the height function h of Σ^n satisfies the following estimate*

$$|h| \leq \frac{\cosh \varrho - 1}{\min H_1}. \tag{3.50}$$

3.5 Height Estimates in GRW Spacetimes

Moreover, estimate (3.50) is sharp in the sense that it is reached by the hyperbolic cap

$$\Sigma_\lambda = \left\{ x \in \mathbb{L}^{n+1} : \langle x, x \rangle = -\lambda^2, \ \lambda \leq x_1 \leq \sqrt{1+\lambda^2} \right\}, \tag{3.51}$$

where λ is the positive constant given by $\lambda = (\cosh \varrho - 1)^{-1/2}$.

Proof By [22, Lemma 1], our assumption that the boundary of Σ^n is contained into the hyperplane $\{0\} \times \mathbb{R}^n$ implies that (after an appropriate choice of the orientation on Σ^n) there exists an elliptic point in Σ^n. Hence the height estimate in (3.50) follows making use of Theorems 3.15 and 3.17.

Finally, it is not difficult to verify that the hyperbolic cap Σ_λ defined in (3.51) is a spacelike hypersurface of the Lorentz-Minkowski spacetime which has constant $(s+1)$-mean curvature given by

$$H_{s+1} = \frac{1}{\lambda^{rs+1}} > 0,$$

for every $0 \leq s \leq n-1$ (if we choose the Gauss map N in the same time-orientation of e_1, for the case in which s is even).

Moreover, the hyperbolic image of Σ_λ is contained in the geodesic ball of center $e_1 \in \mathbb{H}^{n+1}$ and radius

$$\varrho = \cosh^{-1} \sqrt{1 + \frac{1}{\lambda^2}}.$$

Thus, the height function of Σ_λ is given by

$$h = \frac{\cosh \varrho - 1}{\min_{\Sigma_\lambda} H_1},$$

showing that the estimate in (3.50) is sharp. □

We point out that for a spacelike hypersurface with constant $(s+1)$-mean curvature H_{s+1}, Theorem 3.18 improves the estimate obtained in [136, Theorem 4.2]. Indeed, the de Lima's result says that

$$|h| \leq \frac{\cosh \varrho - 1}{H_{s+1}^{1/(s+1)}}.$$

On the other hand, by Lemma 1.1 it follows that

$$\frac{\cosh \varrho - 1}{\min H_1} \leq \frac{\cosh \varrho - 1}{H_{s+1}^{1/(s+1)}}$$

for every $s = 0, \ldots, n-1$.

3.6 Half-Space Theorems and Topology at Infinity in GRW Spacetimes

The purpose of this section is to recover some of the half-space theorems given in Sect. 3.4 for the case of complete spacelike hypersurfaces immersed into a GRW spacetime $-\mathbb{R} \times M^n$. Following [188], our approach is based on the generalized version of the Omori-Yau's maximum principle for trace type differential operators given in Lemma 1.8.

It is worth pointing out that our results give an improvement of those obtained in [188] for hypersurfaces having some constant higher order mean curvature in a GRW spacetime with non-decreasing warping function; see Theorem 3.20 below. Moreover, we are able also to consider the case in which the warping function is non-increasing; see Theorem 3.21 below.

Before, let us recall the following definition, which in the Lorentzian setting was first introduced in [188].

We say that a spacelike hypersurface in a GRW spacetime $-\mathbb{R} \times_f M^n$ lies in an upper or lower half-space if it is, respectively, contained into a region of $-\mathbb{R} \times_f M^n$ of the form

$$[a, +\infty) \times M^n \quad \text{or} \quad (-\infty, a] \times M^n,$$

for some real number $a \in \mathbb{R}$.

We also recall that a GRW spacetime $-\mathbb{R} \times_f M^n$ is said spatially closed if its fiber M^n is compact. In this setting, as an application of Proposition 3.2, we get the following result, which is a generalization of Theorem 26 in [188] for the case in which the warping function is increasing and the Gauss map is future-pointing.

In particular, information on the topology at infinity of these hypersurfaces are given.

Theorem 3.19 *Let* $-\mathbb{R} \times_f M^n$ *be a spatially closed GRW spacetime and let* $\psi : \Sigma^n \to -\mathbb{R} \times_f M^n$ *be a properly immersed complete spacelike hypersurface with positive* $(s+1)$*-mean curvature, for some* $0 \leq s \leq n-1$. *The following holds:*

(a) *If* $f'(h) > 0$ *and, for some* $0 \leq r \leq s$, *we have*

$$\sum_{k=r}^{s} b_k H_{k+1} \geq \sum_{k=r}^{s} b_k \sup[(\log f)']^{k+1}$$

for certain nonnegative constants b_k and, when $s \geq 2$, there exists an elliptic point in Σ^n with respect the future-pointing Gauss map, then Σ^n cannot lie in a lower half-space. In particular, Σ^n must have at least one top end.

(b) If $f'(h) < 0$ and, for some $0 \leq r \leq s$, we have

$$\sum_{k=r}^{s} b_k H_{k+1} \geq \sum_{k=r}^{s} b_k \sup[-(\log f)']^{k+1}$$

for certain nonnegative constants b_k and, when $s \geq 2$, there exists an elliptic point in Σ^n with respect to the past-pointing Gauss map, then Σ^n cannot lie in an upper half-space. In particular, Σ^n must have at least one bottom end.

Proof Since the proof is similar to the Riemannian case, it is sufficient to prove, for instance, item (b). To this aim, arguing by contradiction, let us assume that Σ^n lie in an upper half-space, that is

$$\Sigma^n \subset [a, +\infty) \times M^n$$

for some $a \in \mathbb{R}$.

For any number $t_0 > a$, we denote by Σ_{t_0} the hypersurface

$$\Sigma_{t_0} = \{(t, p) \in \Sigma^n : t \leq t_0\}.$$

Then, Σ_{t_0} is a compact spacelike hypersurface with boundary contained into the slice $\{t_0\} \times M$, since M^n is compact and the immersion is proper.

Therefore, by Proposition 3.2, we get $h \geq t_0$. Since t_0 is arbitrary, we have an absurd. The conclusion is achieved. □

Our aim now is to study the case in which the fiber is not necessarily compact. More precisely, following some ideas already presented in the Riemannian setting, given a GRW spacetime $-\mathbb{R} \times M^n$, we are interest on half-space theorems for noncompact generalized Weingarten spacelike hypersurfaces immersed in these ambient spaces.

A first result in this direction reads as follows.

Theorem 3.20 *Let* $-\mathbb{R} \times_f M^n$ *be a GRW spacetime satisfying the strong NCC and with non-decreasing warping function. Let* $\psi : \Sigma^n \to -\mathbb{R} \times_f M^n$ *be a complete noncompact* (r, s)-*linear Weingarten spacelike hypersurface with positive* $(s + 1)$-*mean curvature, for some* $1 \leq r \leq s \leq n - 1$, *and*

$$d > \sum_{k=r}^{s} b_k \sup[(\log f)']^{k+1}.$$

Suppose that

$$\inf \frac{f''(h)}{f(h)} > -\infty,$$

and, if $s \geq 2$, there exists an elliptic point in Σ^n with respect the future-pointing Gauss map. Assume further that $\sup |H_r| < +\infty$ and the shape operator satisfies $|A| \leq G(r_o)$, where $G \in C^1([0, +\infty))$ is such that

$$G(0) > 0, \quad G'(t) \geq 0 \quad \text{and} \quad \frac{1}{G(t)} \notin L^1(+\infty)$$

and r_o is the distance function from a reference point in Σ^n. Then Σ^n cannot lie in a lower half-space.

Proof As in the proof of Theorem 3.11, we states that the assumptions of Lemma 1.8 of Sect. 1.6 holds. Indeed, we notice that, by Lemma 3.2

$$\overline{R}(U, V)W = R_M(U^*, V^*)W^* + [(\log f)']^2(\langle U, W\rangle V - \langle V, W\rangle U)$$
$$-(\log f)''\langle W, \partial_t\rangle(\langle U, \partial_t\rangle V - \langle V, \partial_t\rangle U)$$
$$-(\log f)''(\langle U, W\rangle\langle V, \partial_t\rangle - \langle V, W\rangle\langle U, \partial_t\rangle)\partial_t,$$

for every U, V, W tangent to $-\mathbb{R} \times_f M^n$.

In particular, for an orthonormal base $\{X, Y\}$ of an arbitrary plane tangent to Σ^n, we get

$$\overline{K}(X, Y) = \frac{1}{f^2(h)} K_M(X^*, Y^*)|X^* \wedge Y^*|^2 + (\log f)'(h)^2$$
$$-(\log f)''(h)(\langle X, \nabla h\rangle^2 + \langle Y, \nabla h\rangle^2).$$

By the strong NCC and the fact that $|X^* \wedge Y^*|^2 = 1 + \langle X, \nabla h\rangle^2 + \langle Y, \nabla h\rangle^2$ we find

$$\overline{K}(X, Y) \geq (\log f)''(h) + (\log f)'(h)^2 = \frac{f''(h)}{f(h)}.$$

On the other hand, by using the Gauss equation (1.1) and the previous inequality, we infer that the sectional curvature K_{Σ^n} of Σ^n satisfies

$$K_{\Sigma^n} = \overline{K}(X, Y) - \langle AX, X\rangle\langle AY, Y\rangle + \langle AX, Y\rangle^2$$
$$\geq \frac{f''(h)}{f(h)} - |AX||AY| \geq \frac{f''(h)}{f(h)} - |A|^2.$$

3.6 Half-Space Theorems and Topology at Infinity in GRW Spacetimes

Hence, the assumption on the warping function and the definition of the shape operator imply that K_{Σ^n} satisfies (1.15) of Lemma 1.8 of Sect. 1.6, proving the claim.

From now on, we argue by contradiction. Let us suppose that Σ^n lies in a lower half-space. In others words, the height function of Σ^n satisfies $h^* = \sup h < +\infty$.

Let $\mathfrak{L}_k : C^\infty(\Sigma^n) \to C^\infty(\Sigma^n)$ be the second order linear differential operator, for each $k = r, \ldots, s$, given by

$$\mathfrak{L}_k = \frac{1}{(-\Theta)^k} \sum_{i=0}^{k} (-1)^i \frac{c_k}{c_i} (\log f)'(h)^{k-i} \Theta^i L_i = \mathrm{tr}(\mathfrak{P}_k \circ \mathrm{Hess}), \tag{3.52}$$

where the tensor $\mathfrak{P}_k : \mathfrak{X}(\Sigma^n) \to \mathfrak{X}(\Sigma^n)$ is defined by

$$\mathfrak{P}_k = \frac{1}{(-\Theta)^k} \sum_{i=0}^{k} (-1)^i \frac{c_k}{c_i} (\log f)'(h)^{k-i} \Theta^i P_i.$$

We notice that

$$\mathfrak{L}_k = \frac{1}{(-\Theta)^k} \mathcal{L}_k^+ \quad \text{and} \quad \mathfrak{P}_k = \frac{1}{(-\Theta)^k} \mathcal{P}_k^+,$$

where $\mathcal{L}_k^+$ and $\mathcal{P}_k^+$ are given by (3.34) and (3.35), respectively.

On the other hand, we easily have that $\mathfrak{L}_k$ is a semi-elliptic operator or, equivalently, $\mathfrak{P}_k$ is a positive semi-definite tensor.

Moreover, since $d > b_k \sup[(\log f)']^{k+1}$ for every $k = r, \ldots, s$, by Lemma 1.1 it follows that

$$\mathrm{tr}(\mathfrak{P}_k) \leq \frac{c_k}{(-\Theta)^k} \sum_{i=0}^{k} (-\Theta)^i \left(\frac{d}{b_k}\right)^{(k-i)/(k+1)} H_r^{i/r} \leq c_k \sum_{i=0}^{k} \left(\frac{d}{b_k}\right)^{(k-i)/(k+1)} H_r^{i/r},$$

which implies that $\sup \mathrm{tr}(\mathfrak{P}_k) < +\infty$.

We set the second order linear differential operator $\mathfrak{L} : C^\infty(\Sigma^n) \to C^\infty(\Sigma^n)$ by

$$\mathfrak{L} = \sum_{k=r}^{s} b_k c_k^{-1} \mathfrak{L}_k^+ = \mathrm{tr}(\mathfrak{P} \circ \mathrm{Hess}), \tag{3.53}$$

where the tensor $\mathfrak{P} : \mathfrak{X}(\Sigma^n) \to \mathfrak{X}(\Sigma^n)$ is given by

$$\mathfrak{P} = \sum_{k=r}^{s} b_k c_k^{-1} \mathfrak{P}_k^+.$$

Then $\mathfrak{P}$ is a positive semi-definite symmetric tensor with $\sup \mathrm{tr}(\mathfrak{P}) < +\infty$. In particular, $\mathcal{L}$ is an semi-elliptic operator.

Consequently, by Lemma 1.8 the Omori-Yau's maximum principle holds on Σ^n for the operator $\mathcal{L}$.

Hence, since $\sigma(h) \leq \sigma(h^*) < +\infty$, there exists a sequence of points $\{p_j\} \subset \Sigma^n$ with the following properties

$$\lim_j \sigma(h)(p_j) = \sigma(h^*), \quad |\nabla \sigma(h)(p_j)| = f(h)(p_j)|\nabla h(p_j)| < \frac{1}{j},$$

and

$$\mathcal{L}\sigma(h)(p_j) < \frac{1}{j}.$$

Then, Eq. (3.39) yields

$$\frac{1}{j} > \sum_{k=r}^{s} \frac{f(h(p_j))}{(-\Theta(p_j))^k} b_k \left(-[(\log f)'(h(p_j))]^{k+1} + (-\Theta(p_j))^{k+1} H_{k+1}(p_j) \right).$$

By relation $|\nabla h|^2 = \Theta^2 - 1$, making $j \to +\infty$ we get

$$d \leq \sum_{k=r}^{s} b_k \sup[(\log f)']^{k+1},$$

obtaining a contradiction. □

Let us recall the following remark already mentioned in a similar way after Theorem 3.11. By

$$|A|^2 = n^2 H_1^2 - n(n-1)H_2,$$

the condition $\sup |A|^2 < +\infty$ is equivalent to $\sup |H_1| < +\infty$, requiring that $\inf H_2 > -\infty$. In general, if there exists an elliptic point for an appropriate choice of the Gauss map and H_{s+1} does not change sign on Σ^n for some $s = 2, \ldots, n-1$, then, by Lemma 1.1, the condition $\sup |A|^2 < +\infty$ is equivalent to $\sup H_1 < +\infty$.

Moreover, if $\inf \dfrac{f''(h)}{f(h)} > -\infty$, arguing as in the proof of Theorem 3.20, we get that the sectional curvature of the hypersurface Σ^n is bounded from below, provided that $\sup |H_1| < +\infty$.

The following result holds.

3.6 Half-Space Theorems and Topology at Infinity in GRW Spacetimes

Corollary 3.13 *Let* $-\mathbb{R} \times_f M^n$ *be a GRW spacetime satisfying the strong NCC and with non-decreasing warping function. Moreover, let* $\psi : \Sigma^n \to -\mathbb{R} \times_f M^n$ *be a complete* (r, s)*-linear Weingarten spacelike hypersurface with positive* $(s+1)$*-mean curvature, for some* $0 \leq r \leq s \leq n-1$*, and*

$$d > \sum_{k=r}^{s} b_k \sup[(\log f)']^{k+1}.$$

Suppose that

$$\sup |H_1| < +\infty \quad \text{and} \quad \inf \frac{f''(h)}{f(h)} > -\infty,$$

and, if $s \geq 2$*, there exists an elliptic point in* Σ^n *with respect the future-pointing Gauss map. Then,* Σ^n *cannot lie in a lower half-space.*

We notice that, if Σ^n is a (s, s)-linear Weingarten hypersurface in Corollary 3.13, i.e. Σ^n has constant $(s+1)$-mean curvature, we recover [188, Theorem 35 (i)].

More precisely, one has the next result.

Corollary 3.14 *Let* $-\mathbb{R} \times_f M^n$ *be a GRW spacetime satisfying the strong NCC and with non-decreasing warping function. Moreover, let* $\psi : \Sigma^n \to -\mathbb{R} \times_f M^n$ *be a complete spacelike hypersurface with constant* $(s+1)$*-mean curvature satisfying*

$$H_{s+1}^{1/(s+1)} > \sup(\log f)'$$

for some $0 \leq s \leq n-1$*. Suppose that*

$$\sup |H_1| < +\infty \quad \text{and} \quad \inf \frac{f''(h)}{f(h)} > -\infty,$$

and, if $s \geq 2$*, there exists an elliptic point in* Σ^n *with respect the future-pointing Gauss map. Then,* Σ^n *cannot lie in a lower half-space.*

We observe that, as showed in [188, Theorem 32], if the hypersurface has constant mean curvature we do not need to assume that f' does not change sign in Corollary 3.14. Finally, we close this section by proving a suitable version of the previous theorem settled on a GRW spacetime satisfying the strong NCC and with non-increasing warping function.

Theorem 3.21 *Let* $-\mathbb{R} \times_f M^n$ *be a GRW spacetime satisfying the strong NCC and with non-increasing warping function. Moreover, let* $\psi : \Sigma^n \to -\mathbb{R} \times_f M^n$ *be a complete* (r, s)*-linear Weingarten spacelike hypersurface with positive* $(s+1)$*-mean curvature, for* some $1 \leq r \leq s \leq n-1$, and

$$d > \sum_{k=r}^{s} b_k \sup[-(\log f)']^{k+1}.$$

Suppose that

$$\inf \frac{f''(h)}{f(h)} > -\infty,$$

and, if $s \geq 2$, there exists an elliptic point in Σ^n with respect the past-pointing Gauss map. Assume further that $\sup |H_r| < +\infty$ and the shape operator satisfies $|A| \leq G(r_o)$, where $G \in C^1([0, +\infty))$ is such that

$$G(0) > 0, \quad G'(t) \geq 0 \quad \text{and} \quad \frac{1}{G(t)} \notin L^1(+\infty)$$

and r_o is the distance function from a reference point in Σ^n. Then Σ^n cannot lie in an upper half-space.

Proof Let us assume by contradiction that Σ^n lies in an upper half-space, i.e. the height function of Σ^n satisfies $h_* = \inf h > -\infty$. Furthermore, for each $k = r, \ldots, s$, let $\mathcal{L}_k$ be the operator given in (3.52).

We observe that $\mathcal{L}_k$ can be written as follows

$$\mathcal{L}_k = \frac{1}{\Theta^k} \sum_{i=0}^{k} (-1)^{k-i} \frac{c_k}{c_i} (\log f)'(h)^{k-i} \Theta^i L_i = \frac{1}{\Theta^k} \mathcal{L}_k^-,$$

where $\mathcal{L}_k^-$ is defined in (3.37). Hence, in this case, $\mathcal{L}_k$ is a semi-elliptic operator. In particular, $\mathcal{L}$ defined in (3.53) is a semi-elliptic operator too. Moreover, the Omori-Yau's maximum principle holds on Σ^n for the semi-elliptic operator $\mathcal{L}$.

Now, since $\sigma(h) \geq \sigma(h_*)$, there exists a sequence of points $\{q_j\} \subset \Sigma^n$ satisfying

$$\lim_j \sigma(h)(q_j) = \sigma(h_*), \quad |\nabla \sigma(h)(q_j)| < \frac{1}{j} \quad \text{and} \quad \mathcal{L}\sigma(h)(q_j) > -\frac{1}{j}.$$

Then, on account of (3.40), one has

$$-\frac{1}{j} < \sum_{k=r}^{s} \frac{f(h(q_j))}{\Theta(q_j)^k} b_k \left([-(\log f)'(h(q_j))]^{k+1} - \Theta(q_j)^{k+1} H_{k+1}(q_j) \right).$$

3.6 Half-Space Theorems and Topology at Infinity in GRW Spacetimes

Therefore, letting $j \to +\infty$ we get

$$d \leq \sum_{k=r}^{s} b_k \sup[-(\log f)']^{k+1},$$

obtaining a contradiction. The proof is now complete. □

A special case of Theorem 3.21 reads as follows.

Corollary 3.15 *Let* $-\mathbb{R} \times_f M^n$ *be a GRW spacetime satisfying the strong NCC and with non-increasing warping function. Moreover, let* $\psi : \Sigma^n \to -\mathbb{R} \times_f M^n$ *be a complete* (r, s)-*linear Weingarten spacelike hypersurface with positive* $(s + 1)$-*mean curvature, for some* $0 \leq r \leq s \leq n - 1$, *and*

$$d > \sum_{k=r}^{s} b_k \sup[-(\log f)']^{k+1}.$$

Suppose that

$$\sup |H_1| < +\infty \quad \text{and} \quad \inf \frac{f''(h)}{f(h)} > -\infty,$$

and, if $s \geq 2$, *there exists an elliptic point in* Σ^n *with respect the past-pointing Gauss map. Then,* Σ^n *cannot lie in an upper half-space.*

The last result of this chapter is valid for hypersurfaces with some constant higher order mean curvature.

Corollary 3.16 *Let* $-\mathbb{R} \times_f M^n$ *be a GRW spacetime satisfying the strong NCC and with non-increasing warping function. Moreover, let* $\psi : \Sigma^n \to -\mathbb{R} \times_f M^n$ *be a complete spacelike hypersurface with constant* $(s + 1)$-*mean curvature satisfying*

$$H_{s+1}^{1/(s+1)} > \sup[-(\log f)'],$$

for some $0 \leq s \leq n - 1$. *Suppose that*

$$\sup |H_1| < +\infty \quad \text{and} \quad \inf \frac{f''(h)}{f(h)} > -\infty,$$

and, if $s \geq 2$, *there exists an elliptic point in* Σ^n *with respect the past-pointing Gauss map. Then,* Σ^n *cannot lie in an upper half-space.*

Spacelike Hypersurfaces in Standard Static Spacetimes

4.1 Introduction

The study of spacelike hypersurfaces immersed with constant mean curvature in a Lorentz manifold has attracted the interest of a considerable group of geometers as it is evidenced by the amount of works which was generated in the last decades.

This is due not only to its mathematical interest, but also to its relevance in General Relativity. For example, constant mean curvature spacelike hypersurfaces are particularly suitable for studying the propagation of gravity radiation. Moreover, as it was observed by Bartnik [68], they have been used to prove positivity of mass [273], analyze the space of solutions of Einstein's equations [177] and in numerical integration schemes for Einstein's equations [166, 257]. Further references can also be found in Choquet-Bruhat and York [118], Marsden [236] and Stumbles [281].

From the mathematical point of view, these hypersurfaces exhibit interesting Bernstein-type properties, and one can truly say that the first remarkable results in this branch were the rigidity theorems of Calabi in [92] and Cheng and Yau in [112], who showed (the former for $n \leq 4$, and the latter for general n) that the only maximal (that is, with zero mean curvature) complete noncompact spacelike hypersurfaces of the Lorentz-Minkowski space $\mathbb{L}^{n+1}$ are the spacelike hyperplanes.

However, in the case that the mean curvature is a positive constant, Treibergs [283] astonishingly showed that there are many entire solutions of the corresponding constant mean curvature equation in $\mathbb{L}^{n+1}$, which he was able to classify by their projective boundary values at infinity.

Our purpose in this chapter is to study the geometry of complete constant mean curvature spacelike hypersurfaces immersed in a standard static spacetime, that is, a Lorentzian manifold endowed with a globally defined timelike Killing vector field.

In this setting, supposing that the ambient space is a warped product of the type $M^n \times_\rho \mathbb{R}_1$ whose Riemannian base M^n has nonnegative sectional curvature and the warping function ρ is convex on M^n, we use the generalized maximum principle of Omori-Yau in order to establish rigidity results concerning these spacelike hypersurfaces.

We also study the parabolicity of maximal spacelike surfaces in $M^2 \times_\rho \mathbb{R}_1$ and we obtain uniqueness results for entire Killing graphs constructed over M^n.

The results presented in this chapter are mainly based on the papers [150, 158].

4.2 Standard Static Spacetimes

Let $\overline{M}^{n+1}$ be a $(n + 1)$-dimensional Lorentz manifold endowed with a timelike Killing vector field K. Suppose that the distribution $\mathcal{D}$ orthogonal to K is of constant rank and integrable. We denote by $\Psi : M^n \times \mathbb{I} \to \overline{M}^{n+1}$ the flow generated by K, where M^n is an arbitrarily fixed spacelike integral leaf of $\mathcal{D}$ labeled as $t = 0$, which we will suppose to be connected, and $\mathbb{I}$ is the maximal interval of definition. Without lost of generality, in what follows we will also consider $\mathbb{I} = \mathbb{R}$.

In this setting, $\overline{M}^{n+1}$ can be viewed as the *standard static spacetime* $M^n \times_\rho \mathbb{R}_1$, that is, the product manifold $M^n \times \mathbb{R}$ endowed with the warping metric

$$\langle \cdot , \cdot \rangle = \pi_M^* (\langle \cdot , \cdot \rangle_M) - (\rho \circ \pi_M)^2 \pi_\mathbb{R}^* \left(dt^2 \right), \tag{4.1}$$

where π_M and $\pi_\mathbb{R}$ denote the canonical projections from $M \times \mathbb{R}$ onto each factor, $\langle \cdot , \cdot \rangle_M$ is the induced Riemannian metric on the fiber M^n, $\mathbb{R}_1$ is the manifold $\mathbb{R}$ endowed with the metric $-dt^2$ and the warping function $\rho \in C^\infty$ is defined by

$$\rho = |K| = \sqrt{-\langle K, K \rangle},$$

in which $|\cdot|$ denotes the norm of a vector field on $\overline{M}^{n+1}$.

Remark 4.1 The importance of standard static spacetimes comes from the fact that they include some classical spacetimes. Some meaningful prototypes are listed below:

(a) The Lorentz-Minkowski space $\mathbb{L}^{n+1}$, which is isometric to the warped product

$$\left(\mathbb{R}^n \times \mathbb{R}_1 \, , \, \pi_{\mathbb{R}^n}^* (g_{\mathbb{R}^n}) + \pi_\mathbb{R}^*(-dt^2) \right).$$

4.2 Standard Static Spacetimes

(b) The Einstein static universe

$$\left(\mathbb{S}^n \times \mathbb{R}_1, \pi^*_{\mathbb{S}^n}(g_{\mathbb{S}^n}) + \pi^*_{\mathbb{R}}(-dt^2)\right)$$

is a standard static space; see [76, Example 5.11].

(c) The exterior Schwarzschild spacetime, which is defined as follows. Let $\mathbb{R}^4$ be given coordinates (t, r, θ, φ), where (r, θ, φ) are the usual spherical coordinates on $\mathbb{R}^3$. Given a positive constant m, the exterior Schwarzschild spacetime is defined on the subset $S = \{(t, r, \theta, \varphi) \in \mathbb{R}^4 : r > 2m\}$ of $\mathbb{R}^4$, which is topologically equivalent to $\mathbb{R}^2 \times \mathbb{S}^2$. The Schwarzschild metric for the region S is given in (t, r, θ, φ) coordinates by

$$ds^2 = -\left(1 - \frac{2m}{r}\right)dt^2 + \left(1 - \frac{2m}{r}\right)^{-1}dr^2 + r^2\left(d\theta^2 + \sin^2\theta d\varphi^2\right).$$

Since the metric for this spacetime is invariant under time translations $t \to t + a$, the coordinate vector field $\partial/\partial t$ is a (globally defined) timelike Killing vector field; see [76, Section 5.2] or [251, Chapter 13]. Consequently, the exterior Schwarzschild spacetime is a standard static spacetime.

(d) The Reissner-Nordström spacetime, whose metric in (t, r, θ, φ) coordinates admits the representation

$$ds^2 = -\left(1 - \frac{2m}{r} + \frac{e^2}{r^2}\right)dt^2 + \left(1 - \frac{2m}{r} + \frac{e^2}{r^2}\right)^{-1}dr^2 + r^2\left(d\theta^2 + \sin^2\theta d\varphi^2\right).$$

This model also presents static regions (which appeared shortly after the Schwarzschild spacetime). More precisely, the metric above has singularities in $r = 0$, $r = r_+$ and $r = r_-$, where $r_\pm = m \pm (m^2 - e^2)^{1/2}$, with $m > 0$ and the Reissner-Nordström spacetime is static in the regions

$$R = \{(t, r, \theta, \varphi) \in \mathbb{R}^4 : r_+ < r < +\infty\},$$

and

$$T = \{(t, r, \theta, \varphi) \in \mathbb{R}^4 : 0 < r < r_-\}.$$

See [200, Section 5.5].

Now, let us consider a connected spacelike hypersurface $\psi : \Sigma^n \to \overline{M}^{n+1}$ immersed in a standard static spacetime $\overline{M}^{n+1} = M^n \times_\rho \mathbb{R}_1$, that is, the metric induced on Σ^n via ψ is a Riemannian metric.

As usual, we also denote for $\langle \cdot, \cdot \rangle$ the metric of Σ^n induced via ψ. Since K is a globally defined timelike vector field on $\overline{M}^{n+1}$, it follows that there exists a unique unitary timelike normal vector field N globally defined on Σ^n which is in same time-orientation as K. By using the inverse Cauchy-Schwarz inequality, we get

$$\langle N, K \rangle \leq -\rho < 0 \quad \text{on} \quad \Sigma^n. \tag{4.2}$$

We will refer to that normal vector field N as the *future-pointing Gauss map* of Σ^n. Throughout this chapter, N will always denote the future-pointing Gauss map of a spacelike hypersurface $\psi : \Sigma^n \to \overline{M}^{n+1}$.

Let $\overline{\nabla}$, ∇ and D denote the Levi-Civita connections in $\overline{M}^{n+1}$, Σ^n and M^n, respectively. Then, the Gauss and Weingarten formulas for the spacelike hypersurface $\psi : \Sigma^n \to \overline{M}^{n+1}$ are given by

$$\overline{\nabla}_X Y = \nabla_X Y - \langle AX, Y \rangle N \tag{4.3}$$

and

$$AX = -\overline{\nabla}_X N, \tag{4.4}$$

for all the tangent vector fields $X, Y \in \mathfrak{X}(\Sigma^n)$, where A stands for the shape operator of Σ^n with respect to its future-pointing Gauss map N.

On the other hand, as in [251], the curvature tensor R of the spacelike hypersurface Σ^n is given by

$$R(X, Y)Z = \nabla_{[X,Y]} - [\nabla_X, \nabla_Y]Z,$$

where $[\cdot, \cdot]$ denotes the Lie bracket and $X, Y, Z \in \mathfrak{X}(\Sigma^n)$.

It is well-known that the curvature tensor R of the spacelike hypersurface Σ^n can be described in terms of the shape operator A and of the curvature tensor $\overline{R}$ of the ambient space $\overline{M}^{n+1} = M^n \times_\rho \mathbb{R}_1$ by the Gauss equation given by

$$R(X, Y)Z = (\overline{R}(X, Y)Z)^\top - \langle AX, Z \rangle AY + \langle AY, Z \rangle AX \tag{4.5}$$

for every $X, Y, Z \in \mathfrak{X}(\Sigma^n)$, where $(\cdot)^\top$ denotes the tangential component of a vector field in $\mathfrak{X}(\overline{M}^{n+1})$ along Σ^n. In particular, when $n = 2$, we have that

$$K_{\Sigma^2} = \overline{K}_{\Sigma^2} - \det A, \tag{4.6}$$

where K_{Σ^2} denotes the Gaussian curvature of Σ^2 and $\overline{K}_{\Sigma^2}$ stands for the sectional curvature in $\overline{M}^3 = M^2 \times_\rho \mathbb{R}_1$ of the plane tangent to Σ^2.

4.3 Some Auxiliary Results

In this setting, we will consider two particular smooth functions on a connected spacelike hypersurface $\psi : \Sigma^n \to \overline{M}^{n+1}$ immersed in $\overline{M}^{n+1} = M^n \times_\rho \mathbb{R}_1$, namely, the (vertical) height function $h = \pi_\mathbb{R} \circ \psi$ and the angle function $\Theta = \langle N, K \rangle$, where we recall that N denotes the future-pointing Gauss map of Σ^n. By (4.2), we note that Θ will be always a negative function.

From the decomposition $K = K^\top - \Theta N$, we obtain

$$\nabla h = -\frac{1}{\rho^2} K^\top \quad \text{and} \quad |\nabla h|^2 = \frac{\Theta^2 - \rho^2}{\rho^4}. \tag{4.7}$$

Moreover, assuming that the mean curvature function is constant, $H = -\frac{1}{n}\text{trace}(A)$, by [7, Proposition 1] we have the following formula

$$\Delta \Theta = \left(\overline{\text{Ric}}(N, N) + |A|^2\right) \Theta, \tag{4.8}$$

where $\overline{\text{Ric}}$ denotes the Ricci tensor of $\overline{M}^{n+1}$ and $|A|$ stands for the Hilbert-Schmidt norm of the shape operator A of Σ^n with respect to its future-pointing Gauss map N.

To close this section, we also recall the following algebraic relation

$$|A|^2 = nH^2 + n(n-1)(H^2 - H_2), \tag{4.9}$$

where

$$H_2 = \frac{2}{n(n-1)} S_2$$

is the mean value of the second elementary symmetric function S_2 on the eigenvalues of A.

4.3 Some Auxiliary Results

The auxiliary results presented here will be use full in the next section in order to prove some uniqueness theorems.

We first study some sufficient conditions in order to guarantee the boundness from below of the Ricci curvature of a spacelike hypersurface $\psi : \Sigma^n \to \overline{M}^{n+1}$ immersed in a standard static spacetime $\overline{M}^{n+1} = M^n \times_\rho \mathbb{R}_1$. To our scope some computations are given below.

Let U, V, W tangent vector fields to $\overline{M}^{n+1}$. We can write

$$U = U^* + U^\perp,$$

where U^* and $U^\perp$ are the orthogonal projections of U onto TM and $T\mathbb{R}_1$, respectively.

Thus
$$U^\perp = \frac{\langle U, K\rangle}{\langle K, K\rangle}K = -\frac{\langle U, K\rangle}{\rho^2}K.$$

On the other hand, a direct computation ensures that

$$\overline{R}(U,V)W = R_M(U^*, V^*)W^* + \frac{\langle V, K\rangle}{\rho^2}\overline{R}(K, U^*)W^* + \frac{\langle V, K\rangle\langle W, K\rangle}{\rho^4}\overline{R}(U^*, K)K$$
$$- \frac{\langle U, K\rangle}{\rho^2}\overline{R}(K, V^*)W^* - \frac{\langle U, K\rangle\langle W, K\rangle}{\rho^4}\overline{R}(V^*, K)K.$$

Then by [251, Lemma 7.34 and Proposition 7.42], we get

$$\overline{R}(U,V)W = R_M(U^*, V^*)W^* + \frac{\langle V, K\rangle\mathrm{Hess}_M\rho(U^*, W^*)}{\rho^3}K$$
$$+ \frac{\langle V, K\rangle\langle W, K\rangle\langle K, K\rangle}{\rho^5}\overline{\nabla}_{U^*}\overline{\nabla}(\rho\circ\pi_M)$$
$$- \frac{\langle U, K\rangle\mathrm{Hess}_M\rho(V^*, W^*)}{\rho^3}K - \frac{\langle U, K\rangle\langle W, K\rangle\langle K, K\rangle}{\rho^5}\overline{\nabla}_{V^*}\overline{\nabla}(\rho\circ\pi_M)$$
$$= R_M(U^*, V^*)W^* + \frac{\langle V, K\rangle\mathrm{Hess}_M\rho(U^*, W^*)}{\rho^3}K$$
$$- \frac{\langle V, K\rangle\langle W, K\rangle}{\rho^3}D_{U^*}D\rho$$
$$- \frac{\langle U, K\rangle\mathrm{Hess}_M\rho(V^*, W^*)}{\rho^3}K + \frac{\langle U, K\rangle\langle W, K\rangle}{\rho^3}D_{V^*}D\rho,$$

where Hess_M is the Hessian on M^n.

In particular, taking a local orthonormal frame $\{E_1, \ldots, E_n\}$ tangent to Σ^n and X be a vector field tangent to Σ^n, we can take $U = W = X$ and $V = E_i$ in the last equation to obtain

$$\overline{R}(X, E_i)X = R_M(X^*, E_i^*)X^* + \frac{\langle E_i, K\rangle\mathrm{Hess}_M\rho(X^*, X^*)}{\rho^3}K$$
$$- \frac{\langle E_i, K\rangle\langle X, K\rangle}{\rho^3}D_{X^*}D\rho$$
$$- \frac{\langle X, K\rangle\mathrm{Hess}_M\rho(E_i^*, X^*)}{\rho^3}K + \frac{\langle X, K\rangle^2}{\rho^3}D_{E_i^*}D\rho.$$

4.3 Some Auxiliary Results

Hence, we conclude that

$$\langle \overline{R}(X, E_i)X, E_i \rangle = \langle R_M(X^*, E_i^*)X^*, E_i \rangle + \frac{\langle E_i, K \rangle^2}{\rho^3} \text{Hess}_M \rho(X^*, X^*)$$

$$- \frac{\langle E_i, K \rangle \langle X, K \rangle}{\rho^3} \langle D_{X^*} D\rho, E_i \rangle$$

$$- \frac{\langle E_i, K \rangle \langle X, K \rangle}{\rho^3} \text{Hess}_M \rho(E_i^*, X^*)$$

$$+ \frac{\langle X, K \rangle^2}{\rho^3} \langle D_{E_i^*} D\rho, E_i \rangle$$

$$= \langle R_{M^n}(X^*, E_i^*)X^*, E_i^* \rangle + \frac{\langle E_i, K \rangle^2}{\rho^3} \text{Hess}_M \rho(X^*, X^*)$$

$$- \frac{\langle E_i, K \rangle \langle X, K \rangle}{\rho^3} \text{Hess}_M \rho(X^*, E_i^*)$$

$$- \frac{\langle E_i, K \rangle \langle X, K \rangle}{\rho^3} \text{Hess}_M \rho(X^*, E_i^*)$$

$$+ \frac{\langle X, K \rangle^2}{\rho^3} \text{Hess}_M(E_i^*, E_i^*).$$

Consequently, one has

$$\langle \overline{R}(X, E_i)X, E_i \rangle = K_M(X^*, E_i^*) \left(\langle X^*, X^* \rangle \langle E_i^*, E_i^* \rangle - \langle X^*, E_i^* \rangle^2 \right)$$

$$+ \frac{\langle E_i, K \rangle^2}{\rho^3} \text{Hess}_M \rho(X^*, X^*)$$

$$- 2 \frac{\langle E_i, K \rangle \langle X, K \rangle}{\rho^3} \text{Hess}_M \rho(X^*, E_i^*)$$

$$+ \frac{\langle X, K \rangle^2}{\rho^3} \text{Hess}_M \rho(E_i^*, E_i^*).$$

That is

$$\langle \overline{R}(X, E_i)X, E_i \rangle = K_M(X^*, E_i^*) \left(\langle X^*, X^* \rangle \langle E_i^*, E_i^* \rangle - \langle X^*, E_i^* \rangle^2 \right)$$

$$+ \frac{1}{\rho} \text{Hess}_M \rho(\widetilde{X}_i^*, \widetilde{X}_i^*)$$

$$- \frac{2}{\rho} \text{Hess}_M \rho(\widetilde{X}_i^*, \widetilde{E}_i^*) + \frac{1}{\rho} \text{Hess}_M \rho(\widetilde{E}_i^*, \widetilde{E}_i^*),$$

where

$$\widetilde{X}_i^* = \frac{\langle E_i, K\rangle}{\rho} X^* \quad \text{and} \quad \widetilde{E}_i^* = \frac{\langle X, K\rangle}{\rho} E_i^*.$$

Hence, one has

$$\langle \overline{R}(X, E_i)X, E_i\rangle = K_M(X^*, E_i^*)\left(\langle X^*, X^*\rangle\langle E_i^*, E_i^*\rangle - \langle X^*, E_i^*\rangle^2\right)$$
$$+ \frac{1}{\rho}\mathrm{Hess}_M \rho(\widetilde{X}_i^* - \widetilde{E}_i^*, \widetilde{X}_i^* - \widetilde{E}_i^*). \tag{4.10}$$

Therefore, we obtain that

$$\sum_{i=1}^n \langle \overline{R}(X, E_i)X, E_i\rangle = \sum_{i=1}^n K_M(X^*, E_i^*)\left(\langle X^*, X^*\rangle\langle E_i^*, E_i^*\rangle - \langle X^*, E_i^*\rangle^2\right)$$
$$+ \sum_{i=1}^n \frac{1}{\rho}\mathrm{Hess}_M \rho(\widetilde{X}_i^* - \widetilde{E}_i^*, \widetilde{X}_i^* - \widetilde{E}_i^*). \tag{4.11}$$

Now, we are in position to prove the following result.

Proposition 4.1 *Let* $\overline{M}^{n+1} = M^n \times_\rho \mathbb{R}_1$ *be a standard static spacetime whose Riemannian base* M^n *has nonnegative sectional curvature* K_M *and warping function* ρ *convex on* M^n. *Moreover, let* $\psi : \Sigma^n \to \overline{M}^{n+1}$ *be a spacelike hypersurface with bounded mean curvature H. Then, the Ricci curvature* Ric *of* Σ^n *is bounded from below.*

Proof By using the Gauss equation (4.5) and taking a local orthonormal frame $\{E_1, \ldots, E_n\}$ tangent to Σ^n, we have that the Ricci curvature Ric of Σ^n is given by

$$\mathrm{Ric}(X, X) = \sum_{i=1}^n \langle \overline{R}(X, E_i)X, E_i\rangle + nH\langle AX, X\rangle + |AX|^2, \tag{4.12}$$

for any vector field X tangent to Σ^n.

Since K_{M^n} is nonnegative and ρ is convex, we can write

$$nH\langle AX, X\rangle + |AX|^2 = \left|AX + \frac{nH}{2}X\right|^2 - \frac{n^2 H^2}{4}|X|^2.$$

4.4 Rigidity and Parabolicity Results

Hence, by (4.11) and (4.12), one has

$$\mathrm{Ric}(X, X) \geq -\frac{n^2 H^2}{4} |X|^2, \qquad (4.13)$$

for any tangent vector field $X \in \mathfrak{X}(\Sigma^n)$.

Therefore, since H is bounded, we conclude that the Ricci curvature of Σ^n is bounded from below. □

Now, let us recall two maximum principles due to Yau. More precisely, in [298], Yau obtained the following version of the celebrated Stokes' Theorem on an n-dimensional, complete noncompact Riemannian manifold Σ^n:

If $\omega \in \Omega^{n-1}(\Sigma^n)$ is an integrable $(n-1)$-differential form on Σ^n, then there exists a sequence B_i of domains on Σ^n such that $B_i \subset B_{i+1}$, $\Sigma^n = \bigcup_{i \geq 1} B_i$ and

$$\lim_{i \to +\infty} \int_{B_i} d\omega = 0.$$

By applying this result to $\omega = \iota_{\nabla f} d\Sigma^n$, where $f : \Sigma^n \to \mathbb{R}$ is a smooth function, ∇f denotes its gradient and $\iota_{\nabla f} d\Sigma^n$ the contraction of the volume element $d\Sigma^n$ of Σ^n in the direction of ∇f, Yau established an extension of the classical Hopf's Theorem on a complete Riemannian manifold.

Let $\mathcal{L}^1_g(\Sigma^n)$ be the space of Lebesgue integrable functions on Σ^n. The next result holds.

Lemma 4.1 *Let Σ^n be a complete Riemannian manifold and let $f : \Sigma^n \to \mathbb{R}$ be a smooth function. If f is a subharmonic (or superharmonic) function with $|\nabla f| \in \mathcal{L}^1_g(\Sigma^n)$, then f must actually be harmonic.*

Finally, the next lemma is just a consequence of an extension of a Liouville's Theorem due to Yau [297].

Lemma 4.2 *The only harmonic semi-bounded functions defined on an n-dimensional complete Riemannian manifold whose Ricci curvature is nonnegative are the constant ones.*

4.4 Rigidity and Parabolicity Results

This section is devoted to present our uniqueness results concerning complete spacelike hypersurfaces immersed in a Lorentz Killing warped product.

So, we state and prove our first result along this direction.

Theorem 4.1 *Let $\overline{M}^{n+1} = M^n \times_\rho \mathbb{R}_1$ be a standard static spacetime whose (not necessarily complete) Riemannian base M^n has nonnegative sectional curvature K_M and warping function ρ convex on M^n. Moreover, let $\psi : \Sigma^n \to \overline{M}^{n+1}$ be a complete spacelike hypersurface with constant mean curvature H. Suppose that the angle function Θ of Σ^n is bounded. Then Σ^n is maximal. Moreover, when $n = 2$ we have that Σ^2 is totally geodesic. In addition, if K_M is positive at some point $p_0 \in \Sigma^2$, then Σ^2 is a slice of $\overline{M}^3$.*

Proof By Proposition 4.1, there exists a sequence of points $\{p_k\} \subset \Sigma^n$ such that

$$\lim_k \Theta(p_k) = \inf_{\Sigma^n} \Theta \quad \text{and} \quad \liminf_k \Delta\Theta(p_k) \geq 0.$$

Now, by [251, Corollary 7.43], we get

$$\overline{\mathrm{Ric}}(N, N) = \overline{\mathrm{Ric}}(N^*, N^*) + \overline{\mathrm{Ric}}(N^\perp, N^\perp)$$

$$= \mathrm{Ric}_M(N^*, N^*) - \frac{1}{\rho}\mathrm{Hess}_M\rho(N^*, N^*) - \langle N^\perp, N^\perp\rangle \frac{\Delta_M \rho}{\rho}$$

$$= \mathrm{Ric}_M(N^*, N^*) - \frac{1}{\rho}\mathrm{Hess}_M\rho(N^*, N^*) + \frac{\Theta^2}{\rho^3}\Delta_M\rho, \qquad (4.14)$$

where Hess_M and Δ_M are, respectively, the Hessian and the Laplacian on M^n.

Consequently, by (4.8) and (4.14) we obtain the following formula

$$\Delta\Theta = \left(\mathrm{Ric}_M(N^*, N^*) - \frac{1}{\rho}\mathrm{Hess}_M\rho(N^*, N^*) + \frac{\Theta^2}{\rho^3}\Delta_M\rho + |A|^2\right)\Theta. \qquad (4.15)$$

In particular, at the points where N^* is zero, by using the convexity of ρ we have that

$$\Delta\Theta \leq \left(\mathrm{Ric}_M(N^*, N^*) + |A|^2\right)\Theta.$$

On the other hand, at the points where N^* is non zero, we can write

$$\frac{1}{\rho}\mathrm{Hess}_M\rho(N^*, N^*) = \frac{|N^*|^2}{\rho}\mathrm{Hess}_M\rho\left(\frac{N^*}{|N^*|}, \frac{N^*}{|N^*|}\right)$$

$$= \frac{\Theta^2 - \rho^2}{\rho^3}\mathrm{Hess}_M\rho\left(\frac{N^*}{|N^*|}, \frac{N^*}{|N^*|}\right),$$

and, taking a local orthonormal frame

$$\left\{E_1 = \frac{N^*}{|N^*|}, E_2, \ldots, E_n\right\}$$

4.4 Rigidity and Parabolicity Results

tangent to M^n, it follows that

$$\frac{\Theta^2}{\rho^3}\Delta_M\rho = \frac{\Theta^2}{\rho^3}\text{Hess}_M\rho\left(\frac{N^*}{|N^*|}, \frac{N^*}{|N^*|}\right) + \frac{\Theta^2}{\rho^3}\sum_{i=2}^{n}\text{Hess}_M\rho(E_i, E_i).$$

Consequently, one has

$$-\frac{1}{\rho}\text{Hess}_M\rho(N^*, N^*) + \frac{\Theta^2}{\rho^3}\Delta_M\rho = \frac{1}{\rho}\text{Hess}_M\rho\left(\frac{N^*}{|N^*|}, \frac{N^*}{|N^*|}\right) + \frac{\Theta^2}{\rho^3}\sum_{i=2}^{n}\text{Hess}_M\rho(E_i, E_i).$$

Now, since ρ is convex and $\Theta \leq 0$, the above equation and (4.15) yield

$$\Delta\Theta \leq \left(\text{Ric}_M(N^*, N^*) + |A|^2\right)\Theta. \tag{4.16}$$

Thus in both the cases (4.16) holds. Then, taking into account relation (4.9), by (4.16) it follows that

$$0 \leq \liminf_{k}\Delta\Theta(p_k) \leq \lim_{k}\left(\text{Ric}_M(N^*, N^*) + |A|^2\right)\Theta(p_k)$$

$$\leq \lim_{k}\left(\text{Ric}_M(N^*, N^*) + nH^2\right)\Theta(p_k) \leq 0.$$

Consequently, since Ric_M is nonnegative, we conclude that $H = 0$, that is, the hypersurface Σ^n is maximal.

Now, let us consider the case $n = 2$. In this setting, $|A|^2 = -2\det A$ and the Gauss equation (4.6) assume the form

$$K_{\Sigma^2} = \overline{K}_{\Sigma^2} + \frac{1}{2}|A|^2, \tag{4.17}$$

where K_{Σ^2} denotes the Gaussian curvature of Σ^2 and $\overline{K}_{\Sigma^2}$ stands for the sectional curvature in $\overline{M}^3$ of the tangent plane to Σ^2.

On the other hand, taking a local orthonormal frame $\{E_1, E_2\}$ tangent to Σ^2, by (4.10) we obtain that

$$\overline{K}_{\Sigma^2} = \langle \overline{R}(E_1, E_2)E_1, E_2\rangle$$

$$= K_M(E_1^*, E_2^*)\left(\langle E_1^*, E_1^*\rangle\langle E_2^*, E_2^*\rangle - \langle E_1^*, E_2^*\rangle^2\right)$$

$$+ \frac{1}{\rho}\text{Hess}_M\rho(\widetilde{E}_1^* - \widetilde{E}_2^*, \widetilde{E}_1^* - \widetilde{E}_2^*)$$

$$= K_M(E_1^*, E_2^*)(1 + \rho^2|\nabla h|^2) + \frac{1}{\rho}\text{Hess}_M\rho(\widetilde{E_1^*} - \widetilde{E_2^*}, \widetilde{E_1^*} - \widetilde{E_2^*})$$

$$= K_M(E_1^*, E_2^*)\frac{\Theta^2}{\rho^2} + \frac{1}{\rho}\text{Hess}_M\rho(\widetilde{E_1^*} - \widetilde{E_2^*}, \widetilde{E_1^*} - \widetilde{E_2^*}),$$

where

$$\widetilde{E_1^*} = \frac{\langle E_2, K \rangle}{\rho}E_1^* \quad \text{and} \quad \widetilde{E_2^*} = \frac{\langle E_1, K \rangle}{\rho}E_2^*.$$

Then, by (4.17), it easily seen that

$$K_{\Sigma^2} = K_M(E_1^*, E_2^*)\frac{\Theta^2}{\rho^2} + \frac{1}{\rho}\text{Hess}_M\rho(\widetilde{E_1^*} - \widetilde{E_2^*}, \widetilde{E_1^*} - \widetilde{E_2^*}) + \frac{1}{2}|A|^2. \qquad (4.18)$$

By using again the convexity of the function ρ in addition to our assumption on K_M, we obtain that the Gaussian curvature of Σ^2 is nonnegative.

By a classical result due to Ahlfors [2] and Blanc-Fiala-Huber [208], a complete surface of nonnegative Gaussian curvature is parabolic in the sense that any bounded superharmonic function on the surface must be constant.

Then, by (4.16) it follows that Θ is a bounded superharmonic function on Σ^2. Hence, Θ must be constant on Σ^2 and, thanks to (4.16) we conclude that $|A|^2 = 0$, that is, Σ^2 is totally geodesic.

Now, we argue as in [28, Corollary 3] obtaining that

$$\nabla\Theta = \frac{1}{\rho}\left(\Theta\overline{\nabla}\rho - \langle N, \overline{\nabla}\rho\rangle K\right),$$

where as before K denotes the timelike Killing vector field on the ambient space.

Since Θ is constant and $\overline{\nabla}\rho$ and K are linearly independent, it follows that ρ is constant. So, if K_M is positive at some point $p_0 \in \Sigma^2$, by (4.16) it follows that $\text{Ric}_M(N^*, N^*)(p_0) = 0$ and, consequently, $N^*(p_0) = 0$.

However, it is easily seen that

$$|\nabla h|^2 = \frac{1}{\rho^2}|N^*|_M^2 = \frac{1}{\rho^2}\left(\frac{\Theta^2}{\rho^2} - 1\right). \qquad (4.19)$$

Therefore, by (4.19) we conclude that Σ^2 is a slice of $\overline{M}^3$. $\square$

4.4 Rigidity and Parabolicity Results

An extension of [48, Theorem 1] is given below.

Theorem 4.2 *Let $\overline{M}^{n+1} = M^n \times_\rho \mathbb{R}_1$ be a standard static spacetime whose (not necessarily complete) Riemannian base M^n has nonnegative sectional curvature K_{M^n} and warping function ρ convex on M^n. Moreover, let $\psi : \Sigma^n \to \overline{M}^{n+1}$ be a complete spacelike hypersurface which lies between two spacelike slices, with constant mean curvature and such that $|\nabla \rho|$ is bounded on it. Then Σ^n is maximal. Moreover, if $\langle \nabla \rho, \nabla h \rangle$ does not change sign on Σ^n and $|\nabla h| \in \mathcal{L}_g^1(\Sigma^n)$, then Σ^n is a slice of $\overline{M}^{n+1}$.*

Proof From the decomposition $K = K^\top - \Theta N$, we have that

$$\overline{\nabla}_X K = \overline{\nabla}_X K^\top - X(\Theta)N - \Theta \overline{\nabla}_X N, \qquad (4.20)$$

for every vector field X tangent to Σ^n. Then applying the Gauss (4.3) and Weingarten (4.4) formulas and (4.20), we get

$$\nabla_X K^\top = (\overline{\nabla}_X K)^\top - \Theta AX.$$

It follows that

$$\begin{aligned}
\nabla_X \nabla h &= X\left(-\frac{1}{\rho^2}\right) K^\top - \frac{1}{\rho^2} \nabla_X K^\top \\
&= \frac{X(\rho^2)}{\rho^4} K^\top - \frac{1}{\rho^2}(\overline{\nabla}_X K)^\top + \frac{1}{\rho^2}\Theta AX \\
&= \frac{2X(\rho)}{\rho^3} K^\top - \frac{1}{\rho^2}(\overline{\nabla}_X K)^\top + \frac{1}{\rho^2}\Theta AX \\
&= \frac{2\langle \nabla\rho, X \rangle}{\rho^3} K^\top - \frac{1}{\rho^2}(\overline{\nabla}_X K)^\top + \frac{1}{\rho^2}\Theta AX, \qquad (4.21)
\end{aligned}$$

for every vector field X tangent to Σ^n.

Since K is a Killing vector field, taking a local orthonormal frame $\{E_1, \ldots, E_n\}$ of $\mathfrak{X}(\Sigma^n)$, by (4.21) we get

$$\Delta h = \sum_{i=1}^n \langle \nabla_{E_i} \nabla h, E_i \rangle$$

$$= \sum_{i=1}^n \frac{2}{\rho^3} \langle \nabla\rho, E_i \rangle \langle K^\top, E_i \rangle - \sum_{i=1}^n \frac{1}{\rho^2} \langle (\overline{\nabla}_{E_i} K)^\top, E_i \rangle + \sum_{i=1}^n \frac{1}{\rho^2} \Theta \langle AE_i, E_i \rangle$$

$$= \frac{2}{\rho^3}\langle\nabla\rho, K^\top\rangle - \sum_{i=1}^{n}\frac{1}{\rho^2}\langle\overline{\nabla}_{E_i}K, E_i\rangle - \frac{nH\Theta}{\rho^2}$$

$$= -\frac{2}{\rho}\langle\nabla\rho, \nabla h\rangle - \frac{nH\Theta}{\rho^2}, \tag{4.22}$$

that is

$$nH\Theta = -\left(2\rho\langle\nabla\rho, \nabla h\rangle + \rho^2\Delta h\right). \tag{4.23}$$

Let us assume that $H \geq 0$. Since $\Theta \leq -\rho$, by (4.23) it follows that

$$nH \leq 2\langle\nabla\rho, \nabla h\rangle + \rho\Delta h \tag{4.24}$$

Now, by Proposition 4.1, there exists a sequence of points $\{p_k\} \subset \Sigma^n$ such that

$$\lim_k |\nabla h(p_k)| = 0 \quad \text{and} \quad \limsup_k \Delta h(p_k) \leq 0.$$

Thus, on account of (4.24), it easily seen that

$$0 \leq nH \leq \limsup_k (2\langle\nabla\rho(p_k), \nabla h(p_k)\rangle + \rho(p_k)\Delta h(p_k))$$

$$= \limsup_k \rho(p_k)\Delta h(p_k) \leq 0,$$

and, consequently, $H = 0$. For the case $H \leq 0$, we observe that, since $\Theta \leq -\rho$,

$$nH \geq 2\langle\nabla\rho, \nabla h\rangle + \rho\Delta h. \tag{4.25}$$

By Proposition 4.1, there exists a sequence of points $\{q_k\} \subset \Sigma^n$ satisfying

$$\lim_k |\nabla h(q_k)| = 0 \quad \text{and} \quad \liminf_k \Delta h(q_k) \geq 0.$$

Hence, by (4.25) we have that

$$0 \geq nH \geq \liminf_k (2\langle\nabla\rho(q_k), \nabla h(q_k)\rangle + \rho(q_k)\Delta h(q_k))$$

$$= \liminf_k \rho(q_k)\Delta h(q_k) \geq 0,$$

and again Σ^n must be maximal.

Now, assuming that $\langle\nabla\rho, \nabla h\rangle$ does not change sign on Σ^n, by (4.22) we conclude that Δh also does not change sign on Σ^n.

4.4 Rigidity and Parabolicity Results

Thus, since $|\nabla h| \in \mathcal{L}_g^1(\Sigma^n)$, we can apply Lemma 4.1 to get that h is a harmonic function on Σ^n.

Moreover, by (4.13) we also get that Σ^n has nonnegative Ricci curvature. Therefore, since Σ^n lies between two slices, Lemma 4.2 guarantees that h must be constant on Σ^n, that is, Σ^n is a slice of $\overline{M}^{n+1}$. □

By Theorem 4.2 the next result immediately holds.

Corollary 4.1 *Let $\overline{M}^{n+1} = M^n \times_\rho \mathbb{R}_1$ be a standard static spacetime whose (not necessarily complete) Riemannian base M^n has nonnegative sectional curvature K_M and warping function ρ convex on M^n. Moreover, let $\psi : \Sigma^n \to \overline{M}^{n+1}$ be a complete maximal spacelike hypersurface which lies between two spacelike slices. If $\langle \nabla \rho, \nabla h \rangle$ does not change sign on Σ^n and $|\nabla h| \in \mathcal{L}_g^1(\Sigma^n)$, then Σ^n is a slice of $\overline{M}^{n+1}$.*

We end this section investigating the parabolicity of maximal spacelike surfaces in $M^2 \times_\rho \mathbb{R}_1$. The following result is achieved.

Theorem 4.3 *Let $\overline{M}^3 = M^2 \times_\rho \mathbb{R}_1$ be a standard static spacetime whose (not necessarily complete) Riemannian base M^2 has nonnegative Gaussian curvature K_M and warping function ρ convex on M^2. Let $\psi : \Sigma^2 \to \overline{M}^3$ be a maximal spacelike surface which is complete with respect to the metric induced from the Riemannian warped product $M^2 \times_\rho \mathbb{R}$. Suppose that the warping function satisfies $\inf_{\Sigma^2} \rho > 0$ and the angle function Θ of Σ^2 is bounded. Then Σ^2 is totally geodesic. In addition, if K_M is positive at some point $p_0 \in \Sigma^2$, then M^2 is complete and Σ^2 is a slice of $\overline{M}^3$.*

Proof Since Θ is bounded, we can consider a constant $\alpha < 2\inf_{\Sigma^2} \Theta$. Now, let us introduce on Σ^2 the conformal metric

$$\tilde{g}(X, X) = (\Theta - \alpha)^2 g(X, X),$$

where $g = \langle \cdot, \cdot \rangle$ stands for the Riemannian metric induced on Σ^2 by the ambient space $\overline{M}^3$. Then, for every tangent vector field X on Σ^2, we have

$$\tilde{g}(X, X) = (\Theta - \alpha)^2 g(X, X) \geq \Theta^2(|X^*|^2 - \rho^2 X(h)^2), \quad (4.26)$$

where, by (4.7), we can write

$$X = X^* - \frac{\langle X, K \rangle}{\rho^2} K = X^* + \langle X, \nabla h \rangle K = X^* + X(h)K$$

and $|\cdot|$ denotes the norm with respect to the Riemannian metric induced on M^2 by the ambient space $\overline{M}^3$. If we set

$$N = N^* - \frac{\Theta}{\rho^2} K,$$

it easily seen that

$$0 = \langle X, N \rangle = \langle N^*, X^* \rangle + \Theta X(h).$$

The Cauchy-Schwarz immediately yields

$$\Theta^2 X(h)^2 \leq |N^*|^2 |X^*|^2 = \frac{\Theta^2 - \rho^2}{\rho^2} |X^*|^2.$$

By (4.26) it follows that

$$\tilde{g}(X, X) = \Theta^2(|X^*|^2 - \rho^2 X(h)^2) \geq \rho^2 |X^*|^2. \tag{4.27}$$

Let us denote by g' the Riemannian metric induced on Σ^2 by the Riemannian warped product $M^2 \times_\rho \mathbb{R}$. Thus

$$g(X, X) = |X^*| - \rho^2 X(h)^2 \quad \text{and} \quad g'(X, X) = |X^*|^2 + \rho^2 X(h)^2,$$

and, consequently, one has

$$|X^*|^2 = \frac{1}{2}\left(g(X, X) + g'(X, X)\right) \geq \frac{1}{2} g'(X, X).$$

Hence, by (4.27), we obtain

$$\tilde{g}(X, X) \geq \rho^2 |X^*|^2 \geq \frac{\rho^2}{2} g'(X, X),$$

for every tangent vector field X on Σ^2.

Now, the main hypothesis on ρ ensures that there exists $\beta > 0$ such that

$$\tilde{g}(X, X) \geq \beta g'(X, X)$$

for every tangent vector field X on Σ^2.

Since we are assuming that Σ^2 is complete with respect to the metric g', we have that $(\Sigma^2, \tilde{g})$ is also complete.

Moreover, by standard arguments, the Gaussian curvature $\tilde{K}_{\Sigma^2}$ of $(\Sigma^2, \tilde{g})$ is given by

$$(\Theta - \alpha)^2 \tilde{K}_{\Sigma^2} = K_{\Sigma^2} - \Delta \log(\Theta - \alpha), \tag{4.28}$$

where Δ is the Laplacian on Σ^2.

The main assumption on K_M and (4.16) ensures that

$$\Delta \log(\Theta - \alpha) = \frac{\Delta \Theta}{\Theta - \alpha} - \frac{|\nabla \Theta|^2}{(\Theta - \alpha)^2}$$

$$\leq \left(K_M(\pi) |N^*|^2 + |A|^2 \right) \frac{\Theta}{\Theta - \alpha} \leq 0$$

Hence, by (4.18) and (4.28), we conclude that

$$(\Theta - \alpha)^2 \tilde{K}_{\Sigma^2} \geq 0,$$

that is, $\tilde{K}_{\Sigma^2} \geq 0$.

Thus $(\Sigma^2, \tilde{g})$ is a complete Riemannian surface with nonnegative Gaussian curvature. Then, again by a classical result due to Ahlfors [2] and Blanc-Fiala-Huber [208], we conclude that $(\Sigma^2, \tilde{g})$ is parabolic.

Since superharmonic is preserved under a conformal change of metric, we have that (Σ^2, g) is also parabolic.

Now, arguing as in Theorem 4.1 and, by using (4.16), the conclusion follows. □

4.5   A Parabolicity Criterion

We begin this section recall that every connected manifold Σ has an universal covering, that is, there exist a simply connected manifold $\tilde{\Sigma}$ (called a universal covering of Σ) and a smooth map $\kappa : \tilde{\Sigma} \to \Sigma$ (called a covering map) such that each point $p \in \Sigma$ has a connected neighborhood U that is evenly covered by κ, that is, κ maps each component of $\kappa^{-1}(U)$ diffeomorphically onto U; see [251, Appendix A] for more details.

Moreover, if Σ is a Riemannian manifold, then there exist an universal covering such that the covering map $\kappa : \tilde{\Sigma} \to \Sigma$ is a local isometry. In this case, $\tilde{\Sigma}$ is said the universal Riemannian covering of Σ; see [163, p. 152].

In [260], the authors studied the parabolicity of complete spacelike hypersurfaces in GRW spacetimes whose Riemannian fiber has a parabolic universal Riemannian covering. In this setting, they were able to guarantee the parabolicity of complete spacelike hypersurfaces, under suitable boundedness assumptions on the warping function and on the hyperbolic angle function of these hypersurfaces.

Our aim in this section is just to obtain an extension of this parabolicity criterion to the context of standard static spacetimes.

More precisely, we get the following result.

Theorem 4.4 *Let $\overline{M}^{n+1} = M^n \times_\rho \mathbb{R}_1$ be a standard static spacetime whose Riemannian base M^n has parabolic universal Riemannian covering. If $\psi : \Sigma^n \to \overline{M}^{n+1}$ is a complete spacelike hypersurface such that the function $\eta = \Theta/\rho$ is bounded on it, then M^n is complete and Σ^n is parabolic.*

Proof The proof is based on two facts:

(*i*) Parabolicity is invariant under a quasi-isometry (cf. [194, 216]);
(*ii*) If the universal Riemannian covering $\widetilde{\Sigma}$ of Σ^n is parabolic, then Σ^n is also parabolic.

Set $\pi = \pi_M \circ \psi : \Sigma^n \to M^n$. For any tangent vector $v \in T\Sigma^n$, we have

$$\langle v, v \rangle = \langle \pi_* v, \pi_* v \rangle_M - \rho^2 \langle h_* v, h_* v \rangle_\mathbb{R} \leq c \langle \pi_* v, \pi_* v \rangle_M,$$

where $c = \sup_{\Sigma^n} \eta^2 \geq 1$.

In particular, it easily seen that $\pi_{*,p} : T_p \Sigma^n \to T_{\pi(p)} M^n$ is an isomorphism for every $p \in \Sigma^n$. Then, the Inverse Function Theorem ensures that π is a local diffeomorphism and by [163, Lemma 7.3.3] we conclude that π is a covering map and that M^n is complete; see also [218, Lemma 8.8.1].

On the other hand, the Cauchy-Schwartz inequality ensures that

$$\langle \nabla h, v \rangle^2 \leq \langle \nabla h, \nabla h \rangle \langle v, v \rangle.$$

Consequently, since $h_* v = dh(v) = \langle \nabla h, v \rangle$, we have

$$\langle v, v \rangle = \langle \pi_* v, \pi_* v \rangle_M - \rho^2 \langle h_* v, h_* v \rangle_\mathbb{R}$$
$$= \langle \pi_* v, \pi_* v \rangle_M - \rho^2 \langle \nabla h, v \rangle^2 \geq \langle \pi_* v, \pi_* v \rangle_M - \rho^2 |\nabla h|^2 \langle v, v \rangle,$$

that is

$$\langle v, v \rangle \left(1 + \rho^2 |\nabla h|^2\right) \geq \langle \pi_* v, \pi_* v \rangle_M.$$

The definition of η and (4.7) give

$$\langle v, v \rangle \geq \frac{1}{\eta^2} \langle \pi_* v, \pi_* v \rangle_M.$$

4.5 A Parabolicity Criterion

By our hypothesis we conclude that

$$c^{-1}\langle \pi_*v, \pi_*v\rangle_M \leq \langle v, v\rangle \leq c\langle \pi_*v, \pi_*v\rangle_M. \qquad (4.29)$$

So, let $\widetilde{\Sigma}$ be the universal Riemannian covering of Σ^n with projection $\pi_{\Sigma^n} : \widetilde{\Sigma} \to \Sigma^n$. Then, the map $\pi_0 = \pi \circ \pi_{\Sigma^n} : \widetilde{\Sigma} \to M^n$ is a covering map. Now, if $\widetilde{M}$ is the universal Riemannian covering of M^n with projection $\widetilde{\pi} : \widetilde{M} \to M^n$, then there exists a diffeomorphism $\varphi : \widetilde{\Sigma} \to \widetilde{M}$ such that $\widetilde{\pi} \circ \varphi = \pi_0$. Moreover, φ is a quasi-isometry.

Indeed, if $v \in T\widetilde{\Sigma}$, by (4.29) we have that

$$\begin{aligned}
\langle \varphi_*v, \varphi_*v\rangle_{\widetilde{M}} &= \langle \widetilde{\pi}_*(\varphi_*v), \widetilde{\pi}_*(\varphi_*v)\rangle_M \\
&= \langle (\pi_0)_*v, (\pi_0)_*v\rangle_M \\
&= \langle \pi_*((\pi_{\Sigma^n})_*v), \pi_*((\pi_{\Sigma^n})_*v)\rangle_M \\
&\leq c\langle (\pi_{\Sigma^n})_*v, (\pi_{\Sigma^n})_*v\rangle_{\Sigma^n} = c\langle v, v\rangle_{\widetilde{\Sigma}}.
\end{aligned}$$

Analogously, we obtain

$$\langle \varphi_*v, \varphi_*v\rangle_{\widetilde{M}} \geq c^{-1}\langle v, v\rangle_{\widetilde{\Sigma}}.$$

Therefore, since the universal Riemannian covering of M^n is parabolic, it follows that the universal Riemannian covering of Σ^n has this property. Hence, Σ^n must be also parabolic. □

As a direct consequence of Theorem 4.4, we have the following result.

Corollary 4.2 *Let $\overline{M}^{n+1} = M^n \times_\rho \mathbb{R}_1$ be a standard static spacetime whose Riemannian base M^n is complete, simply connected and parabolic. If $\psi : \Sigma^n \to \overline{M}^{n+1}$ is a complete spacelike hypersurface such that η is bounded on it, then Σ^n is parabolic.*

As well as, the next property holds.

Corollary 4.3 *Let $\overline{M}^{n+1} = M^n \times \mathbb{R}_1$ be a Lorentzian product whose Riemannian base M^n is complete and has parabolic universal Riemannian covering. If $\psi : \Sigma^n \to \overline{M}^{n+1}$ is a complete spacelike hypersurface whose its angle function Θ is bounded, then Σ^n is parabolic.*

4.6 Further Rigidity Results

In this section, we will apply Theorem 4.4 in order to obtain rigidity results for complete spacelike hypersurfaces in $M^n \times_\rho \mathbb{R}_1$. We will assume that these hypersurfaces are connected. Our first rigidity theorem reads as follows.

Theorem 4.5 *Let $\overline{M}^{n+1} = M^n \times_\rho \mathbb{R}_1$ be a standard static spacetime with nonnegative Ricci curvature and whose Riemannian base M^n has parabolic universal Riemannian covering. Let $\psi : \Sigma^n \to \overline{M}^{n+1}$ be a complete spacelike hypersurface with constant mean curvature such that its angle function Θ is bounded and the warping function ρ satisfies $\inf_{\Sigma^n} \rho > 0$. Then, Σ^n is totally geodesic. In addition, if Ric_M is positive at some point $p_0 \in \Sigma^n$, then Σ^n is a slice of $\overline{M}^{n+1}$.*

Proof By [7, Proposition 1] we have the following formula

$$\Delta \Theta = \left(\overline{\mathrm{Ric}}(N, N) + |A|^2\right) \Theta, \tag{4.30}$$

where $|A|$ stands for the Hilbert-Schmidt norm of A; see also [67, Proposition 3.1].

It follows that Θ is a bounded superharmonic function on Σ^n. By Theorem 4.4, Σ^n is parabolic and thus Θ is constant on it. So, thanks to (4.30), we obtain that $|A|^2 = 0$, that is, Σ^n is totally geodesic.

Now, we claim that ρ is constant. In fact, for any $X \in T\Sigma^n$, we can write

$$X = X^* - \frac{\langle X, K \rangle}{\rho^2} K,$$

where X^* denotes the orthogonal projection of X onto TM^n. Since Σ^n is totally geodesic, by [251, Proposition 7.35], we have that

$$X(\Theta) = \langle N, \overline{\nabla}_X K \rangle$$
$$= \langle N, \overline{\nabla}_{X^*} K \rangle - \frac{\langle X, K \rangle}{\rho^2} \langle N, \overline{\nabla}_K K \rangle$$
$$= \frac{1}{\rho} \langle X, \overline{\nabla}\rho \rangle \langle N, K \rangle - \frac{1}{\rho} \langle X, K \rangle \langle N, \overline{\nabla}\rho \rangle.$$

Thus, the least equation ensures that

$$\nabla \Theta = \frac{1}{\rho} \left(\Theta \overline{\nabla}\rho - \langle N, \overline{\nabla}\rho \rangle K\right) \in TM \oplus T\mathbb{R}_1.$$

4.6 Further Rigidity Results

Since Θ is constant, it follows that ρ is constant. In particular, $\overline{\mathrm{Ric}}(N, N) = \mathrm{Ric}_{M^n}(N^*, N^*)$, where N^* denotes the orthogonal projection of N onto TM^n.

So, if Ric_M is positive at some point $p_0 \in \Sigma^n$, by (4.30) it follows that $\mathrm{Ric}_M(N^*, N^*)(p_0) = 0$ and, consequently, $N^*(p_0) = 0$.

However, it easily seen that

$$|\nabla h|^2 = \frac{1}{\rho^2}|N^*|_M^2 = \frac{1}{\rho^2}\left(\frac{\Theta^2}{\rho^2} - 1\right), \tag{4.31}$$

which means that Σ^n is a slice of $\overline{M}^{n+1}$. □

Taking into account a classical result due to Ahlfors [2] and Blanc-Fiala-Huber [208] which asserts that a complete Riemannian surface of nonnegative Gaussian curvature must be parabolic, arguing as in the proof of Theorem 4.5, the following result holds.

Corollary 4.4 *Let $\overline{M}^3 = M^2 \times_\rho \mathbb{R}_1$ be a standard static spacetime with nonnegative Ricci curvature and whose Riemannian base M^2 has nonnegative Gaussian curvature K_M. Let $\psi : \Sigma^2 \to \overline{M}^3$ be a complete spacelike surface with constant mean curvature such that its angle function Θ is bounded and the warping function ρ satisfies $\inf_{\Sigma^n} \rho > 0$. Then, Σ^2 is totally geodesic. In addition, if K_M is positive at some point $p_0 \in \Sigma^2$, then Σ^2 is a slice of $\overline{M}^3$.*

In the next result, the mean curvature H of the spacelike hypersurface is not supposed to be constant. We just require that H does not change sign.

Theorem 4.6 *Let $\overline{M}^{n+1} = M^n \times_\rho \mathbb{R}_1$ be a standard static spacetime whose Riemannian base M^n has parabolic universal Riemannian covering. Moreover, let $\psi : \Sigma^n \to \overline{M}^{n+1}$ be a complete spacelike hypersurface such that η and h are bounded on it. If the mean curvature H and the function $\langle \nabla \rho, \nabla h \rangle$ have opposite signs, then Σ^n is a slice of $\overline{M}^{n+1}$.*

Proof Taking into account our hypothesis on H and $\langle \nabla \rho, \nabla h \rangle$, by (4.22) we conclude that Δh does not change sing. Therefore, since Theorem 4.4 guarantees the parabolicity of Σ^n, h must be constant and, consequently, Σ^n is a slice of $\overline{M}^{n+1}$. □

By using again [2, 208], Theorem 4.6 yields the following result.

Corollary 4.5 *Let $\overline{M}^3 = M^2 \times_\rho \mathbb{R}_1$ be a standard static spacetime whose Riemannian base M^2 has nonnegative Gaussian curvature. Moreover, let $\psi : \Sigma^2 \to \overline{M}^3$ be a complete spacelike surface such that η and h are bounded on it. If the mean curvature H and the function $\langle \nabla \rho, \nabla h \rangle$ have opposite signs, then Σ^2 is a slice of $\overline{M}^3$.*

We recall that a spacelike hypersurface Σ^n is said *maximal* if its mean curvature vanishes identically on it.

In this setting, by Theorem 4.6 the next result can be easily achieved.

Corollary 4.6 *Let $\overline{M}^{n+1} = M^n \times_\rho \mathbb{R}_1$ be a standard static spacetime whose Riemannian base M^n has parabolic universal Riemannian covering. Let $\psi : \Sigma^n \to \overline{M}^{n+1}$ be a complete maximal spacelike hypersurface such that η and h are bounded on it. If the function $\langle \nabla \rho, \nabla h \rangle$ does not change sign, then Σ^n is a slice of $\overline{M}^{n+1}$.*

Remark 4.2 In each of the exact solutions to Einstein's equations which are presented as warped product manifolds, the warped product decomposition emerges as a natural mathematical expression for physical symmetries. Moreover, formulas for warped product curvatures (see [251, Chapter 7]) indicate that any semi-Riemannian manifold $(\overline{N}, g)$ must possess certain measures of symmetry and flatness in order to be (locally or globally) isometric to a warped product of the form $B \times_f F$. In this context, geometric conditions on a semi-Riemannian manifold $(\overline{N}, g)$ which are necessary and sufficient to ensure that $(\overline{N}, g)$ is locally isometric to a warped product $B \times_f F$ are of extreme interest. For instance, [76, Lemma 3.78] guarantees that the Lorentzian manifold $\overline{M}^{n+1}$ can be viewed as the standard static spacetime, under the validity of the main technical assumptions. This fact shows the "dual character" of the main results in relation to the GRW spacetimes.

4.7 Entire Killing Graphs

According to [131], we define the *entire Killing graph* $\Sigma(u)$ associated to a smooth function $u \in C^\infty(M^n)$ as the hypersurface given by

$$\Sigma(u) = \{\Psi(x, u(x)) : x \in M^n\} \subset M^n \times_\rho \mathbb{R}_1,$$

where $\Psi : M^n \times \mathbb{I} \to \overline{M}^{n+1}$ is the flow generated by the timelike Killing vector field K. The metric induced on M^n by the Lorentzian metric (4.1) via $\Sigma(u)$ is given by

$$\langle \cdot, \cdot \rangle_u = \langle \cdot, \cdot \rangle_M - \rho^2 du^2. \tag{4.32}$$

Moreover, $\Sigma(u)$ is spacelike if and only if $\rho^2 |Du|_M^2 < 1$, where Du denotes the gradient of a function u with respect to the metric $\langle \cdot, \cdot \rangle_M$ of M^n. Indeed, if $\Sigma(u)$ is spacelike, then

$$0 < \langle Du, Du \rangle_u = \langle Du, Du \rangle_{M^n} - \rho^2 \langle Du, Du \rangle_M^2,$$

4.7 Entire Killing Graphs

and, hence, we conclude that $\rho^2|Du|_M^2 < 1$. Conversely, if $\rho^2|Du|_M^2 < 1$ and X is a vector field tangent to $\Sigma(u)$, the Cauchy-Schwarz inequality ensures that

$$\langle X, X \rangle_u = \langle X^*, X^* \rangle_M - \rho^2 \langle Du, X^* \rangle_M^2 \geq \langle X^*, X^* \rangle_M (1 - \rho^2|Du|_M^2), \qquad (4.33)$$

where X^* is the orthogonal projection of X onto TM^n. Thus, by (4.33) we get that $\langle X, X \rangle_u \geq 0$ and $\langle X, X \rangle_u = 0$ if and only if $X = 0$.

Now, the function $g : M^n \times \mathbb{R}_1 \to \mathbb{R}$ given by $g(x, t) = u(x) - t$ is such that $\Sigma(u) = \Psi(g^{-1}(0))$. Thus, for any vector field X tangent to $M^n \times_\rho \mathbb{R}_1$, we have

$$X(g) = X^*(g) - \frac{1}{\rho^2} \langle X, \partial_t \rangle \partial_t(g) = \left\langle \frac{1}{\rho^2} \partial_t + Du, X \right\rangle.$$

Hence, the field

$$\overline{\nabla} g = \frac{1}{\rho^2} \partial_t + Du$$

is a normal vector field on $g^{-1}(0)$ and, consequently

$$N_0 = \Psi_*(\overline{\nabla} g) = \frac{1}{\rho^2} K + \Psi_*(Du),$$

is a normal timelike vector field on $\Sigma(u)$.

Since

$$|N_0| = \frac{(1 - \rho^2|Du|_{M^n}^2)^{1/2}}{\rho},$$

it follows that

$$N = \frac{N_0}{|N_0|} = \frac{1}{\rho(1 - \rho^2|Du|_M^2)^{1/2}} (K + \rho^2 \Psi_*(Du))$$

defines the future-pointing Gauss map of $\Sigma(u)$ such that its angle function is given by

$$\Theta = \langle N, K \rangle = -\frac{\rho}{(1 - \rho^2|Du|_M^2)^{1/2}} < 0. \qquad (4.34)$$

Moreover, for any vector field X tangent to M^n, the shape operator A of $\Sigma(u)$, with respect to N, is given by

$$AX = -\frac{\rho}{(1 - \rho^2|Du|_M^2)^{1/2}} D_X Du - \frac{\rho^3 \langle D_X Du, Du \rangle}{(1 - \rho^2|Du|_M^2)^{3/2}} Du$$

$$-\frac{\rho^2 \langle D\rho, X \rangle |Du|_M^2}{(1-\rho^2|Du|_M^2)^{3/2}} Du - \frac{\langle D\rho, X \rangle}{(1-\rho^2|Du|_M^2)^{1/2}} Du$$
$$-\frac{\langle Du, X \rangle}{(1-\rho^2|Du|_M^2)^{1/2}} D\rho. \tag{4.35}$$

So, by (4.35) it follows that the mean curvature H_u of a spacelike entire Killing graph $\Sigma(u)$ is given by

$$nH_u = \mathrm{Div}\left(\frac{\rho Du}{(1-\rho^2|Du|_M^2)^{1/2}}\right) + \frac{\langle Du, D\rho \rangle}{(1-\rho^2|Du|_M^2)^{1/2}},$$

where Div stands for the divergence operator on M^n with respect to the metric $\langle \cdot, \cdot \rangle_M$. In particular, an entire Killing graph $\Sigma(u)$ is maximal if and only if the function $u \in C^\infty(M^n)$ satisfies the following elliptic partial differential equation of divergence form

$$\begin{cases} \mathrm{Div}\left(\dfrac{\rho Du}{(1-\rho^2|Du|_M^2)^{1/2}}\right) + \dfrac{\langle Du, D\rho \rangle}{(1-\rho^2|Du|_M^2)^{1/2}} = 0, \quad \text{in } M^n \\ \rho^2|Du|_M^2 < 1. \end{cases}$$

We also notice that, since

$$N^* = N - N^\perp = \frac{\rho \Psi_*(Du)}{(1-\rho^2|Du|_M^2)^{1/2}},$$

then

$$|N^*|_{M^n}^2 = \frac{\rho^2|Du|_M^2}{1-\rho^2|Du|_M^2}. \tag{4.36}$$

Thus, by (4.19) and (4.36) we get the following relation

$$|\nabla h|^2 = \frac{|Du|_M^2}{1-\rho^2|Du|_M^2}. \tag{4.37}$$

We point out that, in contrast to the case of graphs into a Riemannian space endowed with a Killing vector field, an entire Killing graph in a standard static spacetime is not necessarily complete. In other words, the induced Riemannian metric (4.32) is not necessary complete on its Riemannian base; see, for instance, [4, Examples 3.1 and 3.3], as well as [5, Example 5.3].

4.7 Entire Killing Graphs

For this reason, it is of interest to establish sufficient conditions ensuring that a graph to be complete. Working along this direction, as a consequence of Theorem 4.1, we obtain the next result.

Theorem 4.7 *Let $\overline{M}^{n+1} = M^n \times_\rho \mathbb{R}_1$ be a standard static spacetime whose Riemannian base M^n is complete with nonnegative sectional curvature K_M and warping function ρ convex on M^n. Let $\Sigma(u)$ be an entire Killing graph of a function $u \in C^\infty(M^n)$, with constant mean curvature. Suppose that $\sup_{\Sigma(u)} \rho < +\infty$ and, for some positive constant $\lambda < 1$,*

$$\sup_{\Sigma(u)} \rho^2 |Du|_M^2 \leq \lambda. \tag{4.38}$$

Then $\Sigma(u)$ is maximal.

Proof In order to apply Theorem 4.1, we show that Θ is bounded and that $\Sigma(u)$ is complete. To this goal, let us observe that by (4.34) and (4.38), we have that

$$\Theta \geq -\frac{\rho}{(1-\lambda)^{1/2}} \geq -\frac{\sup_{\Sigma(u)} \rho}{(1-\lambda)^{1/2}},$$

that is Θ is bounded.

Moreover, by (4.33) and (4.38) we have

$$\langle X, X \rangle_u \geq (1-\lambda)\langle X^*, X^* \rangle_M.$$

The above inequality implies that

$$L_u(\gamma) \geq (1-\lambda)^{1/2} L_M(\gamma^*),$$

where $L_u(\gamma)$ stands for the length of a curve γ on $\Sigma(u)$ with respect to the induced metric (4.32) and $L_{M^n}(\gamma^*)$ denotes the length of the projection γ^* of γ onto M^n with respect to its metric $\langle \cdot, \cdot \rangle_M$.

Consequently, since the projection onto M^n of divergent curves on $\Sigma(u)$ gives divergent curves on M^n and bearing in mind that the metric $\langle \cdot, \cdot \rangle_M$ is complete, we can apply the Hopf-Rinow Theorem to conclude that the induced metric (4.32) is complete. Therefore, by Theorem 4.1 the conclusion follows. □

Remark 4.3 In [68], the author proved the existence of asymptotically flat maximal surfaces in asymptotically flat spacetimes satisfying a uniformity condition in the interior. In particular he showed that the Dirichlet problem in nonflat spacetimes is solvable and proved the existence of constant mean curvature surfaces in cosmological spacetimes.

Taking into account (4.37) and arguing as in the proof of Theorem 4.7, by Theorem 4.2 we have the following result.

Theorem 4.8 *Let $\overline{M}^{n+1} = M^n \times_\rho \mathbb{R}_1$ be a standard static spacetime whose Riemannian base M^n is complete with nonnegative sectional curvature K_M and warping function ρ convex on M^n. Let $\Sigma(u)$ be an entire Killing graph of a bounded function $u \in C^\infty(M^n)$, with constant mean curvature. Suppose that $|\nabla \rho|$ is bounded on $\Sigma(u)$ and, for some positive constant $\lambda < 1$,*

$$\sup_{\Sigma(u)} \rho^2 |Du|_M^2 \leq \lambda.$$

Then $\Sigma(u)$ is maximal. Moreover, if $\langle \nabla \rho, \nabla h \rangle$ does not change sign on $\Sigma(u)$ and $|Du|_M \in \mathcal{L}_g^1(M^n)$, then u is constant on M^n.

By Corollary 4.1 the next result can be achieved.

Corollary 4.7 *Let $\overline{M}^{n+1} = M^n \times_\rho \mathbb{R}_1$ be a standard static spacetime whose Riemannian base M^n is complete with nonnegative sectional curvature K_M and warping function ρ convex on M^n. Let $\Sigma(u)$ be a maximal entire Killing graph of a bounded function $u \in C^\infty(M^n)$. Suppose that for some positive constant $\lambda < 1$,*

$$\sup_{\Sigma(u)} \rho^2 |Du|_M^2 \leq \lambda.$$

If $\langle \nabla \rho, \nabla h \rangle$ does not change sign on $\Sigma(u)$ and $|Du|_M \in \mathcal{L}_g^1(M^n)$, then u is constant on M^n.

As a consequence of the above result, the following theorem holds.

Corollary 4.8 *Let $\overline{M}^{n+1} = M^n \times_\rho \mathbb{R}_1$ be a standard static spacetime whose Riemannian base M^n is complete with nonnegative sectional curvature K_{M^n} and warping function ρ convex on M^n. Suppose that $u \in C^\infty(M^n)$ is a bounded solution to the problem*

$$\begin{cases} \operatorname{Div}\left(\dfrac{\rho Du}{(1 - \rho^2 |Du|_M^2)^{1/2}}\right) + \dfrac{\langle Du, D\rho \rangle}{(1 - \rho^2 |Du|_M^2)^{1/2}} = 0, & \text{in } M^n \\[2mm] \sup_M \rho^2 |Du|_M^2 < 1 \\[2mm] |Du|_M \in \mathcal{L}^1(M^n). \end{cases}$$

4.7 Entire Killing Graphs

If the height function h of the entire Killing graph associate to u is such that $\langle \nabla \rho, \nabla h \rangle$ does not change sign and $|\nabla \rho|$ is bounded on it, then u is constant on M^n.

Remark 4.4 According to [138, Example 4.4], taking $0 < |a| < 1$, we have that the entire Killing graph

$$\Sigma(u) = \{(x, y, a \ln y) : y > 0\} \subset \mathbb{H}^2 \times \mathbb{R}_1$$

is such that

$$|Du|^2_{\mathbb{H}^2} = |a|^2$$

and, hence, $\Sigma(u)$ is a complete spacelike surface in $\mathbb{H}^2 \times \mathbb{R}_1$. Moreover, a direct computation ensures that $\Sigma(u)$ has constant mean curvature $H_u = -\dfrac{a}{2\sqrt{1-a^2}}$ and angle function given by

$$\Theta = -\frac{1}{\sqrt{1-|a|^2}}.$$

Therefore, we conclude that our previous theorems are sharp in the sense that they do not hold when the base of the ambient space has negative sectional curvature; see [4, Example 3.2] for entire Killing maximal graphs.

In context of entire Killing graphs, we can rewrite Theorem 4.3 as follows.

Theorem 4.9 *Let $\overline{M}^3 = M^2 \times_\rho \mathbb{R}_1$ be a standard static spacetime whose (not necessarily complete) Riemannian base M^2 has nonnegative Gaussian curvature K_M and warping function ρ convex on M^2. Let $\Sigma(u)$ be an entire maximal Killing graph which is complete with respect to the metric induced from the Riemannian warped product $M^2 \times_\rho \mathbb{R}$. Suppose that the warping function satisfies $\sup_{\Sigma(u)} \rho < +\infty$ and $\inf_{\Sigma(u)} \rho > 0$. Then $\Sigma(u)$ is totally geodesic. In addition, if K_M is positive at some point $p_0 \in \Sigma(u)$, then M^2 is complete and u is constant on M^2.*

Arguing as in the proof of Theorem 4.7, we also obtain the following nonparametric version of Theorem 4.5.

Theorem 4.10 *Let $\overline{M}^{n+1} = M^n \times_\rho \mathbb{R}_1$ be a standard static spacetime with nonnegative Ricci curvature and whose Riemannian base M^n is complete having parabolic universal Riemannian covering. Let $\Sigma(u)$ be an entire Killing graph in $\overline{M}^{n+1}$ with constant mean* curvature, such that $\rho|_{\Sigma(u)}$ is bounded and $\inf_{\Sigma(u)} \rho > 0$. Suppose that, for some positive constant $\alpha < 1$, one has

$$\sup_{\Sigma(u)} \rho^2 |Du|_M^2 \leq \alpha.$$

Then, $\Sigma(u)$ is totally geodesic. In addition, if Ric_M is positive at some point $p_0 \in \Sigma(u)$, then $\Sigma(u)$ is a slice of $\overline{M}^{n+1}$.

As a consequence of Theorem 4.10 we obtain the following result.

Corollary 4.9 *Let $\overline{M}^{n+1} = M^n \times_\rho \mathbb{R}_1$ be a standard static spacetime with nonnegative Ricci curvature and whose base M^n is complete having parabolic universal Riemannian covering and positive Ricci curvature. Suppose that the warping function ρ is bounded on M^n and $\inf_M \rho > 0$. Let $u \in C^\infty(M^n)$ be a solution to the problem*

$$\begin{cases} \mathrm{Div}\left(\dfrac{\rho Du}{(1 - \rho^2 |Du|_M^2)^{1/2}}\right) + \dfrac{\langle Du, D\rho \rangle}{(1 - \rho^2 |Du|_M^2)^{1/2}} = 0, & \text{in } M^n \\ \sup_M \rho^2 |Du|_M^2 < 1. \end{cases}$$

Then, u is constant on M^n.

We can also apply Theorem 4.6 to get the next result.

Theorem 4.11 *Let $\overline{M}^{n+1} = M^n \times_\rho \mathbb{R}_1$ be a standard static spacetime whose Riemannian base M^n is complete having parabolic universal Riemannian covering. Let $\Sigma(u)$ be an entire Killing graph in $\overline{M}^{n+1}$ of a bounded function $u \in C^\infty(M^n)$ such that for some positive constant $\alpha < 1$,*

$$\sup_{\Sigma(u)} \rho^2 |Du|_M^2 \leq \alpha.$$

If the mean curvature H_u and the function $\langle \nabla \rho, \nabla h \rangle$ have opposite sings, then u is constant on M^n.

Proof Let us observe that the function

$$\eta = -\frac{1}{(1 - \rho^2 |Du|_M^2)^{1/2}}$$

is bounded on $\Sigma(u)$. Therefore, the result it follows by Theorem 4.6. □

4.7 Entire Killing Graphs

We end this chapter with the following consequence of Theorem 4.11.

Corollary 4.10 *Let $\overline{M}^{n+1} = M^n \times_\rho \mathbb{R}_1$ be a standard static spacetime whose Riemannian base M^n is complete having parabolic universal Riemannian covering. Let $u \in C^\infty(M^n)$ be a bounded solution to the problem*

$$\begin{cases} \mathrm{Div}\left(\dfrac{\rho Du}{(1-\rho^2|Du|_M^2)^{1/2}}\right) + \dfrac{\langle Du, D\rho \rangle}{(1-\rho^2|Du|_M^2)^{1/2}} = 0, & \text{in } M^n \\ \sup_{M^n} \rho^2|Du|_M^2 < 1. \end{cases}$$

If the height function h of the entire Killing graph associate to u is such that $\langle \nabla \rho, \nabla h \rangle$ does not change sign, then u is constant on M^n.

Part II

Riemannian Immersions in Weighted Semi-Riemannian Warped Products

Basic Facts Concerning Weighted Manifolds 5

5.1 Introduction

In this chapter, for the sake of clarity we introduce several useful definitions and notations that will appear throughout the forthcoming ones of Part II of this book.

Furthermore, we prove key lemmas which will be crucial to establish the main results of the next two chapters of Part II.

5.2 The Bakry-Émery-Ricci Tensor and the Drift Laplacian

Let (Σ^n, g) be an n-dimensional complete Riemannian manifold. The Laplace operator on Σ^n, $-\Delta$, can be defined as the differential operator associated to the standard Dirichlet form

$$Q(u) = \int_{\Sigma^n} |\nabla u|^2 d\Sigma^n, \quad u \in C_c^\infty(\Sigma^n) \subset L^2(d\Sigma^n),$$

where $|\cdot|$ is the norm induced by the Riemannian inner product $g = \langle \cdot, \cdot \rangle$, and $d\Sigma^n$ is the volume element on Σ^n. Now, let $\varphi \in C^\infty(\Sigma^n)$, which will be referred to as a *weight function*. If we replace the measure $d\Sigma^n$ with the weighted measure $d\mu = e^{-\varphi} d\Sigma^n$ in the definition of Q, we obtain a new quadratic form Q_φ, and we shall denote by Δ_φ the elliptic operator on $C_c^\infty(\Sigma^n) \subset L^2(d\mu)$ induced by Q_φ. In this sense, Δ_φ arises as a natural generalization of the Laplacian. This operator that is clearly symmetric and positive can be naturally extended to a positive operator on $L^2(d\mu)$. Moreover, one has

$$\Delta_\varphi u = \Delta u - \langle \nabla u, \nabla \varphi \rangle, \tag{5.1}$$

by Stokes' Theorem.

Thus, introducing a weight factor is the first step towards decoupling the leading term and the lower order terms of the operator, which in the case of the Laplace operator are completely determined by the metric of Σ^n.

The triple $(\Sigma^n, g, d\mu)$ and the operator Δ_φ defined above and acting over $C^\infty(\Sigma^n)$ will be called, respectively, the *weighted manifold*, Σ^n_φ, associated with (Σ^n, g) and φ, and the corresponding *drift Laplacian* or *φ-Laplacian*.

We also recall that a notion of curvature for weighted manifolds goes back to Lichnerowicz [231], and it was later developed by Bakry and Émery in their seminal work [59], where they introduced the following modified Ricci curvature

$$\text{Ric}_\varphi = \text{Ric} + \text{Hess}\,\varphi. \tag{5.2}$$

According to the current literature, we shall refer to this tensor as being the *Bakry-Émery-Ricci tensor* of Σ^n. We notice that the interplay between the geometry of Σ^n and the behavior of the weight function φ is mostly taken into account by means of its Bakry-Émery-Ricci tensor Ric_φ.

It is interesting to point out that weighted manifolds are closely related to some classical mathematical concepts, as they can be used as a powerful mathematical tool in order to obtain new results related to them. Specifically, in the case where Ric_φ is constant, we can induce on Σ^n a structure of a gradient Ricci soliton. Its mathematical relevance is due to the Perelman's solution of the Poincaré conjecture since gradient Ricci solitons correspond to self-similar solutions to the Hamilton's Ricci flow and often arise as limits of dilations of singularities developed along the Ricci flow. For an overview of results in this scope one can consult [259].

Weighted manifolds have also been considered when studying harmonic heat flows and heat kernels. For instance, Grigor'yan and Saloff-Coste established in [196] a result which relates the heat kernel on a complete, noncompact Riemannian manifold Σ^n with the Dirichlet heat kernel on the exterior of a compact set of Σ^n. For further results of geometric investigations concerning these manifolds, we also refer the reader to the articles of Morgan [245] and Wei-Wylie [294].

We also note that the Bakry-Émery-Ricci curvature tensor relates to φ-Laplacian via the following version Bochner's formula (see [294]),

$$\frac{1}{2}\Delta_\varphi |\nabla u|^2 = |\text{Hess}\,u|^2 + 2\langle \nabla u, \nabla \Delta_\varphi u \rangle + \text{Ric}_\varphi(\nabla u, \nabla u), \tag{5.3}$$

which holds for all $u \in C^\infty(\Sigma^n)$.

Furthermore, it is well-known the validity of the following weak version of the Omori-Yau's generalized maximum principle for the φ-Laplacian on a complete weighted manifold $(\Sigma^n, g, e^{-\varphi}d\Sigma^n)$; see, for instance, [259, Chapter 2, Remark 2.18].

Lemma 5.1 *Let $(\Sigma^n, g, e^{-\varphi}d\Sigma^n)$ be an n-dimensional complete weighted manifold whose Bakry-Émery-Ricci curvature tensor is bounded from below, and let $u \in C^\infty(\Sigma^n)$ be a function bounded from above (resp. bounded from below) on Σ^n. Then, there exists a sequence of points $\{p_k\} \subset \Sigma^n$ such that*

$$\lim_k u(p_k) = \sup_{\Sigma^n} u \quad (\text{resp.}, = \inf_\Sigma u)$$

and

$$\limsup_k \Delta_\varphi u(p_k) \leq 0 \quad (\text{resp.}, \liminf_k \Delta_\varphi u(p_k) \geq 0).$$

5.3 Weighted Riemannian Warped Products

Let us denote by $\overline{M}_\varphi^{n+1}$ the weighted manifold $(\overline{M}^{n+1}, \overline{g}, e^{-\varphi}d\overline{M}^{n+1})$. Motivated by the Cheeger-Gromoll splitting type theorems due to Lichnerowicz [231, 232] (see also [294]), and Fang et al. [176], in Chap. 6 we consider the *weighted warped products* $\overline{M}_\varphi^{n+1} = I \times_f M_\varphi^n$ whose weight function φ does not depend on the parameter $t \in I$, that is $\langle \overline{\nabla}\varphi, \partial_t \rangle = 0$.

In this setting, for a two-sided hypersurface Σ^n immersed in $\overline{M}_\varphi^{n+1}$, the φ-*divergence operator* of a tangent vector field X on Σ^n is defined by

$$\mathrm{div}_\varphi(X) = e^\varphi \mathrm{div}(e^{-\varphi} X).$$

So, as quoted in (5.1), given a smooth function $u: \Sigma^n \to \mathbb{R}$, its φ-*Laplacian* is defined by

$$\Delta_\varphi u = \mathrm{div}_\varphi(\nabla u) = \Delta u - \langle \nabla u, \nabla \varphi \rangle. \tag{5.4}$$

Finally, inspired by Gromov [197], we define the φ-*mean curvature* H_φ of Σ^n by

$$nH_\varphi = nH + \langle \overline{\nabla}\varphi, N \rangle, \tag{5.5}$$

where H denotes the standard mean curvature of Σ^n with respect to its orientation N. In analogy to the case of the standard mean curvature, the φ-mean curvature H_φ on Σ^n is related to the variation of the weighted area functional

$$\mathrm{vol}_\varphi(\Sigma^n) = \int_{\Sigma^n} e^{-\varphi} d\Sigma^n.$$

We observe that, under our assumption on the weight function of the ambient space, the φ-mean curvature of a slice $\Sigma_t \subset \overline{M}_\varphi^{n+1}$, with respect to the orientation $N = -\partial_t$, is given by

$$H_\varphi(t) = H(t) = \frac{f'}{f}(t).$$

In the sequel, we shall establish two preparatory results which will be useful studying the uniqueness of two-sided hypersurfaces $\psi : \Sigma^n \to \overline{M}_\varphi^{n+1}$.

More precisely, in the first result we compute the expression of the φ-Laplacian of three functions naturally attached to Σ^n; namely, the height function and its primitive and the angle function.

While, in the second lemma we give a lower bound for the Bakry-Émery-Ricci curvature of Σ_φ^n.

Lemma 5.2 *Let Σ^n be a two-sided hypersurface immersed in a weighted warped product space $\overline{M}_\varphi^{n+1}$, with orientation N, height function h, and angle function Θ. Then,*

(i) $\Delta_\varphi \sigma(h) = n\left(f'(h) + f(h)\Theta H_\varphi\right)$, where

$$\sigma(t) = \int_{t_0}^{t} f(s)ds$$

and H_φ is the φ-mean curvature of Σ^n related to N;

(ii) $\Delta_\varphi h = (\log f)'(h)(n - |\nabla h|^2) + n\Theta H_\varphi$;

(iii) *If A is the shape operator of Σ^n related to N, $|A|$ denotes its Hilbert-Schmidt norm and Ric^M stands for the Ricci curvature tensor of the fiber M^n,*

$$\Delta_\varphi(f(h)\Theta) = -nf(h)\langle \nabla H_\varphi, \partial_t \rangle - nf'(h)H_\varphi - f(h)\Theta|A|^2$$
$$- f(h)\Theta \,\overline{\mathrm{Hess}}\,\varphi(N, N)$$
$$- f(h)\Theta \left(\mathrm{Ric}^M(N^M, N^M) + (n-1)(\log f)''(h)|\nabla h|^2\right),$$

where $(\cdot)^M$ is the projection of a vector field onto M^n and $\overline{\mathrm{Hess}}$ is the Hessian operator of $\overline{M}^{n+1}$.

5.3 Weighted Riemannian Warped Products

Proof Items (i) and (ii) are proved in [105, Lemma 1]. Now, by [28, Proposition 6], with the aid of [251, Corollary 7.43], we get the following expression for the Laplacian of the function $f(h)\Theta$ as follows

$$\Delta(f(h)\Theta) = -nf(h)\langle \nabla H, \partial_t\rangle - nf'(h)H - f(h)\Theta|A|^2$$
$$- f(h)\Theta \left(\operatorname{Ric}^M(N^M, N^M) + (n-1)(\log f)''(h)|\nabla h|^2\right). \quad (5.6)$$

By definition of the φ-mean curvature (see (5.5)), one has

$$nf(h)\langle \nabla H, \partial_t\rangle = nf(h)\langle \nabla H_\varphi, \partial_t\rangle - f(h)\partial_t^\top \langle \overline{\nabla}\varphi, N\rangle.$$

Moreover, a direct computation yield

$$\partial_t^\top \langle \overline{\nabla}\varphi, N\rangle = -\frac{f'}{f}(h)\langle \overline{\nabla}\varphi, N\rangle - \Theta\overline{\operatorname{Hess}}\,\varphi(N, N) - \langle \overline{\nabla}\varphi, A\partial_t^\top\rangle,$$

and

$$\nabla(f(h)\Theta) = -f(h)A\partial_t^\top.$$

So (5.6) can be written as

$$\Delta(f(h)\Theta) = -nf(h)\langle \nabla H_\varphi, \partial_t\rangle - nf'(h)H_\varphi - f(h)\Theta|A|^2$$
$$- f(h)\Theta\overline{\operatorname{Hess}}\,\varphi(N, N) + \langle \overline{\nabla}\varphi, \nabla(f(h)\Theta)\rangle$$
$$- f(h)\Theta \left(\operatorname{Ric}^M(N^M, N^M) + (n-1)(\log f)''(h)|\nabla h|^2\right). \quad (5.7)$$

Finally, item (iii) follows by (5.7) bearing in mind the definition of φ-Laplacian given in (5.4). □

The next step is to prove some lower bounds for the Bakry-Émery-Ricci tensor of Σ^n. From now on, we will require that $\overline{M}_\varphi^{n+1}$ satisfies the following convergence condition,

$$K^M \geq \sup_I \left(f'^2 - ff''\right), \quad (5.8)$$

where K^M stands for the sectional curvature of the fiber M^n.

Lemma 5.3 *Let $\overline{M}_\varphi^{n+1} = I \times_f M_\varphi^n$ be a weighted warped product obeying (5.8), and such that the Hessian of the weighted function φ is bounded from below. Moreover, let $\psi : \Sigma^n \to \overline{M}_\varphi^{n+1}$ be a two-sided hypersurface. Suppose that both the second fundamental* form and the φ-mean curvature of Σ^n are bounded. If the ratio $\frac{f'''}{f}(h)$ is also bounded on Σ^n, then the Ricci-Bakry-Émery tensor of Σ^n, Ric_φ, is bounded from below.

Proof We recall that the curvature tensor R of Σ^n can be described in terms of the shape operator A and the curvature tensor $\overline{R}$ of $\overline{M}_\varphi^{n+1}$ by the Gauss equation given by

$$R(X,Y)Z = (\overline{R}(X,Y)Z)^\top + \langle AX, Z\rangle AY - \langle AY, Z\rangle AX, \tag{5.9}$$

for all tangent vector fields $X, Y, Z \in \mathfrak{X}(\Sigma^n)$.

Here, as in [251], the curvature tensor R is given by

$$R(X,Y)Z = \nabla_{[X,Y]}Z - [\nabla_X, \nabla_Y]Z,$$

where $[\cdot,\cdot]$ denotes the Lie bracket, and $X, Y, Z \in \mathfrak{X}(\Sigma^n)$.

Let us consider $X \in \mathfrak{X}(\Sigma^n)$ and a local orthonormal frame $\{E_1, \ldots, E_n\}$ of $\mathfrak{X}(\Sigma^n)$. Then, by using the Gauss equation (5.9), it follows that

$$\mathrm{Ric}(X,X) = \sum_{i=1}^n \langle \overline{R}(X, E_i)X, E_i\rangle + nH\langle AX, X\rangle - |AX|^2. \tag{5.10}$$

Moreover, we have that

$$\overline{R}(X,Y)Z = R^M(X^M, Y^M)Z^M - ((\log f)'(h))^2(\langle X, Z\rangle Y - \langle Y, Z\rangle X)$$
$$+ (\log f)''(h)\langle Z, \partial_t\rangle(\langle Y, \partial_t\rangle X - \langle X, \partial_t\rangle Y)$$
$$- (\log f)''(h)(\langle Y, \partial_t\rangle\langle X, Z\rangle - \langle X, \partial_t\rangle\langle Y, Z\rangle)\partial_t,$$

where R^M denotes the curvature tensor of the fiber (M^n, g_{M^n}); see [251, Proposition 7.42] for more details.

Hence, we get

$$\langle \overline{R}(X, E_i)X, E_i\rangle = f(h)^2 K^M(X^M, E_i^M)(\langle X^M, X^M\rangle_M\langle E_i^M, E_i^M\rangle_M$$
$$- \langle X^M, E_i^M\rangle_M^2) - ((\log f)'(h))^2(|X|^2 - \langle X, E_i\rangle^2)$$
$$+ (\log f)''(h)\langle X, \nabla h\rangle(\langle \nabla h, E_i\rangle\langle X, E_i\rangle - \langle X, \nabla h\rangle)$$
$$- (\log f)''(h)(\langle \nabla h, E_i\rangle|X|^2$$
$$- \langle X, \nabla h\rangle\langle X, E_i\rangle)\langle \nabla h, E_i\rangle, \tag{5.11}$$

where $g_M = \langle \cdot, \cdot\rangle_M$.

5.4 Weighted GRW Spacetimes

On the other hand, it is easily seen that

$$\langle X^M, X^M \rangle_M \langle E_i^M, E_i^M \rangle_M - \langle X^M, E_i^M \rangle_M^2 = \frac{1}{f(h)^4}(|X|^2 - \langle X, \nabla h \rangle^2$$
$$- |X|^2 \langle \nabla h, E_i \rangle^2 - \langle X, E_i \rangle^2$$
$$+ 2\langle X, \nabla h \rangle \langle X, E_i \rangle \langle \nabla h, E_i \rangle),$$

which, in addition to (5.8) and (5.11), implies the following lower bound

$$\sum_{i=1}^{n} \langle \overline{R}(X, E_i)X, E_i \rangle \geq \frac{f''}{f}(h)(n-1)|X|^2, \tag{5.12}$$

for every $X \in \mathfrak{X}(\Sigma^n)$. Thus, by (5.10) and (5.12), we get

$$\mathrm{Ric}(X, X) \geq \frac{f''}{f}(h)(n-1)|X|^2 + nH\langle AX, X \rangle - |AX|^2. \tag{5.13}$$

Since the Hessian of φ is bounded from below, we have

$$\mathrm{Hess}\,\varphi(X, X) = \overline{\mathrm{Hess}}\,\varphi(X, X) + \langle \overline{\nabla}\varphi, N \rangle \langle AX, X \rangle$$
$$\geq \beta |X|^2 + \langle \overline{\nabla}\varphi, N \rangle \langle AX, X \rangle, \tag{5.14}$$

for a certain positive constant β.

Therefore, by sing the definition of Ric_φ, on account of (5.5), (5.13), and (5.14), we obtain

$$\mathrm{Ric}_\varphi(X, X) \geq \left(\frac{f''}{f}(h)(n-1) + \beta \right) |X|^2 + nH_\varphi \langle AX, X \rangle - |AX|^2, \tag{5.15}$$

for any $X \in \mathfrak{X}(\Sigma^n)$.

Therefore, since the ratio $\frac{f''}{f}(h)$, the second fundamental form, as well as the φ-mean curvature of Σ^n, are bounded, the conclusion is achieved. $\square$

5.4 Weighted GRW Spacetimes

Motivated by a splitting theorem due to Case (see [101, Theorem 1.2]), throughout Chap. 7 we will deal with *spatially weighted GRW spacetimes* $\overline{M}_\varphi^{n+1} = -I \times_f M_\varphi^n$, which means that the weight function φ does not depend on the parameter $t \in I$, that is, $\langle \overline{\nabla}\varphi, \partial_t \rangle = 0$.

In this setting, for a spacelike hypersurface Σ^n immersed in $\overline{M}_\varphi^{n+1} = -I \times_f M_\varphi^n$, the φ-*divergence operator* on Σ^n of a tangent vector field $X \in \mathfrak{X}(\Sigma^n)$, $\mathrm{div}_\varphi(X)$, and the φ-Laplacian of a function $u \in C^\infty(\Sigma^n)$, $\Delta_\varphi u$, are defined, respectively, as in (5.3) and (5.4). Moreover, following Gromov [197], the φ-*mean curvature* H_φ of Σ^n is given by

$$nH_\varphi = nH - \langle \overline{\nabla}\varphi, N \rangle, \tag{5.16}$$

where H denotes the standard mean curvature of Σ^n with respect to its future-pointing Gauss map N. As in the Riemannian case, the φ-mean curvature H_φ on a spacelike hypersurface Σ^n is related to the variation of the weighted area functional

$$\mathrm{vol}_\varphi(\Sigma^n) = \int_\Sigma e^{-\varphi} d\Sigma^n.$$

We also observe that, under our assumption on the weight function of the ambient spacetime, the φ-mean curvature of a slice $\Sigma_t \subset \overline{M}_\varphi^{n+1}$, with respect to the orientation $N = \partial_t$, is given by

$$H_\varphi(t) = H(t) = \frac{f'}{f}(t).$$

Now, we compute the expression of the φ-Laplacian of the height function h, as well as of the composition of the height with a primitive of the warping function f and, finally, of the angle function Θ of a spacelike hypersurface Σ^n immersed in a spatially weighted GRW spacetime $\overline{M}_\varphi^{n+1} = -I \times_f M_\varphi^n$; see Lemma 5.4 below.

Afterwards, we shall also give some conditions under which we can guarantee that the Bakry-Émery-Ricci tensor has a lower bound; see Lemma 5.5 below.

Lemma 5.4 *Let Σ^n be a spacelike hypersurface with future-pointing Gauss map N immersed in a spatially weighted GRW spacetime $\overline{M}_\varphi^{n+1} = -I \times_f M_\varphi^n$ with weight function φ. Then, we have*

(i)

$$\Delta_\varphi h = -(\log f)'(h)(n + |\nabla h|^2) - n\Theta H_\varphi; \tag{5.17}$$

(ii)

$$\Delta_\varphi \mathcal{F}(h) = -n(f'(h) + f(h)\Theta H_\varphi), \tag{5.18}$$

5.4 Weighted GRW Spacetimes

where $\mathcal{F}(t) = \int_{t_0}^{t} f(s)ds$, and

(iii)

$$\Delta_\varphi(f(h)\Theta) = nf(h)\langle \nabla H_\varphi, \partial_t \rangle + nf'(h)H_\varphi + f(h)\Theta|A|^2$$
$$+ f(h)\Theta\overline{\mathrm{Hess}}\,\varphi(N, N)$$
$$+ f(h)\Theta\left(\mathrm{Ric}^M(N^M, N^M)\right)$$
$$- (n-1)(\log f)''(h)|\nabla h|^2\right), \quad (5.19)$$

where Ric^M stands for the Ricci curvature tensor of the Riemannian fiber M^n, A is the shape operator of Σ^n related to N and $|A|$ denotes its Hilbert-Schmidt norm.

Proof Equations (5.17) and (5.18) correspond to [104, Lemma 1]. Now, we present a proof of (5.19).

In [19, Corollary 8.2] it is proved that

$$\Delta(f(h)\Theta) = nf(h)\langle \nabla H, \partial_t \rangle + nf'(h)H + f(h)\Theta|A|^2$$
$$+ f(h)\Theta\left(\mathrm{Ric}^M(N^M, N^M) - (n-1)(\log f)''(h)|\nabla h|^2\right). \quad (5.20)$$

Taking into account the definition of the φ-mean curvature (see (5.16)), we get

$$nf(h)\langle \nabla H, \partial_t \rangle = nf(h)\langle \nabla H_\varphi, \partial_t \rangle + f(h)\partial_t^\top \langle \overline{\nabla}\varphi, N \rangle.$$

Moreover, a direct computation yields

$$\partial_t^\top \langle \overline{\nabla}\varphi, N \rangle = -\frac{f'}{f}(h)\langle \overline{\nabla}\varphi, N \rangle + \Theta\,\overline{\mathrm{Hess}}\,\varphi(N, N) - \langle \overline{\nabla}\varphi, A\partial_t^\top \rangle,$$

and $\nabla(f(h)\Theta) = -f(h)A\partial_t^\top$. So (5.20) can be written as

$$\Delta(f(h)\Theta) = nf(h)\langle \nabla H_\varphi, \partial_t \rangle - f'(h)\langle \overline{\nabla}\varphi, N \rangle + f(h)\Theta\overline{\mathrm{Hess}}\,\varphi(N, N)$$
$$+ \langle \overline{\nabla}\varphi, \nabla(f(h)\Theta) \rangle + nf'(h)H + f(h)\Theta|A|^2$$
$$+ f(h)\Theta\left(\mathrm{Ric}^M(N^M, N^M) - (n-1)(\log f)''(h)|\nabla h|^2\right). \quad (5.21)$$

Finally, Eq. (5.19) follows by (5.21) and (5.4). □

We recall that a *slab*

$$[t_1, t_2] \times M_\varphi^n = \{(t, q) \in -I \times_f M_\varphi^n : t_1 \leq t \leq t_2\}$$

is called a *timelike bounded region* of the spatially weighted GRW spacetime $\overline{M}_\varphi^{n+1}$.

Following the terminology established by Alías and Colares [19], we say that $\overline{M}_\varphi^{n+1}$ obeys the *strong null convergence condition* (SNCC) when the sectional curvatures K^M of its Riemannian fiber M^n satisfy the following inequality

$$K^M \geq \sup_I (f^2 (\log f)''). \tag{5.22}$$

The following lemma gives some sufficient conditions for which the Bakry-Émery-Ricci curvature of a spacelike hypersurface Σ^n of $\overline{M}_\varphi^{n+1}$ is bounded from below.

Lemma 5.5 *Let $\overline{M}_\varphi^{n+1} = -I \times_f M_\varphi^n$ be a spatially weighted GRW spacetime obeying the SNCC given in (5.22), and such that the Hessian of the weight function φ is bounded from below, that is,*

$$\overline{\text{Hess}}\,\varphi(U, U) \geq \beta |U|^2$$

for some real constant β and for all $U \in \mathfrak{X}(\overline{M}_\varphi^{n+1})$. Let Σ^n be a spacelike hypersurface that lies in a timelike bounded region of $\overline{M}_\varphi^{n+1}$. Suppose that the φ-mean curvature H_φ is bounded on Σ^n. Then the Bakry-Émery-Ricci curvature Ric_φ of Σ^n is bounded from below.

Proof We recall that the curvature tensor R of a spacelike hypersurface Σ^n can be described in terms of its shape operator A and of the curvature tensor $\overline{R}$ of $-I \times_f M_\varphi^n$ by the so-called Gauss equation given by

$$R(X, Y)Z = (\overline{R}(X, Y)Z)^\top - \langle AX, Z \rangle AY + \langle AY, Z \rangle AX, \tag{5.23}$$

for all the tangent vector fields $X, Y, Z \in \mathfrak{X}(\Sigma^n)$. Here, as in [251], the curvature tensor R is given by

$$R(X, Y)Z = \nabla_{[X,Y]} Z - [\nabla_X, \nabla_Y] Z,$$

where $[\cdot, \cdot]$ denotes the Lie bracket, and $X, Y, Z \in \mathfrak{X}(\Sigma^n)$.

5.4 Weighted GRW Spacetimes

Let us consider $X \in \mathfrak{X}(\Sigma^n)$ and a local orthonormal frame $\{E_1, \ldots, E_n\}$ of $\mathfrak{X}(\Sigma^n)$. Then, it follows by the Gauss equation (5.23) that

$$\operatorname{Ric}(X, X) = \sum_{i=1}^{n} \langle \overline{R}(X, E_i)X, E_i \rangle + nH\langle AX, X \rangle + |AX|^2. \tag{5.24}$$

Moreover, we have that

$$\overline{R}(X, Y)Z = R^M(X^M, Y^M)Z^M + ((\log f)'(h))^2(\langle X, Z \rangle Y - \langle Y, Z \rangle X)$$
$$+ (\log f)''(h)\langle Z, \partial_t \rangle(\langle Y, \partial_t \rangle X - \langle X, \partial_t \rangle Y)$$
$$- (\log f)''(h)(\langle Y, \partial_t \rangle \langle X, Z \rangle - \langle X, \partial_t \rangle \langle Y, Z \rangle)\partial_t,$$

where R^M is the curvature tensor of M^n and $(\cdot)^M$ denotes the projection of a vector field onto M^n; see [251, Proposition 7.42] for more details.

Hence, one has

$$\langle \overline{R}(X, E_i)X, E_i \rangle = f(h)^2 K^M(X^M, E_i^M)(|X^M|_M^2 |E_i^M|_M^2 - \langle X^M, E_i^M \rangle_{M^n}^2)$$
$$+ ((\log f)'(h))^2(|X|^2 - \langle X, E_i \rangle^2)$$
$$+ (\log f)''(h)\langle X, \nabla h \rangle(\langle \nabla h, E_i \rangle \langle X, E_i \rangle - \langle X, \nabla h \rangle)$$
$$- (\log f)''(h)(\langle \nabla h, E_i \rangle |X|^2$$
$$- \langle X, \nabla h \rangle \langle X, E_i \rangle)\langle \nabla h, E_i \rangle, \tag{5.25}$$

where $\langle \cdot, \cdot \rangle_M$ is the metric tensor of M^n and $|\cdot|_M$ is the norm induced by $\langle \cdot, \cdot \rangle_M$.

On the other hand, it is not difficult to verify that

$$|X^M|_M^2 |E_i^M|_M^2 - \langle X^M, E_i^M \rangle_M^2 = \frac{1}{f(h)^4}(|X|^2 + \langle X, \nabla h \rangle^2 + |X|^2 \langle \nabla h, E_i \rangle^2$$
$$- \langle X, E_i \rangle^2 - 2\langle X, \nabla h \rangle \langle X, E_i \rangle \langle \nabla h, E_i \rangle).$$

The above expression, in addition to (5.22) and (5.25), implies the following lower bound

$$\sum_{i=1}^{n} \langle \overline{R}(X, E_i)X, E_i \rangle \geq \frac{f''}{f}(h)(n-1)|X|^2, \tag{5.26}$$

for any $X \in \mathfrak{X}(\Sigma^n)$. Thus, by (5.32) and (5.26), we get

$$\operatorname{Ric}(X, X) \geq \frac{f''}{f}(h)(n-1)|X|^2 + nH\langle AX, X \rangle + |AX|^2. \tag{5.27}$$

Since the Hessian of φ is bounded from below, we have

$$\text{Hess}\,\varphi(X,X) = \overline{\text{Hess}}\,\varphi(X,X) - \langle \overline{\nabla}\varphi, N\rangle\langle AX, X\rangle$$
$$\geq \beta|X|^2 - \langle \overline{\nabla} f, N\rangle\langle AX, X\rangle. \tag{5.28}$$

Therefore, by (5.2), (5.16), (5.27) and (5.28), we obtain

$$\text{Ric}_\varphi(X,X) \geq \left(\frac{f''}{f}(h)(n-1) + \beta\right)|X|^2 + nH_\varphi\langle AX, X\rangle + |AX|^2. \tag{5.29}$$

Now, we can write

$$nH_\varphi\langle AX, X\rangle + |AX|^2 = \left|AX + \frac{nH_\varphi}{2}X\right|^2 - \frac{n^2 H_\varphi^2}{4}|X|^2. \tag{5.30}$$

Consequently, inequality (5.29) becomes

$$\text{Ric}_\varphi(X,X) \geq \left(\frac{f''}{f}(h)(n-1) + \beta\right)|X|^2 + \left|AX + \frac{nH_\varphi}{2}X\right|^2 - \frac{n^2 H_\varphi^2}{4}|X|^2, \tag{5.31}$$

for any $X \in \mathfrak{X}(\Sigma^n)$.

Finally, since H_φ is bounded and Σ^n is contained in a timelike bounded region of $-I \times_f M_\varphi^n$, one has that the Bakry-Émery-Ricci curvature Ric_φ of Σ^n is bounded from below. □

If the ambient spacetime is a spatially weighted Lorentzian product space $-I \times M_\varphi^n$, under a different set of assumptions, we are able to prove a lower bound for the Bakry-Émery-Ricci tensor of a spacelike hypersurface $\psi : \Sigma^n \to -I \times M_\varphi^n$.

Lemma 5.6 *Let Σ^n be a spacelike hypersurface immersed in a spatially weighted Lorentzian product space $-I \times M_\varphi^n$ such that both the sectional curvatures K^M of the Riemannian fiber M^n and the Hessian of the weighted function φ are bounded from below. If the φ-mean curvature H_φ and the angle function Θ are bounded on Σ^n, then the Bakry-Émery-Ricci tensor Ric_φ of Σ^n is bounded from below.*

Proof Let us consider $X \in \mathfrak{X}(\Sigma^n)$ and a local orthonormal frame $\{E_1, \ldots, E_n\}$ of $\mathfrak{X}(\Sigma^n)$. By using the Gauss equation (5.23) it follows that

$$\text{Ric}(X,X) = \sum_{i=1}^{n}\langle \overline{R}(X,E_i)X, E_i\rangle + nH\langle AX, X\rangle + |AX|^2. \tag{5.32}$$

5.4 Weighted GRW Spacetimes

Moreover, we have that

$$\langle \overline{R}(X, E_i)X, E_i \rangle = \langle R(X^M, E_i^M)X^M, E_i^M \rangle_M$$
$$= K^M(X^M, E_i^M)\left(\langle X^M, X^M \rangle_M \langle E_i^M, E_i^M \rangle_M \right.$$
$$\left. - \langle X^M, E_i^M \rangle_M^2 \right). \tag{5.33}$$

On the other hand, since $X^M = X + \langle X, \partial_t \rangle \partial_t$, $E_i^M = E_i + \langle E_i, \partial_t \rangle \partial_t$ and $\nabla h = -\partial_t^\top$, after a straightforward computation we have

$$\langle X^M, X^M \rangle_M \langle E_i^M, E_i^M \rangle_M = (1 + \langle E_i, \nabla h \rangle^2)(|X|^2 + \langle X, \nabla h \rangle^2)$$

and

$$\langle X^M, E_i^M \rangle_M^2 = \langle X, E_i \rangle^2 + 2\langle X, \nabla h \rangle \langle E_i, \nabla h \rangle \langle X, E_i \rangle$$
$$+ \langle X, \nabla h \rangle \langle E_i, \nabla h \rangle^2.$$

Hence, since $K^M \geq -\kappa$ for some positive constant κ, by (5.33) we obtain

$$\sum_{i=1}^n \langle \overline{R}(X, E_i)X, E_i \rangle \geq -\kappa \left((n-1)|X|^2 + (n-2)\langle X, \nabla h \rangle^2 + |X|^2 |\nabla h|^2 \right). \tag{5.34}$$

Consequently, by (1.12) and (5.34), we get

$$\sum_{i=1}^n \langle \overline{R}(X, E_i)X, E_i \rangle \geq -(n-1)\kappa \Theta^2 |X|^2. \tag{5.35}$$

Thus, by (5.32) and (5.35) we obtain

$$\mathrm{Ric}(X, X) \geq -(n-1)\kappa \Theta^2 |X|^2 + nH\langle AX, X \rangle + |AX|^2. \tag{5.36}$$

On the other hand, taking into account that the Hessian of φ is bounded from below, we have

$$\mathrm{Hess}\,\varphi(X, X) \geq -\beta |X|^2 - \langle \overline{\nabla} f, N \rangle \langle AX, X \rangle, \tag{5.37}$$

for every $X \in \mathfrak{X}(\Sigma^n)$ and some positive constant β.

Now, by (5.2), (5.36) and (5.37) it follows that

$$\mathrm{Ric}_\varphi(X, X) \geq -\left((n-1)\kappa \Theta^2 + \beta\right)|X|^2 + nH_\varphi \langle AX, X \rangle + |AX|^2. \tag{5.38}$$

Thus, by (5.38) and (5.30) we get the following lower bound for Ric_φ,

$$\mathrm{Ric}_\varphi(X, X) \geq - \left((n-1)\kappa\Theta^2 + \beta + \frac{n^2 H_\varphi^2}{4} \right) |X|^2 \tag{5.39}$$

for any $X \in \mathfrak{X}(\Sigma^n)$.

Therefore, since both H_φ and Θ are supposed to be bounded on Σ^n, by (5.31) we obtain that Ric_φ is bounded from below. □

6 Two-Sided Hypersurfaces in Weighted Riemannian Warped Products

6.1 Introduction

Cavalcante, de Lima and Santos in [105], through an application of some generalized maximum principles, investigated Bernstein type properties related to complete two-sided hypersurfaces immersed in a weighted Riemannian warped product space.

In this setting, supposing a natural comparison inequality between the weighted mean curvatures of the hypersurface and those of the slices of the slab where the hypersurface is supposed to be contained, they established sufficient conditions which guarantee that such a hypersurface must be a slice. Furthermore, they also obtained several rigidity results concerning the slices of weighted product spaces.

Subsequently, de Lima et al. [151] approached the problem of the rigidity of entire graphs defined over the fiber of a weighted product space whose Bakry-Émery-Ricci tensor is nonnegative. Supposing that the weighted mean curvature is constant and assuming suitable constraints on the norm of the gradient of the smooth function which determines such a graph, they used a weak version of the Omori-Yau's generalized maximum principle to show that this function must be constant.

Afterwards, de Lima et al. [153], extending the ideas of [105], also studied the uniqueness of complete two-sided hypersurfaces immersed in a suitable class of weighted Riemannian warped product spaces, via application of some maximum principles. In particular, they proved Moser-Bernstein type results concerning entire graphs in these ambient spaces.

Meanwhile, de Lima and Santos [143] obtained a height estimate concerning compact hypersurfaces with nonzero constant weighted mean curvature and whose boundary is contained into a slice of a weighted Riemannian product space with nonnegative Bakry-Émery-Ricci tensor and used it to get a half-space type result related to complete noncompact two-sided hypersurfaces properly immersed in this ambient space.

In this chapter, we study the uniqueness of slices $\Sigma_t = \{t\} \times M^n$, $t \in I$, among two-sided hypersurfaces Σ^n immersed in a Riemannian weighted warped product of the type $\overline{M}_\varphi^{n+1} = I \times_f M_\varphi^n$, where the weight function φ does not depend on the parameter $t \in I$, that is, $\langle \overline{\nabla}\varphi, \partial_t \rangle = 0$.

Towards this aim, we consider a variety of assumptions on the height function and on the angle function of such a Σ^n, as well as on the φ-mean curvature and on geometric quantities related to Σ^n, and employ analytic tools, such as a weak version of the Omori-Yau maximum principle, $L^{1 \leq p < \infty}$-conditions, and φ-parabolicity criteria, in order to guarantee that $\Sigma^n = \Sigma_t$ for some $t \in I$.

Moser-Bernstein type results concerning entire graphs

$$\Sigma(u) = \{(u(p), p) : p \in M_\varphi^n\} \subset \overline{M}_\varphi^{n+1}$$

of functions $u \in C^\infty(M)$ such that $u(M) \subseteq I$ are also derived.

Finally, we present a height estimate and a half-space type result in weighted Riemannian product spaces.

6.2 Studying Two-Sided Hypersurfaces in Weighted Riemannian Warped Products

With the aid of some suitable analytical tools, we will establish our uniqueness results concerning slices among complete two-sided hypersurfaces immersed in the class of weighted warped product described in Sect. 5.3. The results presented here are based on the paper [153].

6.2.1 Via Weak Omori-Yau Maximum Principle

To study the geometry of a two-sided hypersurface Σ^n immersed in a weighted warped product space of the type $\overline{M}_\varphi^{n+1} = I \times_f M_\varphi^n$. Initially, we will apply the weak Omori-Yau maximum principle related to the φ-Laplacian; see Lemma 5.1.

We recall that by a slab of $\overline{M}_\varphi^{n+1}$, we mean a region of the form $[t_1, t_2] \times M_\varphi^n$, where $t_1, t_2 \in I$ with $t_1 < t_2$.

Theorem 6.1 Let $\overline{M}_\varphi^{n+1}$ be a weighted warped product space obeying (5.8), and such that the Hessian of the weight function φ is bounded by below. Moreover, let $\psi : \Sigma^n \to \overline{M}_\varphi^{n+1}$ be a complete two-sided hypersurface which lies in a slab of $\overline{M}_\varphi^{n+1}$, with bounded second

6.2 Studying Two-Sided Hypersurfaces in Weighted Riemannian...

fundamental form, and angle function Θ *bounded away from zero. Suppose that either the* φ-*mean curvature* H_φ *of* Σ^n *satisfies*

$$0 \leq H_\varphi \leq \inf_{\Sigma^n}(\log f)'(h), \qquad (6.1)$$

or

$$\alpha \leq H_\varphi \leq \inf_{\Sigma^n}(\log f)'(h) \quad \text{and} \quad \inf_{\Sigma^n}(\log f)'(h) \geq 0 \qquad (6.2)$$

for some constant $\alpha \in \mathbb{R}$. *If*

$$|\nabla h| \leq \inf_{\Sigma^n}((\log f)'(h) - H_\varphi), \qquad (6.3)$$

then Σ^n *is a slice* $\Sigma_t \subset \overline{M}^{n+1}$.

Proof By Lemma 5.2, if (6.1) holds, we have

$$\Delta_\varphi \sigma(h) = nf(h)\left((\log f)'(h) + \Theta H_\varphi\right)$$
$$\geq nf(h)\left((\log f)'(h) - H_\varphi\right) \geq 0.$$

Since $\sigma(h)$ is bounded from above (because Σ^n is contained in a slab of $\overline{M}_\varphi^{n+1}$), and, by Lemma 5.3, Ric_φ is bounded from below, the hypotheses of the weak Omori-Yau maximum principle are satisfied for Σ^n and $\sigma(h)$.

Thus, we can take a sequence of points $\{p_k\} \subset \Sigma^n$ such that

$$\limsup_k \Delta_\varphi \sigma(h(p_k)) \leq 0.$$

So, up to a subsequence, we have

$$0 \geq \limsup_k \Delta_\varphi \sigma(h(p_k)) \geq \lim_k \left[nf(h)\left((\log f)'(h) - H_\varphi\right)\right](p_k) \geq 0.$$

Also by the fact that Σ^n is contained in a slab of $\overline{M}_\varphi^{n+1}$, it follows that there exists a positive constant C such that $f(h(p)) \geq C$ for avery $p \in \Sigma^n$.

Thus, we have that

$$\lim_k \left[(\log f)'(h) - H_\varphi\right](p_k) = 0,$$

and, taking into account (6.3), we infer that Σ^n is a slice.

If (6.2) holds, then again by Lemma 5.2, we have that

$$\Delta_\varphi \sigma(h) = nf(h)\left((\log f)'(h) + \Theta H_\varphi\right)$$
$$\geq nf(h)\Theta\left(H_\varphi - (\log f)'(h)\right) \geq 0.$$

Now, arguing as before, we conclude that

$$\lim_k \left\{nf(h)\Theta\left(H_\varphi - (\log f)'(h)\right)\right\}(p_k) = 0.$$

Therefore, since $f(h(p)) \geq C$ for every $p \in \Sigma^n$ and for some positive constant C, taking into account that Θ is bounded away from zero, (6.3) guarantees that Σ^n is a slice. □

We recall that the traceless symmetric tensor

$$\Phi = A - HI$$

is called *traceless second fundamental form*, where I stands for the identity operator on $\mathfrak{X}(\Sigma^n)$. Observe that

$$|\Phi|^2 = \text{tr}(\Phi^2) = |A|^2 - nH^2 \geq 0,$$

where equality holds if and only if Σ^n is totally umbilical.

For this reason, Φ is called the *total umbilicity tensor* of Σ^n.

Theorem 6.2 *Let $\overline{M}_\varphi^{n+1}$ be a weighted warped product space obeying (5.8) with convex weight function φ, that is, $\overline{\text{Hess}\,\varphi} \geq 0$.[1] Moreover, let $\psi : \Sigma^n \to \overline{M}_\varphi^{n+1}$ be a complete two-sided hypersurface which lies in a slab of $\overline{M}_\varphi^{n+1}$, with bounded second fundamental form A, and angle function Θ bounded away from zero. Suppose that φ-mean curvature is constant and satisfies*

$$H_\varphi \sup_{\Sigma^n}(\log f)'(h) \leq H^2(p) \tag{6.4}$$

and

$$H_\varphi(\log f)'(h)(p) \leq -\Theta(p)H_\varphi \sup_{\Sigma^n}(\log f)'(h). \tag{6.5}$$

---

[1] For a detailed discussion concerning convex functions on Riemannian manifolds, we refer the reader to [285, Chapter 3].

6.2 Studying Two-Sided Hypersurfaces in Weighted Riemannian...

If either

$$|\nabla h|^2(p) \leq \inf_{\Sigma^n} H^2 - H_\varphi \sup_{\Sigma^n} (\log f)'(h), \qquad (6.6)$$

or

$$|\nabla h| \leq \inf_{\Sigma^n} |\Phi|^2, \qquad (6.7)$$

then Σ^n is a slice $\Sigma_t \subset \overline{M}^{n+1}$.

Proof By inequality (5.8) we get that

$$\mathrm{Ric}^M(N^M, N^M) + (n-1)(\log f)''(h)|\nabla h|^2 \geq 0.$$

Therefore, by Lemma 5.2 it follows that

$$\Delta_\varphi(f(h)\Theta) \geq -nf'(h)H_\varphi - f(h)\Theta|A|^2. \qquad (6.8)$$

Now, by (6.8) and (6.5), one has

$$\Delta_\varphi(f(h)\Theta) \geq f(h)\Theta \left(nH_\varphi \sup_{\Sigma^n} (\log f)'(h) - |A|^2 \right).$$

On the other hand, $|A|^2 = n^2 H^2 - n(n-1)H_2$, where H_2 is the second order mean curvature defined by

$$\binom{n}{2} H_2 = \sum_{i<j} k_i k_j,$$

being where k_i, $i = 1, \ldots, n$ are the main curvatures of Σ^n. Therefore, one has

$$\Delta_\varphi(f(h)\Theta) \geq f(h)\Theta \left(nH_\varphi \sup_{\Sigma^n} (\log f)'(h) - n^2 H^2 + n(n-1)H_2 \right)$$

$$= nf(h)\Theta \left(H_\varphi \sup_{\Sigma^n} (\log f)'(h) - H^2 \right)$$

$$+ n(n-1)f(h)\Theta(H_2 - H^2) \geq 0, \qquad (6.9)$$

thanks to (6.4) and observing that $H^2 - H_2 \geq 0$.

Since Θ is bounded and Σ^n lies in a slab of $\overline{M}_\varphi^{n+1}$, we have that the function $f(h)\Theta$ is bounded from above.

On the other hand, by Lemma 5.3, we know that Ric_φ is bounded from below. Hence, we can apply the weak Omori-Yau maximum principle for the φ-Laplacian to conclude that there exists a sequence of points $\{p_k\} \subset \Sigma^n$ such that

$$\lim_k f(h(p_k))\Theta(p_k) = \sup_{\Sigma^n} f(h)\Theta, \quad \text{and} \quad \limsup_k \Delta_\varphi f(h(p_k))\Theta(p_k) \leq 0.$$

Thus, up to a subsequence, by (6.9) one has

$$0 \geq \limsup_k \Delta_\varphi f(h(p_k))\Theta(p_k)$$

$$\geq n \lim_k \left[f(h(p_k))\Theta(p_k) \left(H_\varphi \sup_{\Sigma^n} (\log f)'(h) - H^2(p_k) \right) \right]$$

$$+ n(n-1) \lim_k \left[f(h(p_k))\Theta(p_k)(H_2 - H^2)(p_k) \right] \geq 0,$$

so, in particular

$$\lim_k H^2(p_k) - H_\varphi \sup_{\Sigma^n} (\log f)'(h) = 0,$$

since Θ is bounded away from zero. Consequently, by (6.4) we get

$$\inf_{\Sigma^n} H^2 - H_\varphi \sup_{\Sigma^n} (\log f)'(h) = 0,$$

and the result follows by (6.6).

We notice also that by (6.8), (6.4), (6.5), and $\Theta \leq 0$, it follows that

$$\Delta_\varphi(f(h)\Theta) \geq -nf'(h)H_\varphi - f(h)\Theta|A|^2$$

$$\geq nf(h)\Theta H_\varphi \sup_{\Sigma^n}(\log f)'(h) - f(h)\Theta|A|^2$$

$$\geq nf(h)\Theta H^2 - f(h)\Theta|A|^2$$

$$= f(h)\Theta(nH^2 - |A|^2) = -f(h)\Theta|\Phi|^2 \geq 0.$$

Finally, if (6.7) holds, then the conclusion follows once more by applying the weak Omori-Yau maximum principle for the φ-Laplacian to the function $f(h)\Theta$. □

6.2.2 Via Integrability Constraints

Our next theorems give characterizations of slices under some $L^{1 \leq p < \infty}$ conditions.

6.2 Studying Two-Sided Hypersurfaces in Weighted Riemannian...

Theorem 6.3 *Let $\overline{M}_\varphi^{n+1} = I \times_f M_\varphi^n$ be a weighted warped product. Moreover, let Σ^n be a complete two-sided hypersurface immersed in $\overline{M}_\varphi^{n+1}$. Suppose that H_φ and $f'(h)$ satisfy $H_\varphi f'(h) \leq 0$, and that f is log-convex along Σ^n. Suppose also that*

$$\{p \in \Sigma^n : (\log f)''(h(p)) = 0\}$$

is a set of isolated points. Let $d\mu = e^{-\varphi} d\Sigma^n$, where $d\Sigma^n$ stands for the volume element on Σ^n. If $|\nabla h| \in L^1(d\mu)$ and $f'(h)$ is bounded on Σ^n, then Σ^n is a totally geodesic slice of $\overline{M}_\varphi^{n+1}$. Moreover, if $f(h) \in L^p(d\mu)$ for some $p \in (1, +\infty)$, and the weight function is bounded and convex, then Σ^n is a compact totally geodesic slice of $\overline{M}^{n+1}$. In this case, the fiber M^n must be compact.

Proof By Lemma 5.2, and thanks to our assumptions, we have

$$\Delta_\varphi f(h) = f'(h) \Delta_\varphi h + f''(h) |\nabla h|^2$$
$$= n f'(h) \left((\log f)'(h) + \Theta H_\varphi \right)$$
$$+ f(h) (\log f)''(h) |\nabla h|^2 \geq 0. \qquad (6.10)$$

If $|\nabla h| \in L^1(d\mu)$ and $f'(h)$ is bounded on Σ^n, then

$$|\nabla f(h)| = |f'(h)| \cdot |\nabla h| \in L^1(d\mu).$$

In this case, we claim that $f(h)$ is φ-harmonic, that is, $\Delta_\varphi f(h) = 0$.

This claim can be verified by using the following extension of a result due to Yau [298]. More precisely, one has

Let $(\Sigma^n, g, d\mu = e^{-\varphi} d\Sigma^n)$ be a complete n-dimensional weighted Riemannian manifold, and let $u \in C^\infty(\Sigma^n)$ be such that $\Delta_\varphi u$ does not change sign on Σ^n. If $|\nabla u| \in L^1(d\mu)$, then u is φ-harmonic.

We refer to [97, Proposition 2.1] for a detailed proof.

Now, thanks to (6.10), we obtain that

$$f'(h) \equiv 0 \quad \text{and} \quad (\log f)''(h) |\nabla h|^2 \equiv 0 \quad \text{on } \Sigma^n,$$

which implies that Σ^n is a totally geodesic slice, bearing in mind that

$$\{p \in \Sigma^n : (\log f)''(h(p)) = 0\}$$

is supposed to be a set of isolated points.

If $f(h) \in L^p(d\mu)$ for some $p \in (1, +\infty)$, then $f(h)$ is constant on Σ^n. In fact, since $\Delta_\varphi f(h)$ does not change sign on Σ^n, this claim is a consequence of an extension of a Yau's result [298] to the weighted setting.

More precisely, we have:

Let u be a nonnegative smooth φ-subharmonic function on a weighted complete Riemannian manifold $(\Sigma, g, d\mu = e^{-\varphi} d\Sigma^n)$. If $u \in L^p(d\mu)$, for some $p \in (1, +\infty)$, then u is constant.

We refer to [255, Theorem 1.1] for a detailed proof.

Arguing as above, we conclude that Σ^n is a totally geodesic slice $\{t_0\} \times M_\varphi^n$. Furthermore, since $f(h) \equiv C \in L^p(d\mu)$, we have

$$\mathrm{vol}_\varphi(\Sigma^n) = \int_{\Sigma^n} d\mu < +\infty.$$

Since $(\log f)''(t_0) \geq 0$, it follows that $f''(t_0) \geq 0$. Consequently, by the lower bound for the Bakry-Émery-Ricci tensor given in Lemma 5.3 (see inequality (5.15)), one has that $\mathrm{Ric}_\varphi \geq 0$.

However, according to Wei and Wylie [294, Theorem 1.2], a complete noncompact Riemannian manifold with nonnegative Bakry-Émery-Ricci tensor for some bounded weight function ψ have at least linear φ-volume growth.

Thus, we infer that Σ^n, as well as M^n, are compact. □

The following result holds.

Theorem 6.4 *Let $\overline{M}_\varphi^{n+1} = I \times M_\varphi^n$ be a weighted product manifold with convex weight function, and whose fiber M_φ^n has nonnegative sectional curvatures. Let Σ^n be a complete two-sided hypersurface immersed in $\overline{M}_\varphi^{n+1}$. Suppose that Θ has strict sign, H_φ does not change sign, and $|A|$ is bounded on Σ^n. If $|\nabla h| \in L^1(d\mu)$, then Σ^n is a slice of $\Sigma_t \subset \overline{M}^{n+1}$.*

Proof Our hypotheses imply that $H_\varphi \equiv 0$ on Σ^n. In fact, this claim is a consequence of the aforementioned weighted version of Yau's result (see the proof of Theorem 6.3), since $\Delta_\varphi h = nH_\varphi \Theta$ does not change sign on Σ^n (see Lemma 5.2) and we are assuming that $|\nabla h| \in L^1(d\mu)$.

Consequently, again by Lemma 5.2, we get that

$$\Delta_\varphi \Theta = -\Theta \left(|A|^2 + \mathrm{Ric}_\varphi^M(N^M, N^M) \right)$$

does not change sign. Since $|\nabla \Theta| \leq |A| |\nabla h| \in L^1(d\mu)$, it follows that Θ is a φ-harmonic function. Thus, we infer that $|A|^2 \equiv 0$ on Σ^n.

By Lemma 5.3, it follows, in particular, that the Bakry-Émery-Ricci tensor of Σ^n is nonnegative.

Now, we claim that h must be bounded on Σ^n. As in the proof of [212, Theorem 0.16] we argue by contradiction. Then, $\Sigma^n \cap \Sigma_t \neq 0, t \in I$. For a fixed $t \in I$, let

$$\Sigma_t = \{p \in \Sigma : h(p) \geq t\}.$$

By Sard's Theorem, we can suppose that t is a regular value of $h|_{\text{int } \Sigma^n}$, so that Σ_t is a smooth complete manifold with boundary $\partial \Sigma_t = \{p \in \Sigma^n : h(p) = t\}$ and exterior unit normal $\nu_t = -\nabla h/|\nabla h|$. For any $\epsilon > 0$, define on Σ_t

$$h_\epsilon = \max\{h, t + \epsilon\}.$$

Then h_ϵ is a φ-harmonic on Σ_t. In fact, set

$$\Sigma_1 = \{p \in \Sigma_t : h(p) > t + \epsilon\} \quad \text{and} \quad \Sigma_2 = \{p \in \Sigma_t : h(p) = t + \epsilon\},$$

as well as

$$\Sigma_3 = \{p \in \Sigma_t : t < h(p) < t + \epsilon\}.$$

Now, $h_\epsilon = h$ on Σ_1 and h_ϵ is constant (equals to $t + \epsilon$) on Σ_3, so $\Delta_\varphi h_\epsilon = 0$ on both Σ_1 and Σ_3. Since the transversality of Σ^n to $\xi = f(t)\partial_t$ guarantees that h_ϵ is smooth on Σ_2 and $\partial \Sigma_t$, we also have that $\Delta_\varphi h_\epsilon = 0$ on both Σ_2 and $\partial \Sigma_t$ by continuity.

We notice that $h_\epsilon \equiv t + \epsilon$ on $\partial \Sigma_t$, consequently, by the maximum principle for the φ-Laplacian (see [259, Theorem 2.13]), we obtain that $t \leq h \leq t + \epsilon$ on Σ_t. Since this holds for every $\epsilon > 0$, we conclude that $h \equiv t$ on Σ_t, against the boundedness assumption on h.

Finally, the conclusion follows by a Liouville-type theorem due to Brighton [91], which asserts that a bounded φ-harmonic function, on a complete manifold (Σ^n, g) endowed with a weight function φ and whose corresponding Bakry-Émery-Ricci tensor is nonnegative, must be constant. □

Remark 6.1 Another proof of Theorem 6.4 can be achieved as follows. By the weighted Bochner's formula (5.3), we have

$$\frac{1}{2}\Delta_\varphi |\nabla h|^2 = |\text{Hess } h|^2 + g_{\Sigma^n}(\nabla h, \nabla \Delta_\varphi h) + \text{Ric}_\varphi(\nabla h, \nabla h). \tag{6.11}$$

Since h is a φ-harmonic function, $\Theta \equiv \Theta_0 \in \mathbb{R}$, Σ^n is totally geodesic, and $\mathrm{Ric}_\varphi \geq 0$. By Lemma (5.2) and (6.11) we get

$$0 = \frac{1}{2} \Delta_\varphi |\nabla h|^2 \geq |\mathrm{Hess}\, h|^2 \geq 0.$$

By the Cauchy-Schwarz inequality, $|\mathrm{Hess}\, h|^2 \geq (\Delta h)^2/n$. Thus, h is harmonic. Consequently, the maximum principle implies that h is bounded. Now, since Σ^n is totally geodesic and $K^M \geq 0$, then Σ^n has nonnegative Ricci curvature. The conclusion follows by the strong Liouville property; see [297], as well as [229].

6.2.3 Via a φ-Parabolicity Criterion

According to the classical terminology adopted in Linear Potential Theory, a weighted manifold $(P, g, d\mu = e^{-\varphi} dP)$ is called φ-parabolic if there is no nonconstant bounded functions whose φ-Laplacian is globally either nonpositive or nonnegative.

For any compact subset $K \subset P$, we define the φ-capacity of K as

$$\mathrm{cap}_\psi(K) = \inf \left\{ \int_P |\nabla u|^2 d\mu : u \in \mathrm{Lip}_0(P) \text{ and } u\big|_K \equiv 1 \right\},$$

where $\mathrm{Lip}_0(P)$ is the set of all compactly supported Lipschitz functions on P. The notion of φ-capacity is related to the φ-parabolicity by the fact that a weighted manifold $(P, g, d\mu = e^{-\varphi} dP)$ is φ-parabolic if and only if $\mathrm{cap}_\psi(K) = 0$ for any compact $K \subset P$; see [192, Proposition 3], as well as [195, Proposition 2.1].

Let us recall that, given two Riemannian manifolds (P, g) and (P', g'), a diffeomorphism f from P onto P' is called a *quasi-isometry* if there exists a constant $c \geq 1$ such that

$$c^{-1} |v|_g \leq |df(v)|_{g'} \leq c |v|_g$$

for every $v \in T_p P$, $p \in P$; see [215] for details.

Suppose that P and P' are endowed with the same weight function φ. Arguing as in [194, Section 5] it easily seen that the φ-capacity changes under a quasi-isometry at most by a constant factor. So, we have that if (P, g) and (P', g') are two quasi-isometric Riemannian manifolds endowed with the same weight function ψ, then P and P' are either φ-parabolic or not, simultaneously.

Inspired by [264], we can state now the following criterion of parabolicity.

Theorem 6.5 *Let $\psi : \Sigma^n \to \overline{M}_\varphi^{n+1} = I \times_f M_\varphi^n$ be a complete two-sided hypersurface. Suppose that the fiber M^n of $\overline{M}^{n+1}$ has φ-parabolic universal Riemannian covering. If the*

6.2 Studying Two-Sided Hypersurfaces in Weighted Riemannian...

angle function Θ of Σ^n is bounded away from zero, and the restriction $f(h)$ on Σ^n of the warping function f of $\overline{M}_\varphi^{n+1}$ satisfies

(i) $\sup_{\Sigma^n} f(h) < +\infty$ and
(ii) $\inf_{\Sigma^n} f(h) > 0$,

then Σ^n is φ-parabolic.

Proof We first observe that we can introduce on the universal Riemannian covering $(\tilde{M}, g_{\tilde{M}})$ of the fiber (M^n, g_M) the notion of φ-parabolicity in a natural way.

Let $\pi = \pi_M \circ x$. By the Cauchy-Schawrz inequality, for any tangent vector field $V \in T\Sigma^n$ we have

$$g_{\Sigma^n}(V, V) = g_I(h_*V, h_*V) + f(h)^2 g_M(\pi_*V, \pi_*V)$$
$$\leq g_{\Sigma^n}(\nabla h, \nabla h) g_{\Sigma^n}(V, V) + f(h)^2 g_M(\pi_*V, \pi_*V),$$

so that

$$\Theta^2 g_{\Sigma^n}(V, V) \leq f(h)^2 g_M(\pi_*V, \pi_*V).$$

Thus, we have

$$g_{\Sigma^n}(V, V) \leq \frac{f(h)^2}{\Theta^2} g_M(\pi_*V, \pi_*V) \leq C_1 g_M(\pi_*V, \pi_*V),$$

where $C_1 = \max\left\{1, \sup_{\Sigma^n} \frac{f(h)^2}{\Theta^2}\right\}$.

Now, let us observe that π is a covering map, since π increases the distances between (S, g_{Σ^n}) and $(M, \sqrt{C_1} g_M)$. Consequently, the manifold M^n is complete; see [218, Lemma 8.8.1]. On the other hand, it is clear that

$$g_{\Sigma^n}(V, V) \geq f(h)^2 g_M(\pi_*V, \pi_*V) \geq \left(\inf_{\Sigma^n} f(h)^2\right) g_M(\pi_*V, \pi_*V).$$

If we choose $C = \max\{C_1, 1/\inf_{\Sigma^n} f(h)^2\}$, then

$$C^{-1} g_M(\pi_*V, \pi_*V) \leq g_{\Sigma^n}(V, V) \leq C g_M(\pi_*V, \pi_*V). \tag{6.12}$$

Let $(\tilde{\Sigma}, g_{\tilde{\Sigma}})$ be the universal Riemannian covering of (S, g_{Σ^n}), and denote by $\tilde{\pi}_{\Sigma^n} : \tilde{\Sigma} \to S$ the corresponding Riemannian covering map. A standard result on covering spaces can be used now to obtain a lift $\tilde{h} : \tilde{\Sigma} \to \tilde{M}$ of the map $h = \pi \circ \tilde{\pi}_{\Sigma^n} : \tilde{\Sigma} \to M^n$; see, for instance, [199].

It is easy to check that $\tilde{h}$ is, in fact, a diffeomorphism by $\tilde{\Sigma}$ to $\tilde{M}$. Note that (6.12) gives

$$C^{-1} g_{\tilde{M}}(d\tilde{h}(\tilde{v}), d\tilde{h}(\tilde{v})) \leq g_{\tilde{\Sigma}}(\tilde{v}, \tilde{v}) \leq C g_{\tilde{M}}(d\tilde{h}(\tilde{v}), d\tilde{h}(\tilde{v})),$$

for any $\tilde{v} \in T_{\tilde{p}} \tilde{\Sigma}, \tilde{p} \in \tilde{\Sigma}$. Hence, the map $\tilde{h}$ is a quasi-isometry from $(\tilde{\Sigma}, g_{\tilde{\Sigma}})$ onto $(\tilde{M}, g_{\tilde{M}})$.

Finally, let u be a nonnegative φ-superharmonic function on Σ^n, and put $\tilde{u} = u \circ \tilde{\pi}_{\Sigma^n}$. The function $\tilde{u}$ is a nonnegative φ-superharmonic function on the φ-parabolic Riemannian manifold $\tilde{\Sigma}$.

Therefore, $\tilde{u}$ must be constant, and, consequently, u is also constant. □

As a direct consequence of Theorem 6.5, we get the following corollary.

Corollary 6.1 *Let Σ^n be a complete two-sided hypersurface in a weighted warped product $\overline{M}_\varphi^{n+1}$ with weight function φ and whose fiber M^n is simply connected and φ-parabolic. If the angle function Θ of Σ^n is bounded away from zero and the warping function on Σ^n, $f(h)$, is bounded and it satisfies $\inf_{\Sigma^n} f(h) > 0$, then Σ^n is φ-parabolic.*

As well as, the next result holds.

Corollary 6.2 *Let Σ^n be a complete two-sided hypersurface in a weighted product $\overline{M}_\varphi^{n+1} = I \times M_\varphi$ with weight function φ and whose fiber M^n has φ-parabolic universal Riemannian covering. If the angle function Θ of Σ^n is bounded away from zero, then Σ^n is φ-parabolic.*

A hypersurface Σ^n immersed in a weighted warped product $\overline{M}_\varphi$ is called φ-minimal if $H_\varphi = 0$.

In this context, we obtain the following characterization of φ-minimal complete two-sided hypersurfaces.

Theorem 6.6 *Let $\overline{M}_\varphi^{n+1}$ be a weighted warped product whose fiber M^n has φ-parabolic universal Riemannian covering, and such that the warping function f is monotone. The only φ-minimal complete two-sided hypersurfaces contained in a slab of $\overline{M}_\varphi^{n+1}$ and with angle function Θ bounded away from zero are the slices $\Sigma_{t_0} \subset \overline{M}^{n+1}$, where $t_0 \in I$ is such that $f'(t_0) = 0$.*

Proof Let Σ^n be such a hypersurface. By Lemma 5.2, we obtain

$$\Delta_\varphi \sigma(h) = n f'(h).$$

6.2 Studying Two-Sided Hypersurfaces in Weighted Riemannian...

Consequently, the monotonicity of f implies that $\Delta_\varphi \sigma(h)$ is globally either nonpositive or nonnegative signed. Since Σ^n is contained in a slab of $\overline{M}_\varphi^{n+1}$, the function $\sigma(h)$ is clearly bounded on Σ^n.

By Theorem 6.5 we know that Σ^n is φ-parabolic, so $\sigma(h)$ is constant on Σ^n. Hence, the function h must also be constant on Σ^n. $\square$

The next rigidity result is achieved.

Theorem 6.7 *Let $\overline{M}_\varphi^{n+1}$ be a weighted warped product satisfying (5.8), with convex weight function φ and whose fiber M^n has φ-parabolic universal Riemannian covering. Let Σ^n be a φ-minimal complete two-sided hypersurface immersed in $\overline{M}_\varphi^{n+1}$, with angle function Θ bounded away from zero and such that the restriction $f(h)$ on Σ^n of the warping function f of $\overline{M}_\varphi^{n+1}$ satisfies*

(i) $\sup f(h) < +\infty$ and
(ii) $\inf f(h) > 0$.

Then Σ^n is totally geodesic. In addition, if either the inequality (5.8) is strict for all nonzero vector fields on M^n or φ is strictly convex on M, then Σ^n is a slice $\Sigma_{t_0} \subset \overline{M}_\varphi^{n+1}$, where $t_0 \in I$ is such that $f'(t_0) = 0$.

Proof By Lemma 5.2, we have that the φ-Laplacian of the bounded function $f(h)\Theta$ is given by

$$\Delta_\varphi(f(h)\Theta) = -f(h)\Theta|A|^2 - f(h)\Theta\overline{\text{Hess}}\,\varphi(N,N)$$
$$-f(h)\Theta\left(\text{Ric}^M(N^M, N^M) + (n-1)(\log f)''(h)|\nabla h|^2\right). \quad (6.13)$$

Since φ is supposed to be convex and (5.8) holds, it follows that $\Delta_\varphi(f(h)\Theta) \geq 0$. Theorem 6.5 assures that Σ^n is φ-parabolic, so $f(h)\Theta$ must be constant.

Therefore, thanks to (6.13), we infer that $|A|^2 \equiv 0$, that is, Σ^n is totally geodesic,

$$\overline{\text{Hess}}\,\varphi(N,N) = \text{Hess}^M \varphi(N^M, N^M) = 0, \quad (6.14)$$

and

$$\text{Ric}^M(N^M, N^M) + (n-1)(\log f)''(h)|\nabla h|^2 = 0.$$

Consequently, if the inequality (5.8) is strict, or if φ is strictly convex on M^n, then (6.13) also gives that $|N^M| = |\nabla h| = 0$ on Σ^n, that is Σ^n is a slice. $\square$

If the ambient space is simply a product manifold, then the following result holds.

Theorem 6.8 *Let $\overline{M}_\varphi^{n+1}$ be a weighted product manifold, whose fiber M^n has φ-parabolic universal Riemannian covering and such that its Bakry-Émery-Ricci tensor Ric_φ^M is nonnegative. Let Σ^n be a complete two-sided hypersurface immersed in $\overline{M}_\varphi^{n+1}$ with constant φ-mean curvature H_φ. If the angle function Θ of Σ^n is bounded away from zero, then Σ^n is totally geodesic. In addition, if Ric^M is definite positive at some point of Σ^n, then Σ^n is a slice $\Sigma_t \subset \overline{M}^{n+1}$.*

Proof By Lemma 5.2, the definition of Ric_φ (see (5.2)) and (6.14) we get that

$$\Delta_\varphi \Theta = -(\mathrm{Ric}_\varphi^M(N^M, N^M) + |A|^2)\Theta. \tag{6.15}$$

Consequently, since we are supposing that Ric_φ^M is nonnegative and that Θ is negative and bounded away from zero on Σ^n, we can apply Corollary 6.2 to conclude that Θ is constant on Σ^n.

Thus, thanks to (6.15) we get that $|A| \equiv 0$, that is, Σ^n is totally geodesic. Moreover, if Ric_φ^M is definite positive at some $p \in M$, considering once more equation (6.15), we conclude that $\Theta \equiv -1$ on Σ^n, which means that Σ^n is a slice of $\overline{M}^{n+1}$. □

When $\overline{M}^{n+1}$ belongs to a class of manifolds called in [282] *pseudo-hyperbolic spaces*, that is, warped products of the type $I \times_{e^t} M^n$, we get the following result.

Theorem 6.9 *Let $\overline{M}_\varphi^{n+1}$ be a weighted pseudo-hyperbolic space, whose fiber M^n has nonnegative sectional curvatures, and let Σ^n be a complete two-sided hypersurface which lies in a slab of $\overline{M}_\varphi^{n+1}$. Suppose that $|\overline{\nabla}\varphi|$ is bounded on Σ^n. If the φ-mean curvature H_φ of Σ^n is bounded and does not change sign, and $|A|$ is bounded on Σ^n, then $H_\varphi = 1$. If, in addition, the angle function Θ is bounded away from zero and the universal Riemannian covering of M^n is φ-parabolic, then Σ^n is a slice $\Sigma_t \subset \overline{M}^{n+1}$.*

Proof Since H_φ and $|\overline{\nabla}\varphi|$ are supposed to be bounded, it follows immediately by (5.5) that the mean curvature H of Σ^n is bounded. Under the hypotheses of Theorem 6.9, by estimate (5.13) we conclude that the Ricci curvature of Σ^n is bounded by below.

By the Omori-Yau's generalized maximum principle [250, 297] applied to the bounded function h, there exists a sequence of points $\{p_k\} \subset \Sigma^n$ such that

$$\lim_k h(p_k) = \sup_{\Sigma^n} h < +\infty \quad \text{and} \quad |\nabla h|^2(p_k) = 1 - \Theta^2(p_k) < 1/k^2,$$

as well as

$$\Delta_\varphi h(p_k) = \left[n - |\nabla h|^2 + n\Theta H_\varphi\right](p_k) < 1/k.$$

Therefore $\lim_k \Theta(p_k) = -1$ and taking limits in the last inequality we get $H_\varphi \geq 1$.

In a similar way, by applying the Omori-Yau maximum principle to the bounded function $-h$, we obtain another sequence $\{q_k\}$ in Σ^n such that

$$\lim_k h(q_k) = \inf_{\Sigma^n} h > -\infty \quad \text{and} \quad |\nabla h|^2(q_k) = 1 - \Theta^2(q_k) < 1/k^2,$$

as well as

$$\Delta_\varphi h(p_k) = \left[n - |\nabla h|^2 + n\Theta H_\varphi\right](q_k) > -1/k.$$

Thus $\lim_k \Theta(p_k) = -1$ and again taking limits in the last inequality we get $H_\varphi \leq 1$. As a consequence, $H_\varphi = 1$.

Furthermore, by Lemma 5.2, we have

$$\Delta_\varphi e^h = ne^h(1 + \Theta) \geq 0.$$

Assuming that the universal Riemannian covering of M is φ-parabolic and that the angle function is bounded away from zero, Theorem 6.5 assures that Σ^n is φ-parabolic.

Then, we infer that e^h is constant on Σ^n, which implies that h is constant, and Σ^n is a slice. $\square$

6.3 Moser-Bernstein Type Results in Weighted Warped Products

In this section, we shall use the theorems of Sect. 6.2 to establish Moser-Bernstein type results concerning entire graphs in a weighted warped product. To recall here some basic facts that will be useful in the sequel.

Let $\Omega \subseteq M^n$ be a domain. Then, each function $u \in C^\infty(\Omega)$ such that $u(\Omega) \subseteq I$ defines a vertical graph in the Riemannian warped product $I \times_f M^n$. In such a case, $\Sigma(u)$ will denote the graph over Ω determined by u, that is

$$\Sigma(u) = \{(u(p), p) : p \in \Omega\} \subset \overline{M}^{n+1} = I \times_f M^n.$$

The graph is said to be *entire* if $\Omega = M$. Observe that $h(u(p), p) = u(p)$, $p \in \Omega$. Hence, h and u can be identified in a natural way. The metric induced on Ω by the Riemannian metric of the ambient space via $\Sigma(u)$ is given by

$$g_{\Sigma(u)} = du^2 + f(u)^2 g_M.$$

If M is complete and $\inf_{\Sigma(u)} f(u) > 0$, then $\Sigma(u)$ endowed with the metric $g_{\Sigma(u)}$ is also complete. The unit vector field

$$N(p) = -\frac{f(u(p))}{\sqrt{f(u(p))^2 + |Du(p)|_M^2}}\left(\partial_t|_{(u(p),p)} - \frac{Du(p)}{f(u(p))^2}\right), \quad p \in \Omega, \tag{6.16}$$

where Du stands for the gradient of u in M^n and $|Du|_M = g_M(Du, Du)^{1/2}$, gives an orientation of $\Sigma(u)$ with respect to which we have $\bar{g}(N, \partial_t) < 0$, so that the assumption of transversality to the vector field $\xi = f(t)\partial_t$ is not necessary here.

The corresponding shape operator is given by

$$AX = -\frac{1}{f(u)\sqrt{f(u)^2 + |Du|_M^2}} D_X Du + \frac{f'(u)}{\sqrt{f(u)^2 + |Du|_M^2}} X$$

$$- \left(\frac{-g_M(D_X Du, Du)}{f(u)\left(f(u)^2 + |Du|_M^2\right)^{3/2}} - \frac{f'(u)g_M(Du, X)}{\left(f(u)^2 + |Du|_M^2\right)^{3/2}}\right) Du, \tag{6.17}$$

for any vector field X tangent to Ω, where D denotes the Levi-Civita connection in the manifold M^n.

Consequently, if $\Sigma(u)$ is a vertical graph over a domain $\Omega \subseteq M^n$ of a warped product M^n endowed with a weight function φ, it is not difficult to verify that, by (5.5) and (6.17), the φ-mean curvature function $H_\varphi(u)$ of $\Sigma(u)$ is given by

$$nH_\varphi(u) = -\mathrm{div}_\varphi^M\left(\frac{Du}{f(u)\sqrt{f(u)^2 + |Du|_M^2}}\right) + \frac{f'(u)}{\sqrt{f(u)^2 + |Du|_M^2}}\left(n - \frac{|Du|_M^2}{f(u)^2}\right).$$

The non-linear elliptic PDE in φ-divergence form $H_\varphi(u) = 0$ is called the φ-*minimal hypersurface equation* in M^n, and its solutions provide φ-minimal graphs in M^n. Here, div_φ^M is the φ-divergence operator computed in the metric g_M.

We notice that if $u \in C^\infty(\Omega)$ is a function such that $u(\Omega) \subseteq I$, then its shape operator A is bounded provided that u has finite C^2 norm; namely,

$$\|u\|_{C^2(M^n)} = \sup_{|\gamma|\leq 2} |D^\gamma u|_{L^\infty(M^n)} < +\infty.$$

Moreover, we point out that the finiteness of the C^2 norm of u implies, in particular, that u is bounded, which, in turn, guarantees that $\inf_{\Sigma(u)} f(u) > 0$. We shall use this fact without any additional comment.

The condition $\bar{g}(N, \partial_t)$ bounded away from zero is equivalent to $|Du|_M \leq Cf(u)$ for some positive constant C.

In this context, we obtain a nonparametric version of Theorem 6.1.

6.3 Moser-Bernstein Type Results in Weighted Warped Products

Theorem 6.10 *Let $\overline{M}_\varphi^{n+1}$ be a weighted warped product space with complete fiber M^n, obeying (5.8), and such that the Hessian of the weight function φ is bounded by below. Let $\Sigma(u)$ be an entire graph in $\overline{M}^{n+1}$ determined by a function $u \in C^\infty(M^n)$ with finite C^2 norm. Suppose that the φ-mean curvature, H_φ, of $\Sigma(u)$ satisfies either*

$$0 \leq H_\varphi \leq \inf_{\Sigma(u)} (\log f)'(u),$$

or

$$\alpha \leq H_\varphi \leq \inf_{\Sigma(u)} (\log f)'(u), \quad \text{and} \quad \inf_{\Sigma(u)} (\log f)'(u) \geq 0$$

for some constant α, and that $|Du|_M \leq Cf(u)$ for some positive constant C. If

$$|Du|_M^2 \leq \frac{\inf_{\Sigma(u)} f(u)^2 \inf_{\Sigma(u)} \big((\log f)'(u) - H_\varphi\big)}{1 - \inf_{\Sigma(u)} \big((\log f)'(u) - H_\varphi\big)}, \tag{6.18}$$

then $u \equiv t_0$ for some $t_0 \in I$.

Proof This result follows by Theorem 6.1 by noting that (6.18) implies (6.3), bearing in mind that

$$|\nabla h|^2 = \frac{|Du|_M^2}{f(u)^2 + |Du|_M^2}. \tag{6.19}$$

The proof is complete. □

Similarly, a nonparametric version of Theorem 6.2 can be stated and proved.

Theorem 6.11 *Let $\overline{M}_\varphi^{n+1}$ be a weighted warped product space obeying (5.8) with complete fiber M^n and convex weight function φ. Let $\Sigma(u)$ be an entire graph in $\overline{M}^{n+1}$ determined by a function $u \in C^\infty(M^n)$ with finite C^2 norm and with constant φ-mean curvature satisfying*

$$H_\varphi \sup_{\Sigma(u)} (\log f)'(u) \leq H^2(p),$$

and

$$H_\varphi (\log f)'(u)(p) \leq -\Theta(p) H_\varphi \sup_{\Sigma(u)} (\log f)'(u).$$

If

$$|Du(p)|_M^2 \leq \frac{\inf_{\Sigma(u)} f(u)^2 \left(\inf_{\Sigma(u)} H^2 - H_\varphi \sup_{\Sigma(u)} (\log f)'(h)\right)}{1 - \left(\inf_{\Sigma(u)} H^2 - H_\varphi \sup_{\Sigma(u)} (\log f)'(h)\right)},$$

then $u \equiv t_0$ for some $t_0 \in I$.

A nonparametric version of Theorem 6.3 reads as follows.

Theorem 6.12 *Let $\overline{M}_\varphi^{n+1} = I \times_f M_\varphi^n$ be a weighted warped product with complete fiber M^n. Let $\Sigma(u)$ be an entire graph in $\overline{M}^{n+1}$ determined by a bounded function $u \in C^\infty(M^n)$. Suppose that H_φ and $f'(u)$ satisfy $H_\varphi f'(u) \leq 0$, and that f is log-convex along $\Sigma(u)$. Suppose also that*

$$\{p \in M : (\log f)''(u(p)) = 0\}$$

is a set of isolated points. Let $d\mu = e^{-\varphi} dM^n$. If $|Du| \in L^1(d\mu)$ and $f'(u)$ is bounded on $\Sigma(u)$, then $u \equiv t_0$, where $t_0 \in I$ is such that $f'(t_0) = 0$. If $f(u) \in L^p(d\mu)$ for some $p \in (1, +\infty)$ and the weight function is bounded and convex, then $u \equiv t_0$, where $t_0 \in I$ is such that $f'(t_0) = 0$, and M is compact.

Proof Taking into account (6.19), we can see that

$$|Du|_M \in L^1(d\mu) \quad \text{implies} \quad |\nabla h| \in L^1(d\mu_u),$$

where $d\mu_u = e^{-\varphi} d\Sigma(u)$. Then, the first part of Theorem 6.12 follows immediately by Theorem 6.3. For the second part, we note that the volume element of $\Sigma(u)$ is given by

$$d\Sigma^n(u) = f(u)^{n-1} \sqrt{f(u)^2 + |Du|_M^2} \, dM^n.$$

Consequently, since $f(u) \in L^p(d\mu)$ for some $p \in (1, +\infty)$, and $f(h(q)) = f(u(p))$ for every $q = (u(p), p) \in \Sigma(u)$, we infer that $f(h) \in L^p(d\mu_u)$.

The conclusion follows arguing as in Theorem 6.3. □

Similarly, we obtain the following result.

Theorem 6.13 *Let $\overline{M}_\varphi^{n+1} = I \times M_\varphi^n$ be a weighted product with convex weight function and whose fiber is complete has nonnegative sectional curvatures. Let $\Sigma(u)$ be an entire*

6.3 Moser-Bernstein Type Results in Weighted Warped Products

graph in $\overline{M}^{n+1}$ determined by a function $u \in C^\infty(M^n)$ with finite C^2 norm. Suppose that H_φ does not change sign on $\Sigma(u)$. If $|Du|_M \in L^1(d\mu)$, then $u \equiv t_0$ for some $t_0 \in I$.

Now, we shall apply the geometric theorems in Sect. 6.2.3 to its PDE counterparts in order to obtain some Bernstein-Moser type results.

For instance, the following result holds.

Theorem 6.14 Let $\overline{M}_\varphi^{n+1}$ be a weighted warped product whose fiber M^n is complete with φ-parabolic universal Riemannian covering, and such that the warping function f is monotone. The only entire bounded solutions $u \in C^\infty(M^n)$ to

$$\mathrm{div}_\varphi^M \left(\frac{Du}{f(u)\sqrt{f(u)^2 + |Du|_M^2}} \right) = \frac{f'(u)}{\sqrt{f(u)^2 + |Du|_M^2}} \left(n - \frac{|Du|_M^2}{f(u)^2} \right),$$

with $|Du|_M \leq Cf(u)$ for some positive constant C, are the constant functions.

As well as

Theorem 6.15 Let $\overline{M}_\varphi^{n+1}$ be a weighted warped product satisfying (5.8), with convex weight function φ and whose fiber M^n is complete with φ-parabolic universal Riemannian covering. Let $u \in C^\infty(M^n)$ be an entire bounded solution to

$$\mathrm{div}_\varphi^M \left(\frac{Du}{f(u)\sqrt{f(u)^2 + |Du|_M^2}} \right) = \frac{f'(u)}{\sqrt{f(u)^2 + |Du|_M^2}} \left(n - \frac{|Du|_M^2}{f(u)^2} \right),$$

with $|Du|_M \leq Cf(u)$ for some positive constant C. Then $\Sigma(u)$ is totally geodesic. In addition, if either the inequality (5.8) is strict for all non-zero vector fields on M^n or φ is strictly convex on M^n, then $u \equiv t_0$ for some $t_0 \in I$.

Proof It suffices to notice that since u is bounded, the restriction $f(u)$ on $\Sigma(u)$ of the warping function f of $\overline{M}^{n+1}$ satisfies $\sup f(u) < +\infty$ and $\inf f(u) > 0$. □

Remark 6.2 If the Riemannian universal covering of the fiber is not φ-parabolic, some counterexamples can be shown. By following [264, Counterexample 10], let us consider the manifold

$$\left(M^2, g_M \right) = \left(\mathbb{R} \times_k \mathbb{R}, dx^2 + k(x)^2 dy^2 \right),$$

where $k(x) = \sqrt{1 + \cosh^4 x}$. Here, we assume M^2 endowed with the weight function $\varphi(x, y) = f(y) \in C^\infty(\mathbb{R})$. This manifold is complete by [251, Lemma 7.40] but it is not φ-parabolic, indeed the function $v(x) = -1/\cosh^2 x$, $x \in \mathbb{R}$ satisfies

$$\Delta_\varphi v = 2[(\cosh^2 x - 1)^2 + 2]/[\cosh^4 x (1 + \cosh^4 x)] > 0.$$

The function $u(x) = \tanh x$ defines a minimal graph on (M^2, g_{M^2}) in the ambient space $\overline{M}^2 = \mathbb{R} \times M^2$. In $\overline{M}^2$, we consider the weight function $\varphi(t, x, y) = f(y)$. By using the identity

$$\mathrm{div}_\varphi(jX) = \mathrm{div}(X) - g_M(X, Dj),$$

for every $j \in C^\infty(M^2)$ and any $X \in \mathfrak{X}(M^2)$, we have that $\Sigma(u)$ is a φ-minimal graph. Note that it is trivially bounded, and $\overline{g}(N, \partial_t)$ is bounded away from zero. Let us recall that minimal graphs are φ-minimal hypersurfaces for $\varphi = $ constant.

Finally, we point out some remarks concerning the warping function. Let

$$\left(W^2, g_W\right) = \left(\mathbb{R} \times_\varrho \mathbb{R}, dx^2 + \varrho(x)^2 dy^2\right),$$

where $\varrho(x) = (\sqrt{2x^4 + 6x^2 + 5})/(x^2 + 2)$. Since $\sqrt{5}/2 \leq \varrho(x) \leq \sqrt{2}$, the Riemannian manifold W^2 is quasi-isometric to the Euclidean plane, and therefore, W^2 is parabolic.

The function $w(x) = x + \arctan x$ determines a minimal graph on (W^2, g_W), with bounded length of its gradient. We notice that w is unbounded neither by below nor by above and $\overline{g}_W(N, \partial_t)$ is bounded away from zero.

Finally, if the ambient space is simply a product manifold, we obtain the following result.

Theorem 6.16 *Let $\overline{M}_\varphi^{n+1}$ be a weighted product manifold, whose fiber M^n is complete with φ-parabolic universal Riemannian covering and such that its Bakry-Émery-Ricci tensor Ric_φ^M is nonnegative. Let $u \in C^\infty(M^n)$ be an entire bounded solution of the equation*

$$nH_\varphi(u) = -\mathrm{div}_\varphi^M\left(\frac{Du}{f(u)\sqrt{f(u)^2 + |Du|_M^2}}\right) + \frac{f'(u)}{\sqrt{f(u)^2 + |Du|_M^2}}\left(n - \frac{|Du|_M^2}{f(u)^2}\right),$$

with $H_\varphi(u)$ being constant, and $|Du|_M \leq Cf(u)$ for some positive constant C. Then $\Sigma(u)$ is totally geodesic. In addition, if Ric_φ^M is definite positive at some point of $\Sigma(u)$, then $u \equiv t_0$ for some $t_0 \in I$.

6.4 Entire Graphs in Weighted Riemannian Product Spaces

Our aim in this section is to study the rigidity of entire graphs defined over the fiber of a weighted product space $\overline{M}_\varphi^{n+1} = I \times M_\varphi^n$ whose Bakry-Émery-Ricci tensor is nonnegative.

We prove that u must be constant by assuming that the weighted mean curvature is constant and by requiring some conditions on the norm of the gradient of the smooth function u that produces such a graph $\Sigma(u)$.

Our approach is based on the formula for the φ-Laplacian of the angle function attached to a hypersurface immersed in $\overline{M}_\varphi^{n+1}$ (see Lemma 5.2) in addition to a weak version of the Omori-Yau maximum principle; see Sect. 6.2.1.

The results which will be presented in this section can be found in [151]. Along this direction, the following Moser type result.

Theorem 6.17 Let $\overline{M}_\varphi^{n+1} = I \times M_\varphi^n$ be a weighted product space such that its fiber M^n is complete with sectional curvatures bounded by below and nonnegative Bakry-Émery-Ricci tensor Ric_φ^M, and the Hessian of the weight function φ is bounded by below. Let $\Sigma(u) \subset \overline{M}_\varphi^{n+1}$ be an entire graph over M^n with constant φ-mean curvature H_φ and bounded second fundamental form A. If

$$|Du|_M \leq \alpha |A|^\beta, \qquad (6.20)$$

for some positive constants α and β, then $u \equiv t_0$ for some $t_0 \in I$.

Proof First, we observe that $\Sigma(u)$ is complete. Indeed, an entire vertical graph is properly immersed into the Riemannian product space $I \times M^n$, which is obviously complete when the fiber M^n is complete.

Consider on $\Sigma(u)$ the orientation given by

$$N = \frac{1}{\sqrt{1 + |Du|_M^2}} (\partial_t - Du). \qquad (6.21)$$

Note that, with respect to this orientation, we have $\overline{g}(N, \partial_t) > 0$. Since H_φ is supposed to be constant, by Lemma 5.2 we obtain

$$\Delta_\varphi \Theta = -\left(\mathrm{Ric}_\varphi^M(N^M, N^M) + |A|^2\right) \Theta. \qquad (6.22)$$

On the other hand, by Eq. (1.11) it is easily seen that $(N^M)^\top = \Theta \nabla u$ and $|\nabla u|^2 = g_M(N^M, N^M)$. Thus, by (6.21) we obtain that

$$|\nabla u|^2 = \frac{|Du|_M^2}{1 + |Du|_M^2}. \qquad (6.23)$$

Letting $C = \alpha \sup_{p \in \Sigma(u)} |A(p)|^\beta$, by (1.12), (6.20) and 6.23 we have that

$$\Theta \geq \frac{1}{\sqrt{1+C^2}} > 0. \tag{6.24}$$

Since by hypothesis H_φ is constant, $\sup_{p \in \Sigma(u)} |A(p)| < +\infty$, and the fiber M^n has sectional curvatures bounded from below, by Lemma 5.3 we have that the Bakry-Émery-Ricci curvature of $\Sigma(u)$ is also bounded from below.

Hence, we can apply Lemma 5.1 to the function Θ, obtaining a sequence of points $\{p_k\} \subset \Sigma^n(u)$ such that $\lim_k \Theta(p_k) = \inf_{\Sigma(u)} \Theta$ and $\liminf_k \Delta_\varphi \Theta(p_k) \geq 0$.

Consequently, since the Bakry-Émery-Ricci curvature of M^n is nonnegative, by (6.22) and (6.24), up to a subsequence, we have that

$$0 \leq \liminf_k \Delta_\varphi \Theta(p_k) = \liminf_k \left[-\left(\mathrm{Ric}_\varphi^M \left(N^M, N^M \right) + |A|^2 \right) \Theta \right] (p_k)$$

$$= \lim_k \left(\mathrm{Ric}_\varphi^M \left(N^M, N^M \right) + |A|^2 \right) (p_k) \cdot$$

$$\cdot \left(- \inf_{p \in \Sigma(u)} \Theta(p) \right) \leq 0. \tag{6.25}$$

Thus, since $\inf_{p \in \Sigma(u)} \Theta(p) > 0$, by (6.25) we get that $\lim_k |A(p_k)| = 0$. So, taking into account our hypothesis (6.20), by (6.23) we get that $\lim_k |\nabla u(p_k)|^2 = 0$.

Therefore, by (1.12) we infer that $\inf_{p \in \Sigma(u)} \Theta(p) = 1$, and, hence, we must have that $u \equiv t_0$ for some $t_0 \in I$. □

Next, we establish a Moser type result assuming that the fiber of the ambient space has positive Bakry-Émery-Ricci tensor.

Theorem 6.18 *Let $\overline{M}_\varphi^{n+1} = I \times M_\varphi$ be a weighted product space, whose fiber M^n is complete, has sectional curvatures bounded by below, and its Bakry-Émery-Ricci tensor satisfies $\mathrm{Ric}_\varphi^M \geq c$ for some positive constant c, and such that the Hessian of the weight function φ is bounded by below. Let $\Sigma(u) \subset \overline{M}_\varphi^{n+1}$ be an entire graph over M^n with constant φ-mean curvature H_φ and bounded second fundamental form A. If $|Du|_M \leq C$ for some positive constant C, then $u \equiv t_0$ for some $t_0 \in I$.*

Proof We orient $\Sigma(u)$ by choosing N as in (6.21). We notice that our assumption on the Bakry-Émery-Ricci tensor of M^n ensures that

$$\mathrm{Ric}_\varphi^M \left(N^M, N^M \right) \geq c \left| N^M \right|_M^2 = c|\nabla u|^2. \tag{6.26}$$

Consequently, by (6.25) and (6.26) we get that $\lim_k |\nabla u(p_k)|^2 = 0$.

6.4 Entire Graphs in Weighted Riemannian Product Spaces

Therefore, taking into account (1.12), we conclude that $\inf_{p \in \Sigma(u)} \Theta(p) = 1$, and, hence, we must have $u \equiv t_0$ for some $t_0 \in I$. □

Remark 6.3 Related to Theorem 6.18, it is worth pointing out that a complete weighted manifold $(\Sigma, g, e^{-\varphi}dV)$ whose Bakry-Émery-Ricci tensor satisfies $\mathrm{Ric}_\varphi \geq c$, for some positive constant c, is not necessarily compact. A simple example is given by the Gaussian space $\mathbb{G}^n$; see Corollary 6.3 below. On the other hand, under the additional hypothesis that the weight function φ is bounded, an extension of a classical Myers' Theorem due to Wei and Willie (see [294, Theorem 1.4]) guarantees the compactness of Σ.

We recall that, according to the classical terminology in Linear Potential Theory, a weighted manifold $(\Sigma, g, e^{-\varphi}dV)$ is said to be φ-parabolic if any bounded solution f of $\Delta_\varphi f \geq 0$ must be constant.

In this setting, we have the following result.

Theorem 6.19 *Let $\overline{M}_\varphi^{n+1} = I \times M_\varphi^n$ be a weighted product space, whose fiber M^n is complete and such that its Bakry-Émery-Ricci tensor Ric_φ^M is nonnegative. Let $\Sigma(u) \subset \overline{M}_\varphi$ be an entire graph over M^n with constant φ-mean curvature H_φ. If $\Sigma(u)$ is φ-parabolic, then $\Sigma(u)$ is totally geodesic. Moreover, if Ric_φ^M is strictly positive, then $u \equiv t_0$ for some $t_0 \in I$.*

Proof Let $\Sigma(u)$ be oriented by (6.21). We also consider on $\Sigma(u)$ the bounded function $\theta_- = -\Theta$. By Lemma 5.2, we get

$$\Delta_\varphi \theta_- = \left(\mathrm{Ric}_\varphi^M \left(N^M, N^M\right) + |A|^2\right) \Theta \geq 0. \tag{6.27}$$

Consequently, since $\Sigma(u)$ is φ-parabolic, by (6.27) we obtain that θ_- is constant on $\Sigma(u)$. Thus, since $\Theta > 0$ on $\Sigma(u)$, thanks to (6.27) we have that $|A| \equiv 0$, that is, $\Sigma(u)$ is totally geodesic.

Moreover, if Ric_φ^M is strictly positive, we conclude that N^M vanishes identically on $\Sigma(u)$, which means that $N = \partial_t$ on $\Sigma(u)$. So, in this case, $u \equiv t_0$ for some $t_0 \in I$. □

Recently, considering the weighted product space $\mathbb{R} \times \mathbb{G}^n$, which is just the Euclidean space $\mathbb{R}^{n+1} = \mathbb{R} \times \mathbb{R}^n$ endowed with the Euclidean-Gaussian density

$$e^{-\varphi} = (2\pi)^{-\frac{n}{2}} e^{-\frac{|x|}{2}}, \tag{6.28}$$

Hieu and Nam extended the classical Bernstein's Theorem [78] showing that the only weighted minimal graphs $\Sigma(u)$ (that is, with identically zero weighted mean curvature) of smooth functions $u(x_2, \ldots, x_{n+1}) = x_1$ over $\mathbb{G}^n$ are the affine hyperplanes $x_1 = const.$; see [203, Theorem 4].

The *Gaussian space* $\mathbb{G}^n$ is the Euclidean space $\mathbb{R}^n$ with the Gaussian probability density (6.28). Furthermore, we also note that [203, Corollary 3] assures that entire weighted minimal graphs in $\mathbb{R} \times \mathbb{G}^n$ have finite φ-volume.

Moreover, as it was observed by Impera and Rimoldi in [211, Remark 3], the φ-parabolicity holds if $\Sigma(u)$ has finite φ-volume. Consequently, thanks to this fact, these graphs are φ-parabolic.

Hence, by Theorem 6.19 we obtain the following extension of [203, Theorem 4].

Corollary 6.3 *The affine hyperplanes $x_1 = $ const. of $\mathbb{R} \times \mathbb{G}^n$ are the only entire graphs of smooth functions $u(x_2, \ldots, x_{n+1}) = x_1$ over the Gaussian space $\mathbb{G}^n$ that have constant φ-mean curvature and that are φ-parabolic.*

6.5 Uniqueness and Nonexistence of Entire Translating Graphs

According to [162, Definition 2], we say that an entire graph $\Sigma(u) \subset \mathbb{R} \times M_\varphi^n$ constructed over the base $(M^n, \langle \cdot, \cdot \rangle_M)$ is an *entire translating graph* with *soliton constant* $c \in \mathbb{R}$ if its mean curvature function satisfies

$$H(u) = c\Theta = \frac{c}{\sqrt{1 + |Du|_M^2}}. \tag{6.29}$$

We observe that the slices $\{t\} \times M^n$ are trivial translating graphs with soliton constant $c = 0$.

In what follows, based on reference [71], we present uniqueness and nonexistence results concerning entire translating graphs $\Sigma(u) \subset \mathbb{R} \times M_\varphi^n$.

Theorem 6.20 *Let $\mathbb{R} \times M_\varphi^n$ be a weighted product space whose base M^n is complete, with φ-parabolic universal Riemannian covering and nonnegative Bakry-Émery-Ricci tensor, being positive at some point of M^n. If $\Sigma(u) \subset \mathbb{R} \times M_\varphi^n$ is an entire translating graph with soliton constant $|c| \leq 1$, constant φ-mean curvature and such that $|Du|_M^2 \leq H(u)^2$, then $u \equiv t_0$ for some $t_0 \in \mathbb{R}$.*

Proof Let us prove that the soliton constant c of such an entire translating graph must be zero, which means that $H = H(u)$ is identically zero. Hence, since $|Du|_M^2 \leq H^2$, the conclusion follows.

Arguing by contradiction, let us suppose that $c \neq 0$. By item (c) of Proposition 1.1 we have that

$$\Delta\Theta = -(\widetilde{\mathrm{Ric}}(N^*, N^*) + |A|^2)\Theta - \langle \nabla H, \partial_t \rangle, \tag{6.30}$$

6.5 Uniqueness and Nonexistence of Entire Translating Graphs

where $\widetilde{\text{Ric}}$ is the standard Ricci tensor of M^n and $N^* = N + \Theta \partial_t$ denotes the orthonormal projection of N onto M^n.

Thus, since H_φ is constant, by (5.5) and (6.30) we obtain

$$\frac{1}{2}\Delta\Theta^2 = -(\widetilde{\text{Ric}}(N^*, N^*) + |A|^2)\Theta^2 - \langle \nabla H, \partial_t\rangle\Theta + |\nabla\Theta|^2$$

$$= -(\widetilde{\text{Ric}}(N^*, N^*) + |A|^2)\Theta^2 + \partial_t^\top(\langle \overline{\nabla}\varphi, N\rangle)\Theta + |\nabla\Theta|^2. \qquad (6.31)$$

On the other hand, [251, Proposition 7.35] gives that

$$\overline{\nabla}_X \partial_t = 0, \qquad (6.32)$$

for any tangent vector field X on the ambient space.

Then, by (1.11) and (6.32) we have that

$$X(\Theta) = \langle \overline{\nabla}_X N, \partial_t\rangle = \langle -A(\nabla h), X\rangle.$$

Consequently,

$$\nabla\Theta = -A(\nabla h). \qquad (6.33)$$

Thus, by (6.32), (1.11) and (6.33), we get that

$$\partial_t^\top(\langle \overline{\nabla}\varphi, N\rangle) = \langle \overline{\nabla}_{\partial_t^\top}\overline{\nabla}\varphi, N\rangle + \langle \overline{\nabla}\varphi, \overline{\nabla}_{\partial_t^\top}N\rangle$$

$$= -\overline{\text{Hess}}\varphi(N, N)\Theta + \langle \overline{\nabla}\varphi, \nabla\Theta\rangle. \qquad (6.34)$$

Moreover, by (6.34) and (6.31), we obtain

$$\frac{1}{2}\Delta\Theta^2 = -(\widetilde{\text{Ric}}(N^*, N^*) + |A|^2)\Theta^2 - \overline{\text{Hess}}\varphi(N, N)\Theta^2$$

$$+ \frac{1}{2}\langle \overline{\nabla}\varphi, \nabla(\Theta^2)\rangle + |\nabla\Theta|^2. \qquad (6.35)$$

Consequently, since $\overline{\text{Hess}}\varphi(N, N) = \widetilde{\text{Hess}}\varphi(N^*, N^*)$, where $\widetilde{\text{Hess}}$ stands for the Hessian computed in the metric of M^n, by using (5.2), (5.1) and (6.35) we have the following equation

$$\frac{1}{2}\Delta_\varphi\Theta^2 = -(\widetilde{\text{Ric}}_\varphi(N^*, N^*) + |A|^2)\Theta^2 + |\nabla\Theta|^2, \qquad (6.36)$$

where $\widetilde{\text{Ric}}_\varphi$ stands for the Bakry-Émery-Ricci tensor of M^n.

On the other hand, by (1.12) and (6.16), one has

$$|\nabla h|^2 = \frac{|Du|_M^2}{1+|Du|_M^2}. \tag{6.37}$$

Thus, since the soliton constant c satisfies $|c| \leq 1$, by (6.29), (6.33), (6.36) and (6.37) the following inequality holds

$$\frac{1}{2}\Delta_\varphi H^2 = -(\widetilde{\mathrm{Ric}}_\varphi(N^*, N^*) + |A|^2)H^2 + c^2|\nabla\Theta|^2$$
$$\leq -(\widetilde{\mathrm{Ric}}_\varphi(N^*, N^*) + |A|^2)H^2 + |A|^2|\nabla h|^2$$
$$= -\widetilde{\mathrm{Ric}}_\varphi(N^*, N^*)H^2 - |A|^2\left(H^2 - \frac{|Du|_M^2}{1+|Du|_M^2}\right)$$
$$\leq -\widetilde{\mathrm{Ric}}_\varphi(N^*, N^*)H^2 - |A|^2(H^2 - |Du|_M^2). \tag{6.38}$$

Taking into our assumptions on $\widetilde{\mathrm{Ric}}_\varphi$ and $|Du|_M$, by (6.38) we get that H^2 is a nonnegative, φ-superharmonic function defined on $\Sigma(u)$.

Moreover, since the boundedness of $|Du|_M$ implies that

$$\Theta^2 \geq \frac{1}{1+c^2},$$

we can apply Corollary 6.2 obtaining that $\Sigma(u)$ is φ-parabolic.

Hence, we have that H must be constant on $\Sigma(u)$. But, since we have that $\widetilde{\mathrm{Ric}}_\varphi$ is positive at some point $x \in M^n$, thanks to (6.38) we conclude that either $H = 0$ or $|N^*(x)| = |\nabla h(x)| = 0$. Thus, we get that $Du(x) = 0$ and, taking into account (6.38), we obtain

$$A(u(x), x) = 0.$$

Consequently $H = 0$, which gives a contradiction. □

Taking into account the second inequality in (6.38), it is not difficult to verify that the following corollary holds.

Corollary 6.4 *Let $\mathbb{R} \times M_\varphi^n$ be a weighted product space whose base M^n is complete, with φ-parabolic universal Riemannian covering and nonnegative Bakry-Émery-Ricci tensor, being positive at some point of M^n. If $\Sigma(u) \subset \mathbb{R} \times M_\varphi^n$ is an entire translating graph with soliton constant $|c| < 1$, constant φ-mean curvature and such that*

$$|Du|_M^2 \leq \frac{H(u)^2}{1-H(u)^2},$$

then $u \equiv t_0$ for some $t_0 \in \mathbb{R}$.

6.5 Uniqueness and Nonexistence of Entire Translating Graphs

As well as, the next property holds.

Corollary 6.5 *Let $\mathbb{R} \times M_\varphi^n$ be a weighted product space whose base M^n is complete, with φ-parabolic universal Riemannian covering and nonnegative Bakry-Émery-Ricci tensor. There does not exist an entire translating graph $\Sigma(u) \subset \mathbb{R} \times M_\varphi^n$ with soliton constant $|c| \leq 1$, constant φ-mean curvature and such that $H(u)^2 \geq 1$.*

Since $\mathrm{vol}_\varphi(\mathbb{G}^n) = 1$, where φ is the Gaussian probability measure defined in (6.28), by [211, Remark 3.8] we conclude that $\mathbb{G}^n$ is φ-parabolic.

Hence, we obtain the following application of Theorem 6.20 and Corollary 6.4.

Corollary 6.6 *The only entire translating graphs $\Sigma(u) \subset \mathbb{R} \times \mathbb{G}^n$ of the mean curvature flow with respect to ∂_t with soliton constant c, having constant φ-mean curvature (where φ is the Gaussian probability measure defined in (6.28)) and satisfying one of the following conditions*

(i) $|c| \leq 1$ and $|Du|_{\mathbb{G}^n}^2 \leq H(u)^2$;

(ii) $|c| < 1$ and $|Du|_{\mathbb{G}^n}^2 \leq \dfrac{H(u)^2}{1 - H(u)^2}$,

are the slices $\{t\} \times \mathbb{G}^n$.

By Corollary 6.5 we also get the following consequence.

Corollary 6.7 *There is no an entire translating graph $\Sigma(u) \subset \mathbb{R} \times \mathbb{G}^n$ with soliton constant $|c| \leq 1$, constant φ-mean curvature (where φ is the Gaussian probability measure defined in (6.28)) and such that $H(u)^2 \geq 1$.*

By Lemmas 5.1 and 5.3 we obtain the proposition below which gives sufficient conditions to guarantee that the drift Laplacian on a complete hypersurface of a weighted Riemannian product space satisfies the weak Omori-Yau's maximum principle. For our purposes we assume that the Hessian of the weight function is bounded from below.

Proposition 6.1 *Let $\overline{M}_\varphi^{n+1} = \mathbb{R} \times M_\varphi^n$ be a weighted Riemannian product space such that the sectional curvature of the base M^n and the Hessian of the weight function φ are bounded from below. If Σ^n is a complete hypersurface immersed in $\overline{M}_\varphi^{n+1}$ with bounded second fundamental form and constant φ-mean curvature, then the drift Laplacian Δ_φ on Σ^n satisfies the weak Omori-Yau's maximum principle.*

Next, we apply Proposition 6.1 to get the following nonexistence result.

Theorem 6.21 Let $\overline{M}_\varphi^{n+1} = \mathbb{R} \times M_\varphi^n$ be a weighted Riemannian product space such that the sectional curvature of the fiber M^n and the Hessian of the weight function φ are bounded by below. Suppose in addition that the Bakry-Émery-Ricci tensor of M^n satisfies $\widetilde{\mathrm{Ric}}_\varphi \geq \alpha$, for some positive constant α. Then there is no an entire translating graph $\Sigma(u) \subset \overline{M}_\varphi^{n+1}$ with soliton constant $0 < |c| < 1$, constant φ-mean curvature, finite C^2 norm and such that

$$\beta \leq |Du|_M^2 \leq \frac{H(u)^2}{1 - H(u)^2},$$

for some positive constant β.

Proof Let us suppose by contradiction that there exists an entire translating graph $\Sigma(u)$ with the desired properties. Since u has finite C^2 norm, by (6.17) we have that $|A|$ is bounded.

Thus, thanks to the main assumptions, we can apply Proposition 6.1 obtaining a sequence of points $\{p_k\} \subset \Sigma(u)$ such that

$$H(u)^2(p_k) < \inf_{\Sigma(u)} H(u)^2 + \frac{1}{k} \quad \text{and} \quad \Delta_\varphi H(u)^2(p_k) > -\frac{1}{k},$$

for every $k \in \mathbb{N} \setminus \{0\}$.

On the other hand, by using the main hypothesis on the Bakry-Émery-Ricci tensor of M^n in addition to (6.37), we have

$$\widetilde{\mathrm{Ric}}_\varphi(N^*, N^*) \geq \alpha |N^*|^2 = \alpha |\nabla h|^2 = \frac{|Du|_M^2}{1 + |Du|_M^2}. \tag{6.39}$$

Hence, since

$$|Du|_M^2 \leq \frac{H(u)^2}{1 - H(u)^2},$$

which is equivalent to

$$H(u)^2 \geq \frac{|Du|_M^2}{1 + |Du|_M^2},$$

and bearing in mind that the soliton constant c satisfies $0 < c^2 < 1$ and $H(u)^2 \leq c^2$, it is easily seen that

$$0 \leq \frac{1}{2} \liminf_k \Delta_\varphi H(u)^2(p_k) \leq -\alpha(1 - c^2)^2 \liminf_k |Du(p_k)|_M^4 \leq 0, \tag{6.40}$$

on account of (6.38) and (6.39).

Therefore, by (6.40) we conclude that $\liminf_k |Du(p_k)| = 0$, which contradicts the hypothesis $0 < \beta \leq |Du|_M^2$. □

We close this section with the following application of Theorem 6.21.

Corollary 6.8 *There does not exist an entire translating graph $\Sigma(u) \subset \mathbb{R} \times \mathbb{G}^n$ with soliton constant $0 < |c| < 1$, constant φ-mean curvature (where φ is the Gaussian probability measure defined in (6.28)), finite C^2 norm and such that*

$$\beta \leq |Du|_{\mathbb{G}^n}^2 \leq \frac{H(u)^2}{1 - H(u)^2},$$

for some positive constant β.

6.6 A Height Estimate and a Half-Space Type Result in Weighted Riemannian Product Spaces

The results of this section correspond to [143, Section 3]. Firstly, we state and prove a height estimate for compact hypersurfaces immersed in a weighted Riemannian product space.

Theorem 6.22 *Let $\mathbb{R} \times M_\varphi^n$ be a weighted Riemannian product space with nonnegative Bakry-Émery-Ricci tensor $\overline{\mathrm{Ric}}_\varphi$ and let Σ^n be a compact hypersurface with boundary contained into the slice $\{t\} \times M^n$, for some $t \in \mathbb{R}$, and whose angle function Θ does not change sign. If Σ^n has nonzero constant φ-mean curvature such that $nH_\varphi^2 \leq |A|^2$, where A denotes the Weingarten operator of Σ^n with respect to its unit normal vector field N, then the height function h of Σ^n satisfies*

$$|h - t| \leq \frac{1}{|H_\varphi|}. \tag{6.41}$$

Proof We define on Σ^n the function

$$\zeta = H_\varphi h + \Theta. \tag{6.42}$$

By (6.42) and items (*i*) and (*iii*) of Lemma 5.2 we get that

$$\Delta_\varphi \zeta = -\Theta(|A|^2 - nH_\varphi^2 + \widetilde{\mathrm{Ric}}_\varphi(N^*, N^*)). \tag{6.43}$$

Consequently, since $\widetilde{\mathrm{Ric}}_\varphi(N^*, N^*) = \overline{\mathrm{Ric}}_\varphi(N, N) \geq 0, nH_\varphi^2 \leq |A|^2$, by choosing N such that $-1 \leq \Theta \leq 0$ and by (6.43), we get that $\Delta_\varphi \zeta \geq 0$. Thus, by the maximum principle we conclude that $\zeta \leq \zeta_{|\partial \Sigma}$.

Hence, by (6.42) we have that

$$H_\varphi h - 1 \leq H_\varphi h + \Theta \leq H_\varphi t. \tag{6.44}$$

Thus, two possible cases occur. In the case $H_\varphi > 0$, by item (i) of Lemma 5.2 we have $\Delta_\varphi h \leq 0$ and, by Hopf's maximum principle, $h \geq t$ on Σ^n. Thus, by (6.44) we conclude that

$$h - t \leq \frac{1}{H_\varphi}. \tag{6.45}$$

In the case $H_\varphi < 0$, by item (i) of Lemma 5.2 we have $\Delta_\varphi h \geq 0$ and, again by Hopf's maximum principle, $h \leq t$ on Σ^n. Thus, by (6.44) we must have

$$t - h \leq -\frac{1}{H_\varphi}. \tag{6.46}$$

Therefore, estimate (6.41) follows by (6.45) and (6.46). □

By Theorem 6.22 we obtain the following half-space type result

Theorem 6.23 *Let $\mathbb{R} \times M_\varphi^n$ be a weighted Riemannian product space with nonnegative Bakry-Émery-Ricci tensor $\overline{\mathrm{Ric}}_\varphi$ and M^n compact. Let Σ^n be a complete noncompact two-sided hypersurface properly immersed in $\mathbb{R} \times M_\varphi^n$, whose angle function Θ does not change sign. If Σ^n has nonzero constant φ-mean curvature such that $nH_\varphi^2 \leq |A|^2$, then Σ^n cannot lie in a half-space of $\mathbb{R} \times M$. In particular, Σ^n must have at least one top and one bottom end.*

Proof Suppose by contradiction that, for instance, $\Sigma^n \subset (-\infty, \tau] \times M^n$, for some $\tau \in \mathbb{R}$. Thus, for each $t_* < \tau$ we define

$$\Sigma_{t_*}^+ = \{(t, x) \in \Sigma^n : t \geq t_*\}.$$

Since M^n is compact and Σ^n is properly immersed in $\mathbb{R} \times M_\varphi^n$, we have that $\Sigma_{t_*}^+$ is a compact hypersurface contained in a slab of width $\tau - t_*$ and whose boundary is contained in $\{t_*\} \times M^n$.

Thus, we can apply Theorem 6.22 to get that $\Sigma_{t_*}^+$ is contained in a slab of width $1/|H_\varphi|$, so that it must be $\tau - t_* \leq 1/|H_\varphi|$. Consequently, by choosing t_* sufficiently small we violate this estimate, reaching a contradiction.

6.6 A Height Estimate and a Half-Space Type Result in Weighted...

Analogously, if we suppose that $\Sigma^n \subset [\tau, +\infty) \times M^n$ with $\tau \in \mathbb{R}$, for each $t^* > \tau$ we define $\Sigma_{t^*}^-$ by

$$\Sigma_{t^*}^- = \{(t, x) \in \Sigma^n : t \leq t^*\}.$$

Hence, since $\Sigma_{t^*}^-$ is a compact hypersurface with boundary contained in $\{t^*\} \times M^n$, similar to the above arguments ensure the conclusion. $\square$

7 Spacelike Hypersurfaces in Weighted GRW Spacetimes

7.1 Introduction

The investigation of the uniqueness and nonexistence of spacelike hypersurfaces immersed in a weighted Lorentzian manifold through the application of maximum principles for the drift Laplacian Δ_φ constitutes an interesting thematic. In this setting, Cavalcante et al. [104] obtained new Calabi-Bernstein type results concerning complete spacelike hypersurfaces immersed in a spatially weighted GRW spacetime $-I \times_f M_\varphi^n$.

Assuming a natural comparison inequality between the weighted mean curvature H_φ of the spacelike hypersurface and those of the slices of the timelike bounded region where the hypersurface is supposed to be contained, they established sufficient conditions which guarantee that such a hypersurface must be a slice. Furthermore, they also treated the case when the ambient spacetime is a spatially weighted static GRW spacetime $-I \times M_\varphi^n$.

Later on, Albujer et al. [10] also applied some maximum principles in order to study the rigidity of complete spacelike hypersurfaces immersed in a spatially weighted GRW spacetime, which is supposed to obey the so-called strong null convergence condition. Under natural constraints on the weight function and on the weighted mean curvature, they also obtained sufficient conditions to infer that such a hypersurface must be a slice of the ambient spacetime.

Meanwhile, Oliveira and de Lima [249] applied a mean-value inequality for positive subsolutions of the φ-heat operator, obtained from a Sobolev embedding, to prove a nonexistence result concerning complete noncompact φ-maximal spacelike hypersurfaces in a weighted Lorentzian product space of the type $-I \times M_\varphi^n$.

In [143], de Lima and Santos proved a height estimate concerning compact spacelike hypersurfaces with nonzero constant weighted mean curvature and whose boundary is contained into a slice of a weighted Lorentzian product space with nonnegative Bakry-

Émery-Ricci tensor and applied it to obtain a half-space type result related to complete noncompact spacelike hypersurfaces immersed in such an ambient space.

Afterwards, the same authors in [11] extended a technique due to Romero et al. [260, 261] obtaining sufficient conditions to guarantee the parabolicity of complete spacelike hypersurfaces immersed in a spatially weighted GRW spacetime $-I \times_f M_\varphi^n$ whose Riemannian fiber M^n has φ-parabolic universal Riemannian covering. As some applications of this criteria, they proved uniqueness results concerning spacelike hypersurfaces immersed in a suitable spacetime.

In this chapter, we study the uniqueness of spacelike slices $\Sigma_t = \{t\} \times M^n$, $t \in I$, among spacelike hypersurfaces Σ^n immersed in a spatially weighted GRW spacetime of the type $\overline{M}_\varphi^{n+1} = -I \times_f M_\varphi^n$, where the weight function φ does not depend on the parameter $t \in I$.

As in Chap. 6 we introduce suitable abstract conditions in order to guarantee that Σ^n is isometric to Σ_t for some $t \in I$.

Moreover, Calabi-Bernstein type results concerning entire graphs

$$\Sigma(u) = \{(u(p), p) : p \in M_\varphi^n\} \subset \overline{M}_\varphi$$

of functions $u \in C^\infty(M_\varphi^n)$ such that $u(M_\varphi^n) \subseteq I$ are also established.

Finally, we present a height estimate and a half-space type result in weighted Lorentzian product spaces.

7.2 Uniqueness Results for Spacelike Hypersurfaces in Spatially Weighted GRW Spacetimes

In this section we prove some uniqueness results concerning spacelike slices among spacelike hypersurfaces immersed in a spatially weighted generalized Robertson-Walker spacetime via suitable maximum principles.

7.2.1 Comparison Inequalities Between Curvatures in the Presence of a Omori-Yau Maximum Principle

The results presented here are partially contained in [10]. In order to present them, we recall the following version of the Omori-Yau maximum principle for the φ-Laplacian; see [259, Remark 2.18].

7.2 Uniqueness Results for Spacelike Hypersurfaces in Spatially...

Lemma 7.1 *Let $(\Sigma^n, g, e^{-\varphi}dV)$ be a complete weighted manifold whose Bakry-Émery-Ricci curvature tensor is bounded from below, and let $u : \Sigma^n \to \mathbb{R}$ be a smooth function bounded from above (resp. bounded from below) on Σ^n. Then, there exists a sequence of points $\{p_k\} \subset \Sigma^n$ such that*

$$\lim_k u(p_k) = \sup_{\Sigma^n} u \quad (\text{resp.,} = \inf_{\Sigma^n} u)$$

and

$$\limsup_k \Delta_\varphi u(p_k) \leq 0 \quad (\text{resp.,} \liminf_k \Delta_\varphi u(p_k) \geq 0).$$

We know that the φ-mean curvature of a slice $\Sigma_t = \{t\} \times M^n$ in a spatially weighted GRW spacetime $-I \times_f M_\varphi^n$ is given by $H_\varphi(t) = (\log f)'(t)$ with respect to the orientation $N = \partial_t$. The results presented in this subsection use the previous lemma to study the uniqueness of spacelike slices under some comparison inequalities involving their mean curvature and the φ-mean curvature function of a given spacelike hypersurface Σ^n. We also require some bounds on the norm of the gradient of the height function, which can be interpreted as a measure of how much a spacelike hypersurface is no longer a slice. A control on the behaviour of H_φ involving geometric inequalities have already been employed in the literature; see, for instance, the paper [185].

The next result can be view as a generalization of [95, Theorem 4.3] and [104, Theorem 2].

Theorem 7.1 *Let $\overline{M}_\varphi^{n+1} = -I \times_f M_\varphi^n$ be a spatially weighted GRW spacetime obeying the SNCC (5.22) and such that the Hessian of the weight function φ is bounded from below. Moreover, let $\psi : \Sigma^n \to \overline{M}_\varphi^{n+1}$ be a complete spacelike hypersurface that lies in a timelike bounded region of $\overline{M}_\varphi^{n+1}$. Suppose that the φ-mean curvature H_φ of Σ^n satisfies*

$$(\log f)'(h) \leq H_\varphi \leq \alpha \quad \text{and} \quad H_\varphi \geq 0 \tag{7.1}$$

for some positive constant α. If

$$|\nabla h| \leq \beta \inf_{\Sigma^n} |H_\varphi - (\log f)'(h)|^\gamma \tag{7.2}$$

for some constants $\beta > 0$ and $\gamma \neq 0$, then Σ^n is a slice Σ_t for some $t \in I$.

Proof By Lemma 5.4-(ii), for the φ-Laplacian of the function

$$\mathcal{F}(t) = \int_{t_0}^t f(s)ds$$

we get

$$\Delta_\varphi \mathcal{F}(h) = -nf(h)\left((\log f)'(h) + \Theta H_\varphi\right) \geq nf(h)\left(H_\varphi - (\log f)'(h)\right).$$

Thus, by (7.1), we conclude that $\Delta_\varphi \mathcal{F}(h) \geq 0$ on Σ^n. Since the smooth function $\mathcal{F}$ is bounded from above, and Ric_φ is bounded from below by Lemma 5.5, the hypotheses of Lemma 7.1 are verified.

Hence, we can take a sequence of points $\{p_k\} \subset \Sigma^n$ such that

$$\limsup_k \Delta_\varphi \mathcal{F}(h(p_k)) \leq 0.$$

Up to a subsequence, we have

$$0 \geq \limsup_k \Delta_\varphi \mathcal{F}(h(p_k)) \geq \lim_k \left[nf(h)\left(H_\varphi - (\log f)'(h)\right)\right](p_k) \geq 0.$$

Now, since Σ^n is contained in a timelike bounded region of $\overline{M}_\varphi^{n+1}$, there exists a positive constant C such that $f(h(p)) \geq C$ for all $p \in \Sigma^n$.

Therefore, we have that

$$\lim_k \left(H_f - (\log f)'(h)\right)(p_k) = 0.$$

Thanks to (7.2), the conclusion immediately follows. $\square$

Observing that the φ-mean curvature is simply the typical mean curvature H when the function φ is constant, our next result can be interpreted as a generalization of [9, Theorem 3.3]. Interestingly, despite the fact that [9] deals with hypersurfaces in RW spacetimes, the same results can be reached when GRW spacetimes $-I \times_f M$ are taken into consideration, where the fiber's sectional curvatures are bounded from below, as is the case under the SNCC.

Theorem 7.2 *Let $\overline{M}_\varphi^{n+1} = -I \times_f M_\varphi^n$ be a spatially weighted GRW spacetime obeying the SNCC (5.22) with convex weight function φ, that is, $\overline{\mathrm{Hess}}\,\varphi \geq 0$. Moreover, let $\psi : \Sigma^n \to \overline{M}_\varphi^{n+1}$ be a complete spacelike hypersurface that lies in a timelike bounded region of $\overline{M}_\varphi^{n+1}$ and with constant φ-mean curvature H_φ satisfying*

$$0 \leq H_\varphi \sup_{\Sigma^n} (\log f)'(h) \leq H^2. \tag{7.3}$$

If

$$|\nabla h|^2 \leq \alpha \left(\inf_{\Sigma^n} H^2 - H_\varphi \sup_{\Sigma^n} (\log f)'(h) \right)^\beta \quad (7.4)$$

for some constants $\alpha > 0$ *and* $\beta \neq 0$, *then* $\Sigma^n = \Sigma_t$ *for some* $t \in I$.

Proof By the SNCC (5.22) we get that

$$\mathrm{Ric}^M(N^M, N^M) - (n-1)(\log f)''(h)|\nabla h|^2 \geq 0.$$

Therefore, by Lemma 5.4-(iii), the following holds

$$\Delta_\varphi(f(h)\Theta) \leq nf'(h)H_\varphi + f(h)\Theta|A|^2, \quad (7.5)$$

thanks to the main assumptions.

By using (7.5) and (7.3), we obtain

$$\Delta_\varphi(f(h)\Theta) \leq f(h)\Theta \left(|A|^2 - nH_\varphi \sup_{\Sigma^n} (\log f)'(h) \right).$$

On the other hand, a simple algebraic computation shows that

$$|A|^2 = n^2 H^2 - n(n-1)H_2,$$

where H_2 is the second order mean curvature defined by

$$\binom{n}{2} H_2 = \sum_{i<j} k_i k_j,$$

in which k_i, $i = 1, \ldots, n$ are the principal curvatures of Σ^n.

Therefore, one has

$$\Delta_\varphi(f(h)\Theta) \leq f(h)\Theta \left(n^2 H^2 - n(n-1)H_2 - nH_\varphi \sup_{\Sigma^n} (\log f)'(h) \right)$$

$$= n(n-1)f(h)\Theta(H^2 - H_2)$$

$$+ nf(h)\Theta \left(H^2 - H_\varphi \sup_{\Sigma^n} (\log f)'(h) \right) \leq 0, \quad (7.6)$$

where we use again (7.3) and $H^2 - H_2 \geq 0$.

Now, by (7.4) and (1.12) we have that Θ is bounded. Consequently, since Σ^n lies in a timelike bounded region of $\overline{M}_\varphi^{n+1}$, the function $f(h)\Theta$ is bounded from below.

By Lemma 5.5, we know that Ric_φ is also bounded from below. So, we can apply the Omori-Yau maximum principle for the φ-Laplacian (see Lemma 7.1) to get a sequence of points $\{p_k\} \subset \Sigma^n$ such that

$$\lim_k (f(h(p_k))\Theta(p_k)) = \inf_{\Sigma^n} f(h)\Theta \quad \text{and} \quad \liminf_k \Delta_\varphi (f(h(p_k))\Theta(p_k)) \geq 0.$$

Up to a subsequence (7.6) we have that

$$0 \leq \liminf_k \Delta_\varphi f(h(p_k))\Theta(p_k)$$

$$\leq n(n-1)\lim_k \left(f(h(p_k))\Theta(p_k)(H^2 - H_2)(p_k) \right)$$

$$+n \lim_k \left(f(h(p_k))\Theta(p_k) \left(H^2(p_k) - H_\varphi \sup_{\Sigma^n} (\log f)'(h) \right) \right) \leq 0,$$

so, in particular

$$\lim_k \left(H^2(p_k) - H_\varphi \sup_{\Sigma^n} (\log f)'(h) \right) = 0.$$

Consequently, by (7.3) we get

$$\inf_{\Sigma^n} \left(H^2 - H_\varphi \sup_{\Sigma^n} (\log f)'(h) \right) = 0.$$

The result follows by (7.4). □

7.2.2 Uniqueness Under Certain Integrability Conditions

The result presented in this subsection can be founded in reference [10].

Let $(\Sigma^n, g, d\mu = e^{-\varphi}dV)$ be a weighted manifold. For $1 \leq p < \infty$, let us consider the set

$$L^p(d\mu) = \left\{ u : \Sigma^n \to \mathbb{R} : \int_{\Sigma^n} |u|^p d\mu < +\infty \right\}.$$

We recall the following result, which is a consequence of [256, Theorem 1.1].

Lemma 7.2 *Let u be a nonnegative smooth φ-subharmonic function on a complete Riemannian manifold Σ^n. If $u \in L^p(d\mu)$ for some $1 \leq p < \infty$, then u is constant.*

7.2 Uniqueness Results for Spacelike Hypersurfaces in Spatially...

The next result is due to Wei and Wylie [294].

Lemma 7.3 *Let Σ^n be a noncompact complete Riemannian manifold with nonnegative Bakry-Émery-Ricci curvature for some bounded weight function $\varphi \in C^\infty(\Sigma^n)$. Then Σ^n has at least linear φ-volume growth, i.e., for any $p \in \Sigma^n$, $\mathrm{vol}_\varphi(B(p, R))$ has at least linear growth on R, where $B(p, R)$ is the geodesic ball in Σ^n centered at p with radius R.*

Combining Lemmas 7.2 and 7.3, we can prove the main result of this subsection which reads as follows.

Theorem 7.3 *Let $\overline{M}_\varphi^{n+1} = -I \times_f M_\varphi^n$ be a spatially weighted GRW spacetime such that the Hessian of the weight function φ is bounded from below, and let $\psi : \Sigma^n \to \overline{M}_\varphi^{n+1}$ be a complete spacelike hypersurface. Suppose that $H_\varphi > 0$, that $f'(h) > 0$, and that the following inequalities are satisfied*

$$\frac{n^2}{4}((\log f)'(h))^2 \leq \frac{n^2 H_\varphi^2}{4} \leq (n-1)\frac{f''}{f}(h). \tag{7.7}$$

If $f(h) \in L^p(d\mu)$ for some $1 \leq p < \infty$, where $d\mu = e^{-\varphi}d\Sigma^n$, then Σ^n is a slice of $\overline{M}_\varphi^{n+1}$ with $\mathrm{vol}_\varphi(\Sigma^n) < +\infty$. In addition, if $\overline{M}_\varphi^{n+1}$ obeys the SNCC (5.22) and φ is bounded and convex, then Σ^n is compact.

Proof By Lemma 5.4-(i) we get

$$\Delta_\varphi h = -(\log f)'(h)(n + |\nabla h|^2) - n H_\varphi \Theta. \tag{7.8}$$

Moreover, it is easily seen that hypothesis (7.7) implies that $(\log f)''(h) \geq 0$ on Σ^n.

Consequently, by our assumptions, by (7.8) we get

$$\Delta_\varphi f(h) = f'(h)\Delta_\varphi h + f''(h)|\nabla h|^2$$
$$\geq n f'(h) \left(H_\varphi - (\log f)'(h) \right) \geq 0. \tag{7.9}$$

Thus, since $f(h) \in L^p(d\mu)$, in view of (7.9) we can apply Lemma 7.2 to conclude that $f(h)$ is constant on Σ^n, and so $\mathrm{vol}_\varphi(\Sigma^n) < +\infty$.

Hence, since $f'(h) > 0$ on Σ^n, we get that h is also constant, and, consequently, Σ^n must be a slice of $\overline{M}_\varphi^{n+1}$.

Furthermore, if $\overline{M}_\varphi^{n+1}$ obeys the SNCC and the weight function φ is convex, by (5.31) we obtain

$$\mathrm{Ric}_\varphi(X,X) \geq \left((n-1)\frac{f''}{f}(h) - \frac{n^2 H_\varphi^2}{4}\right)|X|^2 \qquad (7.10)$$

for any $X \in \mathfrak{X}(\Sigma^n)$.

Thus, by (7.7) and (7.10), we have that Ric_φ is nonnegative. Therefore, Lemma 7.3 guarantees that Σ^n must be compact. □

Remark 7.1 By [24, Proposition 3.2], if a GRW spacetime admits a compact spacelike hypersurface, then it is spatially closed. Therefore, Theorem 7.3 implies, in particular, that the fiber M^n must be also compact. Moreover, the compactness of Σ^n could be achieved in a direct way requiring that M^n is compact. In fact, again by [24, Proposition 3.2], if M^n is compact and Σ^n is a complete spacelike hypersurface such that $f(h)$ is bounded, then Σ^n is compact. We emphasize that under the assumptions of Theorem 7.3 we have that $f(h)$ is constant.

Remark 7.2 Following [268], given a spacelike hypersurface Σ^n immersed in $\overline{M}_\varphi^{n+1} = -I \times_f M_\varphi^n$, with future-pointing Gauss map N, for each $q \in \Sigma^n$ we set

$$\xi_q = E(q) N_q + \xi_q^\top,$$

where

$$E(q) = -\overline{g}(\xi_q, N_q) = -f(h(q))\Theta(q) > 0,$$

and $\xi_q^\top$ are, respectively, the energy and the n-momentum that the instantaneous observer N_q measures for ξ_q.

Extending the concept of total energy established in [144], for $1 \leq p < \infty$, we say that Σ^n has finite total (φ, p)-energy if

$$\int_{\Sigma^n} E^p d\mu < +\infty.$$

Finally, since $E(q) \geq f(h(q))$ for every $q \in \Sigma^n$, we observe that the conclusion of Theorem 7.3 still holds if we assume that the hypersurface Σ^n has finite total (φ, p)-energy, instead of $f(h) \in L^p(d\mu)$.

7.2.3 Uniqueness Under φ-Parabolicity Criteria

In this subsection, we extend a technique due to Romero et al. [260–262] establishing sufficient conditions to guarantee the φ-parabolicity of complete spacelike hypersurfaces immersed in a weighted generalized Robertson-Walker spacetime whose fiber has φ-parabolic universal Riemannian covering.

As application of these criteria, we obtain uniqueness results concerning spacelike slices among spacelike hypersurfaces immersed in a spatially weighted generalized Robertson-Walker spacetimes.

The results presented here are contained in reference [11].

We will say that a smooth function u on a weighted manifold $P_\varphi = (P, g, d\mu = e^{-\varphi} dP)$ is φ-superharmonic if $\Delta_\varphi u \leq 0$.

The weighted manifold P_φ is called φ-parabolic if there is no nonconstant, nonnegative, φ-superharmonic function on it.

Moreover, for any compact subset $K \subset P$, we define the φ-capacity of K as

$$\mathrm{cap}_\varphi(K) = \inf \left\{ \int_P |\nabla u|^2 d\mu \,:\, u \in \mathrm{Lip}_0(P) \text{ and } u\big|_K \equiv 1 \right\},$$

where $\mathrm{Lip}_0(P)$ is the set of all compactly supported Lipschitz functions on P.

The following statement describe the link between the notion of φ-capacity and the concept of φ-parabolicity; see [195, Proposition 2.1].

Lemma 7.4 *The weighted manifold* $(P, g, d\mu = e^{-\varphi} dP)$ *is φ-parabolic if and only if* $\mathrm{cap}_\varphi(K) = 0$ *for any compact set* $K \subset P$.

Let us recall that, given two Riemannian manifolds (P, g) and (P', g'), a diffeomorphism ψ from P onto P' is called a *quasi-isometry* if there exists a constant $c \geq 1$ such that

$$c^{-1}|v|_g \leq |d\psi(v)|_{g'} \leq c|v|_g$$

for every $v \in T_p P$, $p \in P$; see [215] for more details.

Now, let P and P' endowed with same weight function φ. Arguing as in [194, Section 5] one has that the φ-capacity changes under a quasi-isometry at most by a constant factor. So, by Lemma 7.4, we can state the following result.

Lemma 7.5 *Let (P, g) and (P', g') be two Riemannian manifolds endowed with the same weight function φ. If P and P' are quasi-isometric, then P and P' are φ-parabolic or not simultaneously.*

Given a spacelike hypersurface $\psi : \Sigma^n \to M^n$ in a GRW spacetime $\overline{M}^{n+1} = -I \times_f M^n$, the following lemma provides sufficient conditions to guarantee that the hypersurface Σ^n and the fiber M^n are quasi-isometric; see [260, Lemma 4.1].

Lemma 7.6 *Let $\psi : \Sigma^n \to M^n$ be a spacelike hypersurface in a GRW spacetime $\overline{M}^{n+1} = -I \times_f M^n$, whose hyperbolic angle function Θ is bounded. If the warping function f on Σ^n satisfies*

(i) $\sup_{\Sigma^n} f(h) < +\infty$ and
(ii) $\inf_{\Sigma^n} f(h) > 0$,

then $\pi = \pi_M \circ \psi$ is a quasi-isometry from Σ^n onto M^n.

We can now present our main criterion of φ-parabolicity.

Theorem 7.4 *Let Σ^n be a complete spacelike hypersurface in a spatially weighted GRW spacetime M_φ with weight function φ, whose fiber M is complete with φ-parabolic universal Riemannian covering. If the angle function Θ of Σ^n is bounded and the restriction $f(h)$ on Σ^n of the warping function f of M satisfies*

(i) $\sup_{\Sigma^n} f(h) < +\infty$ and
(ii) $\inf_{\Sigma^n} f(h) > 0$,

then Σ^n is φ-parabolic.

The proof of Theorem 7.4 follows by using the same arguments of [260, Theorem 4.4]; see also [261, Theorem 1]. Along this directions, the following standard result on covering spaces will be useful for our purposes; see [199] for additional comments and remarks.

Lemma 7.7 *Let $\rho : (\tilde{E}, \tilde{x}_0) \to (E, x_0)$ be a covering space and let $h : (W, y_0) \to (E, x_0)$ be a continuous map, where W is a path connected and locally path connected topological space. Then, there exists a lift $\tilde{h} : (W, y_0) \to (\tilde{E}, \tilde{x}_0)$ of h if and only if $h_*(\pi_1(W, y_0)) \subset \rho_*(\pi_1(\tilde{E}, \tilde{x}_0))$.*

Proof of (Theorem 7.4) Let us introduce on the universal Riemannian covering $(\tilde{M}, g_{\tilde{M}})$ of the fiber (M, g_M) the notion of φ-parabolicity as usual. Now, by using [24, Lemma 3.1], we know that the projection on the fiber, $\pi : \tilde{M} \to M$, is a covering map.

Moreover, by Lemma 7.6, we can find a constant $c \geq 1$ such that

$$c^{-1} g_M(d\pi(v), d\pi(v)) \leq g_{\Sigma^n}(v, v) \leq c\, g_M(d\pi(v), d\pi(v)) \tag{7.11}$$

7.2 Uniqueness Results for Spacelike Hypersurfaces in Spatially...

for every $v \in T_p \Sigma^n$ and any $p \in \Sigma^n$.

Let $(\tilde{\Sigma}, g_{\tilde{\Sigma}})$ be the universal Riemannian covering of (Σ^n, g_{Σ^n}), and denote by $\tilde{\pi}_{\Sigma^n} : \tilde{\Sigma} \to \Sigma^n$ the corresponding Riemannian covering map.

By Lemma 7.7 we conclude that there exists a lift $\tilde{h} : \tilde{\Sigma} \to \tilde{M}$ of the map $h = \pi \circ \tilde{\pi}_{\Sigma^n} : \tilde{\Sigma} \to M$.

It is easy to check that $\tilde{h}$ is, in fact, a diffeomorphism from $\tilde{\Sigma}$ to $\tilde{M}$. Note that (7.11) implies

$$c^{-1} g_{\tilde{M}}(d\tilde{h}(\tilde{v}), d\tilde{h}(\tilde{v})) \leq g_{\tilde{\Sigma}}(\tilde{v}, \tilde{v}) \leq c\, g_{\tilde{M}}(d\tilde{h}(\tilde{v}), d\tilde{h}(\tilde{v}))$$

for any $\tilde{v} \in T_{\tilde{p}} \tilde{\Sigma}$, $\tilde{p} \in \tilde{\Sigma}$, which means that $\tilde{h}$ is a quasi-isometry from $(\tilde{\Sigma}, g_{\tilde{\Sigma}})$ onto $(\tilde{M}, g_{\tilde{M}})$.

Finally, let u be a nonnegative φ-superharmonic function on Σ^n, and put $\tilde{u} = u \circ \tilde{\pi}_{\Sigma^n}$. The function $\tilde{u}$ is a nonnegative φ-superharmonic function on the φ-parabolic Riemannian manifold $\tilde{\Sigma}$.

Therefore, $\tilde{u}$ must be constant, and, consequently, u is also constant. □

As a direct consequence of Theorem 7.4 we get the following corollaries.

Corollary 7.1 *Let Σ^n be a complete spacelike hypersurface in a spatially weighted GRW spacetime M_φ with weight function φ and whose fiber M is complete, simply connected and φ-parabolic. If the angle function Θ of Σ^n is bounded, and the warping function on Σ^n, $f(h)$, is bounded and satisfies $\inf_{\Sigma^n} f(h) > 0$, then Σ^n is φ-parabolic.*

We recall that a GRW is said to be *static* if its warping function is constant, which, without loss of generality, can be supposed equal to 1.

Corollary 7.2 *Let Σ^n be a complete spacelike hypersurface in a static spatially weighted GRW spacetime M_φ with weight function φ and whose fiber M is complete with φ-parabolic universal Riemannian covering. If the angle function Θ of Σ^n is bounded, then Σ^n is φ-parabolic.*

As an application of Theorem 7.4, we will prove some uniqueness results concerning spacelike slices among spacelike hypersurfaces immersed in a spatially weighted GRW spacetime.

Theorem 7.5 *Let M_φ be a spatially weighted GRW spacetime whose fiber M is complete with φ-parabolic universal Riemannian covering, and such that the warping function f is monotone. The only φ-maximal complete spacelike hypersurfaces contained in a timelike bounded region of M and with bounded angle function Θ are the slices $\{t_0\} \times M$, where $t_0 \in I$ is such that $f'(t_0) = 0$.*

Proof Let Σ^n be such a spacelike hypersurface. By Lemma 5.4, we obtain

$$\Delta_\varphi \mathcal{F}(h) = -n f'(h).$$

Consequently, the monotonicity of f implies that $\Delta_\varphi \mathcal{F}(h)$ is globally either nonpositive or nonnegative.

Since Σ^n is contained in a timelike bounded region of M and the warping function f is monotone, the function $\mathcal{F}(h)$ is clearly bounded on Σ^n.

By Theorem 7.4 we know that Σ^n is φ-parabolic, so $\mathcal{F}(h)$ is constant in Σ^n. Consequently, h must also be constant in Σ^n. □

Now, let us assume that the ambient space obeys the so-called *null convergence condition* (NCC), that is

$$\mathrm{Ric}^M \geq (n-1) f^2 (\log f)'' g_M. \tag{7.12}$$

The above condition is equivalent to the fact that the Ricci curvature of M is nonnegative on null or lightlike directions; see [243] for additional comments and remarks.

In our next result we will also assume that the weight function φ is convex, that is, $\overline{\mathrm{Hess}}\,\varphi \geq 0$.

Theorem 7.6 *Let M_φ be a spatially weighted GRW spacetime satisfying (7.12), with convex weight function φ, and whose fiber M is complete with φ-parabolic universal Riemannian covering. Let Σ^n be a complete φ-maximal spacelike hypersurface immersed in M, with bounded angle function Θ and such that the restriction $f(h)$ on Σ^n of the warping function f of M satisfies*

(i) $\sup f(h) < +\infty$ *and*
(ii) $\inf f(h) > 0.$

Then Σ^n is totally geodesic. In addition, if either the inequality (7.12) is strict for all nonzero vector fields on M or φ is strictly convex on M, then Σ^n is a slice Σ_{t_0}, where $t_0 \in I$ is such that $f'(t_0) = 0$.

Proof By Lemma 5.4, we have that the φ-Laplacian of the bounded function $f(h)\Theta$ is given by

$$\Delta_\varphi(f(h)\Theta) = f(h)\Theta|A|^2 + f(h)\Theta\overline{\mathrm{Hess}}\,\varphi(N, N)$$
$$+ f(h)\Theta \left(\mathrm{Ric}^M(N^M, N^M) - (n-1)(\log f)''(h)|\nabla h|^2 \right). \tag{7.13}$$

7.2 Uniqueness Results for Spacelike Hypersurfaces in Spatially...

Since φ is convex and the null convergence condition (7.12) holds, it follows that $\Delta_\varphi(f(h)\Theta) \leq 0$. Now, Theorem 7.4 assures that Σ^n is φ-parabolic, so $f(h)\Theta$ must be constant.

Therefore, thanks to (7.13), we infer that $|A| \equiv 0$, that is, Σ^n is totally geodesic,

$$\overline{\mathrm{Hess}}\,\varphi(N, N) = \mathrm{Hess}^M \varphi(N^M, N^M) = 0, \tag{7.14}$$

and

$$\mathrm{Ric}^M(N^M, N^M) - (n-1)(\log f)''(h)|\nabla h|^2 = 0.$$

Consequently, if the inequality (7.12) is strict, or if φ is strictly convex on M, then (7.13) also gives that $|N^M| = |\nabla h| = 0$ on Σ^n, that is, Σ^n is a slice. □

We recall that the Gaussian space $\mathbb{G}^n$ corresponds to the Euclidean space $\mathbb{R}^n$ endowed with the Gaussian probability density

$$e^{-\varphi(x)} = (2\pi)^{-\frac{n}{2}} e^{-\frac{|x|^2}{2}}.$$

By [203, Corollary 3] we have that $\mathbb{G}^n$ has finite φ-volume. Consequently, taking into account of [211, Remark 3.8], we conclude that $\mathbb{G}^n$ is φ-parabolic. On the other hand, it is not difficult to verify that the weight function φ of $\mathbb{G}^n$ is strictly convex.

Consequently, by Theorem 7.6 we get the next property.

Corollary 7.3 *The only complete φ-maximal spacelike hypersurfaces of $-\mathbb{R} \times \mathbb{G}^n$ having bounded angle function are the spacelike hyperplanes $\{t\} \times \mathbb{G}^n$.*

Now, considering as ambient space a static GRW spacetime, we obtain the following result.

Theorem 7.7 *Let $\overline{M}_\varphi$ be a static spatially weighted GRW spacetime, whose fiber M is complete with φ-parabolic universal Riemannian covering and such that its Bakry-Émery-Ricci tensor Ric^M_φ is nonnegative. Let Σ^n be a complete spacelike hypersurface immersed in $\overline{M}_\varphi$ with constant φ-mean curvature, H_φ. If the angle function Θ of Σ^n is bounded, then Σ^n is totally geodesic. In addition, if Ric^M_φ is definite positive at some point of Σ^n, then Σ^n is a slice $\Sigma_t \subset \overline{M}_\varphi$.*

Proof By Lemma 5.4, in addition to (5.2) and (7.14), we get that

$$\Delta_\varphi \Theta = (\mathrm{Ric}^M_\varphi(N^M, N^M) + |A|^2)\Theta. \tag{7.15}$$

Consequently, since Ric_φ^M is nonnegative and Θ is negative and bounded on Σ^n, we can apply Corollary 7.2 to conclude that Θ is constant on Σ^n. Thus, thanks to (7.15), we get $|A| \equiv 0$, that is, Σ^n is totally geodesic.

Moreover, if Ric_φ^M is definite positive at some $p \in M$, considering once more equation (7.15), we conclude that $\Theta \equiv -1$ on Σ^n, which means that Σ^n is a slice of the manifold M. □

We recall that Xin [295] and Aiyama [3] proved simultaneously and independently that the only spacelike hypersurfaces of the Lorentz-Minkowski space $\mathbb{L}^{n+1} = -\mathbb{R} \times \mathbb{R}^n$, with constant mean curvature and having bounded angle function, are the spacelike hyperplanes. Hence, by Theorem 7.7 we get the following extension of this Xin-Aiyama result.

Corollary 7.4 *The only complete spacelike hypersurfaces of $-\mathbb{R} \times \mathbb{G}^n$, with constant φ-mean curvature and having bounded angle function, are the spacelike hyperplanes $\{t\} \times \mathbb{G}^n$.*

Proceeding, we will use a Bochner's formula due to Wei and Wylie [294] to obtain the following theorem.

Theorem 7.8 *Let M_φ be a static spatially weighted GRW spacetime endowed with a convex weight function φ and whose fiber M is complete, with nonnegative sectional curvature, and such that its universal Riemannian covering is φ-parabolic. Let Σ^n be a complete spacelike hypersurface lying in a semi-space of M and with constant φ-mean curvature H_φ. If the angle function Θ is bounded, then Σ^n is a slice Σ_t.*

Proof Since M has nonnegative sectional curvature, it follows by (3.3) and (3.4) of [142] that

$$\mathrm{Ric}(X, X) \geq nH g_{\Sigma^n}(AX, X) + |AX|^2. \tag{7.16}$$

On the other hand, taking into account that

$$\mathrm{Hess}\, \varphi(X, X) = \overline{\mathrm{Hess}}\, \varphi(X, X) - \overline{g}(\overline{\nabla}\varphi, N) g_{\Sigma^n}(AX, X),$$

by using the convexity of the weight function φ we get

$$\mathrm{Hess}\, \varphi(X, X) \geq -\overline{g}(\overline{\nabla}\varphi, N) g_{\Sigma^n}(AX, X). \tag{7.17}$$

Now, by (7.16) and (7.17) we have the following lower bound for Ric_φ,

$$\mathrm{Ric}_\varphi(X, X) \geq nH_\varphi g_{\Sigma^n}(AX, X) + |AX|^2. \tag{7.18}$$

7.2 Uniqueness Results for Spacelike Hypersurfaces in Spatially...

Inequality (7.18) yields

$$\text{Ric}_\varphi(\nabla h, \nabla h) \geq n H_\varphi g_{\Sigma^n}(A(\nabla h), \nabla h) + |A(\nabla h)|^2. \tag{7.19}$$

Since H_φ is constant, we have

$$\nabla \Delta_\varphi h = -n H_\varphi A(\nabla h). \tag{7.20}$$

On the other hand, by Bochner's formula (5.3),

$$\frac{1}{2}\Delta_\varphi|\nabla h|^2 = |\text{Hess } h|^2 + g_{\Sigma^n}(\nabla h, \nabla \Delta_\varphi h) + \text{Ric}_\varphi(\nabla h, \nabla h). \tag{7.21}$$

Consequently, by (1.12), (7.19), (7.20) and (7.21) we get

$$\frac{1}{2}\Delta_\varphi \Theta^2 = \frac{1}{2}\Delta_\varphi|\nabla h|^2 \geq |\text{Hess } h|^2 \geq 0. \tag{7.22}$$

Thus, by Corollary 7.2, we have that Θ is constant, and, thanks to (7.22), we get $|\text{Hess } h|^2 = 0$ in Σ^n.

Then, since $n|\text{Hess } h|^2 \geq (\Delta h)^2$, we have that h is harmonic. So, since $\Delta h = -nH\Theta$, we also get that $H = 0$ in Σ^n, and by (7.16) we have that Σ^n has nonnegative Ricci curvature.

Therefore, since Σ^n lies in a semi-space of M, we can apply the strong Liouville property due to Yau [297] (see also [229, Theorem 4.8]) to conclude that h must be constant and, hence, Σ^n is a slice of M. $\square$

Inspired by [6], we now consider *spatially weighted steady state type spacetimes*, that is, GRW spacetimes of the type $-\mathbb{R} \times_{e^t} M$ whose fiber M is endowed with a weight function φ.

In this setting, the next result is an extension of [6, Theorem 8].

Theorem 7.9 *Let M_φ be a spatially weighted steady state type spacetime, whose fiber M is complete, with nonnegative sectional curvature, and let Σ^n be a complete spacelike hypersurface that lies in a timelike bounded region of M. Suppose that $|\overline{\nabla}\varphi|$ is bounded on Σ^n. If the φ-mean curvature H_φ of Σ^n is constant and the angle function Θ is bounded, then $H_\varphi = 1$. In addition, if the universal Riemannian covering of M is φ-parabolic, then Σ^n is a slice $\{t\} \times M$.*

Proof We claim that the mean curvature H of Σ^n is bounded. Indeed, since $\overline{g}(\overline{\nabla}\varphi, \partial_t) = 0$, by (5.16) we have that

$$n|H| \leq n|H_\varphi| + |\overline{g}(\overline{\nabla}\varphi, N)| = n|H_\varphi| + |\overline{g}(\overline{\nabla}\varphi, N^M)|. \tag{7.23}$$

Thus, by (7.23) we get

$$n|H| \leq n|H_\varphi| + |\overline{\nabla}\varphi|(\Theta^2 - 1). \tag{7.24}$$

Consequently, since H_φ is constant, and both $\overline{\nabla}\varphi$ and Θ are supposed to be bounded on Σ^n, it follows by (7.24) that H is also bounded on Σ^n.

On the other hand, by (16) of [6] we have that the Ricci curvature of Σ^n satisfies

$$\mathrm{Ric}(X, X) \geq n - 1 - \frac{n^2 H^2}{4}.$$

So, we conclude that the Ricci curvature of Σ^n is bounded from below.

Thus, we can apply the maximum principle of Omori [250] and Yau [297] to guarantee that there exists a sequence $\{p_k\} \subset \Sigma^n$ such that

$$\lim_k h(p_k) = \sup_{\Sigma^n} h, \quad \lim_k |\nabla h(p_k)| = 0 \quad \text{and} \quad \limsup_k \Delta h(p_k) \leq 0.$$

Now, by (1.12) and (5.4), we also get that

$$\lim_k \Theta(p_k) = -1 \quad \text{and} \quad \limsup_k \Delta_\varphi h(p_k) \leq 0.$$

Hence, taking into account formula (5.17), arguing as in the proof of [6, Theorem 8], we obtain that $H_\varphi = 1$. Furthermore, by (5.18) we also have that

$$\Delta_\varphi e^h = -ne^h(1 + \Theta) \geq 0.$$

Therefore, assuming now that the universal Riemannian covering of M is φ-parabolic, we can apply Theorem 7.4 to conclude that h is constant on Σ^n. □

The next result extends [121, Theorem 5.3].

Theorem 7.10 *Let $\overline{M}_\varphi$ be a spatially weighted steady state type spacetime whose fiber M is complete, with φ-parabolic universal Riemannian covering, and let Σ^n be a complete spacelike hypersurface that lies in a timelike bounded region of M, with $1 \leq H_\varphi \leq \alpha$ for a certain constant $\alpha \geq 1$. If the angle function Θ of Σ^n satisfies $-\Theta \leq H_\varphi$, then Σ^n is a slice Σ_t.*

Proof By (5.17) we get

$$\Delta_\varphi e^{-h} = e^{-h}\left(|\nabla h|^2 - \Delta_\varphi h\right)$$

$$\leq ne^{-h}\left(|\nabla h|^2 + 1 + H_\varphi \Theta\right) = ne^{-h}\Theta\left(\Theta + H_\varphi\right).$$

7.2 Uniqueness Results for Spacelike Hypersurfaces in Spatially... 207

Hence, our hypothesis on Θ guarantees that the function e^{-h} is a φ-superharmonic positive function on Σ^n. Therefore, we can apply once more Theorem 7.4 to get that h is constant on Σ^n. □

7.2.4 Constant φ-Mean Curvature Spacelike Hypersurfaces

Our aim in this subsection is to study the uniqueness of spacelike slices among complete spacelike hypersurfaces immersed in a weighted Lorentzian product space of type $-\mathbb{R} \times M_\varphi^n$, whose Riemannian fiber M^n has nonnegative Bakry-Émery-Ricci tensor and such that the Hessian of the weight function φ is bounded from below.

By assuming that the weighted mean curvature H_φ of a given spacelike hypersurface Σ^n is constant, as well as some appropriated constraints on the norm of the gradient of the height function of Σ^n, we prove that such a hypersurface must be a slice of the ambient spacetime.

The results presented here are contained in [149].

Theorem 7.11 *Let $-\mathbb{R} \times M_\varphi^n$ be a weighted Lorentzian product space, whose Riemannian fiber M^n is complete, has sectional curvatures bounded from below, and its Bakry-Émery-Ricci tensor satisfies $\mathrm{Ric}_\varphi^M \geq c$ for some positive constant c. Suppose that the Hessian of the weighted function φ is bounded from below. Moreover, let Σ^n be a complete spacelike hypersurface immersed in $-\mathbb{R} \times M_\varphi^n$ with constant φ-mean curvature H_f. If Θ is bounded on Σ^n, then Σ^n is a slice Σ_t for some $t \in \mathbb{R}$.*

Proof Since H_φ is constant, Lemma 5.4 gives

$$\Delta_\varphi \Theta = (\mathrm{Ric}_\varphi^M(N^M, N^M) + |A|^2)\Theta. \tag{7.25}$$

Moreover, since Θ is bounded and the fiber M^n has sectional curvatures bounded from below, by Lemma 5.5 we have that the Bakry-Émery-Ricci curvature of Σ^n is also bounded from below.

Thus, we can apply Lemma 7.1 to the function Θ to obtain a sequence of points $\{p_k\} \subset \Sigma^n$ such that

$$\lim_k \Theta(p_k) = \inf_{\Sigma^n} \Theta \quad \text{and} \quad \liminf_k \Delta_\varphi \Theta(p_k) \geq 0.$$

Consequently, since the Bakry-Émery-Ricci curvature Ric_φ^M of M^n is positive, taking into account that $\Theta \leq -1$, by (7.25), up to a subsequence, we get

$$0 \leq \liminf_k \Delta_\varphi \Theta(p_k) = \lim_k \left[\mathrm{Ric}_\varphi^M(N^M, N^M) + |A|^2\right](p_k) \inf_{p \in \Sigma^n} \Theta(p) \leq 0. \tag{7.26}$$

Furthermore, the main assumption on the Bakry-Émery-Ricci tensor of M^n yields

$$\mathrm{Ric}_\varphi^M(N^M, N^M) \geq c|N^M|_M^2 = c|\nabla h|^2. \tag{7.27}$$

Thus, by (7.26) and (7.27) we get that $\lim_k |\nabla h(p_k)| = 0$. Therefore, taking into account (1.12), we conclude that

$$\inf_{p \in \Sigma^n} \Theta(p) = -1.$$

Hence, Σ^n must be a slice $\Sigma_t = \{t\} \times M^n$ for some $t \in \mathbb{R}$. □

Remark 7.3 We emphasize that a complete weighted manifold $(\Sigma^n, g, e^{-\varphi}dV)$ whose Bakry-Émery-Ricci tensor satisfies $\mathrm{Ric}_\varphi \geq c$, for some positive constant c, is not necessarily compact. For instance, the Gaussian space $\mathbb{G}^n$ gives a simple prototype for which our claim is true. On the other hand, under the additional hypothesis that the weighted function φ is bounded, an extension of a Myers' Theorem due to Wei and Willie guarantees the compactness of Σ^n; see [294, Theorem 1.4].

The following result holds.

Theorem 7.12 *Let $-\mathbb{R} \times M_\varphi^n$ be a weighted Lorentzian product space. Suppose that the fiber M^n is complete and has sectional curvatures bounded from below and nonnegative Bakry-Émery-Ricci tensor, Ric_φ^M, and that the Hessian of the weight function φ is bounded from below. Moreover, let Σ^n be a complete spacelike hypersurface immersed in $-\mathbb{R} \times M_\varphi^n$ with constant φ-mean curvature H_φ and bounded second fundamental form A. If*

$$|\nabla h| \leq \alpha |A|^\beta, \tag{7.28}$$

for some positive constants α and β, then Σ^n is a slice $\Sigma_t = \{t\} \times M_\varphi^n$ for some $t \in \mathbb{R}$.

Proof By (7.28) the function Θ is bounded on Σ^n. Hence, arguing as in the proof of Theorem 7.11 one has that (7.26) still holds. Consequently, we get that $\lim_k |A(p_k)| = 0$. Thus, thanks to (7.28), we obtain that $\lim_k |\nabla h(p_k)| = 0$. Therefore, by (1.12) we conclude that $\inf_{p \in \Sigma^n} \Theta(p) = -1$. Hence, Σ^n must be a slice $\Sigma_t = \{t\} \times M_\varphi^n$ for some $t \in \mathbb{R}$, as desired. □

The next result can be achieved requiring that Σ^n is φ-parabolic.

Theorem 7.13 *Let $-\mathbb{R} \times M_\varphi^n$ be a weighted Lorentzian product space whose fiber M^n is complete and has nonnegative Bakry-Émery-Ricci tensor, Ric_φ^M. Moreover, let Σ^n be a complete spacelike hypersurface immersed in $-\mathbb{R} \times M_\varphi^n$ with constant φ-mean curvature*

H_φ. If Σ^n is φ-parabolic and Θ is bounded on Σ^n, then Σ^n is totally geodesic. In addition, if Ric_φ^M is strictly positive, then Σ^n is a slice Σ_t for some $t \in \mathbb{R}$.

Proof Let us consider on Σ^n the bounded function $u = -\Theta$. By Lemma 5.4 we have that

$$\Delta_\varphi u = -(\mathrm{Ric}_\varphi^M(N^M, N^M) + |A|^2)\Theta \geq 0. \tag{7.29}$$

Since Σ^n is φ-parabolic, by (7.29), we get that u is constant.

Thus, since $\Theta < 0$ on Σ^n, and taking into account that $\mathrm{Ric}_\varphi^M \geq 0$, thanks to (7.29) we get that $|A| \equiv 0$, that is, Σ^n is totally geodesic.

Moreover, in the case that Ric_φ^M is strictly positive, we conclude that N^M vanishes identically on Σ^n, which means that $N = \partial_t$ on Σ^n. So, in this case, Σ^n must be a slice Σ_t for some $t \in \mathbb{R}$. □

Remark 7.4 As observed in [211, Remark 3], a sufficient condition for a weighted manifold Σ^n_φ to be φ-parabolic is that Σ^n_φ is geodesically complete and

$$\mathrm{vol}_\varphi(\partial B_r(o))^{-1} \notin L^1_\varphi(\Sigma^n),$$

for some origin $o \in \Sigma^n$. Here, $B_r(o)$ denotes the geodesic ball of Σ^n centered at o and of radius $r > 0$, and $L^1_\varphi(\Sigma^n)$ stands for the set of integrable functions on Σ^n with respect to the weighted volume element $d\mu = e^{-\varphi}d\Sigma^n$. Furthermore, φ-parabolicity is also guaranteed if we assume the stronger condition

$$\mathrm{vol}_\varphi(B_r) = O(r^2), \quad \text{as} \quad r \to +\infty.$$

In particular, φ-parabolicity holds if Σ^n has finite φ-volume. Finally, if Σ^n is complete noncompact with $\mathrm{Ric}_\varphi \geq 0$, [293, Theorem 1.1] asserts that a sufficient condition for Σ^n to have finite φ-volume is that the space of L^2_φ harmonic one-forms on Σ^n is nontrivial.

7.3 On the Existence of φ-Maximal Spacelike Hypersurfaces

In this section, we prove the following nonexistence result firstly showed in [249].

Theorem 7.14 Let $\overline{M}_\varphi^{n+1} = -I \times M_\varphi^n$ be a weighted Lorentzian product space whose fiber M^n is noncomplete with nonnegative sectional curvature and such that the weight function φ is bounded and convex. Then there is no complete noncompact φ-maximal spacelike hypersurfaces immersed in M_φ^n.

As a consequence of Theorem 7.14, which will be proved in Sect. 7.3.2, we get the following Calabi-Bernstein type result.

Theorem 7.15 *Let $\overline{M}^{n+1} = -I \times M^n$ be a Lorentzian product space whose fiber M^n is complete with nonnegative sectional curvature. The only maximal noncompact spacelike hypersurfaces of $\overline{M}^{n+1}$ are the slices.*

7.3.1 Additional Key Lemmas

In this subsection, we present two key lemmas which will be used to prove Theorem 7.14.

Let $(\Sigma, g, d\mu = e^{-\varphi}dV)$ be an n-dimensional weighted complete Riemannian manifold. Take any point $x \in \Sigma$ and denote the volume form in geodesic polar coordinates centered at x by

$$dV\big|_{\mathrm{Exp}_x(r\zeta)} = J(x, r, \zeta) dr d\zeta,$$

where $r > 0$ and $\zeta \in T_x\Sigma$ is a unitary tangent vector at x. It is well-known that if $y \in \Sigma$ is any point such that $y = \mathrm{Exp}_x(r\zeta)$, then

$$\Delta_\varphi d(x, y) = \frac{J'_\varphi(x, r, \zeta)}{J_\varphi(x, r, \zeta)},$$

where $J_\varphi(x, r, \zeta) = e^{-\varphi} J(x, r, \zeta)$ is the φ-volume form in geodesic polar coordinates. For a fixed point $p \in \Sigma$ and $R > 0$, define

$$A(R) = \sup_{B_p(3R)} |\varphi(x)|. \tag{7.30}$$

For a subset $\Omega \subseteq \Sigma$, we will denote by $V(\Omega)$ the volume of Ω with respect to the usual volume form dV, and by $V_\varphi(\Omega)$ the φ-volume of Ω, namely $V_\varphi(\Omega) = \int_\Omega d\mu$. If Σ has nonnegative Bakry-Émery-Ricci curvature, then along any minimizing geodesic starting from $x \in B_p(R)$, we have

$$\frac{J_\varphi(x, r_2, \zeta)}{J_\varphi(x, r_1, \zeta)} \leq e^{4A} \left(\frac{r_2}{r_1}\right)^{n-1} \tag{7.31}$$

for any $0 < r_1 < r_2 < R$; in particular, for any $0 < r_1 < r_2 < R$,

$$\frac{V_\varphi(B_x(r_2))}{V_\varphi(B_x(r_1))} \leq e^{4A} \left(\frac{r_2}{r_1}\right)^n. \tag{7.32}$$

7.3 On the Existence of φ-Maximal Spacelike Hypersurfaces

where $A = A(R)$ is defined in (7.30); see [247, Lemma 2.1].

If Σ is noncompact, the comparison inequality (7.31) guarantees that there exist constants $\nu > 2$, C_1 and C_2, depending only on n, such that for all $u \in C_0^\infty(B_x(r))$ the following local Sobolev inequality holds,

$$\left(\int_{B_x(r)} |u|^{\frac{2\nu}{\nu-2}} d\mu\right)^{\frac{\nu-2}{\nu}} \leq C_1 \frac{e^{C_2 A_r 2}}{V_\varphi(B_x(r))^{\frac{2}{\nu}}} \int_{B_x(r)} \left(|\nabla u|^2 + r^{-2}|u|^2\right) d\mu, \tag{7.33}$$

where $x \in B_p(R)$ and $0 < r < R$; see [106, Lemma 2.3].

Such a family of Sobolev inequalities can be used to obtain a mean value inequality for subsolutions to the φ-heat equation as in [270, Theorem 5.2.9]; see [106, Lemma 2.5].

Lemma 7.8 *Let* $(\Sigma, g, d\mu = e^{-\varphi} dV)$ *be an n-dimensional weighted complete Riemannian manifold which satisfies the local Sobolev inequality (7.33) for all* $u \in C_0^\infty(B_o(\rho))$ *and* $0 < \rho \leq R$, *where* $o \in \Sigma$ *is a fixed origin. Fix* $q \in (0, +\infty)$, *and let* v *be a positive subsolution of the φ-heat equation, that is,*

$$Lv \geq 0,$$

in the cylinder $Q = B_o(r) \times (s - r^2, s)$ *for some* $s \in \mathbb{R}$ *and* $0 < r < R$, *where*

$$L = \Delta_\varphi + \frac{\partial}{\partial t}.$$

Then, for any $0 < \delta < \delta' \leq 1$, *there exist constants* $C_3 = C_3(n, \sup|\varphi|, \nu, q)$ *and* $C_4 = C_4(n, \nu, q)$ *such that*

$$\sup_{Q_\delta} v^q \leq C_3 \frac{e^{C_4 A(R)}}{(\delta' - \delta)^{\nu+2} r^2 V_\varphi(B_o(r))} \int_{Q_{\delta'}} v^q d\mu dt,$$

where $Q_z = B_o(zr) \times (s - zr^2, s)$.

Our second key lemma gives sufficient conditions to guarantee that the Bakry-Émery-Ricci curvature of a φ-maximal spacelike hypersurface immersed in a weighted Lorentzian product is nonnegative.

Lemma 7.9 *Let* $\overline{M}_\varphi^{n+1} = -I \times M_\varphi^n$ *be a Lorentzian weighted product whose fiber* M^n *has nonnegative sectional curvatures and with convex weight function. Moreover, let* $\psi : (\Sigma^n, g_{\Sigma^n}) \to (\overline{M}_\varphi^{n+1}, \overline{g})$ *be an φ-maximal spacelike hypersurface with orientation* N *and shape operator* A. *Then the Bakry-Émery-Ricci curvature* Ric_φ *of* Σ^n *is nonnegative.*

Proof The Gauss equation (5.23) yields

$$\mathrm{Ric}(X, X) = \sum_{i=1}^{n} \overline{g}(\overline{R}(X, E_i)X, E_i) + nH g_{\Sigma^n}(AX, X) + g_{\Sigma^n}(AX, AX) \tag{7.34}$$

for every $X, Y, Z \in \mathfrak{X}(\Sigma^n)$, where Ric and $\overline{R}$ stand for the Ricci curvature and the curvature tensor of $\overline{M}^{n+1}$, respectively, and H is the mean curvature function of Σ^n.

Moreover, we have

$$\overline{R}(X, Y)Z = R^M(X^M, Y^M)Z^M,$$

where R^M and $(\cdot)^M$ denote, respectively, the curvature tensor and the projection of a vector field onto the fiber M^n; see [251, Proposition 7.42].

Hence

$$\overline{g}(\overline{R}(X, E_i)X, E_i) = K^M(X^M, E_i^M) \left(g_M(X^M, X^M) g_M(E_i^M, E_i^M) - g_M(X^M, E_i^M) \right),$$

where K^M stands for the sectional curvature of M^n. Since M^n is non-negatively curved, the above equation and (7.34) yield

$$\mathrm{Ric}(X, X) \geq nH g_{\Sigma^n}(AX, X) + g_{\Sigma^n}(AX, AX). \tag{7.35}$$

Since φ is convex, we have

$$\mathrm{Hess}\,\varphi(X, X) = \overline{\mathrm{Hess}}\,\varphi(X, X) - \overline{g}(\overline{\nabla}\varphi, N) g_{\Sigma^n}(AX, X)$$

$$\geq -\overline{g}(\overline{\nabla}\varphi, N) g_{\Sigma^n}(AX, X). \tag{7.36}$$

Therefore, by (5.2), (5.16), (7.35) and (7.36), we obtain

$$\mathrm{Ric}_\varphi(X, X) \geq nH_\varphi g_{\Sigma^n}(AX, X) + g_{\Sigma^n}(AX, AX). \tag{7.37}$$

The result follows by (7.37) taking into account the hypothesis that Σ^n is φ-maximal.
$\square$

7.3.2 Proof of Theorem 7.14

Let Σ^n be a complete noncompact spacelike φ-maximal hypersurface immersed in $\overline{M}_\varphi^{n+1} = -I \times M_\varphi^n$. We observe that the height function of Σ^n, $h = \pi_I|_{\Sigma^n}$, is a φ-harmonic function.

7.3 On the Existence of φ-Maximal Spacelike Hypersurfaces

By Lemma 5.4-(i), the φ-Laplacian of h on Σ^n is given by $\Delta_\varphi h = -n H_\varphi \overline{g}(N, \partial_t)$ and the φ-harmonicity of h follows by the fact that Σ^n is supposed to be φ-maximal. We claim that h must be bounded from above.

We argue by contradiction as in the proof of [212, Theorem 0.13]. Suppose that $\Sigma^n \cap \Sigma_t \neq 0$, $\forall t \in I$. For a fixed $t \in I$, let

$$\Sigma_t = \{ p \in \Sigma^n : h(p) \geq t \}.$$

By Sard's Theorem, we can suppose that t is a regular value of $h|_{\text{int } \Sigma^n}$, so that Σ_t is a smooth complete manifold with boundary

$$\partial \Sigma_t = \{ p \in \Sigma^n : h(p) = t \},$$

and exterior unit normal $\nu_t = -\nabla h / |\nabla h|$.

For any $\epsilon > 0$, let us define on Σ_t the function

$$h_\epsilon = \max\{h, t + \epsilon\}.$$

We claim that h_ϵ is φ-harmonic on Σ_t. Indeed, set

$$\Sigma_1 = \{ p \in \Sigma_t : h(p) > t + \epsilon \} \quad \text{and} \quad \Sigma_2 = \{ p \in \Sigma_t : h(p) = t + \epsilon \},$$

as well as

$$\Sigma_3 = \{ p \in \Sigma_t : t < h(p) < t + \epsilon \}.$$

Then, $h_\epsilon = h$ on Σ_1 and h_ϵ is constant (equals to $t + \epsilon$) on Σ_3, so $\Delta_\varphi h_\epsilon = 0$ on both Σ_1 and Σ_3.

Now, we notice that the tranversality of Σ^n and ∂_t, i.e. the function $\overline{g}(N, \partial_t)$ is negative on Σ^n, implies a monotonic behavior of the height function h, which in turn guarantees that h_ϵ is smooth on Σ_2 and on $\partial \Sigma_t$.

So, we also have that $\Delta_\varphi h_\epsilon = 0$ on both Σ_2 and $\partial \Sigma_t$ by continuity.

Whence, on noting that $h_\epsilon \equiv t + \epsilon$ on $\partial \Sigma_t$, by the maximum principle for the φ-Laplacian, we obtain that $t \leq h \leq t + \epsilon$ on Σ_t. Since this holds for every $\epsilon > 0$, we conclude that $h \equiv t$ on Σ_t. Since h is unbounded from above we have a contradiction.

Clearly, the above procedure can be also applied by proving that h is bounded from below.

We now observe that h has sublinear growth, that is,

$$\lim_{p \to \infty} \frac{|h|(p)}{r(p)} = 0, \tag{7.38}$$

where $r(p) = d(p, p_0)$ is the distance from a fixed point $p_0 \in \Sigma^n$.

This fact follows by the noncompactness of Σ^n and taking into account that

$$\alpha = \sup_{\Sigma^n} |h| < +\infty.$$

Indeed

$$0 \leq \frac{|h|(p)}{r(p)} \leq \frac{\alpha}{r(p)} \to 0, \quad \text{as} \quad p \to +\infty.$$

By $\Delta_\varphi h = 0$, Lemma 7.9 and the weighted Bochner formula (7.21), one has that $|\nabla h|^2$ is φ-subharmonic. Now, let us take $q = 1$ in Lemma 7.8, since every φ-subharmonic function is also a subsolution to the φ-heat equation, we get the following mean value inequality

$$\sup_{B_p(R/2)} |\nabla h|^2 \leq \frac{\beta}{R^2 V_f(B_p(R))} \int_{B_p(R)} |\nabla h|^2 d\mu, \tag{7.39}$$

where β is a constant depending only on n and $\sup_{\Sigma^n} |\varphi|$.

Now, we argue exactly as in [247, Theorem 3.2] applying a standard cut-off argument. To this aim, let us consider a cut-off function χ such that $\chi = 1$ on $B_p(R)$, $\chi = 0$ on $\Sigma^n \setminus B_p(2R)$ and $|\nabla \chi| \leq \frac{\beta}{R}$. Integrating by parts and by using $\Delta_\varphi h = 0$, we obtain

$$\int_{\Sigma^n} |\nabla h|^2 \chi^2 d\mu = -2 \int_{\Sigma^n} h\chi \langle \nabla h, \nabla \chi \rangle d\mu$$

$$\leq 2 \int_{\Sigma^n} |h| \chi |\langle \nabla h, \nabla \chi \rangle| d\mu$$

$$\leq 2 \left(\int_{\Sigma^n} |\nabla h|^2 \chi^2 d\mu \right)^{1/2} \left(\int_{\Sigma^n} h^2 |\nabla \chi|^2 d\mu \right)^{1/2}.$$

Thanks to (7.32), it follows that

$$\int_{\Sigma^n} |\nabla h|^2 \chi^2 d\mu \leq 4 \int_{\Sigma^n} h^2 |\nabla \chi|^2 d\mu$$

$$\leq \frac{\beta^2}{R^2} \int_{B_p(2R) \setminus B_p(R)} h^2 d\mu$$

$$\leq \frac{\beta^2}{R^2} \left(\sup_{B_p(2R)} h^2 \right) V_\varphi(B_p(2R)) \leq \frac{\gamma}{R^2} \left(\sup_{B_p(2R)} h^2 \right) V_\varphi(B_p(R)),$$

for a positive constant γ.

Taking into account (7.38), we obtain

$$\lim_{R\to\infty} \frac{\beta}{R^2 V_\varphi(B_p(R))} \int_{B_p(R)} |\nabla h|^2 d\mu = 0,$$

so that, by (7.39), $|\nabla h| = 0$ on Σ^n, that is, Σ^n is a slice of $\overline{M}^{n+1}$. Since the fiber M^n is noncomplete, this fact cannot occur.

Remark 7.5 As a matter of fact, Munteanu and Wang [247] already established that a sublinear growth φ-harmonic function in a complete noncompact weighted manifold, with bounded weight function φ, must be constant. Since the main approach is based on a mean value inequality slightly different by inequality (3.14) in [247], we have decided to present the full argument.

7.4 Calabi-Bernstein Type Results

Let $\Omega \subseteq M^n$ be a connected domain, and let $u \in C^\infty(\Omega)$ be a smooth function. Then, $\Sigma(u)$ will denote the vertical graph over Ω determined by u, that is,

$$\Sigma(u) = \{(u(x), x) : x \in \Omega\} \subset \overline{M}^{n+1} = -I \times_f M^n.$$

The graph is said to be *entire* if $\Omega = M^n$. The metric induced on Ω from the Lorentzian metric of the ambient space via $\Sigma(u)$ is

$$g_{\Sigma(u)} = -du^2 + f^2(u)g_M. \tag{7.40}$$

Now, a graph $\Sigma(u)$ is a spacelike hypersurface if and only if $|Du|_M^2 < f^2(u)$, where Du is the gradient of u in M^n and $|Du|_M$ denotes its norm, both with respect to the metric g_M. Moreover, if M^n is a simply connected manifold, every complete spacelike hypersurface Σ^n immersed in M^n such that the warping function f is bounded on Σ^n is an entire spacelike graph over M^n; see [24, Lemma 3.1].

In particular, this happens for complete spacelike hypersurfaces contained in a timelike bounded region of M^n. It is interesting to observe that, in contrast to the case of graphs into a Riemannian space, an entire spacelike graph $\Sigma(u)$ in a Lorentzian spacetime is not necessarily complete, in the sense that the induced Riemannian metric is not necessarily complete on M^n.

However, it can be proved that if M^n is complete and $|Du|_M^2 \leq f^2(u) - c$ for a certain positive constant c, then $\Sigma(u)$ is complete. Although a particular case of this claim is proved in [9, Theorem 4.1], and the general proof is given in [10, Proposition 9], we will expose it here for the sake of completeness.

Proposition 7.1 Let M^n be a complete Riemannian manifold, and let $\Sigma(u)$ an entire spacelike vertical graph in $\overline{M}^{n+1} = -I \times_f M^n$. If

$$|Du|_M^2 \leq f^2(u) - c$$

for a certain positive constant c, then $\Sigma(u)$ is complete.

Proof By (7.40), the Cauchy-Schwarz inequality and the main assumption yield

$$g_{\Sigma(u)}(X, X) = -g_M(Du, X)^2 + f^2(u)g_M(X, X)$$
$$\geq \left(f^2(u) - |Du|_M^2\right) g_M(X, X) \geq c g_M(X, X),$$

for every $X \in \mathfrak{X}(\Sigma(u))$.

This implies that $L \geq \sqrt{c} L_M$, where L and L_M denote the length of a curve on $\Sigma(u)$ with respect to the Riemannian metrics $g_{\Sigma(u)}$ and g_M, respectively.

As a consequence, since M^n is complete, the induced metric on $\Sigma(u)$ by the metric of M^n is also complete. □

Remark 7.6 Let us observe that a constant c, as in Proposition 7.1, clearly exists if M^n is assumed to be compact and the spacelike graph is entire. Moreover, if M^n is complete and noncompact, the existence on an entire graph of such a constant prevents that the tangent vector field to a divergent curve in Σ^n asymptotically approaches to a lightlike direction in the ambient spacetime.

The future-pointing Gauss map of a spacelike vertical graph $\Sigma(u)$ over Ω is given by the vector field

$$N(x) = \frac{f(u(x))}{\sqrt{f^2(u(x)) - |Du(x)|_M^2}} \left(\partial_t|_{(u(x),x)} + \frac{1}{f^2(u(x))} Du(x)\right), \quad (7.41)$$

for all $x \in \Omega$. Moreover, the shape operator A of $\Sigma(u)$ with respect to the orientation (7.41) is given by

$$AX = -\frac{1}{f(u)\sqrt{f^2(u) - |Du|_M^2}} D_X Du - \frac{f'(u)}{\sqrt{f^2(u) - |Du|_M^2}} X$$
$$+ \left(\frac{-g_M(D_X Du, Du)}{f(u) \left(f^2(u) - |Du|_M^2\right)^{3/2}} + \frac{f'(u)g_M(Du, X)}{\left(f^2(u) - |Du|_M^2\right)^{3/2}}\right) Du \quad (7.42)$$

for any tangent vector field X tangent to Ω.

7.4 Calabi-Bernstein Type Results

Consequently, if $\Sigma(u)$ is a spacelike vertical graph over a domain Ω of the fiber M^n of a spatially weighted GRW spacetime $\overline{M}^{n+1}$ endowed with a weight function φ, it is not difficult to verify that, by (5.16) and (7.42), the φ-mean curvature function $H_\varphi(u)$ of $\Sigma(u)$ is given by

$$nH_\varphi(u) = -\mathrm{div}_\varphi\left(\frac{Du}{f(u)\sqrt{f(u)^2 - |Du|_M^2}}\right) - \frac{f'(u)}{\sqrt{f(u)^2 - |Du|_M^2}}\left(n + \frac{|Du|_M^2}{f(u)^2}\right).$$

The differential equation $H_\varphi(u) = 0$ with the constraint $|Du|_{M^n} < f(u)$ is called the φ-maximal spacelike hypersurface equation on M^n, and its solutions provide φ-maximal spacelike graphs in M^n.

In this context, we can establish a nonparametric version of Theorem 7.1.

Theorem 7.16 *Let M^n be a complete n-dimensional Riemannian manifold, and consider a spatially weighted GRW spacetime $\overline{M}_\varphi^{n+1} = -I \times_f M_\varphi^n$ obeying the SNCC (5.22) and such that the Hessian of the weight function φ is bounded from below. Let $\Sigma(u)$ be an entire vertical graph in M_φ^n determined by a bounded smooth function $u \in C^\infty(M^n)$. Suppose that the φ-mean curvature H_φ of $\Sigma(u)$ satisfies*

$$(\log f)'(u) \leq H_\varphi \leq \alpha \quad \text{and} \quad H_\varphi \geq 0$$

for some positive constant α. If

$$|Du|_M^2 \leq \frac{\beta \inf_{\Sigma(u)} f^2(u) \inf_{\Sigma(u)} |H_\varphi - (\log f)'(u)|^\gamma}{1 + \beta \inf_{\Sigma(u)} |H_\varphi - (\log f)'(u)|^\gamma} \tag{7.43}$$

for some constants $\beta > 0$ and $\gamma \neq 0$, then $u \equiv t_0$ for some $t_0 \in I$.

Proof Let us observe that, under the assumptions of the theorem, $\Sigma(u)$ is complete. In fact, from (7.43) we easily obtain

$$f^2(u) - |Du|_M^2 \geq c = \frac{\inf_{\Sigma(u)} f^2(u)}{1 + \beta \inf_{\Sigma(u)} |H_\varphi - (\log f)'(h)|^\gamma} > 0,$$

and the completeness of $\Sigma^n(u)$ follows by Proposition 7.1.

On the other hand, by (1.12) and (7.41) we get

$$|\nabla h|^2 = \frac{|Du|_M^2}{f^2(u) - |Du|_M^2}. \tag{7.44}$$

Therefore, (7.43) implies (7.2), and the result follows from Theorem 7.1. □

The nonparametric version of Theorem 7.2 can be stated and proved in a similar way as for the previous theorem.

Theorem 7.17 *Let M^n be a complete n-dimensional Riemannian manifold, and consider a spatially weighted GRW spacetime $\overline{M}_\varphi^{n+1} = -I \times_f M_\varphi^n$ obeying the SNCC (5.22) with convex weight function φ. Let $\Sigma(u)$ be an entire vertical graph in M_φ^n determined by a bounded smooth function $u \in C^\infty(M^n)$ with constant φ-mean curvature H_φ satisfying*

$$0 \leq H_\varphi \sup_{\Sigma(u)} (\log f)'(u) \leq H^2.$$

If

$$|Du|_M^2 \leq \frac{\alpha \inf_{\Sigma(u)} f^2(u) \left(\inf H^2 - H_\varphi \sup_{\Sigma(u)} (\log f)'(u) \right)^\beta}{1 + \alpha \left(\inf H^2 - H_\varphi \sup_{\Sigma(u)} (\log f)'(u) \right)^\beta}$$

for some constants $\alpha > 0$ and $\beta \neq 0$, then $u \equiv t_0$ for some $t_0 \in I$.

A non-parametric version of Theorem 7.3 is given below. We emphasize that in this setting we require some additional assumptions in order to recover the completeness for entire graphs.

Theorem 7.18 *Let M^n be a complete n-dimensional Riemannian manifold, and consider a spatially weighted GRW spacetime $\overline{M}_\varphi^{n+1} = -I \times_f M_\varphi^n$. Let $\Sigma(u)$ be an entire vertical graph in M_φ^n of a bounded smooth function $u \in C^\infty(M^n)$, and suppose that $H_\varphi > 0$, $f'(u) > 0$, and that the following inequalities are satisfied,*

$$\frac{n^2}{4} (\log f)'^2(u) \leq \frac{n^2 H_\varphi^2}{4} \leq (n-1) \frac{f''}{f}(u).$$

If $|Du|_M^2 \leq \alpha f^2(u)$ for some constant $0 < \alpha < 1$, and $f(u) \in L^p(d\mu)$, where $d\mu = e^{-\varphi} dM^n$, for some $1 < p < +\infty$, then $u \equiv t_0$ for some $t_0 \in I$, with $\mathrm{vol}_\varphi(\Sigma(u)) < +\infty$. In addition, if M_φ^n obeys (5.22) and the weight function φ is bounded and convex, then $\Sigma(u)$ is compact.

Proof Since u is bounded and $|Du|_M^2 \leq \alpha f^2(u)$ for some constant $0 < \alpha < 1$, it is easily seen that the assumptions of Proposition 7.1 are satisfied, so $\Sigma(u)$ is complete. Moreover,

7.4 Calabi-Bernstein Type Results

by (5.9) of [30] we get that

$$d\Sigma(u) = f^{n-1}(u)\sqrt{f^2(u) - |Du|_M^2}\, dM^n.$$

Consequently, since $f(u) \in L^p(d\mu)$ for $1 < p < +\infty$, and $f(h(q)) = f(u(x))$ for every $q = (u(x), x) \in \Sigma(u)$, we infer that $f(h) \in L^p(d\mu_u)$ $1 < p < +\infty$, where $d\mu_u = e^{-\varphi} d\Sigma(u)$.

The conclusion follows by Theorem 7.3. □

In [38] the authors obtained a Calabi-Bernstein type result for $-\mathbb{R} \times \mathbb{G}^n$ showing that the graph $\Sigma(u)$ of a function $u(x) = t$ over $\mathbb{G}^n$, with $|Du|_{\mathbb{G}^n}$ bounded away from 1, is φ-maximal if and only if u is constant.

An extension of this result is presented below.

Theorem 7.19 *Let M_φ^n be a spatially weighted GRW spacetime, whose fiber M^n is complete with φ-parabolic universal Riemannian covering, and such that the warping function f is monotone. The only entire bounded solutions on M_φ^n to the following modified φ-maximal spacelike hypersurface equation*

$$(E) \begin{cases} \mathrm{div}_\varphi \left(\dfrac{Du}{f(u)\sqrt{f(u)^2 - |Du|_M^2}} \right) = -\dfrac{f'(u)}{\sqrt{f(u)^2 - |Du|_M^2}} \left(n + \dfrac{|Du|_M^2}{f(u)^2} \right) \\ \\ |Du|_M \leq \alpha f(u) \qquad (0 < \alpha < 1) \end{cases}$$

are the constant functions $u = t_0$ with $f'(t_0) = 0$.

Proof The main assumption on $|Du|_{M^n}$ assures the boundedness of the angle function $\Theta(u)$ of $\Sigma(u)$. This fact easily follows by (7.44). Hence, by using (1.12) and (7.44), if

$$|Du|_M \leq \alpha f(u)$$

then

$$\Theta(u) \geq \frac{-1}{\sqrt{1-\alpha^2}}.$$

Moreover, since we are looking for bounded solutions to (E), taking

$$c = (1-\alpha^2) \inf_{\Sigma(u)} f^2(u),$$

we can apply Proposition 7.1 to obtain that such a solution must be complete. Therefore, the result follows by Theorem 7.5. □

Arguing as in the proof of Theorem 7.19, a nonparametric version of the results in Sect. 7.2.3 can be achieved. For instance, a nonparametric version of Theorem 7.10 is presented below.

Theorem 7.20 *Let $\overline{M}_\varphi^{n+1}$ be a spatially weighted steady state type spacetime, whose fiber M^n is complete with φ-parabolic universal Riemannian covering. The only entire bounded solutions on M_φ^n to the equation*

$$(E') \begin{cases} \operatorname{div}_\varphi\left(\dfrac{Du}{e^u\sqrt{e^{2u} - |Du|_M^2}}\right) = -nH_\varphi(u) - \dfrac{e^u}{\sqrt{e^{2u} - |Du|_M^2}}\left(n + \dfrac{|Du|_M^2}{e^{2u}}\right) \\ 1 \leq H_\varphi(u) \leq \alpha \qquad (\alpha \geq 1) \\ |Du|_M \leq \sqrt{1 - \dfrac{1}{H_\varphi^2(u)}} e^u \end{cases}$$

are the constant functions $u = t_0$.

Proof Let us first observe that, under the main assumptions, the solutions of (E') determine complete entire graphs $\Sigma(u)$. Indeed, the last inequality in (E') yields

$$e^{2u} - |Du|_M^2 \geq c = e^{2\inf_{\Sigma(u)} u} \sup_{\Sigma(u)} \frac{1}{H_\varphi^2(u)} > 0.$$

Moreover, the completeness of $\Sigma(u)$ follows by Proposition 7.1. Therefore, by using (7.44) and again the last inequality of (E'), the angle function $\Theta(u)$ of $\Sigma(u)$ satisfies

$$-\Theta(u) \leq H_\varphi(u).$$

The conclusion follows by Theorem 7.10. □

Remark 7.7 An essential characteristic of the Robertson-Walker spacetimes is that they are locally conformally flat, as noted in reference [11]. This property is not true in the generalized Robertson-Walker spacetimes. Therefore, a useful extension of Robertson-Walker spacetimes might be to consider locally conformally flat multiple warped products and instead of reducing the assumption on the fiber's geometry, make a more general assumption on the warped product structure.

7.4 Calabi-Bernstein Type Results

The following result can be viewed as an application of Theorem 7.11.

Theorem 7.21 *Let $\overline{M}_\varphi^{n+1} = -\mathbb{R} \times M_\varphi^n$ be a weighted Lorentzian product space, whose fiber M^n is complete with sectional curvature bounded from below and its Bakry-Émery-Ricci tensor satisfies $\mathrm{Ric}_\varphi^M \geq c$ for some positive constant c, and such that the Hessian of the weight function φ is bounded from below. Let $\Sigma(u) \subset M_\varphi^n$ be an entire graph over M^n with constant φ-mean curvature, H_φ. If $|Du|_M \leq C$ for some constant $0 < C < 1$, then $u \equiv t_0$ for some $t_0 \in \mathbb{R}$.*

Proof Proposition 7.1 guarantees that such $\Sigma(u)$ is a complete spacelike hypersurface of M_φ^n. We observe that the choice of the future-pointing Gauss map of $\Sigma(u)$ given in (7.41) implies that $\Theta \leq -1$. Hence, by (7.44) we have that

$$-\frac{1}{\sqrt{1-C^2}} \leq \Theta \leq -1.$$

Therefore, by Theorem 7.11 we have that $\Sigma(u)$ is a slice Σ_{t_0} for some $t_0 \in \mathbb{R}$, which means that $u \equiv t_0$ as claimed. □

Let us recall that

$$|u|_{C^2(M^n)} = \max_{|\gamma| \leq 2} |D^\gamma u|_{L^\infty(M^n)}.$$

By Theorem 7.12 the next result holds.

Theorem 7.22 *Let $\overline{M}_\varphi^{n+1} = -\mathbb{R} \times M_\varphi^n$ be a weighted Lorentzian product space such that its fiber M^n is complete with sectional curvature bounded from below and nonnegative Bakry-Émery-Ricci tensor Ric_φ^M, and such that the Hessian of the weight function φ is bounded from below. Let $\Sigma(u) \subset M_\varphi^n$ be an entire graph over M^n with constant φ-mean curvature H_φ. If $|u|_{C^2(M^n)} < +\infty$ and $|Du|_M \leq \min\{\alpha|A|^\beta, C\}$, for some positive constants α, β and $C < 1$, then $u \equiv t_0$ for some $t_0 \in \mathbb{R}$.*

Proof By (7.42) the shape operator A of $\Sigma(u)$ with respect to N defined in (7.41) is given by

$$AX = -\frac{1}{\sqrt{1-|Du|_M^2}} D_X Du - \frac{g_M(D_X Du, Du)}{(1-|Du|_M^2)^{3/2}} Du \qquad (7.45)$$

for every tangent vector field $X \in \mathfrak{X}(\Sigma(u))$. Thus, since $|u|_{C^2(M^n)} < +\infty$ and $|Du|_M \leq C$ for some positive constant $C < 1$, by (7.45) we get that $|A|$ is bounded on $\Sigma(u)$. Therefore, arguing as in the proof of Theorem 7.21, the conclusion follows by Theorem 7.12. □

By Theorem 7.13 the following result holds.

Corollary 7.5 *Let $\overline{M}_\varphi^{n+1} = -\mathbb{R} \times M_\varphi^n$ be a weighted Lorentzian product space, whose fiber M^n is complete and such that its Bakry-Émery-Ricci tensor Ric_φ^M is nonnegative. Let $\Sigma(u) \subset M_\varphi^n$ be an entire graph over M^n with constant φ-mean curvature H_φ. If $\Sigma(u)$ is φ-parabolic and $|Du|_M \leq C$ for some positive constant $C < 1$, then $\Sigma(u)$ is totally geodesic. In addition, if Ric_φ^M is strictly positive, then $u \equiv t_0$ for some $t_0 \in \mathbb{R}$.*

Finally, a nonparametric version of Theorem 7.15 is given below.

Corollary 7.6 *Let M^n be an n-dimensional complete noncompact manifold with nonnegative sectional curvatures. The only maximal entire functions into $-I \times M^n$ such that $|Du|_M \leq c$ for some constant $c \in (0, 1)$, are the constant ones.*

7.5 Spacelike Translating Solitons of the Mean Curvature Flow in Weighted Lorentzian Product Spaces

The mean curvature flow $\Psi : [0, T) \times \Sigma^n \to \overline{M}^{n+1}$ of a spacelike hypersurface $\psi : \Sigma^n \to \overline{M}^{n+1}$, satisfies $\Psi(0, \cdot) = \psi(\cdot)$ and the evolution equation

$$\frac{\partial \Psi}{\partial t} = \mathbf{H},$$

where $\mathbf{H}(t, \cdot)$ is the (non-normalized) mean curvature vector field of the spacelike hypersurface $\Sigma_t^n = \Psi(t, \Sigma^n)$ for every $t \in [0, T)$, with $T > 0$.

Roughly speaking, the family of spacelike hypersurfaces $\Sigma_t^n = \Psi(t, \Sigma^n)$ evolves by mean curvature flow Ψ if the velocity $\frac{\partial \Psi}{\partial t}$ coincides with the mean curvature vector $\mathbf{H}$ at every point of $[0, T) \times \Sigma^n$.

Mean curvature flow in a Lorentzian manifold is an important thematic in the scope of Geometric Analysis and it has been extensively studied by several authors; see, among others, the papers [1, 168–171, 186, 210], as well as [213, 214, 225–227, 230].

More recently, Batista et al. [73, 75] studied solitons of the spacelike mean curvature flow in GRW spacetimes $-I \times_f M^n$. More precisely, under suitable constraints on the warping function f and on the curvatures of the Riemannian fiber M^n, they applied suitable maximum principles in order to obtain nonexistence and uniqueness results concerning these solitons. Moreover, they derived applications to standard models of GRW spacetimes, namely, the Einstein-de Sitter spacetime, steady state type spacetimes, de Sitter and anti-de Sitter spaces. Furthermore, they also established some Calabi-Bernstein type results related to entire spacelike mean curvature flow graphs constructed over the Riemannian fiber of the ambient spacetime.

7.5 Spacelike Translating Solitons of the Mean Curvature Flow in...

Afterwards, Molica Bisci et al. [85] investigated the uniqueness and nonexistence of n-dimensional spacelike translating solitons of the mean curvature flow immersed in a GRW spacetime $-I \times_f M^n$ obeying appropriated curvature constraints on the Riemannian fiber M^n which also involve the warping function f, via a sharp Bochner type inequality jointly with several maximum principles related to a drift Laplacian naturally attached to such a spacelike translating soliton. See also [74, 161, 184] for further results concerning the uniqueness and nonexistence of solitons of the spacelike mean curvature flow in a GRW spacetime.

This wide interest in the current literature is mainly due to the fact that *spacelike translating solitons* can be regarded as a natural way of foliating spacetimes by almost null-like hypersurfaces; for more details, see [169]. We also recall here that spacelike translating solitons are spacelike hypersurfaces $\psi : \Sigma^n \to \overline{M}^{n+1}$ such that $\mathbf{H} = c V^{\perp}$ for some constant c, where V stands for a suitable timelike vector field globally defined on $\overline{M}^{n+1}$.

In this setting and according to [72, 73], we say that a spacelike hypersurface $\psi : \Sigma^n \to \overline{M}^{n+1}$ immersed in a Lorentzian product space $\overline{M}^{n+1} = -\mathbb{R} \times M^n$ is a *spacelike translating soliton* of the mean curvature flow with respect to ∂_t and with *soliton constant* $c \in \mathbb{R}$ if its future mean curvature function satisfies the following equation

$$H = c\Theta. \tag{7.46}$$

In particular, we observe that the slices $\{t\} \times M^n$ are spacelike translating solitons of the mean curvature flow with respect to ∂_t and with soliton constant $c = 0$.

Our aim is to investigate the uniqueness of complete spacelike translating solitons in a weighted Lorentzian product space of the form $\overline{M}_{\varphi}^{n+1} = -\mathbb{R} \times M_{\varphi}^n$ under the assumption that the future φ-mean curvature is constant.

Theorem 7.23 *Let $\overline{M}_{\varphi}^{n+1} = -\mathbb{R} \times M_{\varphi}^n$ be a weighted Lorentzian product space such that its Riemannian base M^n is complete, with nonnegative Bakry-Émery-Ricci tensor. Let $\psi : \Sigma^n \to \overline{M}_{\varphi}^{n+1}$ be a complete spacelike translating soliton of the mean curvature flow with respect to ∂_t, having soliton constant c and constant future φ-mean curvature. If $H \in \mathcal{L}^q(d\mu)$, for some $q > 2$, then Σ^n is maximal. Moreover, if in addition M^n has nonnegative sectional curvature and Σ^n lies in a vertical half-space of $\overline{M}_{\varphi}^{n+1}$, then Σ^n must be a slice $\{t_0\} \times M^n$ for some $t_0 \in \mathbb{R}$.*

Proof Let Σ^n be such a spacelike translating soliton. If $c = 0$ the first conclusion is immediate. So, assume that $c \neq 0$. By item (c) of Proposition 1.1 we have the following key formula

$$\Delta \Theta = (\widetilde{\text{Ric}}(N^*, N^*) + |A|^2)\Theta + \langle \nabla H, \partial_t \rangle. \tag{7.47}$$

where $\widetilde{\mathrm{Ric}}$ is the standard Ricci tensor of M^n and $N^* = N + \Theta \partial_t$ denotes the orthonormal projection of N onto M^n.

Thus, since H_φ is constant, by (5.5) and (7.47) we obtain

$$\frac{1}{2}\Delta\Theta^2 = (\widetilde{\mathrm{Ric}}(N^*, N^*) + |A|^2)\Theta^2 + \langle \nabla H, \partial_t \rangle \Theta + |\nabla\Theta|^2$$
$$= (\widetilde{\mathrm{Ric}}(N^*, N^*) + |A|^2)\Theta^2 + \partial_t^\top(\langle \overline{\nabla}\varphi, N \rangle)\Theta + |\nabla\Theta|^2. \quad (7.48)$$

On the other hand, [251, Proposition 7.35] gives that

$$\overline{\nabla}_X \partial_t = 0, \quad (7.49)$$

for every tangent vector field X on the ambient space. Then, by (1.11) and (7.49) we have that

$$X(\Theta) = \langle \overline{\nabla}_X N, \partial_t \rangle = \langle A(\nabla h), X \rangle$$

and, consequently,

$$\nabla\Theta = A(\nabla h). \quad (7.50)$$

Thus, using once more (7.49), by (1.11) and (7.50) we get that

$$\partial_t^\top(\langle \overline{\nabla}\varphi, N \rangle) = \langle \overline{\nabla}_{\partial_t^\top} \overline{\nabla}\varphi, N \rangle + \langle \overline{\nabla}\varphi, \overline{\nabla}_{\partial_t^\top} N \rangle$$
$$= \overline{\mathrm{Hess}}\varphi(N, N)\Theta + \langle \overline{\nabla}\varphi, \nabla\Theta \rangle. \quad (7.51)$$

Now, by (7.51) in (7.48), we obtain

$$\frac{1}{2}\Delta\Theta^2 = (\widetilde{\mathrm{Ric}}(N^*, N^*) + |A|^2)\Theta^2 + \overline{\mathrm{Hess}}\varphi(N, N)\Theta^2 + \frac{1}{2}\langle \overline{\nabla}\varphi, \nabla(\Theta^2) \rangle + |\nabla\Theta|^2. \quad (7.52)$$

Consequently, since $\overline{\mathrm{Hess}}\varphi(N, N) = \widetilde{\mathrm{Hess}}\varphi(N^*, N^*)$, where $\widetilde{\mathrm{Hess}}$ stands for the Hessian computed in the metric of M^n, using (5.2) and (5.1) in (7.52) we reach at the following equation

$$\frac{1}{2}\Delta_\varphi\Theta^2 = (\widetilde{\mathrm{Ric}}_\varphi(N^*, N^*) + |A|^2)\Theta^2 + |\nabla\Theta|^2, \quad (7.53)$$

where $\widetilde{\mathrm{Ric}}_\varphi$ stands for the Bakry-Émery-Ricci tensor of M^n.

7.5 Spacelike Translating Solitons of the Mean Curvature Flow in...

Hence, since the soliton constant c is nonzero, by (7.46) and (7.53) we get the following inequality

$$\frac{1}{2}\Delta_\varphi H^2 \geq (\widetilde{\mathrm{Ric}}_\varphi(N^*, N^*) + |A|^2)H^2. \tag{7.54}$$

Since we are also supposing that $\widetilde{\mathrm{Ric}}_\varphi$ is nonnegative, by (7.54) we obtain that $\Delta_\varphi H^2 \geq 0$. So, since $H \in \mathcal{L}^q(d\mu)$, for some $q > 2$, we can apply Lemma 7.2 for $p = q/2$ to conclude that H is constant.

Thanks to (7.54), we also get that $|A|$ vanishes identically on Σ^n, which means that Σ^n is totally geodesic. Consequently, H must be identically zero and thus Σ^n is maximal and the first conclusion follows.

Moreover, considering $X \in \mathfrak{X}(\Sigma^n)$ and $\alpha = \inf_M K_M \geq 0$, where K_M stands for the sectional curvature of M^n, by Lemma 2.1 we obtain the following suitable lower bound for the Ricci tensor of Σ^n

$$\mathrm{Ric}(X, X) \geq (n-1)\alpha|X|^2 + \alpha|\nabla h|^2|X|^2 + (n-2)\alpha\langle X, \nabla h\rangle^2 + |AX|^2. \tag{7.55}$$

In particular, from (7.55) we get that the Ricci tensor of Σ^n is nonnegative.

Therefore, by item (a) of Proposition 1.1 we have that h is an harmonic function on Σ^n. If we also assume that Σ^n lies in vertical half-space of $\overline{M}_\varphi^{n+1}$ (which means that h is semi-bounded), we can apply Lemma 7.3 to conclude that h is constant, that is, Σ^n is a slice $\{t_0\} \times M^n$ for some $t_0 \in \mathbb{R}$. □

Taking into account that the Bakry-Émery-Ricci tensor of the Gaussian space $\mathbb{G}^n$ is positive, by Theorem 7.23 we obtain the following result.

Corollary 7.7 *The only complete spacelike translating solitons* $\psi : \Sigma^n \to -\mathbb{R} \times \mathbb{G}^n$ *of the mean curvature flow with respect to* ∂_t, *having constant future φ-mean curvature (where φ is the Gaussian probability measure defined in (6.28)), lying in a vertical half-space of* $-\mathbb{R} \times \mathbb{G}^n$ *and such that either H^2 attains its maximum on Σ^n or $H \in \mathcal{L}^q(d\mu)$, for some $q > 2$, are the slices $\{t\} \times \mathbb{G}^n$.*

Now, let $\mathbb{H}^n = \{(x_1, \cdots, x_n) \in \mathbb{R}^n : x_n > 0\}$ be the n-dimensional hyperbolic space endowed with its standard complete metric

$$\langle \cdot, \cdot \rangle_{\mathbb{H}^n} = \frac{1}{x_n^2}(dx_1^2 + \cdots + dx_n^2)$$

and let $\varphi : \mathbb{H}^n \to \mathbb{R}$ be the weight function given by

$$\varphi(x) = (d(x, x_0))^2, \tag{7.56}$$

where $d(x, x_0)$ denotes the distance in $\mathbb{H}^n$ between x and a fixed point $x_0 \in \mathbb{H}^n$.

According to [294, Example 7.2], we have that the Bakry-Émery-Ricci tensor of $\mathbb{H}^n_\varphi$ satisfies $\mathrm{Ric}_\varphi \geq (n-1)$. Thus, the base $\mathbb{H}^n_\varphi$ of the weighted Lorentzian product space $-\mathbb{R} \times \mathbb{H}^n_\varphi$ has nonnegative Bakry-Émery-Ricci tensor.

By Theorem 7.23 we also derive the following result.

Corollary 7.8 *Let $-\mathbb{R} \times \mathbb{H}^n_\varphi$ be the weighted Lorentzian product space with weight function φ defined by (7.56). Let $\psi : \Sigma^n \to -\mathbb{R} \times \mathbb{H}^n_\varphi$ be a complete spacelike translating soliton of the mean curvature flow with respect to ∂_t, having soliton constant c and constant future φ-mean curvature. If either H^2 attains its maximum on Σ^n or $H \in \mathcal{L}^q(d\mu)$, for some $q > 2$, then Σ^n is maximal.*

Since the drift Laplacian is an elliptic operator and taking into account inequality (7.54), we can apply the classical strong maximum principle due to Hopf [206] obtaining the following result; see also the book [258] due to Pucci and Serrin for related topics.

Theorem 7.24 *Let $\overline{M}_\varphi^{n+1} = -\mathbb{R} \times M^n_\varphi$ be a weighted Lorentzian product space such that its Riemannian base M^n is complete, with nonnegative Bakry-Émery-Ricci tensor. Let $\psi : \Sigma^n \to \overline{M}_\varphi^{n+1}$ be a complete spacelike translating soliton of the mean curvature flow with respect to ∂_t, having soliton constant c and constant future φ-mean curvature. If H^2 attains its maximum on Σ^n, then Σ^n is maximal. Moreover, if in addition M^n has nonnegative sectional curvature and Σ^n lies in vertical half-space of $\overline{M}_\varphi^{n+1}$, then Σ^n must be a slice $\{t_0\} \times M^n$ for some $t_0 \in \mathbb{R}$.*

7.6 A Height Estimate and a Half-Space Type Result in Weighted Lorentzian Product Spaces

The results of this section can be found in [143, Section 4]. For our purposes, we firstly prove a height estimate for compact spacelike hypersurfaces immersed in a weighted Lorentzian product space.

Theorem 7.25 *Let $-\mathbb{R} \times M^n_\varphi$ be a weighted Lorentzian product space with nonnegative Bakry-Émery-Ricci tensor $\overline{\mathrm{Ric}}_\varphi$ and let Σ^n be a compact spacelike hypersurface with boundary contained into the slice $\{t\} \times M^n$, for some $s \in \mathbb{R}$. If Σ^n has nonzero constant φ-mean curvature such that $nH^2_\varphi \leq |A|^2$, where A denotes the Weingarten operator of Σ^n with respect to its future-pointing unit normal vector field N, then the height function h of Σ^n satisfies*

$$|h - t| \leq \frac{\max_{\partial \Sigma^n} |\Theta| - 1}{|H_\varphi|}. \tag{7.57}$$

Proof As in the proof of Theorem 6.22, let us define on Σ^n the function

$$\zeta = -H_\varphi h - \Theta. \tag{7.58}$$

By (7.58) and items (i) and (iii) of Lemma 5.4, we get that

$$\Delta_\varphi \zeta = -\Theta(|A|^2 - nH_\varphi^2 + \widetilde{\mathrm{Ric}}_\varphi(N^*, N^*)).$$

Consequently, since $\widetilde{\mathrm{Ric}}_\varphi(N^*, N^*) = \overline{\mathrm{Ric}}_\varphi(N, N) \geq 0$ and $nH_\varphi^2 \leq |A|^2$, by choosing N future-pointing (that is, $\Theta \leq -1$), we get that $\Delta_\varphi \zeta \geq 0$.

Thus, we conclude by using the maximum principle that $\zeta \leq \zeta|_{\partial \Sigma}$. Hence, by (7.58) we have that

$$-H_\varphi h + 1 \leq -H_\varphi h - \Theta \leq -H_\varphi t + \max_{\partial \Sigma^n} |\Theta|. \tag{7.59}$$

Now, if $H_\varphi > 0$, we notice that $\Delta_\varphi h \geq 0$ and, by Hopf's maximum principle, $h \leq t$ on Σ^n. Thus, by (7.59) we conclude that

$$t - h \leq \frac{\max_{\partial \Sigma^n} |\Theta| - 1}{H_\varphi}. \tag{7.60}$$

Otherwise, if $H_\varphi < 0$, we have $\Delta_\varphi h \leq 0$ and, again by the Hopf's maximum principle, $h \geq t$ on Σ^n. Thus, by (7.59) we must have

$$h - t \leq \frac{1 - \max_{\partial \Sigma^n} |\Theta|}{H_\varphi}. \tag{7.61}$$

Therefore, estimate (7.57) follows by (7.60) and (7.61). □

Finally, arguing as in the proof of Theorem 6.23, by using Theorem 7.25 the following half-space type result holds.

Theorem 7.26 *Let* $-\mathbb{R} \times M_\varphi^n$ *be a weighted Lorentzian product space with nonnegative Bakry-Émery-Ricci tensor* $\overline{\mathrm{Ric}}_\varphi$ *and* M^n *compact. Moreover, let* Σ^n *be a complete noncompact spacelike hypersurface properly immersed in* $-\mathbb{R} \times M_\varphi^n$, *with bounded angle function* Θ. *If* Σ^n *has nonzero constant φ-mean curvature such that* $nH_\varphi^2 \leq |A|^2$, *then* Σ^n *cannot lie in a half-space of* $-\mathbb{R} \times M$. *In particular,* Σ^n *must have at least one top and one bottom end.*

8 Spacelike Hypersurfaces in Weighted Standard Static Spacetimes

8.1 Introduction

In this chapter our aim is to obtain new uniqueness results concerning the mean curvature equation in a weighted standard static spacetime $M_\varphi^n \times_\rho \mathbb{R}_1$ having warping function ρ and whose weight function φ does not depend on the parameter $t \in \mathbb{R}$.

At this purpose, we establish a φ-parabolicity criterion in order to study the rigidity of spacelike hypersurfaces immersed in $M_\varphi^n \times_\rho \mathbb{R}_1$ and, in particular, entire Killing graphs constructed over the Riemannian base M^n.

Furthermore, we deduce an extension of a classical maximum principle due to Nishikawa [248] in order to study the rigidity of complete spacelike hypersurfaces immersed in a weighted static spacetime $M_\varphi^n \times \mathbb{R}_1$. Applications to weighted standard static spacetimes of the type $\mathbb{G}^n \times_\rho \mathbb{R}_1$, where $\mathbb{G}^n$ denotes for the so-called Gaussian space, are also given.

The results presented here are mainly based on the papers [51, 159, 160].

8.2 Weighted Standard Static Spacetimes

In the sequel, we will consider an $(n+1)$-dimensional Lorentz manifold $\overline{M}^{n+1}$ with Lorentzian metric $g = g(\cdot, \cdot)$ and endowed with a Killing timelike vector field Y. Here, timelike referred to a vector field means that $Y_p \in T_p \overline{M}^{n+1}$ is a timelike (and so nonzero) vector for each $p \in \overline{M}^{n+1}$. On the other hand, Killing means that the $\mathcal{L}_Y g = 0$, where $\mathcal{L}_Y$ stands for the Lie derivative of g in the direction of Y.

We observe that the distribution $\mathcal{D}$ of all the smooth vector fields of $\overline{M}^{n+1}$ that are orthogonal to Y, defined at each point by

$$\overline{M}^{n+1} \ni p \longmapsto \mathcal{D}(p) = \left\{ v \in T_p \overline{M}^{n+1} : g(v, Y_p) = 0 \right\}, \tag{8.1}$$

is of constant rank and integrable. Given a Riemannian integral leaf M^n of that distribution $\mathcal{D}$, let $\Psi : \mathbb{I} \times M^n \to \overline{M}^{n+1}$ be the flow generated by Y with initial values in M^n, where $\mathbb{I}$ is a maximal interval of definition. Without loss of generality, we consider $\mathbb{I} = \mathbb{R}$.

In this setting, the space $\overline{M}^{n+1}$ can be regarded as the *standard static spacetime* $M^n \times_\rho \mathbb{R}_1$ (cf. [251, Proposition 12.38]), that is, the Lorentzian product manifold $M^n \times \mathbb{R}_1$ endowed with the warping metric

$$\langle \cdot, \cdot \rangle = \pi_M^*(\langle \cdot, \cdot \rangle_M) + (\rho \circ \pi_M)^2 \pi_\mathbb{R}^*(-dt^2), \tag{8.2}$$

where π_M and $\pi_\mathbb{R}$ denote the canonical projections from $M^n \times \mathbb{R}_1$ onto each factor, $\langle \cdot, \cdot \rangle_M$ is the induced Riemannian metric on the base M^n, $\mathbb{R}_1$ is the manifold $\mathbb{R}$ endowed with the metric $-dt^2$ and

$$\rho = |Y| = \sqrt{-\langle Y, Y \rangle} > 0$$

is the warping function.

We denote by $C^\infty(M^n \times_\rho \mathbb{R}_1)$ the ring of real functions of class C^∞ on $M^n \times_\rho \mathbb{R}_1$ and by $\mathfrak{X}(M^n \times_\rho \mathbb{R}_1)$ the $C^\infty(M^n \times_\rho \mathbb{R}_1)$-module of vector fields of class C^∞ on $M^n \times_\rho \mathbb{R}_1$. Finally, the Levi-Civita connections of $M^n \times_\rho \mathbb{R}_1$ and M^n will be denoted by $\overline{\nabla}$ and $\widetilde{\nabla}$, respectively.

Now, in the framework described above, let $(M^n \times_\rho \mathbb{R}_1)_\varphi$ be a *weighted standard static spacetime*, namely, a standard static spacetime $M^n \times_\rho \mathbb{R}_1$ endowed with a weighted volume form $d\overline{\sigma} = e^{-\varphi} d\overline{v}$, where $\varphi \in C^\infty(M^n \times_\rho \mathbb{R}_1)$ is a real-valued function, called *weight function* (or *density function*), and $d\overline{v}$ is the volume element induced by the warping metric $\langle \cdot, \cdot \rangle$ defined in (8.2). Fixed $(M^n \times_\rho \mathbb{R}_1)_\varphi$, we recall that the Bakry-Émery-Ricci tensor $\overline{\mathrm{Ric}}_\varphi$ is defined by

$$\overline{\mathrm{Ric}}_\varphi = \overline{\mathrm{Ric}} + \overline{\mathrm{Hess}}\varphi, \tag{8.3}$$

where $\overline{\mathrm{Ric}}$ and $\overline{\mathrm{Hess}}$ stand for the Ricci tensor and the Hessian operator in $M^n \times_\rho \mathbb{R}_1$, respectively.

Throughout this chapter, we deal with complete spacelike hypersurfaces

$$\psi : \Sigma^n \to (M^n \times_\rho \mathbb{R}_1)_\varphi,$$

8.2 Weighted Standard Static Spacetimes

namely, isometric immersions by an (connected) n-dimensional Riemannian manifold Σ^n into weighted static spacetime $\left(M^n \times_\rho \mathbb{R}_1\right)_\varphi$. In this setting, the Levi-Civita connection of Σ^n will be denoted by ∇. As $\left(M^n \times_\rho \mathbb{R}_1\right)_\varphi$ is time-oriented by the timelike vector field Y and $\psi : \Sigma^n \to \left(M^n \times_\rho \mathbb{R}_1\right)_\varphi$ is a spacelike hypersurface, then Σ^n is orientable (cf. [251, Proposition 5.26]) and one can choose a globally defined unit normal vector field N on Σ^n having the same time-orientation of $\left(M^n \times_\rho \mathbb{R}_1\right)_\varphi$ (cf. [251, Proposition 5.29]), that is

$$\langle Y, N \rangle < 0. \tag{8.4}$$

Such N is said the *future-pointing Gauss map* of Σ^n. Let A denote the shape operator of Σ^n with respect to N. So that at each $p \in \Sigma^n$ A restricts to a self-adjoint linear map

$$A_p : T_p \Sigma^n \to T_p \Sigma^n$$
$$v \mapsto A_p v = -\overline{\nabla}_v N.$$

According to Gromov [197], the *weighted mean curvature* (or simply the φ-mean curvature) H_φ of Σ^n is given by

$$nH_\varphi = nH - \langle \overline{\nabla}\varphi, N \rangle, \tag{8.5}$$

where

$$H = -\frac{1}{n}\mathrm{tr}(A)$$

denotes the standard mean curvature of Σ^n with respect to its orientation N. Moreover, we say that $\psi : \Sigma^n \to \left(M^n \times_\rho \mathbb{R}\right)_\varphi$ is φ-*maximal* when its φ-mean curvature vanishes identically.

The φ-*divergence* on Σ^n is defined by

$$\mathrm{div}_\varphi : C^\infty(\Sigma^n) \to C^\infty(\Sigma^n)$$
$$X \mapsto \mathrm{div}_\varphi(X) = \mathrm{div}X - \langle \nabla\varphi, X \rangle,$$

where $\mathrm{div}(\cdot)$ denotes the standard divergence on Σ^n. We define the φ-*Laplacian* (also called the *drift Laplacian*) of Σ^n by

$$\Delta_\varphi : C^\infty(\Sigma^n) \to C^\infty(\Sigma^n)$$
$$u \mapsto \Delta_\varphi(u) = \mathrm{div}_\varphi(\nabla u) = \Delta u - \langle \nabla\varphi, \nabla u \rangle \tag{8.6}$$

where Δ is the standard Laplacian on Σ^n.

In the sequel, associated to a spacelike hypersurface $\psi : \Sigma^n \to (M^n \times_\rho \mathbb{R}_1)_\varphi$, we will consider two particular smooth functions, namely, the (vertical) *height function*

$$h = (\pi_\mathbb{R})|_{\Sigma^n} : \Sigma^n \to \mathbb{R} \tag{8.7}$$

and the *angle function*

$$\begin{aligned}\Theta : \Sigma^n &\to \mathbb{R} \\ p &\mapsto \Theta(p) = \langle N(p), Y(p)\rangle,\end{aligned} \tag{8.8}$$

where N is the future-pointing Gauss map of Σ^n and Y is the Killing vector field on $(M^n \times_\rho \mathbb{R}_1)_\varphi$. By (8.4), we note that Θ will be always a negative function on Σ^n.

We have that

$$\nabla h = -\frac{1}{\rho^2} Y^\top, \tag{8.9}$$

where $(\cdot)^\top$ denote the projections of a smooth vector field in $\mathfrak{X}(M^n \times_\rho \mathbb{R}_1)$ on $\mathfrak{X}(\Sigma^n)$. Moreover,

$$N^* = N + \frac{1}{\rho^2}\Theta Y, \tag{8.10}$$

where $(\cdot)^*$ denote the projections of a smooth vector field in $\mathfrak{X}(M^n \times_\rho \mathbb{R}_1)$ on $\mathfrak{X}(M^n)$. Hence, by (8.9) and (8.10) it is not difficult to verify that

$$|\nabla h|^2 = \frac{1}{\rho^2}|N^*|^2_{M^n}. \tag{8.11}$$

8.3 A φ-Parabolicity Criterion for Spacelike Hypersurfaces in $(M^n \times_\rho \mathbb{R}_1)_\varphi$

In [260], Romero et al. investigated the parabolicity of complete spacelike hypersurfaces in GRW spacetimes whose Riemannian fiber has a parabolic universal Riemannian covering.

In this setting, they were able to guarantee the parabolicity of complete spacelike hypersurfaces, under suitable boundedness assumptions on the warping function and on the hyperbolic angle function of these hypersurfaces. Our aim in this section is just, following the ideas of [155], to obtain an extension of this parabolicity criterion to the context of standard static spacetimes.

A smooth function u on a weighted manifold Σ^n_φ is said to be φ-*superharmonic* if $\Delta_\varphi u \leq 0$. Taking this into account, the weighted manifold Σ^n_φ is called φ-*parabolic* if there is no nonconstant, nonnegative, φ-superharmonic function on Σ^n.

8.3 A φ-Parabolicity Criterion for Spacelike Hypersurfaces in $(M^n \times_\rho \mathbb{R}_1)_\varphi$

On the other hand, given a weighted manifold Σ_φ^n we define, for any compact subset $K \subset \Sigma^n$, the φ-*capacity* of K as

$$\mathrm{cap}_\varphi(K) = \inf \left\{ \int_{\Sigma^n} |\nabla u|^2 e^{-\varphi} d\Sigma^n : u \in \mathrm{Lip}_0(\Sigma^n) \text{ and } u|_K \equiv 1 \right\},$$

where $\mathrm{Lip}_0(\Sigma^n)$ is the set of all compactly supported Lipschitz functions on Σ^n.

The following statement relates the notion of φ-capacity to the concept of φ-parabolicity; see [195, Proposition 2.1].

Lemma 8.1 *The weighted manifold Σ_φ^n is φ-parabolic if and only if $\mathrm{cap}_\varphi(K) = 0$ for any compact set $K \subset \Sigma^n$.*

Let us recall that given two Riemannian manifolds (Σ', g') and (Σ, g), a diffeomorphism ϑ by Σ' onto Σ is called a *quasi-isometry* if there exists a constant $c \geq 1$ such that

$$c^{-1} |v|_{g'} \leq |d\vartheta(v)|_g \leq c |v|_{g'}$$

for every $v \in T_p \Sigma'$, $p \in \Sigma'$; see [215] for more details.

In this case, given a smooth function $\varphi : \Sigma \to \mathbb{R}$, by arguing as in [194, Section 5], it is easily seen that the $(\varphi \circ \vartheta)$-capacity of the compact subsets in Σ' changes under a quasi-isometry at most by a constant factor of the φ-capacity of the compact subsets in Σ.

With the above notations, Lemma 8.1 of [155] states the following

Lemma 8.2 *The following facts hold:*

(a) *Given a quasi-isometry $\vartheta : \Sigma' \to \Sigma$, Σ is φ-parabolic if and only if Σ' is $(\varphi \circ \vartheta)$-parabolic;*
(b) *Let $\widetilde{\Sigma}$ be the universal Riemannian covering of Σ with canonical projection $\pi_\Sigma : \widetilde{\Sigma} \to \Sigma$. If $\widetilde{\Sigma}$ is $(\varphi \circ \pi_\Sigma)$-parabolic, then Σ is φ-parabolic.*

Recall that every connected manifold Σ has an universal covering, i.e. there exist a simply connected manifold $\widetilde{\Sigma}$ (called a universal covering of Σ) and a smooth map $\widetilde{\pi} : \widetilde{\Sigma} \to \Sigma$ (called a covering map) such that each point $p \in \Sigma$ has a connected neighborhood U that is evenly covered by $\widetilde{\pi}$, that is, $\widetilde{\pi}$ maps each component of $\widetilde{\pi}^{-1}(U)$ diffeomorphically onto U; see [251, Appendix A] for additional comments and remarks.

Moreover, if Σ is a Riemannian manifold, then it is possible to give $\widetilde{\Sigma}$ a Riemannian structure such that the covering map $\widetilde{\pi} : \widetilde{\Sigma} \to \Sigma$ is a local isometry. In this case, $\widetilde{\Sigma}$ is said a universal Riemannian covering of Σ; see [163, p. 152].

From now on, we will denote by $\widetilde{M}$ the universal Riemannian covering of base M^n with projection $\widetilde{\pi} : \widetilde{M} \to M^n$ and $\widetilde{f}$ will denote the composition $f \circ \widetilde{\pi}$.

In this setting, a standard static spacetime $(M^n \times_\rho \mathbb{R}_1)_\varphi$ will be said *spatially φ-parabolic* if the universal Riemannian covering $\tilde{M}$ of its base M^n is $\tilde{\varphi}$-parabolic.

Proposition 8.1 *Let $(M^n \times_\rho \mathbb{R}_1)_\varphi$ be a weighted standard static spacetime which is spatially $\tilde{\varphi}$-parabolic. If $\psi : \Sigma^n \to \overline{M}^{n+1}$ is a spacelike hypersurface such that the function*

$$\eta = \frac{\Theta}{\rho}$$

is bounded on it, then Σ^n is φ-parabolic.

Proof By Lemma 8.2 we have that

(i) φ-parabolicity is invariant under a quasi-isometry;
(ii) If the universal Riemannian covering $\tilde{\Sigma}$ of Σ^n is $(\varphi \circ \pi_{\Sigma^n})$-parabolic, then Σ^n is also φ-parabolic.

Denoting by $\pi = \pi_M \circ \psi : \Sigma^n \to M^n$, for any tangent vector $v \in T\Sigma^n$ we have

$$\langle v, v \rangle = \langle \pi_* v, \pi_* v \rangle_M - \rho^2 \langle h_* v, h_* v \rangle_\mathbb{R} \leq c \langle \pi_* v, \pi_* v \rangle_M,$$

where $c = \sup_{\Sigma^n} \eta^2 \geq 1$. In particular, by previous inequality we see that $\pi_{*,p} : T_p\Sigma^n \to T_{\pi(p)}M^n$ is a isomorphism for every $p \in \Sigma^n$. Then, the Inverse Function Theorem yields that the map π is a local diffeomorphism and by applying [163, Lemma 7.3.3] (see also [218, Lemma 8.8.1]), we can conclude that π is a covering map and that M^n is complete.

On the other hand, by using the Cauchy-Schwartz inequality it follows that

$$\langle \nabla h, v \rangle^2 \leq \langle \nabla h, \nabla h \rangle \langle v, v \rangle.$$

Consequently, since $h_* v = dh(v) = \langle \nabla h, v \rangle$, we have

$$\langle v, v \rangle = \langle \pi_* v, \pi_* v \rangle_M - \rho^2 \langle h_* v, h_* v \rangle$$
$$= \langle \pi_* v, \pi_* v \rangle_M - \rho^2 \langle \nabla h, v \rangle^2 \geq \langle \pi_* v, \pi_* v \rangle_M - \rho^2 |\nabla h|^2 \langle v, v \rangle,$$

that is,

$$\langle v, v \rangle \left(1 + \rho^2 |\nabla h|^2\right) \geq \langle \pi_* v, \pi_* v \rangle_M.$$

8.4 Rigidity Results for Spacelike Hypersurfaces in $M_\varphi^n \times_\rho \mathbb{R}_1$

Now, by using the definition of η and (8.11), one has

$$\langle v, v \rangle \geq \frac{1}{\eta^2} \langle \pi_* v, \pi_* v \rangle_M$$

We conclude that

$$c^{-1} \langle \pi_* v, \pi_* v \rangle_M \leq \langle v, v \rangle \leq c \langle \pi_* v, \pi_* v \rangle_M. \tag{8.12}$$

So, let $\widetilde{\Sigma}$ be the universal Riemannian covering of Σ^n with projection $\pi_{\Sigma^n} : \widetilde{\Sigma} \to \Sigma^n$. Then, the map $\pi_0 = \pi \circ \pi_{\Sigma^n} : \widetilde{\Sigma} \to M^n$ is a covering map. Now, if $\widetilde{M}$ is the universal Riemannian covering of M^n with projection $\widetilde{\pi} : \widetilde{M} \to M^n$, then there exists a diffeomorphism $\varphi : \widetilde{\Sigma} \to \widetilde{M}$ such that $\widetilde{\pi} \circ \varphi = \pi_0$.

Moreover, φ is a quasi-isometry. Indeed, if $v \in T\widetilde{\Sigma}$, by (8.12) we have

$$\langle \varphi_* v, \varphi_* v \rangle_{\widetilde{M}} = \langle \widetilde{\pi}_*(\varphi_* v), \widetilde{\pi}_*(\varphi_* v) \rangle_M = \langle (\pi_0)_* v, (\pi_0)_* v \rangle_M$$
$$= \langle \pi_*((\pi_{\Sigma^n})_* v), \pi_*((\pi_\Sigma)_* v) \rangle_M \leq c \langle (\pi_{\Sigma^n})_* v, (\pi_{\Sigma^n})_* v \rangle_{\Sigma^n} = c \langle v, v \rangle_{\widetilde{\Sigma}}.$$

Similarly, we also have

$$\langle \varphi_* v, \varphi_* v \rangle_{\widetilde{M}} \geq c^{-1} \langle v, v \rangle_{\widetilde{\Sigma}}.$$

Therefore, since the universal Riemannian covering of M^n is parabolic, the same property holds for the universal Riemannian covering of Σ^n. In conclusion, Σ^n must be also parabolic. □

8.4 Rigidity Results for Spacelike Hypersurfaces in $M_\varphi^n \times_\rho \mathbb{R}_1$

According to [101], in a weighted timelike geodesically complete spacetime $\overline{M}_\varphi^{n+1}$ that contains a timelike line, with $\overline{\mathrm{Ric}}_\varphi(X, X) \geq 0$ for all timelike vector field X and whose weight function φ is bounded, the weight function φ must be constant along timelike line of $\overline{M}_\varphi^{n+1}$.

Consequently, in any weighted standard static spacetime $(M^n \times_\rho \mathbb{R}_1)_\varphi$ having nonnegative Bakry-Émery-Ricci tensor for timelike vector fields and with bounded weight function φ, we have that φ does not depend on the parameter of the flow associated to the Killing vector field

$$\frac{\partial}{\partial t} \equiv Y.$$

Motivated by this fact, we consider standard static spacetimes $M^n \times_\rho \mathbb{R}_1$ endowed with a weight function φ such that $\langle \overline{\nabla}\varphi, Y \rangle = 0$. As usual, we denote this ambient space by $M_\varphi^n \times_\rho \mathbb{R}_1$.

In this section, we apply Proposition 8.1 in order to obtain rigidity results for spacelike hypersurfaces in $M^n \times_\rho \mathbb{R}_1$.

To this aim, we compute an explicit formula for the drift Laplacian of the angle function Θ defined in (8.8).

Proposition 8.2 *Let $\psi : \Sigma^n \to M_\varphi^n \times_\rho \mathbb{R}_1$ be an immersed spacelike hypersurface and let $\Theta \in C^\infty(\Sigma^n)$ be the angle function defined in (8.8). Then,*

$$\Delta_\varphi \Theta = nY^\top (H_\varphi) + \left(\widetilde{\mathrm{Ric}}_\varphi(N^*, N^*) - \frac{1}{\rho}\widetilde{\mathrm{Hess}}\,\rho(N^*, N^*) + \Theta^2 \frac{\widetilde{\Delta}_\varphi(\rho)}{\rho^3} + |A|^2 \right) \Theta.$$

Proof Firstly, since Y is a Killing vector field, for any $X \in \mathfrak{X}(\Sigma^n)$, we have

$$\langle \nabla \Theta, X \rangle = X(\langle N, Y \rangle) = \langle \overline{\nabla}_X N, Y \rangle + \langle N, \overline{\nabla}_X Y \rangle = \langle -A(Y^\top) - \overline{\nabla}_N Y, X \rangle,$$

which assures that

$$\nabla \Theta = -A(Y^\top) - (\overline{\nabla}_N Y)^\top. \tag{8.13}$$

On the other hand, by (8.5) we note that

$$nY^\top(H) = Y^\top \left(nH_\varphi + \langle \overline{\nabla}\varphi, N \rangle \right)$$
$$= nY^\top (H_\varphi) + Y^\top (\langle \overline{\nabla}\varphi, N \rangle)$$
$$= nY^\top (H_\varphi) + \langle Y, \overline{\mathrm{Hess}}\varphi(N) \rangle + \Theta \overline{\mathrm{Hess}}\varphi(N, N)$$
$$\quad - \langle A(Y^\top), \overline{\nabla}\varphi \rangle, \tag{8.14}$$

where we used the decomposition $Y = Y^\top - \Theta N$.

Moreover, since φ is supposed to be invariant along the flow determinate by Y, on account of (8.13), one has

$$\langle \nabla \Theta, \overline{\nabla}\varphi \rangle = -\langle A(Y^\top) + (\overline{\nabla}_N Y)^\top, \overline{\nabla}\varphi \rangle$$
$$= -\langle A(Y^\top), \overline{\nabla}\varphi \rangle - \langle \overline{\nabla}_N Y, \overline{\nabla}\varphi \rangle$$
$$= -\langle A(Y^\top), \overline{\nabla}\varphi \rangle + \langle Y, \overline{\nabla}_N \overline{\nabla}\varphi \rangle$$
$$= -\langle A(Y^\top), \overline{\nabla}\varphi \rangle + \langle Y, \overline{\mathrm{Hess}}\varphi(N) \rangle. \tag{8.15}$$

8.4 Rigidity Results for Spacelike Hypersurfaces in $M_\varphi^n \times_\rho \mathbb{R}_1$

Now, by (8.15) and (8.14), we get

$$nY^\top(H) = nY^\top(H_\varphi) + \Theta\,\overline{\mathrm{Hess}}\varphi(N,N) + \langle \nabla\Theta, \overline{\nabla}\varphi\rangle. \tag{8.16}$$

Moreover, by [64, Proposition 2.12], we have

$$\Delta\Theta = nY^\top(H) + \Theta\left(\mathrm{Ric}(N,N) + |A|^2\right). \tag{8.17}$$

Thus, by (8.3), (8.6), (8.17) and (8.16), we obtain that

$$\Delta_\varphi \Theta = nY^\top(H_\varphi) + \left(\overline{\mathrm{Ric}}_\varphi(N,N) + |A|^2\right)\Theta. \tag{8.18}$$

Now, if we consider the decomposition $N = N^* + N^\perp$ of N, where $(\cdot)^\perp$ denote the projection of a vector field in $\mathfrak{X}(M^n \times_\rho \mathbb{R}_1)$ on $\mathfrak{X}(\mathbb{R}_1)$, we have

$$\overline{\mathrm{Hess}}\varphi(N,N) = \langle \overline{\nabla}_N \overline{\nabla}\varphi, N\rangle = \langle \overline{\nabla}_N \widetilde{\nabla}\varphi, N^* + N^\perp\rangle$$

$$= \widetilde{\mathrm{Hess}}\varphi(N^*, N^*) + \frac{1}{\rho}\langle \widetilde{\nabla}\varphi, \widetilde{\nabla}\rho\rangle |N^\perp|^2$$

$$= \widetilde{\mathrm{Hess}}\varphi(N^*, N^*) - \frac{1}{\rho^3}\langle \widetilde{\nabla}\varphi, \widetilde{\nabla}\rho\rangle\Theta^2. \tag{8.19}$$

By [251, Corollary 7.43] we get that

$$\overline{\mathrm{Ric}}(N,N) = \widetilde{\mathrm{Ric}}(N^*, N^*) - \frac{1}{\rho}\widetilde{\mathrm{Hess}}\rho(N^*, N^*) + \Theta^2 \frac{\widetilde{\Delta}(\rho)}{\rho^3}. \tag{8.20}$$

Hence, by (8.3), (8.19) and (8.20), one has

$$\overline{\mathrm{Ric}}_\varphi(N,N) = \widetilde{\mathrm{Ric}}_\varphi(N^*, N^*) - \frac{1}{\rho}\widetilde{\mathrm{Hess}}\rho(N^*, N^*) + \Theta^2 \frac{\widetilde{\Delta}_\varphi(\rho)}{\rho^3} \tag{8.21}$$

Therefore, by (8.21) and (8.18) the conclusion follows. $\square$

We are in position now to present the rigidity theorem given below.

Theorem 8.1 *Let $M_\varphi^n \times_\rho \mathbb{R}_1$ be a weighted standard static spacetime which is spatially $\widetilde{f}$-parabolic. Suppose that $\widetilde{\mathrm{Ric}}_\varphi \geq 0$, the warping function ρ is convex and $\langle \widetilde{\nabla}\varphi, \widetilde{\nabla}\rho\rangle \leq 0$. Moreover, let $\psi : \Sigma^n \to \overline{M}^{n+1}$ be an immersed spacelike hypersurface with constant φ-mean curvature H_φ such that its angle function Θ is bounded and $\inf_{\Sigma^n} \rho > 0$. Then, Σ^n is totally geodesic and ρ is a positive constant. In addition, if $\widetilde{\mathrm{Ric}}_\varphi$ is positive at some point $p_0 \in \Sigma^n$, then Σ^n is contained in a slice $M^n \times \{t_0\}$, for some $t_0 \in \mathbb{R}$.*

Proof Since H_φ is constant, by Proposition 8.2, we have the following formula

$$\Delta_\varphi \Theta = \left(\widetilde{\mathrm{Ric}}_\varphi(N^*, N^*) - \frac{1}{\rho} \widetilde{\mathrm{Hess}} \rho(N^*, N^*) + \Theta^2 \frac{\widetilde{\Delta}_\varphi(\rho)}{\rho^3} + |A|^2 \right) \Theta. \tag{8.22}$$

Let us observe that at the points where N^* is different from zero we have

$$\frac{1}{\rho} \widetilde{\mathrm{Hess}}\, \rho(N^*, N^*) = \frac{|N^*|^2}{\rho} \widetilde{\mathrm{Hess}}\, \rho \left(\frac{N^*}{|N^*|}, \frac{N^*}{|N^*|} \right) = \frac{\Theta^2 - \rho^2}{\rho^3} \widetilde{\mathrm{Hess}}\, \rho \left(\frac{N^*}{|N^*|}, \frac{N^*}{|N^*|} \right).$$

Besides, taking a local orthonormal frame

$$\left\{ E_1 = \frac{N^*}{|N^*|}, E_2, \ldots, E_n \right\}$$

tangent to M^n, we also have

$$\frac{\Theta^2}{\rho^3} \widetilde{\Delta}(\rho) = \frac{\Theta^2}{\rho^3} \widetilde{\mathrm{Hess}}\, \rho \left(\frac{N^*}{|N^*|}, \frac{N^*}{|N^*|} \right) + \frac{\Theta^2}{\rho^3} \sum_{i=2}^n \widetilde{\mathrm{Hess}}\, \rho(E_i, E_i).$$

Then, one has

$$-\frac{1}{\rho} \widetilde{\mathrm{Hess}}\, \rho(N^*, N^*) + \frac{\Theta^2}{\rho^3} \widetilde{\Delta}(\rho) = \frac{1}{\rho} \widetilde{\mathrm{Hess}}\, \rho \left(\frac{N^*}{|N^*|}, \frac{N^*}{|N^*|} \right) + \frac{\Theta^2}{\rho^3} \sum_{i=2}^n \widetilde{\mathrm{Hess}}\, \rho(E_i, E_i).$$

Thus, by using (8.6) and the convexity of ρ, it follows that

$$-\frac{1}{\rho} \widetilde{\mathrm{Hess}}\, \rho(N^*, N^*) + \frac{\Theta^2}{\rho^3} \widetilde{\Delta}_\varphi(\rho) = \frac{1}{\rho} \widetilde{\mathrm{Hess}}\, \rho \left(\frac{N^*}{|N^*|}, \frac{N^*}{|N^*|} \right)$$

$$+ \frac{\Theta^2}{\rho^3} \sum_{i=2}^n \widetilde{\mathrm{Hess}}\, \rho(E_i, E_i)$$

$$- \frac{\Theta^2}{\rho^3} \langle \widetilde{\nabla} \varphi, \widetilde{\nabla} \rho \rangle \geq 0, \tag{8.23}$$

bearing in mind that $\langle \widetilde{\nabla} \varphi, \widetilde{\nabla} \rho \rangle \leq 0$.

On the other hand, the main assumption on $\widetilde{\mathrm{Ric}}_\varphi$ and thanks to (8.23), one has that Θ is a bounded φ-superharmonic function on Σ^n.

Now, by Proposition 8.1, the hypersurface Σ^n is φ-parabolic. Thus, the function Θ is constant on Σ^n. So, thanks to (8.22), we obtain $|A|^2 = 0$, that is, Σ^n is totally geodesic.

8.4 Rigidity Results for Spacelike Hypersurfaces in $M_\varphi^n \times_\rho \mathbb{R}_1$

We claim that ρ is a positive constant. Indeed, for any $X \in T\Sigma^n$, we can write

$$X = X^* - \frac{\langle X, Y \rangle}{\rho^2} Y,$$

where X^* denotes the orthogonal projection of X onto TM^n. Since Σ^n is totally geodesic, by [251, Proposition 7.35], we have that

$$X(\Theta) = \langle N, \overline{\nabla}_X Y \rangle = \langle N, \overline{\nabla}_{X^*} Y \rangle - \frac{\langle X, Y \rangle}{\rho^2} \langle N, \overline{\nabla}_Y Y \rangle$$

$$= \frac{1}{\rho} \langle X, \overline{\nabla}\rho \rangle \langle N, Y \rangle - \frac{1}{\rho} \langle X, Y \rangle \langle N, \overline{\nabla}\rho \rangle.$$

Thus, we conclude that

$$\nabla \Theta = \frac{1}{\rho} \left(\Theta \overline{\nabla}\rho - \langle N, \overline{\nabla}\rho \rangle Y \right).$$

Since Θ is constant, taking into account that $\overline{\nabla}\rho$ and Y are linearly independent, it follows that ρ is a positive constant.

Furthermore, we have, again by (8.22), that $\widetilde{\mathrm{Ric}}_\varphi(N^*, N^*)(p_0) = 0$. So, if $\widetilde{\mathrm{Ric}}_\varphi$ is positive at some point $p_0 \in \Sigma^n$, then $N^*(p_0) = 0$. Consequently, by using (8.11) it is easily seen that

$$|\nabla h|^2 = \frac{1}{\rho^2} |N^*|_M^2 = \frac{1}{\rho^2} \left(\frac{\Theta^2}{\rho^2} - 1 \right) = 0,$$

i.e. Σ^n is contained in a slice $M^n \times \{t_0\}$, for some $t_0 \in \mathbb{R}$. $\square$

In the next result $\widetilde{\mathrm{Ric}}_\varphi$ is not necessarily nonnegative.

Theorem 8.2 *Let $M_\varphi^n \times_\rho \mathbb{R}_1$ be a weighted standard static spacetime which is spatially $\widetilde{f}$-parabolic. Suppose that $\widetilde{\mathrm{Ric}}_\varphi \geq -\kappa$, for some constant $\kappa > 0$, and that ρ is a convex warping function such that $\langle \overline{\nabla}\varphi, \overline{\nabla}\rho \rangle \leq 0$. Moreover, let $\psi : \Sigma^n \to M_\varphi^n \times_\rho \mathbb{R}_1$ be an immersed spacelike hypersurface with constant φ-mean curvature, bounded angle function Θ and such that $\inf_\Sigma \rho > 0$. If the height function h satisfies*

$$|\nabla h|^2 \leq \frac{\alpha}{\kappa \rho^2} |A|^2, \tag{8.24}$$

for some constant $\alpha \in (0, 1)$, then Σ^n is contained in a slice $M^n \times \{t_0\}$, for some $t_0 \in \mathbb{R}$.

Proof Since H_φ is constant, $\Theta < 0$ on Σ^n and taking into account our assumption on $\widetilde{\mathrm{Ric}}_\varphi$, by (8.11) and (8.23), in addition to Proposition 8.2, we obtain

$$\Delta_\varphi \Theta \leq \left(-\kappa \rho^2 |\nabla h|^2 + |A|^2\right) \Theta. \tag{8.25}$$

By using (8.24), by (8.25) we get that

$$\Delta_\varphi (\Theta) \leq (1 - \alpha)|A|^2 \Theta. \tag{8.26}$$

Hence, by (8.26) it follows that Θ is a bounded φ-superharmonic function on Σ^n. Since Proposition 8.1 guarantees that Σ^n is φ-parabolic, Θ must be constant on Σ^n.

So, thanks to (8.26), we have that Σ^n is totally geodesic. Therefore, hypothesis (8.24) assures that h is constant on Σ^n, that is, there exists $t_0 \in \mathbb{R}$ such that $\Sigma^n \subset M^n \times \{t_0\}$. $\square$

In the sequel, we will deal with some specific weight functions that will be defined in terms of the warping function ρ.

The next result will be usefull in the sequel.

Proposition 8.3 *Let $\psi : \Sigma^n \to M^n \times_\rho \mathbb{R}_1$ be an immersed spacelike hypersurface and let $h \in C^\infty(\Sigma^n)$ be the height function. Then*

$$\Delta h = -n\rho^{-2} \Theta \, H_{\log \rho^2}, \tag{8.27}$$

where Θ is the angle function and $H_{\log \rho^2}$ is the $\log \rho^2$-mean curvature of Σ^n.

Proof Let $\{E_1, \ldots, E_n\}$ be an orthonormal frame defined in a neighborhood of some point of Σ^n. By (8.9) the following computation holds

$$\rho^{-2} \mathrm{div}\,(\nabla h) = \rho^{-2} \mathrm{div}\left(-\rho^{-2} Y^\top\right)$$

$$= -\rho^{-2} \langle \nabla \rho^{-2}, Y^\top \rangle - \rho^{-4} \mathrm{div}\left(Y^\top\right)$$

$$= \langle \nabla \rho^{-2}, \nabla h \rangle - \rho^{-4} \mathrm{div}\,(Y + \Theta N)$$

$$= \langle \nabla \rho^{-2}, \nabla h \rangle - \rho^{-4} \sum_{i=1}^{n} \langle \nabla_{E_i} (Y + \Theta N), E_i \rangle$$

$$= \langle \nabla \rho^{-2}, \nabla h \rangle - \rho^{-4} \sum_{i=1}^{n} \langle \overline{\nabla}_{E_i} (Y + \Theta N), E_i \rangle$$

$$= \langle \nabla \rho^{-2}, \nabla h \rangle - \rho^{-4} \underbrace{\sum_{i=1}^{n} \langle \overline{\nabla}_{E_i} Y, E_i \rangle}_{=0} - \rho^{-4} \sum_{i=1}^{n} \langle \overline{\nabla}_{E_i} (\Theta N), E_i \rangle$$

8.4 Rigidity Results for Spacelike Hypersurfaces in $M_\varphi^n \times_\rho \mathbb{R}_1$

$$= \langle \nabla \rho^{-2}, \nabla h \rangle - \rho^{-4} \sum_{i=1}^{n} \underbrace{\langle E_i(\Theta) \langle N, E_i \rangle + \Theta \overline{\nabla}_{E_i} N, E_i \rangle}_{=0}$$

$$= \langle \nabla \rho^{-2}, \nabla h \rangle + \rho^{-4} \Theta \operatorname{tr}(A) = \langle \nabla \rho^{-2}, \nabla h \rangle - n\rho^{-4} H \Theta.$$

Therefore, one has

$$\Delta h = \operatorname{div}(\nabla h) = \rho^2 \langle \nabla \rho^{-2}, \nabla h \rangle - n\rho^{-2} H \Theta$$
$$= \langle \nabla \log \rho^{-2}, -\rho^2 Y^\top \rangle - n\rho^{-2} H \Theta$$
$$= -\rho^{-2} \langle \overline{\nabla} \log \rho^{-2}, Y^\top \rangle - n\rho^{-2} H \Theta$$
$$= -\rho^{-2} \langle \overline{\nabla} \log \rho^{-2}, Y + \Theta N \rangle - n\rho^{-2} H \Theta$$
$$= -\rho^{-2} \underbrace{\langle \overline{\nabla} \log \rho^{-2}, Y \rangle}_{=0} - \rho^{-2} \langle \overline{\nabla} \log \rho^{-2}, N \rangle \Theta - n\rho^{-2} H \Theta$$

$$= -\rho^{-2} \Theta \left\{ nH + \langle \overline{\nabla} (\log \rho^{-2}), N \rangle \right\} = -n\rho^{-2} \Theta H_{\log \rho^2},$$

where the last equality holds thanks to (8.5). □

Let us recall now that a *slab* of a standart static spacetime $M^n \times_\rho \mathbb{R}_1$ is a region of the type

$$M^n \times_\rho [t_1, t_2] = \{(t, q) \in M^n \times_\rho \mathbb{R}_1 : t_1 \leq t \leq t_2\}.$$

Making use of the notion above, in Theorem 8.3 the weighted mean curvature $H_{\log \rho^2}$ of the spacelike hypersurface is not supposed to be constant. We just assume a certain control on the sign of $H_{\log \rho^2}$.

More precisely, the following result holds.

Theorem 8.3 *Let $M_{\log \rho^2}^n \times_\rho \mathbb{R}_1$ be a weighted standard static spacetime which is spatially $\log \widetilde{\rho}^2$-parabolic. Moreover, let $\psi : \Sigma^n \to M_{\log \rho^2}^n \times_\rho \mathbb{R}_1$ be an immersed spacelike hypersurface such that η is bounded. Suppose that the $\log \rho^2$-mean curvature $H_{\log \rho^2}$ and the function $\langle \nabla \rho, \nabla h \rangle$ have opposite signs. If Σ^n lies in a slab, then Σ^n is contained in a slice $M^n \times \{t_0\}$, for some $t_0 \in \mathbb{R}$.*

Proof By (8.6) and Proposition 8.3, we have that

$$\Delta_{\log \rho^2} h = -n\rho^{-2} \Theta H_{\log \rho^2} - \langle \nabla \log \rho^2, \nabla h \rangle = -n\rho^{-2} \Theta H_{\log \rho^2} - \frac{2}{\rho} \langle \nabla \rho, \nabla h \rangle.$$

Taking into account that $H_{\log \rho^2}$ and $\langle \nabla \rho, \nabla h \rangle$ have opposite sign, we conclude that $\Delta_{\log \rho^2} h$ does not change sing.

Therefore, since Proposition 8.1 guarantees the $\log \rho^2$-parabolicity of Σ^n, h must be constant and, consequently, Σ^n is contained in a slice $M^n \times \{t_0\}$, for some $t_0 \in \mathbb{R}$. □

We recall that a spacelike hypersurface Σ^n is said φ-*maximal* if its φ-mean curvature vanishes identically on it.

In this setting, the following meaningfull consequence of Theorem 8.3 can be achieved.

Corollary 8.1 *Let $M^n_{\log \rho^2} \times_\rho \mathbb{R}_1$ be a weighted standard static spacetime which is spatially $\log \widetilde{\rho}^{\,2}$-parabolic. Moreover, let $\psi : \Sigma^n \to M^n_{\log \rho^2} \times_\rho \mathbb{R}_1$ be a $\log \rho^2$-maximal spacelike hypersurface, contained in a slab, such that η is bounded. If the function $\langle \nabla \rho, \nabla h \rangle$ does not change sign, then Σ^n is contained in a slice $M^n \times \{t_0\}$, for some $t_0 \in \mathbb{R}$.*

We also get the next rigidity result.

Theorem 8.4 *Let $M^n_{\log \rho^{-2}} \times_\rho \mathbb{R}_1$ be a weighted standard static spacetime which is spatially $\log \widetilde{\rho}^{\,-2}$-parabolic. Moreover, let $\psi : \Sigma^n \to M^n_{\log \rho^{-2}} \times_\rho \mathbb{R}_1$ be a maximal spacelike hypersurface such that η is bounded and $\inf_{\Sigma^n} \rho > 0$. If $\mathrm{Ric}_{\log \rho^{-2}} \geq \kappa$, for some constant $\kappa > 0$, then Σ^n is contained in a slice $M^n \times \{t_0\}$, for some $t_0 \in \mathbb{R}$.*

Proof Let us observe that, arguing as in the proof of Proposition 8.3, we obtain

$$\Delta h = \mathrm{div}\,(\nabla h) = \rho^2 \langle \nabla \rho^{-2}, \nabla h \rangle - n\rho^{-2} H \Theta = \langle \nabla \log \rho^{-2}, \nabla h \rangle - n\rho^{-2} H \Theta.$$

Therefore, by using (8.6), we get

$$\Delta_{\log \rho^{-2}} h = -n\rho^{-2} H \Theta. \tag{8.28}$$

Now, the Bochner's formula (5.3) yields

$$\frac{1}{2} \Delta_{\log \rho^{-2}} |\nabla h|^2 = |\mathrm{Hess}\, h|^2 + \mathrm{Ric}_{\log \rho^{-2}}(\nabla h, \nabla h) + \langle \nabla \Delta_{\log \rho^{-2}} h, \nabla h \rangle. \tag{8.29}$$

Consequently, taking into account our assumption on $\mathrm{Ric}_{\log \rho^{-2}}$ and bearing in mind that Σ^n is maximal, by (8.28) and (8.29), we obtain that

$$\frac{1}{2} \Delta_{\log \rho^{-2}} |\nabla h|^2 \geq \mathrm{Ric}_{\log \rho^{-2}}(\nabla h, \nabla h) \geq \kappa |\nabla h|^2 \geq 0. \tag{8.30}$$

On the other hand, Proposition 8.1 guarantees that Σ^n is $\log \rho^{-2}$-parabolic. Now, by (8.11), $\inf_{\Sigma^n} \rho > 0$ implies the boundedness of $|\nabla h|$ and, consequently, the boundedness of $|\nabla h|^2$. We conclude that $|\nabla h|^2$ is constant by using the $\log \rho^{-2}$- parabolicity of Σ^n. Hence $\Delta_{\log \rho^2} |\nabla h|^2 = 0$.

Thanks to (8.30), we obtain that $|\nabla h| = 0$ and Σ^n is contained in a slice. □

8.5 Further Rigidity Results in a Weighted Static Spacetime $M_\varphi^n \times \mathbb{R}_1$

We start this section presenting the main analytical tool that we will use in order to prove our rigidity result for complete spacelike hypersurface in a weighted static spacetime $M_\varphi^n \times \mathbb{R}_1$.

The main approach is based on a careful use of the generalized maximum principle of Omori [250] and Yau [297] for the drift Laplacian which we can viewed as a generalization of a Nishikawa's result [248]; see Lemma 2.3.

Proposition 8.4 *Let Σ^n be an n-dimensional complete Riemannian manifold having Ricci curvature bounded from below and let $\varphi : \Sigma^n \to \mathbb{R}$ be a smooth function such that $|\nabla \varphi|$ is bounded on Σ^n. If $u \in C^2(\Sigma^n)$ is nonnegative and satisfies $\Delta_\varphi u \geq a u^b$, for some real numbers $a > 0$ and $b > 1$, then u vanishes identically on Σ^n.*

Proof Let us take the auxiliary function $\zeta = (1+u)^{-\alpha}$, where α is a positive constant to be chosen later, to deduce the following relation

$$\nabla \zeta = -\alpha (1+u)^{-\alpha-1} \nabla u$$

and

$$\Delta \zeta = \left(\frac{\alpha+1}{\alpha}\right) \frac{1}{\zeta} |\nabla \zeta|^2 - \alpha(1+u)^{-\alpha-1} \Delta u.$$

Making use of the above expressions, a direct computation yields

$$\Delta_\varphi \zeta = \left(\frac{\alpha+1}{\alpha}\right) \frac{1}{\zeta} |\nabla \zeta|^2 - \alpha(1+u)^{-\alpha-1} \Delta_\varphi u.$$

Taking $\alpha = (b-1)/2$, it follows that

$$\frac{b-1}{2} \frac{\Delta_\varphi u}{(1+u)^b} = \left(\frac{b+1}{b-1}\right) |\nabla \zeta|^2 - \zeta \Delta_\varphi \zeta. \tag{8.31}$$

Since $\inf_{\Sigma^n} \zeta > -\infty$, we can use the generalized maximum principle of Omori [250] and Yau [297] to obtain a sequence $\{p_k\} \subset \Sigma^n$ such that

$$\zeta(p_k) < \inf_{\Sigma^n} \zeta + \frac{1}{k}, \quad |\nabla\zeta|(p_k) < \frac{1}{k}, \quad \Delta\zeta(p_k) > -\frac{1}{k}.$$

Consequently, since $|\nabla\varphi| < C$ for some constant C, we deduce

$$\Delta_\varphi \zeta(p_k) = \Delta\zeta(p_k) - \langle \nabla\zeta, \nabla\varphi \rangle(p_k)$$
$$> -\frac{1}{k} - |\nabla\zeta|(p_k)|\nabla\varphi|(p_k) > -\frac{1+C}{k}.$$

Hence, thanks to $\Delta_\varphi u \geq au^b$ and (8.31), we have

$$\frac{a(b-1)}{2} \left(\frac{u(p_k)}{1+u(p_k)} \right)^b < \left(\frac{b+1}{b-1} \right) \frac{1}{k^2} + \left(\inf_\Sigma \zeta + \frac{1}{k} \right) \frac{1+C}{k}.$$

Therefore, making $k \to +\infty$, owing to $u(p_k) \to \sup_{\Sigma^n} u$, we conclude that $u \equiv 0$ on the hypersurface Σ^n. □

W are in position now to state and prove the following rigidity result.

Theorem 8.5 *Let $M_\varphi^n \times \mathbb{R}_1$ be a weighted static spacetime, with compact Riemannian base M^n having positive sectional curvature K_M and Bakry-Émery-Ricci tensor satisfying $\widetilde{\mathrm{Ric}}_\varphi \geq c$, for some positive constant c. Suppose that the gradient of the weight function φ is such that $|\widetilde{\nabla}\varphi|^2 \leq 4\alpha$, where $\widetilde{\nabla}\varphi$ stands for the gradient of φ in M^n and $\alpha = \min_M K_M$. If Σ^n is a complete spacelike hypersurface immersed in $M_\varphi^n \times \mathbb{R}_1$ with constant φ-mean curvature and bounded second fundamental form, then Σ^n is a slice $M^n \times \{t_0\}$, for some $t_0 \in \mathbb{R}$.*

Proof By Lemma 2.1, if $X \in \mathfrak{X}(\Sigma^n)$, the following lower bound for the Ricci curvature Ric^Σ of Σ^n holds

$$\mathrm{Ric}^\Sigma(X, X) \geq (n-1)\alpha|X|^2 + \alpha|\nabla h|^2|X|^2 + (n-2)\alpha\langle X, \nabla h \rangle^2$$
$$+ nH\langle AX, X \rangle + |AX|^2.$$

8.5 Further Rigidity Results in a Weighted Static Spacetime $M_\varphi^n \times \mathbb{R}_1$

Thus, by (8.5) we obtain

$$\text{Ric}^\Sigma(X, X) \geq (n-1)\alpha|X|^2 + \alpha|\nabla h|^2|X|^2 + (n-2)\alpha\langle X, \nabla h\rangle^2$$
$$+ \langle \widetilde{\nabla}\varphi, N^*\rangle\langle AX, X\rangle + H_\varphi\langle AX, X\rangle + |AX|^2$$
$$\geq (n-1)\alpha|X|^2 + \left(\alpha - \frac{1}{4}|\widetilde{\nabla}\varphi|^2\right)|\nabla h|^2|X|^2 + (n-2)\alpha\langle X, \nabla h\rangle^2$$
$$- |H_\varphi||A||X|^2 + \left|AX + \frac{\langle \widetilde{\nabla}\varphi, N^*\rangle}{2}X\right|^2. \tag{8.32}$$

Consequently, taking into account our assumption on $|\widetilde{\nabla}\varphi|$, since H_φ is constant and A is bounded, by (8.32) the Ricci curvature of Σ^n is bounded from below.

On the other hand, by Proposition 8.2 we obtain

$$\Delta\Theta = (\widetilde{\text{Ric}}(N^*, N^*) + |A|^2)\Theta + n\langle \nabla H, \partial_t\rangle, \tag{8.33}$$

where $\widetilde{\text{Ric}}$ is the standard Ricci tensor of M^n and $N^* = N + \Theta\partial_t$ is the orthonormal projection of N onto M^n.

Thus, since $\nabla\Theta = A(\nabla h)$ and H_φ is constant, by using (8.11), (8.5) and (8.33) it is easily seen that

$$\frac{1}{2}\Delta|\nabla h|^2 = (\widetilde{\text{Ric}}(N^*, N^*) + |A|^2)\Theta^2 + n\Theta\langle \nabla H, \partial_t\rangle + |A(\nabla h)|^2$$
$$= (\widetilde{\text{Ric}}(N^*, N^*) + |A|^2)\Theta^2 + \Theta\partial_t^\top(\langle \overline{\nabla}\varphi, N\rangle) + |A(\nabla h)|^2. \tag{8.34}$$

Now, we notice that

$$\partial_t^\top(\langle \overline{\nabla}\varphi, N\rangle) = \langle \overline{\nabla}_{\partial_t^\top}\overline{\nabla}\varphi, N\rangle + \langle \overline{\nabla}\varphi, \nabla\Theta\rangle = \Theta\overline{\text{Hess}}\varphi(N, N) + \langle \overline{\nabla}\varphi, \nabla\Theta\rangle.$$

Thus, by (8.34), one has

$$\frac{1}{2}\Delta|\nabla h|^2 = (\widetilde{\text{Ric}}(N^*, N^*) + |A|^2)\Theta^2 + \Theta^2\overline{\text{Hess}}\varphi(N, N) + \frac{1}{2}\langle \nabla\varphi, \nabla|\nabla h|^2\rangle + |A(\nabla h)|^2.$$

Moreover, since $\overline{\text{Hess}}\varphi(N, N) = \widetilde{\text{Hess}}\varphi(N^*, N^*)$, by using (8.3) and (8.6) we obtain that

$$\frac{1}{2}\Delta_\varphi|\nabla h|^2 = (\widetilde{\text{Ric}}_\varphi(N^*, N^*) + |A|^2)\Theta^2 + |A(\nabla h)|^2, \tag{8.35}$$

where $\widetilde{\text{Ric}}_\varphi$ stands for the Bakry-Émery-Ricci tensor of M^n.

Hence, taking into account that $\widetilde{\mathrm{Ric}}_\varphi \geq c > 0$, by (8.35) we deduce

$$\Delta_\varphi |\nabla h|^2 \geq 2c|N^*|_M^2 \Theta^2 = 2c|\nabla h|^2(1 + |\nabla h|^2) \geq 2c|\nabla h|^4.$$

Therefore, we can apply Proposition 8.4 to conclude that $|\nabla h|^2$ vanishes identically on Σ^n and, consequently, Σ^n must be a slice $M^n \times \{t_0\}$, for some $t_0 \in \mathbb{R}$. □

Arguing as in the proof of Theorem 8.5 we also obtain the following rigidity result concerning φ-maximal spacelike hypersurfaces of $M_\varphi^n \times \mathbb{R}_1$, where the boundedness of the second fundamental form is not assumed.

Corollary 8.2 *Let $M_\varphi^n \times \mathbb{R}_1$ be a weighted static spacetime, with compact Riemannian base M^n having positive sectional curvature K_M and Bakry-Émery-Ricci tensor satisfying $\widetilde{\mathrm{Ric}}_\varphi \geq c$, for some positive constant c. Suppose that the gradient of the weight function φ is such that $|\widetilde{\nabla}\varphi|^2 \leq 4\alpha$, where $\widetilde{\nabla}\varphi$ stands for the gradient of φ in M^n and $\alpha = \min_M K_M$. If Σ^n is a complete φ-maximal spacelike hypersurface immersed in $M_\varphi^n \times \mathbb{R}_1$, then Σ^n is a slice $M^n \times \{t_0\}$, for some $t_0 \in \mathbb{R}$.*

Set

$$\mathcal{L}_\varphi^p(\Sigma^n) = \left\{ u : \Sigma^n \to \mathbb{R} : \int_{\Sigma^n} |u|^p e^{-\varphi} d\Sigma^n < +\infty \right\},$$

where $d\Sigma^n$ denotes the standard Riemannian volume element of Σ^n.

The following rigidity result holds.

Theorem 8.6 *Let $M_\varphi^n \times \mathbb{R}_1$ be a weighted static spacetime, with complete noncompact Riemannian base M^n having nonnegative Bakry-Émery-Ricci tensor $\widetilde{\mathrm{Ric}}_\varphi$. Let Σ^n be a complete (noncompact) spacelike hypersurface immersed in $M_\varphi^n \times \mathbb{R}_1$ with constant φ-mean curvature H_φ. If $|\nabla h|^2 \in \mathcal{L}_\varphi^p(\Sigma^n)$ for some $p > 1$, then Σ^n is a slice $M^n \times \{t_0\}$, for some $t_0 \in \mathbb{R}$.*

Proof Since H_φ is constant and $\widetilde{\mathrm{Ric}}_\varphi$ is nonnegative, by (8.35) we have that

$$\frac{1}{2}\Delta_\varphi |\nabla h|^2 \geq (\widetilde{\mathrm{Ric}}_\varphi(N^*, N^*) + |A|^2)\Theta^2 \geq 0.$$

Now, since $|\nabla h|^2 \in \mathcal{L}_\varphi^p(\Sigma^n)$ for some $p > 1$, by Lemma 7.2 we infer that $|\nabla h| \equiv C$ for some nonnegative constant C. Moreover, we notice that, A vanishes identically on Σ^n and then, this allow us to conclude that

$$\mathrm{Ric}_\varphi^\Sigma(X, X) = \widetilde{\mathrm{Ric}}_\varphi(X^*, X^*) \geq 0,$$

for all $X \in \mathfrak{X}(\Sigma^n)$.

Now, if $C \neq 0$ by Lemma 7.3 we obtain that $\mathrm{vol}_\varphi(\Sigma^n) = +\infty$. However, this fact contradicts $|\nabla h|^2 \in \mathcal{L}_\varphi^p(\Sigma^n)$. Hence, we obtain that $|\nabla h|$ vanishes identically on Σ^n. Therefore, Σ^n must be a slice $M^n \times \{t_0\}$, for some $t_0 \in \mathbb{R}$. □

8.6 Entire Killing Graphs

According to [134], we define the *entire Killing graph* $\Sigma(u)$ associated to a smooth function $u \in C^\infty(M^n)$ the hypersurface given by

$$\Sigma(u) = \{\Psi(x, u(x)) : x \in M^n\} \subset M^n \times_\rho \mathbb{R}_1.$$

The metric induced on M^n from the Lorentzian metric (8.2) via $\Sigma(u)$ is given by

$$\langle \cdot, \cdot \rangle_u = \langle \cdot, \cdot \rangle_M - \rho^2 du^2.$$

Moreover, $\Sigma(u)$ is spacelike if and only if $\rho^2 |Du|_M^2 < 1$, where Du denotes the gradient of a function u with respect to the metric $\langle \cdot, \cdot \rangle_M$ of M^n. Indeed, if $\Sigma(u)$ is spacelike, then

$$0 < \langle Du, Du \rangle_u = \langle Du, Du \rangle_M - \rho^2 \langle Du, Du \rangle_M^2$$

and, hence, we conclude that $\rho^2 |Du|_M^2 < 1$.

Conversely, if $\rho^2 |Du|_M^2 < 1$ and X is a vector field tangent to $\Sigma(u)$, we obtain, by Cauchy-Schwarz inequality

$$\langle X, X \rangle_u = \langle X^*, X^* \rangle_M - \rho^2 \langle Du, X^* \rangle_M^2 \geq \langle X^*, X^* \rangle_M (1 - \rho^2 |Du|_M^2),$$

where X^* is the orthogonal projection of X onto TM^n. Thus, $\langle X, X \rangle_u \geq 0$ and $\langle X, X \rangle_u = 0$ if and only if $X = 0$.

The function $g : M^n \times \mathbb{R}_1 \to \mathbb{R}$ given by $g(x, t) = u(x) - t$ is such that $\Sigma(u) = \Psi(g^{-1}(0))$. Thus, for each vector field X tangent to $M^n \times_\rho \mathbb{R}_1$, we have

$$X(g) = X^*(g) - \frac{1}{\rho^2} \langle X, \partial_t \rangle \partial_t(g) = \left\langle \frac{1}{\rho^2} \partial_t + Du, X \right\rangle.$$

Hence,

$$\overline{\nabla} g = \frac{1}{\rho^2} \partial_t + Du$$

is a normal vector field on $g^{-1}(0)$ and, consequently

$$N_0 = \Psi_*(\overline{\nabla} g) = \frac{1}{\rho^2} Y + \Psi_*(Du)$$

is a normal timelike vector field on $\Sigma(u)$.

Since,

$$|N_0| = \frac{(1 - \rho^2 |Du|_M^2)^{1/2}}{\rho},$$

it follows that

$$N = \frac{N_0}{|N_0|} = \frac{1}{\rho(1 - \rho^2 |Du|_M^2)^{1/2}} (Y + \rho^2 \Psi_*(Du)), \qquad (8.36)$$

defines the future-pointing Gauss map of $\Sigma(u)$ such that its angle function is given by

$$\Theta = \langle N, Y \rangle = -\frac{\rho}{(1 - \rho^2 |Du|_M^2)^{1/2}} < 0. \qquad (8.37)$$

Moreover, for each vector field X tangent to M^n, the shape operator A of $\Sigma(u)$ with respect to N is given by

$$\begin{aligned}
AX = & -\frac{\rho}{(1 - \rho^2 |Du|_M^2)^{1/2}} D_X Du \\
& - \frac{\rho^3 \langle D_X Du, Du \rangle}{(1 - \rho^2 |Du|_M^2)^{3/2}} Du - \frac{\rho^2 \langle D\rho, X \rangle |Du|_M^2}{(1 - \rho^2 |Du|_M^2)^{3/2}} Du \\
& - \frac{\langle D\rho, X \rangle}{(1 - \rho^2 |Du|_M^2)^{1/2}} Du - \frac{\langle Du, X \rangle}{(1 - \rho^2 |Du|_M^2)^{1/2}} D\rho,
\end{aligned} \qquad (8.38)$$

where D denotes the Levi-Civita connections in M^n.

8.6.1 The φ-Mean Curvature Equation in $M_\varphi^n \times_\rho \mathbb{R}_1$

By (8.38) it follows that the mean curvature H_u of a spacelike entire Killing graph $\Sigma(u)$ is given by

$$n H(u) = \mathrm{Div}\left(\frac{\rho Du}{(1 + \rho^2 |Du|_M^2)^{1/2}}\right) + \frac{\langle Du, D\rho \rangle}{(1 + \rho^2 |Du|_M^2)^{1/2}},$$

8.6 Entire Killing Graphs

where Div stands for the divergence operator on M^n with respect to the metric $\langle \cdot, \cdot \rangle_M$. A direct computation shows that the φ-mean curvature is given by

$$n(H_u)_\varphi = \text{Div}_\varphi \left(\frac{\rho Du}{(1 - \rho^2 |Du|_M^2)^{1/2}} \right) + \frac{\langle Du, D\rho \rangle}{(1 - \rho^2 |Du|_M^2)^{1/2}}.$$

Consequently, an entire Killing graph $\Sigma(u)$ is spacelike with constant φ-mean curvature C if and only if the function $u \in C^\infty(M^n)$ satisfies the following elliptic partial differential equation of φ-divergence form

$$\begin{cases} \text{Div}_\varphi \left(\dfrac{\rho Du}{(1 - \rho^2 |Du|_M^2)^{1/2}} \right) + \dfrac{\langle Du, D\rho \rangle}{(1 - \rho^2 |Du|_M^2)^{1/2}} = C, & \text{in } M^n \\ \rho^2 |Du|_M^2 < 1. \end{cases} \quad (8.39)$$

We will use the theorems obtained in the previous section on the entire Killing graph setting to obtain uniqueness results for equations of the type given in (8.39).

By Theorem 8.1 the following result holds.

Theorem 8.7 *Let $M_\varphi^n \times_\rho \mathbb{R}_1$ be a weighted standard static spacetime which is spatially $\widetilde{f}$-parabolic with convex warping function ρ, $\langle \widetilde{\nabla}\varphi, \widetilde{\nabla}\rho \rangle \leq 0$ and $\widetilde{\text{Ric}}_\varphi \geq 0$. If the entire Killing graph $\Sigma(u)$ associated to $u \in C^\infty(M^n)$ is such that $\rho|_{\Sigma(u)}$ is bounded and $\widetilde{\text{Ric}}_\varphi$ is positive at some point $p_0 \in \Sigma(u)$, the only solutions of the problem*

$$\begin{cases} \text{Div}_\varphi \left(\dfrac{\rho Du}{(1 - \rho^2 |Du|_M^2)^{1/2}} \right) + \dfrac{\langle Du, D\rho \rangle}{(1 - \rho^2 |Du|_M^2)^{1/2}} = C, & u \in C^\infty(M^n) \\ \sup_{\Sigma(u)} \left(\rho^2 |Du|_M^2 \right) < 1, \end{cases}$$

are the constant ones.

Proof Since $\sup \rho^2 |Du|_M^2 < 1$, by (8.37), the boundedness of $\rho|_{\Sigma(u)}$ is equivalent to the boundedness of Θ. Furthermore, we observe that the condition $\sup \rho^2 |Du|_M^2 < 1$ also implies the boundedness of η. Indeed, by using (8.37), we have that

$$\eta = \frac{1}{(1 - \rho^2 |Du|_M^2)^{1/2}}.$$

Hence, arguing as in Theorem 8.1 the conclusion is achieved. □

An important example of weighted Riemannian manifold is the so-called *Gaussian space* $\mathbb{G}^n$ which corresponds to the Euclidean space $\mathbb{R}^n$ endowed with the Gaussian probability measure

$$e^{-\varphi}dx^2 = (2\pi)^{-\frac{n}{2}}e^{-\frac{|x|^2}{2}}dx^2.$$

In weighted product spaces $\mathbb{G}^n \times \mathbb{R}_1$, in [38, Theorem 4] the authors extended the classical Bernstein's Theorem [77] showing that the only weighted minimal graphs $\Sigma^n(u)$ of functions $u(x_2, \ldots, x_{n+1}) = x_1$ over $\mathbb{G}^n$, with $\sup_{\Sigma(u)} |Du|_\mathbb{G} < 1$, are the hyperplanes $x_1 = \text{const}$.

By Theorem 8.7 an extension of [38, Theorem 4] is given below.

Corollary 8.3 *Consider the weighted standard static spacetime $\mathbb{G}^n \times_\rho \mathbb{R}_1$, where $\mathbb{G}^n$ is the Gaussian space and the warping function ρ is convex with $\langle \widetilde{\nabla}\varphi, \widetilde{\nabla}\rho \rangle \leq 0$. If the entire Killing graph $\Sigma(u)$ associated to $u \in C^\infty(\mathbb{G}^n)$ is such that $\rho|_{\Sigma(u)}$ is bounded, the only solutions of the problem*

$$\begin{cases} \mathrm{Div}_\varphi \left(\dfrac{\rho Du}{(1 - \rho^2 |Du|^2_{\mathbb{G}^n})^{1/2}} \right) + \dfrac{\langle Du, D\rho \rangle}{(1 - \rho^2 |Du|^2_{\mathbb{G}^n})^{1/2}} = C, \quad u \in C^\infty(\mathbb{G}^n) \\ \sup_{\Sigma(u)} \left(\rho^2 |Du|^2_{\mathbb{G}^n} \right) < 1, \end{cases}$$

are the constant ones.

Proof Since $\mathrm{vol}_\varphi(\mathbb{G}^n) = 1$, by [211, Remark 3] one has that $\mathbb{G}^n$ is φ-parabolic. Moreover, a direct computation ensures that $\widetilde{\mathrm{Ric}}_\varphi = 1$. Therefore, since $\mathbb{G}^n$ is also simply connected, the result follows by Theorem 8.7. □

The next result is an application of Theorem 8.2.

Theorem 8.8 *Let $M^n_\varphi \times_\rho \mathbb{R}_1$ be a weighted standard static spacetime which is spatially $\tilde{f}$-parabolic with convex warping function ρ, $\langle \widetilde{\nabla}\varphi, \widetilde{\nabla}\rho \rangle \leq 0$ and $\widetilde{\mathrm{Ric}}_\varphi \geq -\kappa$, for some constant $\kappa > 0$. If the entire Killing graph $\Sigma(u)$ associated to u is such that $\rho|_{\Sigma(u)}$ is bounded and $\alpha \in (0, 1)$ is a constant, the only solutions of the problem*

$$\begin{cases} \mathrm{Div}_\varphi \left(\dfrac{\rho Du}{(1 - \rho^2 |Du|^2_M)^{1/2}} \right) + \dfrac{\langle Du, D\rho \rangle}{(1 - \rho^2 |Du|^2_M)^{1/2}} = C, \quad u \in C^\infty(M^n) \\ \sup_{\Sigma(u)} \left(\rho^2 |Du|^2_M \right) < \dfrac{\alpha |A|^2}{\alpha |A|^2 + \kappa}, \end{cases}$$

are the constant ones.

8.6 Entire Killing Graphs

Proof By (8.36), we have that

$$N^* = N - N^{\perp} = \frac{\rho \Psi_*(Du)}{(1 - \rho^2 |Du|_M^2)^{1/2}}. \tag{8.40}$$

So, by Eq. (8.40) it follows that

$$|N^*|_M^2 = \frac{\rho^2 |Du|_M^2}{1 - \rho^2 |Du|_M^2}. \tag{8.41}$$

Thus, Eqs. (8.11) and (8.41) give us the following relation

$$|\nabla h|^2 = \frac{|Du|_M^2}{1 - \rho^2 |Du|_M^2}. \tag{8.42}$$

Now, by using (8.42) we conclude that the hypothesis (8.24) is equivalent to

$$\rho^2 |Du|_M^2 \le \frac{\alpha |A|^2}{\alpha |A|^2 + \kappa}.$$

Furthermore, since $\kappa > 0$, we have that that

$$\frac{\alpha |A|^2}{\alpha |A|^2 + \kappa} \le 1.$$

The conclusion follows by Theorem 8.2. □

Arguing as in Corollary 8.3, the next result holds.

Corollary 8.4 *Consider the weighted standard static spacetime $\mathbb{G}^n \times_\rho \mathbb{R}_1$, where $\mathbb{G}^n$ is the Gaussian space and the warping function ρ is convex with $\langle \widetilde{\nabla} \varphi, \widetilde{\nabla} \rho \rangle \le 0$. If the entire Killing graph $\Sigma(u)$ associated to $u \in C^\infty(\mathbb{G}^n)$ is such that $\rho|_{\Sigma(u)}$ is bounded, then, for any constants $k > 0$ and $\alpha \in (0, 1)$, the only solutions of the problem*

$$\begin{cases} \mathrm{Div}_\varphi \left(\dfrac{\rho Du}{(1 - \rho^2 |Du|_{\mathbb{G}^n}^2)^{1/2}} \right) + \dfrac{\langle Du, D\rho \rangle}{(1 - \rho^2 |Du|_{\mathbb{G}^n}^2)^{1/2}} = C, \quad u \in C^\infty(\mathbb{G}^n) \\[2mm] \sup_{\Sigma(u)} \left(\rho^2 |Du|_{\mathbb{G}^n}^2 \right) < \dfrac{\alpha |A|^2}{\alpha |A|^2 + \kappa}, \end{cases}$$

are the constant ones.

By Theorem 8.3, the next result holds.

Theorem 8.9 *Let $M^n_{\log \rho^2} \times_\rho \mathbb{R}_1$ be a weighted standard static spacetime which is spatially $\log \tilde{\rho}^2$-parabolic. If the entire Killing graph associate to u is such that $\langle \nabla \rho, \Psi_*(Du) \rangle$ does not change sign, then the only bounded solutions of the problem*

$$\begin{cases} \mathrm{Div}_{\log \rho^2} \left(\dfrac{\rho Du}{(1 - \rho^2|Du|_M^2)^{1/2}} \right) + \dfrac{\langle Du, D\rho \rangle}{(1 - \rho^2|Du|_M^2)^{1/2}} = C, & u \in C^\infty(M^n) \\ \sup_{\Sigma(u)} \left(\rho^2|Du|_M^2 \right) < 1. \end{cases}$$

are the constant ones.

Proof Let us observe that

$$\langle \nabla \rho, \nabla N \rangle = \nabla h(\rho) = -\frac{1}{\rho^2} Y^\top(\rho) = -\frac{1}{\rho^2} Y^\top((-\langle Y, Y \rangle)^{\frac{1}{2}})$$

$$= -\frac{1}{\rho^2} (\frac{1}{2}(-\langle Y, Y \rangle)^{\frac{1}{2}} Y^\top \langle Y, Y \rangle) = -\frac{1}{2\rho^3} Y^\top \langle Y, Y \rangle$$

$$= -\frac{1}{\rho^3} \langle \overline{\nabla}_{Y^\top} Y, Y \rangle = -\frac{1}{\rho^3} \langle \overline{\nabla}_{Y+\Theta N} Y, Y \rangle$$

$$= -\frac{1}{\rho^3} (\underbrace{\langle \overline{\nabla}_Y Y, Y \rangle}_{=0} + \langle \overline{\nabla}_{\Theta N} Y, Y \rangle) = -\frac{1}{\rho^3} \langle \overline{\nabla}_{\Theta N} Y, Y \rangle$$

$$= -\frac{\Theta}{\rho^3} \langle \overline{\nabla}_N Y, Y \rangle = -\frac{\Theta}{2\rho^3} N \langle Y, Y \rangle = -\frac{\Theta}{2\rho^3} N(\rho^2)$$

$$= -\frac{\Theta}{2\rho^3} \cdot 2\rho N^*(\rho) = \frac{\Theta}{\rho^2} \langle \overline{\nabla} \rho, N^* \rangle. \tag{8.43}$$

Thus, by (8.43) and (8.40) we obtain

$$\langle \nabla \rho, \nabla N \rangle = \frac{\Theta}{\rho} \left\langle \overline{\nabla} \rho, \frac{\rho \Psi_*(Du)}{(1 - \rho^2|Du|_M^2)^{1/2}} \right\rangle = \frac{\Theta}{\rho(1 - \rho^2|Du|_M^2)^{1/2}} \langle \overline{\nabla} \rho, \Psi_*(Du) \rangle.$$

Therefore, $\langle \nabla \rho, \nabla N \rangle$ do not change sign if and only if $\langle \overline{\nabla} \rho, \Psi_*(Du) \rangle$ do not change sign. The result follows by Corollary 8.1. □

8.6 Entire Killing Graphs

Taking

$$\rho = \left(e^{\frac{|x|^2}{2} + \log(2\pi)^{n/2}} \right)^{1/2} \tag{8.44}$$

in Theorem 8.9, we obtain the following consequence.

Corollary 8.5 *Consider the weighted standard static spacetime $\mathbb{G}^n \times_\rho \mathbb{R}_1$, where $\mathbb{G}^n$ is the Gaussian space and ρ is defined in (8.44). If the entire Killing graph associate to u is such that $\langle \nabla \rho, \Psi_*(Du) \rangle$ does not change sign, then the only bounded solutions of the problem*

$$\begin{cases} \operatorname{Div}_\varphi \left(\dfrac{\rho Du}{(1 - \rho^2 |Du|_{\mathbb{G}^n}^2)^{1/2}} \right) + \dfrac{\langle Du, D\rho \rangle}{(1 - \rho^2 |Du|_{\mathbb{G}^n}^2)^{1/2}} = C, \quad u \in C^\infty(\mathbb{G}^n) \\ \sup_{\Sigma(u)} \left(\rho^2 |Du|_{\mathbb{G}^n}^2 \right) < 1, \end{cases}$$

are the constant ones.

By Theorem 8.4 the following result holds.

Theorem 8.10 *Let $M^n_{\log \rho^{-2}} \times_\rho \mathbb{R}_1$ be a weighted standard static spacetime which is spatially $\log \widetilde{\rho}^{-2}$-parabolic. If the entire Killing graph associate to u is such that $|Du|_M^2$ is bounded and $\operatorname{Ric}_{\log \rho^{-2}} \geq \kappa$, for some constant $\kappa > 0$, then the only bounded solutions of the problem*

$$\begin{cases} \operatorname{Div} \left(\dfrac{\rho Du}{(1 - \rho^2 |Du|_M^2)^{1/2}} \right) + \dfrac{\langle Du, D\rho \rangle}{(1 - \rho^2 |Du|_M^2)^{1/2}} = 0, \quad u \in C^\infty(M^n) \\ \sup_{\Sigma(u)} \left(\rho^2 |Du|_M^2 \right) < 1, \end{cases} \tag{8.45}$$

are the constant ones.

Proof We observe that if $u \in C^\infty(M)$ is solution of problem (8.45), then the entire Killing graph $\Sigma(u)$ is spacelike and maximal. Moreover using (8.42), we note that the boundedness of $|\nabla h|^2$ follows by the boundedness of $|Du|_M^2$. Then, the result follows by Theorem 8.4. □

Next, considering

$$\rho = \left(e^{\frac{|x|^2}{2} + \log(2\pi)^{n/2}}\right)^{-1/2} \tag{8.46}$$

in Theorem 8.10, we have the next result.

Corollary 8.6 *Consider the weighted standard static spacetime $\mathbb{G}^n \times_\rho \mathbb{R}_1$, where $\mathbb{G}^n$ is the Gaussian space and ρ is defined in (8.46). If the entire Killing graph associate to u is such that $|Du|^2_{\mathbb{G}^n}$ is bounded and $\mathrm{Ric}_{\log \rho^{-2}} \geq \kappa$, for some constant $\kappa > 0$, then the only bounded solutions of the problem*

$$\begin{cases} \mathrm{Div}\left(\dfrac{\rho Du}{(1 - \rho^2 |Du|^2_{\mathbb{G}^n})^{1/2}}\right) + \dfrac{\langle Du, D\rho \rangle}{(1 - \rho^2 |Du|^2_{\mathbb{G}^n})^{1/2}} = 0, \quad u \in C^\infty(\mathbb{G}^n) \\ \sup_{\Sigma(u)} \left(\rho^2 |Du|^2_{\mathbb{G}^n}\right) < 1, \end{cases}$$

are the constant ones.

It is easily seen that Theorem 8.5 can be applied to get the following Calabi-Bernstein type result.

Corollary 8.7 *Let $M^n_\varphi \times \mathbb{R}_1$ be a weighted static spacetime, with compact Riemannian base M^n having positive sectional curvature K_M and Bakry-Émery-Ricci tensor satisfying $\widetilde{\mathrm{Ric}}_\varphi \geq c$, for some positive constant c. Suppose that the gradient of the weight function φ is such that $|\widetilde{\nabla}\varphi|^2 \leq 4\alpha$, where $\alpha = \min_M K_M$. The only entire solutions of the following modified φ-mean curvature equation*

$$\begin{cases} \mathrm{Div}_\varphi\left(\dfrac{Du}{\sqrt{1 - |Du|^2_M}}\right) = C \\ |Du|_M \leq \alpha, \end{cases}$$

where C and $0 < \alpha < 1$ are constant, are the constant functions $u = t_0$.

Finally, we also quote a nonparametric version of Theorem 8.6. To this aim, let us consider the following set

$$\mathcal{L}^p_\varphi(M^n) = \left\{u : M^n \to \mathbb{R} : \int_{M^n} |u|^p e^{-\varphi} dM^n < +\infty\right\},$$

where dM^n stands for the Riemannian volume element of M^n.

8.6 Entire Killing Graphs

Corollary 8.8 *Let $M_\varphi^n \times \mathbb{R}_1$ be a weighted static spacetime, with complete noncompact Riemannian fiber M^n having nonnegative Bakry-Émery-Ricci tensor $\widetilde{\mathrm{Ric}}_\varphi$. The only entire solutions of the following modified φ-mean curvature equation*

$$\begin{cases} \mathrm{div}_\varphi \left(\dfrac{Du}{\sqrt{1 - |Du|_M^2}} \right) = C \\ |Du|_M^2 \in \mathcal{L}_\varphi^p(M^n), \\ |Du|_M \leq \alpha, \end{cases}$$

where C and $0 < \alpha < 1$ are constant and $p > 1$, are the constant functions $u = t_0$.

Part III

Submanifolds Immersed in Semi-Riemannian Warped Products

9 Submanifolds in a Riemannian Warped Product

9.1 Introduction

Conformal Killing vector fields are important objects which have been widely used in order to understand the geometry of submanifolds immersed in Riemannian spaces. In this setting, Montiel [242] studied constant mean curvature compact hypersurfaces immersed in warped products of the type $\mathbb{R} \times_f M^n$ and $\mathbb{S}^1 \times_f M^n$.

We observe that such class of warped products are endowed with a globally defined conformal Killing vector field given by $f \partial_t$, where ∂_t stands for the unit vector field tangent to either $\mathbb{R}$ or $\mathbb{S}^1$. By supposing that such hypersurfaces are locally graphs on M^n, Montiel proved that (up to exceptional well-understood cases) they must be slices of the form $\{t\} \times M^n$.

Later on, this thematic was revisited in [21] by Alías and Dajczer, where they generalized Montiel's results considering complete, not necessarily compact, hypersurfaces immersed in $\mathbb{R} \times_f M^n$. Afterwards, Caminha and de Lima [98] and Aquino and de Lima [45] investigated the uniqueness of complete vertical graphs with constant mean curvature in a warped product $I \times_f M^n$. Under suitable restrictions on the values of the mean curvature and the norm of the gradient of the height function, they obtained uniqueness theorems concerning to such graphs.

In [31], Alías et al. characterized compact and complete hypersurfaces with some constant higher order mean curvature into warped product spaces. Their approach was based on the use of a new version of the Omori-Yau's maximum principle for a trace type operator which seems to be interesting in its own.

When the ambient space is a Riemannian product of the type $M^n \times \mathbb{R}$, it was shown by Rosenberg et al. [267] that if the Ricci curvature of the base M^n is nonnegative and its sectional curvature is bounded from below, then any entire minimal graph over M^n with nonnegative height function must be a slice. This result extends the celebrated theorem due to Bombieri et al. [89] for entire minimal hypersurfaces in the Euclidean space. In [145], de Lima et al. studied complete two-sided hypersurfaces immersed in $M^n \times \mathbb{R}$, whose base is also supposed to have sectional curvature bounded from below.

In this setting, they extended a technique developed in [141] obtaining sufficient conditions which assure that such a hypersurface is a slice of the ambient space, provided that its angle function has some suitable behavior.

Motivated by these works, in this chapter we deal with n-dimensional submanifolds immersed in a slab of a warped product of the type $I \times_f M^{n+p-1}$. Under suitable constraints on the warping function f and assuming that such a submanifold Σ^n is either complete or stochastically complete, we apply some maximum principles in order to show that Σ^n must be contained in a slice of $I \times_f M^{n+p-1}$. In particular, as a byproduct of our results we guarantee the nonexistence of n-dimensional closed minimal submanifolds immersed in $\mathbb{R}^m \times \mathbb{H}^{n+1}$. We also construct a nontrivial duo-graph in $\mathbb{R} \times \mathbb{H}^2 \times \mathbb{R}$ which illustrates the importance of our rigidity results.

Finally, we study n-dimensional complete submanifolds immersed in a weighted warped product of the type $I \times_f M_\varphi^{n+p-1}$, whose warping function f has convex logarithm and φ that does not depend on the real parameter $t \in I$. Assuming the constancy of an appropriate support function involving the φ-mean curvature vector field of the submanifold Σ^n jointly with suitable constraints on the Bakry-Émery-Ricci tensor of Σ^n, we apply a weak Omori-Yau's generalized maximum principle and Liouville-type results for the drift Laplacian in order to prove that Σ^n must be contained in a slice of the ambient space.

As applications, we obtain codimension reductions and Bernstein-type results for complete φ-minimal bounded multi graphs constructed over the n-dimensional Gaussian space.

The results present in this chapter are mainly based on the papers [52–54].

9.2 Recalling Some Basic Facts

In this section we introduce some basic facts and notations that will appear along this chapter.

More precisely, let M^{n+p-1} be a connected $(n+p-1)$-dimensional oriented Riemannian manifold, $I \subset \mathbb{R}$ an open interval and $f : I \to \mathbb{R}$ a positive smooth function. Moreover, in the product manifold $\overline{M}^{n+p} = I \times M^{n+p-1}$ denote by π_I and π_M the projections onto the factors I and M^{n+p-1}, respectively.

The class of manifolds that we consider here is the one obtained by endowing $\overline{M}^{n+p}$ with the metric

$$\langle v, w \rangle_q = \langle (\pi_I)_* v, (\pi_I)_* w \rangle_{\pi_I(q)} + ((f \circ \pi_I)(q))^2 \langle (\pi_M)_* v, (\pi_M)_* w \rangle_{\pi_M(q)}, \quad (9.1)$$

for every $q \in \overline{M}^{n+p}$ and $v, w \in T_q \overline{M}^{n+p}$.

9.2 Recalling Some Basic Facts

In this case, we simply write

$$I \times_f M^{n+p-1}.$$

We observe that the vector field given by

$$K(t, x) = f(t)\partial_t|_{(t,x)}, \quad (t, x) \in I \times_f M^{n+p-1}, \tag{9.2}$$

determines a non-vanishing conformal vector field on $I \times_f M^{n+p-1}$ which is also closed, in the sense that its metrically equivalent 1-form is closed.

Indeed, by (9.2) we deduce that

$$\overline{\nabla}_V K = f'(t) V \tag{9.3}$$

for every vector field V on $I \times_f M^{n+p-1}$, where $\overline{\nabla}$ denotes the Levi-Civita connection of $I \times_f M^{n+p-1}$.

For every $\tau \in I$, the slice $M_\tau = \{\tau\} \times M \subset I \times_f M^{n+p-1}$ is a hypersurface. Actually, the induced metric on M_τ is given by $f(\tau)^2 \langle \cdot, \cdot \rangle_M$, which means that M_τ is homotetic to M with scale factor $f(\tau)$. The restriction of ∂_t to M_τ gives an orientation for it. By (9.2) it follows that

$$\overline{\nabla}_V \partial_t = \overline{\nabla}_V \left(\frac{1}{f(t)} K\right) = -\frac{1}{f(t)^2} \langle V, \overline{\nabla} \overline{f} \rangle K + \frac{1}{f(t)} f'(t) V \tag{9.4}$$

for every $V \in \mathfrak{X}(I \times_f M^{n+p-1})$, where $\overline{\nabla} \overline{f}$ denotes the gradient on $I \times_f M^{n+p-1}$ of $\overline{f}(t, x) = f(t)$. Observe that the gradient on $I \times_f M^{n+p-1}$ of the projection $\pi_I(t, x) = t$ is given by

$$\overline{\nabla} \pi_I = \langle \overline{\nabla} \pi_I, \partial_t \rangle \partial_t = \partial_t. \tag{9.5}$$

Hence, writing $\overline{f} = f \circ \pi_I$, by (9.5) we get

$$\overline{\nabla} \overline{f} = f'(t) \overline{\nabla} \pi_I = f'(t) \partial_t$$

and (9.4) becomes

$$\overline{\nabla}_V \partial_t = \frac{f'(t)}{f(t)} \left(V - \langle V, \partial_t \rangle \partial_t\right), \tag{9.6}$$

for every $V \in \mathfrak{X}(I \times_f M^{n+1})$.

In particular, one has

$$\overline{\nabla}_v \partial_t = \frac{f'(\tau)}{f(\tau)} v,$$

for every tangent vector $v \in T_{(\tau,x)} M_\tau$. This means that M_τ is a totally umbilical hypersurface in $I \times_f M^{n+p}$ with shape operator (with respect to the orientation ∂_t) given by

$$A_\tau v = -\overline{\nabla}_v \partial_t = -\frac{f'(\tau)}{f(\tau)} v,$$

for every $v \in T_{(\tau,x)} M_\tau$.

Therefore, the correspondence $I \ni \tau \mapsto M_\tau \subset I \times_f M^{n+p-1}$ determines a foliation of $I \times_f M^{n+p-1}$ by totally umbilical hypersurface with constant mean curvature given by

$$\mathcal{H}(\tau) = \frac{1}{n+p-1} \mathrm{tr}(A_\tau) = -\frac{f'(\tau)}{f(\tau)}. \tag{9.7}$$

Let Σ^n be a codimension p submanifold immersed into a $I \times_f M^{n+p-1}$. In other words, Σ^n is an n-dimensional connected manifold for which there exists a smooth immersion $\psi : \Sigma^n \to I \times_f M^{n+p-1}$.

As usual, we denote by $\langle \cdot, \cdot \rangle$ the induced metric. In this setting, we denote by $\overline{\nabla}$ and ∇ the Levi-Civita connections of $I \times_f M^{n+p-1}$ and Σ^n, respectively. The Gauss formula of Σ^n in $I \times_f M^{n+p-1}$ is given by

$$\overline{\nabla}_X Y = \nabla_X Y + \alpha(X, Y), \tag{9.8}$$

for every tangent vector fields $X, Y \in \mathfrak{X}(\Sigma^n)$. Here $\alpha : \mathfrak{X}(\Sigma^n) \times \mathfrak{X}(\Sigma^n) \to \mathfrak{X}^\perp(\Sigma^n)$ stands for the vector valued second fundamental form of Σ^n, which is defined by

$$\alpha(X, Y) = (\overline{\nabla}_X Y)^\perp, \tag{9.9}$$

where $(\overline{\nabla}_X Y)^\perp$ denotes the normal component of $\overline{\nabla}_X Y$ along Σ^n.

Moreover, the Weingarten formula is given by

$$\overline{\nabla}_X \eta = -A_\eta X + \nabla_X^\perp \eta, \tag{9.10}$$

for every tangent vector field $X \in \mathfrak{X}(\Sigma^n)$ and normal vector field $\eta \in \mathfrak{X}^\perp(\Sigma)$, where $\nabla^\perp$ is just the normal connection of Σ^n and $A_\eta : \mathfrak{X}(\Sigma^n) \to \mathfrak{X}(\Sigma^n)$ denotes the shape operator with respect to η; that is, the self-adjoint operator on $\mathfrak{X}(\Sigma^n)$ defined by

$$\langle A_\eta X, Y \rangle = \langle \alpha(X, Y), \eta \rangle,$$

for all $X, Y \in \mathfrak{X}(\Sigma^n)$.

9.2 Recalling Some Basic Facts

The mean curvature vector field **H** of Σ^n is defined by

$$\mathbf{H} = \frac{1}{n}\mathrm{tr}(\alpha) = \frac{1}{n}\sum_{i=1}^{n}\alpha(E_i, E_i), \tag{9.11}$$

where $\{E_1, \ldots, E_n\}$ is a local orthonormal frame on Σ^n. Denoting by $\{N_1, N_2, \ldots, N_p\}$ an orthonormal frame of the normal bundle $\mathfrak{X}^\perp(\Sigma^n)$, by (9.11) we get

$$\mathbf{H} = \sum_{i=1}^{p} H_i N_i, \tag{9.12}$$

for some smooth functions $H_1, H_2, \ldots, H_p$ defined on Σ^n.

As for the hypersurfaces, we measure height projecting into the real line. When we restrict it to the submanifold, we get $h : \Sigma^n \to I$ such that $h = \pi_I \circ \psi$. Since the support function depends on the normal that we choose, we consider ξ a normal vector field on Σ^n and we define the support function as $\langle \xi, \partial_t \rangle$, where ξ denotes a normal map of Σ^n.

Let us denote by $\overline{\nabla}$ and ∇ the gradients with respect to the metrics of $I \times_f M^{n+p-1}$ and Σ^n, respectively. Then, a simple computation shows that the gradient of π_I on $I \times_f M^{n+p-1}$ is given by

$$\overline{\nabla}\pi_I = \langle \overline{\nabla}\pi_I, \partial_t \rangle \partial_t = \partial_t, \tag{9.13}$$

so the gradient of h on Σ^n is given by

$$\nabla h = (\overline{\nabla}\pi_I)^\top = \partial_t^\top = \partial_t - \sum_{i=1}^{p}\langle N_i, \partial_t \rangle N_i, \tag{9.14}$$

where $(\cdot)^\top$ denotes the tangential component of a vector field in $\mathfrak{X}(I \times_f M^{n+p-1})$ along Σ^n.

Thus, we get

$$|\nabla h|^2 = 1 - \left(\sum_{i=1}^{p}\langle N_i, \partial_t \rangle^2\right), \tag{9.15}$$

where $|\cdot|$ denotes the norm of a vector field on Σ^n. Finally, denoting $\Theta_i = \langle N_i, \partial_t \rangle$ for every $i \in \{1, 2, \ldots, p\}$, by (9.15) we immediately have

$$|\nabla h|^2 = 1 - \left(\sum_{i=1}^{p}\Theta_i^2\right). \tag{9.16}$$

9.3 Rigidity Results for n-Dimensional Submanifolds in $I \times_f M^{n+p-1}$

This section is devoted to present rigidity results concerning submanifolds of codimension p immersed in $I \times_f M^{n+p-1}$. More precisely, we establish sufficient conditions which guarantee that such submanifold Σ^n be contained in the fiber M^{n+p-1}. The main results are based on some analytic tools.

9.3.1 Rigidity Via a Suitable Integrability Condition

In order to establish our main results, we need some auxiliary results. The first one gives us suitable formulas for the Laplacian of the height function of a submanifold in $I \times_f M^{n+p-1}$.

Lemma 9.1 *Let Σ^n be a submanifold immersed in a warped product $I \times_f M^{n+p-1}$ with height function h and let us consider the function $u = g(h)$, where $g : I \to \mathbb{R}$ is an arbitrary primitive of the warping function f. Then, one has*

(i) $\Delta h = (\log f)'(h)[n - |\nabla h|^2] + n\langle \mathbf{H}, \partial_t \rangle$;
(ii) $\Delta u = n(f'(h) + f(h)\langle \mathbf{H}, \partial_t \rangle)$.

Proof Using a local orthonormal frame $\{e_1, \ldots, e_n\}$ on Σ^n, by (9.3) we obtain that

$$f(h)\mathrm{div}\partial_t = \sum_{i=1}^n \langle f(h)\overline{\nabla}_{e_i}\partial_t, e_i\rangle = \sum_{i=1}^n \langle \overline{\nabla}_{e_i} f(h)\partial_t, e_i\rangle - \sum_{i=1}^n \langle e_i(f(h))\partial_t, e_i\rangle$$

$$= \sum_{i=1}^n \langle f'(h)e_i, e_i\rangle - \sum_{i=1}^n \langle f'(h)e_i(h)\partial_t, e_i\rangle$$

$$= f'(h)(n - \langle \nabla h, \partial_t\rangle)$$

$$= f'(h)\left(n - \langle \nabla h, \partial_t^\top + \sum_{i=1}^p \Theta_i N_i\rangle\right) = f'(h)(n - |\nabla h|^2). \quad (9.17)$$

Then, by (9.17) we get

$$\Delta h = \mathrm{div}(\nabla h) = \mathrm{div}\left(\partial_t^\top\right) = \mathrm{div}\left(\partial_t - \sum_{i=1}^p \Theta_i N_i\right)$$

$$= \mathrm{div}(\partial_t) - \sum_{i=1}^p \mathrm{div}(\Theta_i N_i). \quad (9.18)$$

9.3 Rigidity Results for n-Dimensional Submanifolds in $I \times_f M^{n+p-1}$

Now, for each $i \in \{1, \ldots, p\}$, we have that

$$\mathrm{div}(\Theta_i N_i) = \sum_{j=1}^{n} \langle \overline{\nabla}_{e_j} \Theta_i N_i, e_j \rangle = \sum_{j=1}^{n} \Theta_i \langle \overline{\nabla}_{e_j} N_i, e_j \rangle + \sum_{j=1}^{n} \langle e_j(\Theta_i) N_i, e_j \rangle$$

$$= \sum_{j=1}^{n} \Theta_i \langle -A_{N_i} e_j, e_j \rangle + \langle N_i, \sum_{j=1}^{n} e_j(\Theta_i) e_j \rangle$$

$$= -n \Theta_i H_i + \langle N_i, \nabla \Theta_i \rangle$$

$$= -n \Theta_i H_i. \tag{9.19}$$

Hence, by (9.19) and (9.18) we get

$$\Delta h = (\log f)'(h)(n - |\nabla h|^2) + n \sum_{i=1}^{p} \Theta_i H_i$$

$$= (\log f)'(h)(n - |\nabla h|^2) + n \langle \mathbf{H}, \partial_t \rangle, \tag{9.20}$$

proving item (i).

Let us observe now that

$$\nabla u = f(h) \nabla h = f(h) \nabla \partial_t^\top = K^\top, \tag{9.21}$$

where $K^\top$ denotes the tangential component of the closed conformal vector field K defined in (9.2).

By using (9.8), (9.10) and taking into account that $K = K^\top + K^\perp$, we obtain

$$\overline{\nabla}_X K = \nabla_X K^\top + \alpha(X, K^\top) - A_{K^\perp} X + \nabla_X^\perp K^\perp$$

for every $X \in \mathfrak{X}(\Sigma^n)$. Hence,

$$(\overline{\nabla}_X K)^\top = \nabla_X K^\top - A_{K^\perp} X \tag{9.22}$$

and

$$(\overline{\nabla}_X K)^\perp = \alpha(X, K^\top) + \nabla_X^\perp K^\perp.$$

On the other hand, Eq. (9.3) implies $\overline{\nabla}_X K = f'(h) X$, so that

$$(\overline{\nabla}_X K)^\top = f'(h) X \tag{9.23}$$

and

$$(\overline{\nabla}_X K)^\perp = 0.$$

Thus, by (9.22) and (9.23), we see that

$$\nabla_X K^\top = f'(h)X + A_{K^\perp} X. \tag{9.24}$$

Therefore, by (9.21) and (9.24) we get

$$\nabla_X \nabla u = \nabla_X K^\top = f'(h)X + A_{K^\perp} X.$$

Consequently, we get

$$\Delta u = nf'(h) + \mathrm{tr}(A_{K^\perp}) = n(f'(h) + \langle \mathbf{H}, K \rangle) = n(f'(h) + f(h)\langle \mathbf{H}, \partial_t \rangle). \tag{9.25}$$

In conclusion, item (ii) is proved. □

Lemma 9.2 below is an extension of the Hopf's Theorem on a complete Riemannian manifold due to Yau in [298].

Lemma 9.2 *Let ζ be a smooth function defined on a complete Riemannian manifold Σ^n, such that $\Delta \zeta$ does not change sign on Σ^n. If $|\nabla \zeta| \in \mathcal{L}_g^1(\Sigma^n)$, then $\Delta \zeta$ vanishes identically on Σ^n.*

As usual, $\mathcal{L}_g^1(\Sigma^n)$ denotes the space of Lebesgue integrable functions on Σ^n. We also recall that a *slab* is a bounded region delimited by two slices, that is

$$[t_1, t_2] \times M^{n+p-1} = \left\{ (t,q) \in I \times_f M^{n+p-1} : t_1 \leq t \leq t_2 \right\}.$$

With the above notations we are able to prove our first rigidity result concerning n-dimensional complete submanifolds immersed in $I \times_f M^{n+p-1}$.

Theorem 9.1 *Let Σ^n be a complete submanifold which lies in a slab of a warped product $I \times_f M^{n+p-1}$ whose warping function f is monotone non-decreasing (non-increasing). If $\langle \mathbf{H}, \partial_t \rangle$ is nonnegative (nonpositive) on Σ^n and $|\nabla h| \in \mathcal{L}_g^1(\Sigma^n)$, then Σ^n is contained in a slice $\{t\} \times M^{n+p-1}$.*

Proof Taking into account that f is monotone non-decreasing (non-increasing) and that $\langle \mathbf{H}, \partial_t \rangle$ is nonnegative (nonpositive) on Σ^n, then by item (i) of Lemma 9.1 we obtain that Δh does not change sign on Σ^n.

9.3 Rigidity Results for n-Dimensional Submanifolds in $I \times_f M^{n+p-1}$

Thus, since $|\nabla h| \in \mathcal{L}_g^1(\Sigma^n)$, by Lemma 9.2 we get that $\Delta h = 0$ on Σ^n. On the other hand, since Σ^n lies in a slab of $I \times_f M^{n+p-1}$, we have $|\nabla h^2| \in \mathcal{L}_g^1(\Sigma^n)$. Now, we notice that

$$\Delta h^2 = 2h\Delta h + 2|\nabla h|^2 = 2|\nabla h|^2 \geq 0.$$

Thus, we can apply Lemma 9.2 to obtain that $\Delta h^2 = 0$ and, hence, $|\nabla h| = 0$ on Σ^n. Therefore, h is constant on Σ^n, which means that Σ^n is contained in a slice $\{t\} \times M^{n+p-1}$. □

By Theorem 9.1 the next result holds.

Corollary 9.1 *Let Σ^n be a complete minimal submanifold which lies in a slab of a warped product $I \times_f M^{n+p-1}$ whose warping function f is monotone. If $|\nabla h| \in \mathcal{L}_g^1(\Sigma^n)$, then Σ^n is contained in a slice $\{t\} \times M^{n+p-1}$.*

Taking into account the nonexistence of closed minimal hypersurfaces in the hyperbolic space $\mathbb{H}^{n+1}$ (see, for instance [27, Lemma 8]), we can apply recursively Theorem 9.1 in order to obtain the following nonexistence result

Corollary 9.2 *There do not exist n-dimensional compact minimal submanifolds immersed in $\mathbb{R}^m \times \mathbb{H}^{n+1}$, for all $m \geq 0$.*

The next result is related to a *proper* warped product $I \times_f M^{n+p-1}$, which means that its warping function f is non-locally constant.

Theorem 9.2 *Let Σ^n be a complete submanifold which lies in a slab of a proper warped product $I \times_f M^{n+p-1}$ and such that its mean curvature vector field $\mathbf{H}$ satisfies*

$$(\log f)'(h) \geq |\mathbf{H}|. \tag{9.26}$$

If $|\nabla h| \in \mathcal{L}_g^1(\Sigma^n)$, then Σ^n is contained in a slice $\{t\} \times M^{n+p-1}$.

Proof By Lemma 9.1 and (9.26), we get

$$\Delta u = nf(h)\left((\log f)'(h) + \langle \mathbf{H}, \partial_t \rangle\right) \geq nf(h)\left((\log f)'(h) - |\mathbf{H}|\right). \tag{9.27}$$

Thus, thanks to (9.26), by (9.27) we have that $\Delta u \geq 0$ on the hypersurface Σ^n.
On the other hand, since Σ^n lies in a slab of $I \times_f M^{n+p-1}$, there exists a positive constant C such that

$$|\nabla u| = f(h)|\nabla h| \leq C|\nabla h|.$$

Consequently, since $|\nabla h| \in \mathcal{L}_g^1(\Sigma^n)$, one has $|\nabla u| \in \mathcal{L}_g^1(\Sigma^n)$. So, we can apply Lemma 9.2 to assure that $\Delta u = 0$ on Σ^n. Thus, thanks to (9.27) we get

$$(\log f)'(h) = |\mathbf{H}| = -\langle \mathbf{H}, \partial_t \rangle.$$

Therefore, it holds the equality in Cauchy-Schwarz inequality, that is

$$\mathbf{H} = -(\log f)'(h)\partial_t. \tag{9.28}$$

Now, we claim that $\mathcal{A}_t = \{t\} \times M \cap \Sigma^n$ is open and closed in Σ^n. Clearly, $\mathcal{A}_t$ is closed since it is the intersection of closed sets. It remains to prove that it is open.

To this aim, let us consider two cases.

Case 1 Let $x \in \mathcal{A}_t$ and $f'(t) \neq 0$. In this case, we have that $f'(t(y)) \neq 0$ for any y in a neighborhood of $x \in \mathcal{U} \subset \Sigma^n$. By (9.28) we obtain that ∂_t is orthogonal to $\Sigma^n \cap \mathcal{U}$. Thus, $\nabla h|_{\mathcal{U}} = \partial_t^\top = 0$, that is, $\mathcal{U} \subset \mathcal{A}_t$. Therefore $\mathcal{A}_t$ is open, and $\Sigma^n \subset \mathcal{A}_t$ by connectness.

Case 2 Let $x \in \mathcal{A}_t$ and $f'(t) = 0$. Notice that in this case $\Sigma^n \cap \mathcal{A}_t = \emptyset$ for every $t \in I$ such that $f'(t) \neq 0$, otherwise it would be Case 1. Hence, $f'(t) \equiv 0$ on Σ^n and, since f is not locally constant, we have that Σ^n must be contained in a slice.

We conclude in both the cases that Σ^n is contained in a slice $\{t\} \times M^{n+p-1}$. □

The next result is actually a consequence of the proof of Theorem 9.2.

Corollary 9.3 *Let Σ^n be a complete submanifold which lies in a slab of a warped product $I \times_f M^{n+p-1}$ and such that inequality (9.26) is satisfied. If $|\nabla h| \in \mathcal{L}_g^1(\Sigma^n)$, then Σ^n is either contained in a slice $\{t\} \times M^{n+p-1}$ or it is minimal. In the last case, we must have $f'(t) \equiv 0$ on Σ^n.*

9.3.2 Rigidity Via Omori-Yau Maximum Principle

In the next lemma, we obtain a lower estimate for the Ricci curvature of a codimension p submanifold in $I \times_f M^{n+p-1}$.

To this aim, we will assume that the ambient space obeys a convergence condition which was established by Montiel [242].

Lemma 9.3 *Let Σ^n be a submanifold immersed in a warped product $I \times_f M^{n+p-1}$, which obeys the following convergence condition*

$$K_M \geq \sup_I (f'^2 - ff''). \tag{9.29}$$

9.3 Rigidity Results for n-Dimensional Submanifolds in $I \times_f M^{n+p-1}$

Then, for every $X \in \mathfrak{X}(\Sigma^n)$, the Ricci curvature of Σ^n satisfies the following inequality

$$\mathrm{Ric}(X, X) \geq -n \frac{|f|''}{f} |X|^2 - \sum_{i=1}^{p} \left| A_{N_i} X - \frac{n H_i}{2} X \right|^2 + \frac{n^2 |\mathbf{H}|^2}{4} |X|^2,$$

where $\{N_1, N_2, \ldots, N_p\}$ is a local orthonormal frame of the normal bundle $\mathfrak{X}^\perp(\Sigma^n)$, $\mathbf{H}$ is the mean curvature vector field of Σ^n and the functions $H_1, H_2, \ldots, H_p$ are defined in (9.12).

Proof Let us observe that the curvature tensor R of the submanifold Σ^n can be described in terms of its shape operator and the curvature tensor $\overline{R}$ of the ambient $I \times_f M^{n+p-1}$ via Gauss equation.

More precisely, we have that

$$\langle R(X, Y)Z, W \rangle = \langle \overline{R}(X, Y)Z, W \rangle + \langle \alpha(X, Z), \alpha(Y, W) \rangle$$
$$- \langle \alpha(X, W), \alpha(Y, Z) \rangle, \tag{9.30}$$

for all the tangent vector fields $X, Y, Z \in \mathfrak{X}(\Sigma^n)$, where α stands for the second fundamental form of Σ^n.

Now, let $X \in \mathfrak{X}(\Sigma^n)$ and consider a local orthonormal frame $\{E_1, \ldots, E_n\}$ of $\mathfrak{X}(\Sigma^n)$. By using the Gauss equation (9.30) the Ricci curvature tensor of Σ^n is given by

$$\mathrm{Ric}(X, X) = \sum_{i=1}^{n} \langle \overline{R}(X, E_i)X, E_i \rangle + n \langle \alpha(X, X), \mathbf{H} \rangle - \sum_{i=1}^{n} |\alpha(X, E_i)|^2$$

$$= \sum_{i=1}^{n} \langle \overline{R}(X, E_i)X, E_i \rangle + n \left\langle \sum_{k=1}^{p} A_{N_k} X, X \right\rangle H_k$$

$$- \sum_{i=1}^{n} \left| \sum_{k=1}^{p} \langle A_{N_k} X, E_i \rangle N_k \right|^2,$$

where $\alpha(X, Y) = \sum_{i=1}^{p} \langle A_{N_i} X, Y \rangle N_i$. Consequently, we get

$$\mathrm{Ric}(X, X) = \sum_{i=1}^{n} \langle \overline{R}(X, E_i)X, E_i \rangle - \sum_{i=1}^{p} \left| A_{N_i} X - \frac{n H_i}{2} X \right|^2 + \frac{n^2 |\mathbf{H}|^2}{4} |X|^2. \tag{9.31}$$

Moreover, since condition (9.29) holds, by (4.17) of [47] we have that

$$\sum_i \langle \overline{R}(X, E_i)X, E_i \rangle \geq -n\frac{|f|''}{f}|X|^2. \tag{9.32}$$

The conclusion is achieved by (9.31) and (9.32). □

Next lemma is a generalization of the classical Omori-Yau's maximum principle; see [31, 33, 65, 90, 256] for related topics.

Lemma 9.4 *Let (Σ^n, g) be an n-dimensional complete Riemannian manifold such that $\mathrm{Ric} \geq -G(r)$ for a function $G \in C^1([0, +\infty))$ obeying the following properties*

$$G(0) > 0, \ G' \geq 0 \ \text{and} \ G^{-1/2} \notin L^1[0, +\infty).$$

If $\zeta \in C^2(\Sigma^n)$ with $\sup_{\Sigma^n} \zeta < +\infty$, then there exists a sequence of points $\{p_k\} \subset \Sigma^n$ satisfying

$$\lim_k \zeta(p_k) = \sup_{\Sigma^n} \zeta, \quad \lim_k |\nabla \zeta(p_k)| = 0, \quad \lim_k \Delta \zeta(p_k) \leq 0.$$

The following result is an extension of the main theorem in [45].

Theorem 9.3 *Let Σ^n be a complete submanifold which lies in a slab of a warped product $I \times_f M^{n+p-1}$ obeying the strong null convergence condition (9.29). Suppose that the spatial shape operator of Σ^n satisfies $|S|^2 \leq G(r)$, for some generic radial function $G(r)$ satisfying the properties in Lemma 9.4. If the following inequality is satisfied*

$$|\nabla h| \leq \inf_{\Sigma^n} \left((\log f)'(h) - |\mathbf{H}| \right), \tag{9.33}$$

then Σ^n is contained in a slice $\{t\} \times M^{n+p-1}$.

Proof By item (i) of Lemma 9.1 we have that

$$\Delta h \geq n((\log f)'(h) - |\mathbf{H}|) - (\log f)'(h)|\nabla h|^2. \tag{9.34}$$

Since Σ^n lies in a slab of $I \times_f M^{n+p-1}$, by Lemmas 9.3 and 9.4 there exists a sequence $\{p_k\} \subset \Sigma^n$ such that

$$\lim_k h(p_k) = \sup_{\Sigma^n} h, \quad \lim_k |\nabla h(p_k)| = 0 \quad \text{and} \quad \limsup_k \Delta h(p_k) \leq 0.$$

Hence, by (9.34) we get

$$0 \geq \limsup_k \Delta h(p_k) \geq \lim_k((\log f)'(h) - |\mathbf{H}|)(p_k) \geq 0.$$

Therefore, it follows that

$$\lim_k(|\mathbf{H}| - (\log f)'(h))(p_k) = 0.$$

Taking into account (9.33), the conclusion immediately follows. □

9.3.3 Rigidity Via Stochastic Completeness

According to [256, Chapter 3], a Riemannian manifold Σ^n is said to be *stochastically complete* if, for some (and, hence, for any) $(x, t) \in \Sigma^n \times (0, +\infty)$, the heat kernel $p(x, y, t)$ of the Laplace-Beltrami operator Δ satisfies the conservation property

$$\int_{\Sigma^n} p(x, y, t) dy = 1. \qquad (9.35)$$

From the probabilistic viewpoint, stochastic completeness is the property of a stochastic process to have infinite life time. For the Brownian motion on a manifold, the conservation property (9.35) means that the total probability of a particle to be found in the state space is constantly equal to one; see [173, 193, 194, 280].

In the cited paper, the authors showed that stochastic completeness turns out to be equivalent to the validity of a weak form of the Omori-Yau maximum principle [250, 297]; see [254, Theorem 1.1] or [256, Theorem 3.1]. More precisely, one has

Lemma 9.5 *A Riemannian manifold Σ^n is stochastically complete if and only if for every $\zeta \in C^2(\Sigma^n)$ satisfying $\sup_{\Sigma^n} \zeta < +\infty$, there exists a sequence of points $\{p_k\} \subset \Sigma^n$ such that*

$$\lim_{k \to \infty} \zeta(p_k) = \sup_{\Sigma^n} \zeta \quad \text{and} \quad \limsup_{k \to \infty} \Delta \zeta(p_k) \leq 0.$$

In the next result, Lemma 9.5 has been applied in order to prove the rigidity of stochastically complete submanifolds in $I \times_f M^{n+p-1}$.

Theorem 9.4 *Let Σ^n be a stochastically complete submanifold which lies in a slab of a warped product $I \times_f M^{n+p-1}$, with $f' \leq 0$ ($f' \geq 0$) on Σ^n. If $|\mathbf{H}|$ is bounded and (9.33) is satisfied, then Σ^n is a complete minimal hypersurface of a slice $\{t\} \times_f M^{n+p-1}$, with $f'(t) = 0$.*

Proof Since Σ^n is stochastically complete and lies in a slab of $I \times_f M^{n+p-1}$, by item (ii) of Lemma 9.1 and Lemma 9.5, there exists a sequence $\{p_k\} \subset \Sigma^n$ such that

$$0 \geq \limsup_k \Delta u(p_k) \geq nC \lim_k ((\log f)'(h) - |\mathbf{H}|)(p_k) \geq 0,$$

where $C = \inf_{\Sigma^n} f > 0$. Since $|\mathbf{H}|$ is bounded and $f' \leq 0$, we get $(\log f)'(t) = 0$, for some $t \in I$, as well as $|\mathbf{H}| = 0$. Therefore, by (9.33) the conclusion follows. □

Remark 9.1 We recall that a submanifold has parallel mean curvature vector field $\mathbf{H}$ if it is parallel as a section of the normal bundle. Related to Theorem 9.4, we observe that the hypothesis that $|\mathbf{H}|$ be bounded is weaker than to assume that $\mathbf{H}$ being parallel.

9.4 A Duo-Graph in $\mathbb{R} \times \mathbb{H}^2 \times \mathbb{R}$

This section is devoted to construct an example of submanifold immersed in a warped product, which is not contained in a slice of the ambient space and, hence, illustrates the importance of our previous rigidity results.

Let us consider the smooth functions $\zeta_a, \zeta_b : \mathbb{H}^2 \to \mathbb{R}$ given by $\zeta_a(x, y) = a \ln y$ and $\zeta_b(x, y) = b \ln y$, respectively. Furthermore, let us consider the corresponding entire duo-graph defined as follows

$$\Sigma(a, b) = \{(a \ln y, x, y, b \ln y) : y > 0\} \subset \mathbb{R} \times \mathbb{H}^2 \times \mathbb{R}.$$

We have that $D\zeta_a(x, y) = (0, ay)$ and $D\zeta_b(x, y) = (0, by)$. Hence, $|D\zeta_a(x, y)|^2 = a^2$ and $|D\zeta_b(x, y)|^2 = b^2$. Consequently, $\Sigma(a, b)$ will be a complete surface in the space $\mathbb{R} \times \mathbb{H}^2 \times \mathbb{R}$.

Moreover, considering the coordinates (t, x, y, s) in $\mathbb{R} \times \mathbb{H}^2 \times \mathbb{R}$, the normal vectors are given by

$$N = \frac{\partial_t - D\zeta_a}{\sqrt{1 + a^2}} \quad \text{and} \quad \overline{\nu} = \frac{\partial_s - D\zeta_b}{\sqrt{1 + b^2}}.$$

Indeed, let $G_a : \mathbb{R} \times \mathbb{H}^2 \to \mathbb{R}$ be given by $G_a(t, p) = t - \zeta_a(p)$ and $G_b(p, t) = t - \zeta_b(p)$. Note that $G_a|_{\Sigma(a,b)} = 0$ and $G_b|_{\Sigma(a,b)} = 0$. Let γ be a curve in $\Sigma(a, b)$ with $\gamma : (-\epsilon, \epsilon) \to \Sigma(a, b)$ such that $\gamma(0) = p$ and $\gamma'(0) = v \in T_p\Sigma(a, b)$. Hence

$$v(G_a) = v(G_b) = 0.$$

Moreover, if $v \in T_{(t,p)}\mathbb{R} \times \mathbb{H}^2$, then

$$v(G_a) = \langle -\partial_t + D\zeta_a, v \rangle$$

9.4 A Duo-Graph in $\mathbb{R} \times \mathbb{H}^2 \times \mathbb{R}$

Therefore, one has

$$N = \frac{-\overline{\nabla} G_a}{|\overline{\nabla} G_a|} = \frac{\partial_t - D\zeta_a}{\sqrt{1+a^2}}.$$

Similarly, we have that

$$\bar{\nu} = \frac{\partial_s - D\zeta_b}{\sqrt{1+b^2}}.$$

Despite these unit vectors are perpendicular to $\Sigma(a,b)$ they are not orthogonal to each other. Nonetheless we use Gram-Schmidt process to obtain another normal vector field given by $\tilde{\nu} = \bar{\nu} + \langle N, \bar{\nu} \rangle N$. So, the unit normal vector field will be given by $\nu = \frac{\tilde{\nu}}{|\tilde{\nu}|}$. We have

$$\tilde{\nu} = \frac{\partial_s - D\zeta_b}{\sqrt{1+b^2}} + \left\langle \frac{\partial_t - D\zeta_a}{\sqrt{1+a^2}}, \frac{\partial_s - D\zeta_b}{\sqrt{1+b^2}} \right\rangle \frac{\partial_t - D\zeta_a}{\sqrt{1+a^2}}$$

$$= \frac{1}{\sqrt{1+b^2}} \left(\partial_s + \frac{ab}{1+a^2} \partial_t - \frac{b + 2a^2 b}{1+a^2} D\zeta \right),$$

where

$$\zeta = \frac{1}{a} \zeta_a.$$

We also observe that ∂_t, ∂_s and $D\zeta$ are unit and orthogonal vectors. Hence, with a straightforward computation, ensures that

$$|\tilde{\nu}|^2 = \frac{(1+a^2)^2 + a^2 b^2 + (b + 2a^2 b)^2}{(1+a^2)^2 (1+b^2)}.$$

In the particular case $a = b$, we have

$$|\tilde{\nu}|^2 = \frac{(1+a^2)^2 + a^4 + (a + 2a^3)^2}{(1+a^2)^3}.$$

Moreover, if $ab = 0$ then $|\tilde{\nu}| = 1$.

By Eq. (9.15) the height function h satisfies

$$|\nabla h|^2 = \frac{a^2}{1+a^2} - \frac{1}{|\tilde{\nu}|^2} \langle \tilde{\nu}, \partial_t \rangle^2.$$

Consequently, one has

$$|\nabla h|^2 = \frac{a^2}{1+a^2} - \frac{(1+a^2)(1+b^2)ab}{[(1+a^2)^2 + a^2b^2 + (b+2a^2b)^2](\sqrt{1+b^2})}.$$

Hence, if $a = b$ it follows that

$$|\nabla h|^2 = \frac{a^2}{1+a^2} - \frac{(1+a^2)(1+a^2)a^2}{[(1+a^2)^2 + a^4 + (a+2a^3)^2](\sqrt{1+a^2})}.$$

Moreover, if $b = 0$ we get the same formula for the surface in $\mathbb{R} \times \mathbb{H}^2$, i.e.

$$|\nabla h|^2 = \frac{a^2}{1+a^2}$$

Consequently, if $a = 0$ it follows that $\nabla h \equiv 0$, since it is contained in the slice $\{0\} \times \mathbb{H}^2 \times \mathbb{R}$. Finally, the mean curvature H_N of $\Sigma(a,b)$ is given by

$$2H_N = \mathrm{Div}\left(\frac{aD\zeta}{\sqrt{1+a^2|D\zeta|^2}}\right) = \mathrm{Div}\left(\frac{aD\zeta}{\sqrt{1+a^2}}\right),$$

where Div is the divergence on $\mathbb{H}^2$. So, taking into account that $\mathrm{Div} = \mathrm{Div}_0 + \frac{2}{y}dy$, where Div_0 denotes the usual divergence on $\mathbb{R}^2$, we have that $\Delta u = 1$. Thus

$$H_N = \frac{a}{2\sqrt{1+a^2}}.$$

Moreover, since $2H_v = \mathrm{Div}\left(\frac{\tilde{v}}{|\tilde{v}|}\right)$, we obtain

$$2H_v = \frac{1}{|\tilde{v}|} \frac{b+2a^2b}{\sqrt{1+b^2}(1+a^2)}.$$

Therefore, one has

$$H_v = \frac{1}{2}\frac{b+2a^2}{\sqrt{(1+a^2)^2 + a^2b^2 + (b+2a^2b)^2}}.$$

We claim that $\Sigma(a,b)$ is stochastically complete. To verify this fact, we consider the lineal function $T : \mathbb{R}^2 \to \mathbb{R}^2$ such that $T(a,b) = \lambda(1,0)$ and $T(-b,a) = \lambda(0,1)$, where $\lambda^2 = a^2 + b^2$. It is clear that T is an isometry whenever $\lambda \neq 0$. If $a = b = 0$ the corresponding manifold Σ is trivially isometric to $\mathbb{H}^2$ therefore stochastically complete. For simplicity, let us consider $\mathbb{H}^2 \times \mathbb{R}^2$ as ambient space for $\Sigma(a, b)$.

Now, we observe that the map $\Psi = Id_{\mathbb{H}^2} \oplus T$ is a smooth isometry. Thus

$$\Psi|_{\Sigma(a,b)} : \Sigma(a, b) \to \mathbb{H}^2 \times \mathbb{R}^2$$

is also a smooth isometry. We obtain

$$\Psi(\Sigma(a, b)) = \{(x, y, \sqrt{a^2 + b^2} \ln y, 0) : (x, y) \in \mathbb{H}^2\}.$$

Hence, $\Sigma(a, b)$ is isometric to the graph of Abresch-Rosenberg given by

$$\Xi = \{(x, y, c \ln y), (x, y) \in \mathbb{H}^2\},$$

see [167] and [145].

Therefore, since the graph Ξ has Ricci curvature bounded from below, it follows that $\Sigma(a, b)$ must be stochastically complete. Consequently, we conclude that the hypothesis assumed in Theorem 9.4 that the hypersurface must be contained in a slab of the ambient space is necessary.

9.5 Further Rigidity and Nonexistence Results

Taking into account that the value of the mean curvature $\mathcal{H}(\tau)$ of a slice M_τ is given in (9.7), as an application of (9.25) we have the following result.

Lemma 9.6 *Let* $\psi : \Sigma^n \to I \times_f M^{n+p-1}$ *be a closed submanifold immersed into* $I \times_f M^{n+p-1}$. *Then*

(i) $\min_{\Sigma^n} \langle \mathbf{H}, \partial_t \rangle \leq \mathcal{H}(h^*)$, *where* $h^* = \max_{\Sigma^n} h$, *and*
(ii) $\max_{\Sigma^n} \langle \mathbf{H}, \partial_t \rangle \geq \mathcal{H}(h_*)$, *where* $h_* = \min_{\Sigma^n} h$.

Proof Let us consider on Σ^n the function $u = g(h)$, where $g : I \to \mathbb{R}$ is an arbitrary primitive of the warping function f. Since Σ^n is closed, the function u attains its minimum and maximum at some points $p_{\min}$ and $p_{\max}$.

Since $g' = f > 0$, g is strictly increasing and, at $p_{\min}$, it holds

$$u(p_{\min}) = u_* = \min_{\Sigma^n} = g(h_*),$$

where $h_* = h(p_{\min}) = \min_{\Sigma^n} h$, and

$$0 \leq \Delta u(p_{\min}) = n\left(f'(h_*) + f(h_*)\langle \mathbf{H}, \partial_t\rangle|_{p_{\min}}\right)$$
$$= nf(h_*)\left(\frac{f'(h_*)}{f(h_*)} + \langle \mathbf{H}, \partial_t\rangle|_{p_{\min}}\right).$$

Thus, one has

$$\langle \mathbf{H}, \partial_t\rangle|_{p_{\min}} \geq -\frac{f'(h_*)}{f(h_*)} = \mathcal{H}(h_*).$$

Consequently

$$\max_{\Sigma^n}\langle \mathbf{H}, \partial_t\rangle \geq -\frac{f'(h_*)}{f(h_*)} = \mathcal{H}(h_*).$$

Therefore, item (ii) is proved.

The proof of item (i) is quite similar, working at $p_{\max}$. □

Next, we extend Lemma 9.6 to the noncompact case under the assumption of stochastic completeness.

Lemma 9.7 *Let $\psi : \Sigma^n \to M^{n+p-1}$ be a stochastically complete submanifold immersed in $I \times_f M^{n+p-1}$.*

(i) If Σ^n lies above a slice of $I \times_f M^{n+p-1}$, then

$$\sup_{\Sigma^n}\langle \mathbf{H}, \partial_t\rangle \geq \mathcal{H}(h_*),$$

where $h_* = \inf_{\Sigma^n} h \in I$;

(ii) If Σ^n lies below a slice of $I \times_f M^{n+p-1}$, then

$$\inf_{\Sigma^n}\langle \mathbf{H}, \partial_t\rangle \leq \mathcal{H}(h^*),$$

where $h^* = \sup_{\Sigma^n} h \in I$.

Proof Let us assume that Σ^n lies above a slice of $I \times_f M^{n+p-1}$ and let us apply the weak maximum principle on Σ^n to the function $u = g(h)$, which satisfies $u_* = \inf_{\Sigma^n} u = g(h_*) > -\infty$, where $h_* = \inf_{\Sigma^n} h \geq \tau_* \in I$.

9.5 Further Rigidity and Nonexistence Results

By Lemma 9.5 there exists a sequence of points $\{p_k\} \subset \Sigma^n$ such that

$$u(p_k) < u_* + \frac{1}{k} \quad \text{and} \quad \Delta u(p_k) > -\frac{1}{k}. \tag{9.36}$$

Now, we observe that $\lim_k h(p_k) = h_*$, since g is strictly increasing. So, by (9.25) and (9.36) we obtain that

$$-\frac{1}{nk} < \frac{1}{n}\Delta u(p_k) = f'(h(p_k)) + f(h(p_k))\langle \mathbf{H}, \partial_t \rangle(p_k).$$

Consequently,

$$\langle \mathbf{H}, \partial_t \rangle(p_k) > \frac{1}{f(h(p_k))}\left(-f'(h(p_k)) - \frac{1}{nk}\right). \tag{9.37}$$

Hence, by (9.37) we get

$$\sup_{\Sigma^n}\langle \mathbf{H}, \partial_t \rangle \geq \langle \mathbf{H}, \partial_t \rangle(p_k) > \frac{1}{f(h(p_k))}\left(-f'(h(p_k)) - \frac{1}{nk}\right). \tag{9.38}$$

Therefore, since

$$\lim_k f(h(p_k)) = f(h_*) \quad \text{and} \quad \lim_k f'(h(p_k)) = f'(h_*),$$

making $k \to +\infty$ in (9.38) we prove item (i). The proof of item (ii) is quite similar. □

Next lemma is a meanigfull consequence of a Liouville-type theorem due to Yau; see [297] for additional comments and remarks.

Lemma 9.8 *The only harmonic semi-bounded functions defined on an n-dimensional complete Riemannian manifold whose Ricci curvature is nonnegative are the constant ones.*

We derive now some rigidity results for submanifolds Σ^n immersed in a warped product $I \times_f M^{n+p-1}$ whose warping function has convex logarithm.

Theorem 9.5 *Let $I \times_f M^{n+p-1}$ be a warped product such that $(\log f)'' \geq 0$, and let $\psi : \Sigma^n \to I \times_f M^{n+p-1}$ be a closed submanifold with mean curvature vector field $\mathbf{H}$ such that the support function $\langle \mathbf{H}, \partial_t \rangle$ is constant. Then, $\psi(\Sigma^n)$ is contained in a slice*

$\{\tau\} \times M^{n+p-1}$, for some $\tau \in I$. Moreover, when $p = 2$, $\phi = \pi_M \circ \psi : \Sigma^n \to M^{n+1}$ is a hypersurface with mean curvature H_ϕ satisfying

$$|\mathbf{H}|^2 = \frac{H_\phi^2 + f'(\tau)^2}{f(\tau)^2}. \tag{9.39}$$

Proof By Lemma 9.6 and by using $(\log f)'' \geq 0$, we have

$$\min_{\Sigma^n}\langle \mathbf{H}, \partial_t \rangle \leq \mathcal{H}(h^*) \leq \mathcal{H}(h_*) \leq \max_{\Sigma^n}\langle \mathbf{H}, \partial_t \rangle. \tag{9.40}$$

Since $\langle \mathbf{H}, \partial_t \rangle$ is constant, by (9.40) we get

$$\mathcal{H}(h_*) = \mathcal{H}(h^*) = \langle \mathbf{H}, \partial_t \rangle = const. \tag{9.41}$$

By using again that $(\log f)'' \geq 0$, by (9.41) it follows that $\mathcal{H}(t) = \langle \mathbf{H}, \partial_t \rangle = const.$ on $[h_*, h^*]$. That is, $\mathcal{H}(h) = \langle \mathbf{H}, \partial_t \rangle$ on Σ^n.
So

$$\mathcal{H}(h) = -\frac{f'(h)}{f(h)} = \langle \mathbf{H}, \partial_t \rangle$$

implies that $f'(h) + f(h)\langle \mathbf{H}, \partial_t \rangle = 0$ on Σ^n. Thus, by (9.25) we have that $\Delta u = 0$ on Σ^n. In other words, u is a harmonic function on Σ^n, which is a closed manifold. Hence $u = g(h)$ is constant on Σ^n and, since g is an increasing function, this means that h is itself constant on Σ^n. Consequently, $\psi(\Sigma^n)$ is contained in a slice M_τ.

Now, it is easily seen that the projection $\phi = \pi_M \circ \psi : \Sigma^n \to M^{n+1}$ is an immersed hypersurface for which $\psi(p) = (\tau, \phi(p)) = \phi_\tau(p)$. Moreover, since $\langle \cdot, \cdot \rangle_\tau = \phi_\tau^*(\langle \cdot, \cdot \rangle) = f(\tau)^2 \langle \cdot, \cdot \rangle_{\Sigma^n}$, intrinsically, the manifold $(\Sigma^n, \langle \cdot, \cdot \rangle_\tau)$ is homothetic to $(\Sigma^n, \langle \cdot, \cdot \rangle_{\Sigma^n})$ with scale factor $f(\tau)$.

The main objective now is to express the extrinsic geometry of the codimension two submanifold $\phi_\tau : \Sigma^n \to I \times_f M^{n+1}$ in terms of the extrinsic geometry of the hypersurface $\phi : \Sigma^n \to M^{n+1}$. In order to compute the second fundamental form α_τ of the immersion ϕ_τ, let us denote by N the (locally defined) unit normal vector field of the hypersurface $\phi : \Sigma^n \to M^{n+1}$, with $\langle N, N \rangle_M = 1$.

We notice that

$$\langle N, N \rangle = f(\tau)^2 \langle N, N \rangle_M = f(\tau)^2. \tag{9.42}$$

Thus, it follows that

$$\eta_\tau(p) = \frac{1}{f(\tau)} N(p), \quad \xi_\tau(p) = \partial_t|_{(\tau, \phi(p))}, \quad p \in \Sigma^n,$$

9.5 Further Rigidity and Nonexistence Results

define a local orthonormal frame of vector fields normal along the immersion ϕ_τ, with

$$\langle \eta_\tau, \eta_\tau \rangle = 1, \quad \langle \eta_\tau, \xi_\tau \rangle = 0, \quad \langle \xi_\tau, \xi_\tau \rangle = 1.$$

The second fundamental form α_τ of the immersion ϕ_τ is then written as

$$\begin{aligned}\alpha_\tau(X, Y) &= \langle \alpha_\tau(X, Y), \eta_\tau \rangle \eta_\tau + \langle \alpha_\tau(X, Y), \xi_\tau \rangle \xi_\tau \\ &= \langle A_{\eta_\tau} X, Y \rangle_\tau \eta_\tau + \langle A_{\xi_\tau} X, Y \rangle_\tau \xi_\tau, \end{aligned} \quad (9.43)$$

for every tangent vector fields $X, Y \in \mathfrak{X}(\Sigma^n)$. Now, observe that, for every $X \in \mathfrak{X}(\Sigma^n)$, one has

$$\overline{\nabla}_X \eta_\tau = \frac{1}{f(\tau)} \overline{\nabla}_X N = \frac{1}{f(\tau)} \nabla^*_X N,$$

where ∇^* denotes the Levi-Civita connection of $(M^{n+1}, \langle \cdot, \cdot \rangle_M)$. Therefore, taking into account that

$$AX = \nabla^*_X N,$$

where $A : \mathfrak{X}(\Sigma^n) \to \mathfrak{X}(\Sigma^n)$ stands for the shape operator of the hypersurface $\phi : \Sigma^n \to M^{n+1}$ with respect to N, it follows from here, using the Weingarten formula (9.10), that $\nabla^\perp_X \eta_\tau = 0$ and

$$A_{\eta_\tau} X = \overline{\nabla}_X \eta_\tau = \frac{1}{f(\tau)} AX.$$

On the other hand, since $\langle X, \xi_\tau \rangle = 0$ for every $X \in \mathfrak{X}(\Sigma^n)$, by (9.6) we have

$$\overline{\nabla}_X \xi_\tau = \frac{f'(\tau)}{f(\tau)} X,$$

which yields $\nabla^\perp_X \xi_p = 0$ and

$$A_{\xi_\tau} X = \frac{f'(\tau)}{f(\tau)} X$$

for every $X \in \mathfrak{X}(\Sigma^n)$. By (9.43) we obtain that

$$\alpha_\tau(X, Y) = \frac{1}{f(\tau)^2} \langle AX, Y \rangle_\tau N + \frac{f'(\tau)}{f(\tau)} \langle X, Y \rangle_\tau \xi_\tau$$

for every tangent vector fields $X, Y \in \mathfrak{X}(\Sigma^n)$. Thus, the mean curvature vector field of ψ is given by

$$\mathbf{H}_\tau = \frac{1}{n}\mathrm{tr}(\alpha_\tau) = \frac{1}{n}\sum_{i=1}^{n}\alpha_\tau(E_i, E_i) \tag{9.44}$$

$$= \frac{1}{n}\left(\frac{1}{f(\tau)^2}\sum_{i=1}^{n}\langle AE_i, E_i\rangle_\tau N_\tau + \frac{f'(\tau)}{f(\tau)}\sum_{i=1}^{n}\langle E_i, E_i\rangle_\tau \xi_\tau\right),$$

where $\{E_1, \ldots, E_n\}$ is a local orthonormal frame on Σ, with respect to the metric $\langle \cdot, \cdot \rangle_\tau$. In particular,

$$\sum_{i=1}^{n}\langle E_i, E_i\rangle_\tau = n.$$

On the other hand, by using the metric $\langle \cdot, \cdot \rangle_\tau = \phi_\tau^*(\langle \cdot, \cdot \rangle) = f(\tau)^2 \langle \cdot, \cdot \rangle_{\Sigma^n}$ we have that

$$\langle AE_i, E_i\rangle_\tau = f(\tau)^2 \langle AE_i, E_i\rangle_{\Sigma^n} = \langle Ae_i, e_i\rangle_{\Sigma^n},$$

for every $i = 1, \ldots, n$, where $e_i = f(\tau)E_i$ and $\{e_1, \ldots, e_n\}$ is a local orthonormal frame on Σ^n with respect to the metric $\langle \cdot, \cdot \rangle_{\Sigma^n}$.

Observe that the mean curvature function of the hypersurface $\phi : \Sigma^n \to M^{n+1}$ is given

$$H = \frac{1}{n}\mathrm{tr}(A) = \frac{1}{n}\sum_{i=1}^{n}\langle Ae_i, e_i\rangle_{\Sigma^n} = \frac{1}{n}\sum_{i=1}^{n}\langle AE_i, E_i\rangle_\tau.$$

Putting this into (9.44) we get that

$$\mathbf{H}_\tau = \frac{H}{f(\tau)^2}N + \frac{f'(\tau)}{f(\tau)}\xi_\tau. \tag{9.45}$$

Therefore, by (9.42) and (9.45) we deduce (9.39). □

By (9.39) in Theorem 9.5 we obtain the next nonexistence result.

Corollary 9.4 *There are no closed minimal submanifolds* $\psi : \Sigma^n \to I \times_f M^{n+1}$ *immersed in a warped product* $I \times_f M^{n+1}$ *such that* $(\log f)'' \geq 0$ *and* f' *does not vanish on* I.

9.5 Further Rigidity and Nonexistence Results

We consider now the case in which the submanifold is stochastically complete. As before, a *slab* of a warped product $I \times_f M^{n+p-1}$ means the region between two slices M_{τ_1} and M_{τ_2}, for some $\tau_1 < \tau_2$.

Theorem 9.6 *Let $I \times_f M^{n+p-1}$ be a warped product such that $(\log f)'' \geq 0$, with the equality $(\log f)'' = 0$ holding only at isolated points of I, and let $\psi : \Sigma^n \to I \times_f M^{n+p-1}$ be a stochastically complete submanifold which lies in a slab of $I \times_f M^{n+p-1}$ and with mean curvature vector field $\mathbf{H}$ such that the support function $\langle \mathbf{H}, \partial_t \rangle$ is constant. Then, $\psi(\Sigma)$ is contained in a slice $\{\tau\} \times M^{n+p-1}$, for some $\tau \in I$. Moreover, when $p = 2$, $\phi = \pi_M \circ \psi : \Sigma^n \to M^{n+1}$ is a hypersurface with mean curvature H_ϕ satisfying (9.39).*

Proof Arguing as in the proof of Theorem 9.5, by using now Lemma 9.7 instead of Lemma 9.6, we have that

$$\mathcal{H}(h^*) = \mathcal{H}(h_*) = \langle \mathbf{H}, \partial_t \rangle = \text{const.} \tag{9.46}$$

Our assumption on $\log f$ implies that the function $\mathcal{H}(t)$ is strictly decreasing on I. Hence, by (9.46) we have that $h_* = h^*$ and, consequently, h is constant on Σ^n. Therefore, $\psi(\Sigma^n)$ must be contained in a slice $\{\tau\} \times M^{n+p-1}$. $\square$

By Theorem 9.6 we obtain the following nonexistence property.

Corollary 9.5 *There are no stochastically complete minimal submanifolds $\psi : \Sigma^n \to I \times_f M^{n+1}$ which lie in a slab of a warped product $I \times_f M^{n+1}$ such that $(\log f)'' \geq 0$, with the equality $(\log f)'' = 0$ holding only at isolated points of I, and f' does not vanish on I.*

The next rigidity result reads as follows.

Theorem 9.7 *Let $I \times_f M^{n+p-1}$ be a warped product such that $(\log f)'' \geq 0$ and let $\psi : \Sigma^n \to I \times_f M^{n+p-1}$ be a complete submanifold which lies in a slab of $I \times_f M^{n+p-1}$, having nonnegative Ricci curvature and with mean curvature vector field $\mathbf{H}$ such that the support function $\langle \mathbf{H}, \partial_t \rangle$ is constant. Then, $\psi(\Sigma)$ is contained in a slice $\{\tau\} \times M^{n+p-1}$, for some $\tau \in I$. Moreover, when $p = 2$, $\phi = \pi_M \circ \psi : \Sigma^n \to M^{n+1}$ is a hypersurface with mean curvature H_ϕ satisfying (9.39).*

Proof A Riemannian manifold with Ricci curvature bounded from below is stochastically complete. Since the Omori-Yau maximum principle [250, 297], we can proceed as in the proof of Theorem 9.6 to infer that the function $u = g(h)$ is a harmonic function on Σ^n.

Hence, since $\psi(\Sigma^n)$ lies in a slab of $I \times_f M^{n+p-1}$, we can apply Lemma 9.8 to conclude that u is constant and, consequently, h is constant on Σ^n. Therefore, $\psi(\Sigma^n)$ must be contained in a slice $\{\tau\} \times M^{n+p-1}$. $\square$

By Theorem 9.7 the following result holds.

Corollary 9.6 *There are no complete minimal submanifolds $\psi : \Sigma^n \to I \times_f M^{n+1}$ having nonnegative Ricci curvature and lying in a slab of a warped product $I \times_f M^{n+1}$ such that $(\log f)'' \geq 0$ and f' does not vanish on I.*

9.6 Submanifolds in a Weighted Riemannian Warped Product

We consider here Σ^n immersed in a Riemannian warped product $I \times_f M_\varphi^{n+p-1}$ endowed with a weight function φ which does not depend on the parameter $t \in I$. In this context, by (5.4), (9.21) and (9.25), we get

$$\begin{aligned}
\Delta_\varphi u &= \Delta u - \langle \nabla u, \overline{\nabla}\varphi\rangle \\
&= n(f'(h) + f(h)\langle \mathbf{H}, \partial_t\rangle) - f(h)\langle \partial_t^\top, \overline{\nabla}\varphi\rangle \\
&= n(f'(h) + f(h)\langle \mathbf{H}, \partial_t\rangle) - f(h)\langle \partial_t - \partial_t^\perp, \overline{\nabla}\varphi\rangle \\
&= n(f'(h) + f(h)\langle \mathbf{H}, \partial_t\rangle) + f(h)\langle \partial_t^\perp, (\overline{\nabla}\varphi)^\perp\rangle. \quad (9.47)
\end{aligned}$$

According to Gromov [197] and in a similar way of (5.5), the *weighted mean curvature vector field*, or simply φ-*mean curvature vector field*, $\mathbf{H}_\varphi$ of Σ^n is defined by the following formula

$$\mathbf{H}_\varphi = \mathbf{H} + \frac{1}{n}(\overline{\nabla}\varphi)^\perp, \quad (9.48)$$

where $\mathbf{H}$ denotes the standard mean curvature vector field of Σ^n defined in (9.11) and $(\overline{\nabla}\varphi)^\perp \in \mathfrak{X}^\perp(\Sigma^n)$ stands for the normal component of $\overline{\nabla}\varphi$ along Σ^n.

Thus, by (9.48) and (9.47), we obtain

$$\Delta_\varphi u = n(f'(h) + f(h)\langle \mathbf{H} + \frac{1}{n}(\overline{\nabla}\varphi)^\perp, \partial_t\rangle) = n(f'(h) + f(h)\langle \mathbf{H}_\varphi, \partial_t\rangle). \quad (9.49)$$

Consequently, by (9.49) we have the following preparatory property.

Lemma 9.9 *Let Σ^n be a submanifold immersed in $I \times_f M_\varphi^{n+p-1}$. If $u = g(h)$, where $g : I \to \mathbb{R}$ is an arbitrary primitive of f and h is the height function of Σ^n, then*

$$\Delta_\varphi u = n(f'(h) + f(h)\langle \mathbf{H}_\varphi, \partial_t\rangle).$$

Taking into account Lemma 9.9, by arguing as in the proof of Lemma 9.6 we have the following result.

9.6 Submanifolds in a Weighted Riemannian Warped Product

Lemma 9.10 *Let Σ^n be a closed submanifold immersed in $I \times_f M_\varphi^{n+p-1}$. Then, one has:*

(i) $\min_{\Sigma^n} \langle \mathbf{H}_\varphi, \partial_t \rangle \leq \mathcal{H}_\varphi(h^*)$, where $h^* = \max_{\Sigma^n} h$, and
(ii) $\max_{\Sigma^n} \langle \mathbf{H}_\varphi, \partial_t \rangle \geq \mathcal{H}_\varphi(h_*)$, where $h_* = \min_{\Sigma^n} h$.

We are able to prove now the next rigidity result.

Theorem 9.8 *Let $I \times_f M_\varphi^{n+p-1}$ be a weighted warped product such that $(\log f)'' \geq 0$, and let $\psi : \Sigma^n \to I \times_f M_\varphi^{n+p-1}$ be a closed submanifold with φ-mean curvature vector field $\mathbf{H}_\varphi$ such that the support function $\langle \mathbf{H}_\varphi, \partial_t \rangle$ is constant. Then, $\psi(\Sigma^n)$ is contained in a slice $\{\tau\} \times M^{n+p-1}$, for some $\tau \in I$. Moreover, when $p = 2$, $\phi = \pi_M \circ \psi : \Sigma^n \to M^n$ is a hypersurface with φ-mean curvature $H_{\phi,\varphi}$ satisfying*

$$|\mathbf{H}_\varphi|^2 = \frac{H_{\phi,\varphi}^2 + f'(\tau)^2}{f(\tau)^2}. \tag{9.50}$$

Proof By Lemma 9.10 and by using the fact that $(\log f)'' \geq 0$, we have that

$$\min_{\Sigma^n} \langle \mathbf{H}_\varphi, \partial_t \rangle \leq \mathcal{H}_\varphi(h^*) \leq \mathcal{H}_\varphi(h_*) \leq \max_{\Sigma^n} \langle \mathbf{H}_\varphi, \partial_t \rangle. \tag{9.51}$$

Thus, since $\langle \mathbf{H}_\varphi, \partial_t \rangle$ is constant, by (9.51) we get

$$\mathcal{H}_\varphi(h_*) = \mathcal{H}_\varphi(h^*) = \langle \mathbf{H}_\varphi, \partial_t \rangle = \text{const.} \tag{9.52}$$

By using again $(\log f)'' \geq 0$, by (9.52) it follows that $\mathcal{H}_\varphi(t) = \langle \mathbf{H}_\varphi, \partial_t \rangle = const.$ on $[h_*, h^*]$. That is, $\mathcal{H}_\varphi(h) = \langle \mathbf{H}_\varphi, \partial_t \rangle$ on Σ^n.

Hence

$$\mathcal{H}_\varphi(h) = -\frac{f'(h)}{f(h)} = \langle \mathbf{H}_\varphi, \partial_t \rangle \tag{9.53}$$

implies $f'(h) + f(h)\langle \mathbf{H}_\varphi, \partial_t \rangle = 0$ on Σ^n. Then, by (9.47) one has $\Delta_\varphi u = 0$ on Σ^n. In other words, u is a φ-harmonic function on Σ^n, which is a closed manifold.

Hence, by (5.3) and (5.4), we can apply the Divergence Theorem to infer that $u = g(h)$ is constant on Σ^n. Since g is an increasing function this means that h is itself constant on Σ^n. Consequently, $\psi(\Sigma^n)$ is contained in a slice M_τ.

When $p = 2$, as in the proof of Theorem 9.5, we can consider the (locally defined) unit normal vector field N of the hypersurface $\phi : \Sigma^n \to M^n$, with $\langle N, N \rangle_M = 1$.

Thus, by (9.45) and (9.48), since φ does not depend on the parameter $t \in I$, it is easily seen that the following equation holds

$$\mathbf{H}_\varphi = \frac{H_{\phi,\varphi}}{f(\tau)^2} N + \frac{f'(\tau)}{f(\tau)} \partial_t. \tag{9.54}$$

Indeed, since $\overline{\nabla}\varphi = \frac{1}{f(\tau)^2}\nabla\varphi$ and $\langle N, N \rangle = f(\tau)^2 \langle N, N \rangle_M = f(\tau)^2$, bearing in mind that

$$H_\phi + \frac{1}{n}\langle \overline{\nabla}\varphi, N \rangle = H_\phi + \frac{1}{n}\langle \nabla\varphi, N \rangle_M = H_{\phi,\varphi}, \tag{9.55}$$

it follows that (9.54) is verified.

Finally, by (9.54) relation (9.50) easily follows. □

If the ambient space is a weighted product space of the form $\mathbb{R}^p \times M_\varphi^n$, we can apply p times Theorem 9.8 in order to get the following result.

Corollary 9.7 *The only n-dimensional closed φ-minimal submanifolds immersed in a weighted product space $\mathbb{R}^p \times M_\varphi^n$ are the closed φ-minimal hypersurfaces immersed in M_φ^n.*

By (9.50) in Theorem 9.8 we also obtain the following nonexistence property.

Corollary 9.8 *There do not exist closed φ-minimal submanifolds Σ^n immersed in a weighted warped product $I \times_f M_\varphi^n$ such that $(\log f)'' \geq 0$ and f' does not vanish on I.*

Lemmas 5.1 and 9.9 enable us to obtain an extension of Lemma 9.10.

Lemma 9.11 *Let Σ^n be a complete submanifold immersed in $I \times_f M_\varphi^{n+p-1}$, such that its Bakry-Émery-Ricci tensor is bounded from below. Then, the following facts holds:*

(i) *If Σ^n lies above a slice of $I \times_f M_\varphi^{n+p-1}$, then*

$$\sup_{\Sigma^n} \langle \mathbf{H}_\varphi, \partial_t \rangle \geq \mathcal{H}_\varphi(h_*),$$

where $h_* = \inf_{\Sigma^n} h \in I$;

(ii) *If Σ^n lies below a slice of $I \times_f M_\varphi^{n+p-1}$, then*

$$\inf_{\Sigma^n} \langle \mathbf{H}_\varphi, \partial_t \rangle \leq \mathcal{H}_\varphi(h^*),$$

where $h^* = \sup_{\Sigma^n} h \in I$.

9.6 Submanifolds in a Weighted Riemannian Warped Product

In our next result, we will assume that the ambient space obeys a convergence condition which was established by Montiel [242]. To this aim, we recall that a *slab* of a weighted warped product $I \times_f M_\varphi^{n+p-1}$ is just a region between two slices M_{τ_1} and M_{τ_2}, for some $\tau_1 < \tau_2$.

Theorem 9.9 *Let $I \times_f M_\varphi^{n+p-1}$ be a weighted warped product such that $(\log f)'' \geq 0$, with the equality $(\log f)'' = 0$ holding only at isolated points of I, and which obeys the convergence condition*

$$K_M \geq \sup_I (f'^2 - ff''), \tag{9.56}$$

where K_M stands for the sectional curvature of M^{n+p-1}. Suppose in addition that the Hessian of the weight function φ is bounded from below. Let $\psi : \Sigma^n \to I \times_f M_\varphi^{n+p-1}$ be a complete submanifold which lies in a slab of $I \times_f M_\varphi^{n+p-1}$, with bounded second fundamental form and such that the support function $\langle \mathbf{H}_\varphi, \partial_t \rangle$ is constant. Then, $\psi(\Sigma^n)$ is contained in a slice $\{\tau\} \times M^{n+p-1}$, for some $\tau \in I$. Moreover, when $p = 2$, $\phi = \pi_M \circ \psi :$ $\Sigma^n \to M^n$ is a hypersurface with φ-mean curvature $H_{\phi,\varphi}$ satisfying (9.50).

Proof We start showing that the Bakry-Émery-Ricci tensor of Σ^n is bounded from below. To this aim, we recall that the curvature tensor R of Σ^n can be described in terms of its second fundamental form α and the curvature tensor $\overline{R}$ of the ambient space $I \times_f M_\varphi^{n+p-1}$ by Gauss equation. More precisely, we have that

$$\langle R(X,Y)Z, W \rangle = \langle \overline{R}(X,Y)Z, W \rangle + \langle \alpha(X,Z), \alpha(Y,W) \rangle - \langle \alpha(X,W), \alpha(Y,Z) \rangle,$$

for every tangent vector fields $X, Y, Z \in \mathfrak{X}(\Sigma^n)$.

Taking $X \in \mathfrak{X}(\Sigma^n)$ and a local orthonormal frame $\{E_1, \ldots, E_n\}$ of $\mathfrak{X}(\Sigma^n)$, by the Gauss equation it follows that the Ricci curvature tensor of Σ^n is given by

$$\mathrm{Ric}(X,X) = \sum_{i=1}^{n} \langle \overline{R}(X, E_i)X, E_i \rangle + n \langle \alpha(X,X), \mathbf{H} \rangle - \sum_{i=1}^{n} |\alpha(X, E_i)|^2$$

$$= \sum_{i=1}^{n} \langle \overline{R}(X, E_i)X, E_i \rangle + n \langle \sum_{k=1}^{p+1} A_k X, X \rangle H_k$$

$$- \sum_{i=1}^{n} \left| \sum_{k=1}^{p+1} \langle A_k X, E_i \rangle \eta_k \right|^2,$$

where

$$\alpha(X, Y) = \sum_{i=1}^{p+1} \langle A_i X, Y \rangle \eta_i$$

and $\{\eta_1, \ldots, \eta_{p+1}\}$ is a local orthonormal frame of $\mathfrak{X}^\perp(\Sigma^n)$. Consequently, taking account that $\mathbf{H}$ can be expressed in the following way

$$\mathbf{H} = \sum_{i=1}^{p+1} H_i \eta_i, \tag{9.57}$$

for some smooth functions $H_1, H_2, \ldots, H_{p+1}$ defined on Σ^n, by (9.57) and (9.57) we get

$$\mathrm{Ric}(X, X) = \sum_{i=1}^{n} \langle \overline{R}(X, E_i)X, E_i \rangle - \sum_{i=1}^{p+1} \left| A_i X - \frac{n H_i}{2} X \right|^2 + \frac{n^2 |\mathbf{H}|^2}{4} |X|^2. \tag{9.58}$$

Moreover, since the convergence condition (9.56) holds, by inequality (4.17) of [47] we have that

$$\sum_i \langle \overline{R}(X, E_i)X, E_i \rangle \geq -n \frac{|f''(h)|}{f(h)} |X|^2. \tag{9.59}$$

Thus, by (9.59), (9.58) and (9.57), we have

$$\mathrm{Ric}(X, X) \geq -n \frac{|f''(h)|}{f(h)} |X|^2 - \sum_{i=1}^{p+1} \left| A_i X - \frac{n H_i}{2} X \right|^2 + \frac{n^2 |\mathbf{H}|^2}{4} |X|^2$$

$$\geq -\left(n \frac{|f''(h)|}{f(h)} + |\alpha|^2 \right) |X|^2. \tag{9.60}$$

Now, since Σ^n lies in a slab of $I \times_f M_\varphi^{n+p-1}$, $|\alpha|$ is bounded and Hess φ is bounded from below, by (5.2) and (9.60) we get that the Bakry-Émery-Ricci tensor of Σ^n is bounded from below.

Consequently, as in the proof of Theorem 9.8, by using now Lemma 9.11 instead of Lemma 9.10, we have that

$$\mathcal{H}_\varphi(h^*) = \mathcal{H}_\varphi(h_*) = \langle \mathbf{H}_\varphi, \partial_t \rangle = \mathrm{const}. \tag{9.61}$$

Our assumption on $\log f$ implies that the function $\mathcal{H}_\varphi$ is strictly decreasing on I. Hence, by (9.61) we get that $h_* = h^*$ and, consequently, h is constant on Σ^n. Therefore, $\psi(\Sigma^n)$ must be contained in a slice $\{\tau\} \times M^{n+p-1}$. □

9.6 Submanifolds in a Weighted Riemannian Warped Product

By Theorem 9.9 we obtain the next nonexistence result.

Corollary 9.9 *Let $I \times_f M_\varphi^n$ be a weighted warped product such that $(\log f)'' \geq 0$, with the equality $(\log f)'' = 0$ holding only at isolated points of I, and which obeys the convergence condition (9.56). Suppose in addition that f' does not vanish on I and Hess φ is bounded from below. There do not exist complete φ-minimal submanifolds $\psi : \Sigma^n \to I \times_f M_\varphi^n$ lying in a slab of $I \times_f M_\varphi^n$ and with bounded second fundamental form.*

Now, we recall a Liouville-type result due to Brighton in [91].

Lemma 9.12 *The only φ-harmonic bounded functions defined on an n-dimensional complete weighted Riemannian manifold Σ_φ^n, whose Bakry-Émery-Ricci tensor is nonnegative, are the constant ones.*

As a meaningful consequence of Lemma 9.12, the next result holds.

Theorem 9.10 *Let $I \times_f M_\varphi^{n+p-1}$ be a weighted warped product such that $(\log f)'' \geq 0$ and let $\psi : \Sigma^n \to I \times_f M_\varphi^{n+p-1}$ be a complete submanifold which lies in a slab of $I \times_f M_\varphi^{n+p-1}$, having nonnegative Bakry-Émery-Ricci tensor and such that the support function $\langle \mathbf{H}_\varphi, \partial_t \rangle$ is constant. Then, $\psi(\Sigma^n)$ is contained in a slice $\{\tau\} \times M^{n+p-1}$, for some $\tau \in I$. Moreover, when $p = 2$, $\phi = \pi_M \circ \psi : \Sigma^n \to M^n$ is a hypersurface with φ-mean curvature $H_{\phi,\varphi}$ satisfying (9.50).*

Proof Arguing as in the proof of Theorem 9.9 we can proce that the function $u = g(h)$ is φ-harmonic on Σ^n. Hence, since $\psi(\Sigma)$ lies in a slab of $I \times_f M_\varphi^{n+p-1}$, we can apply Lemma 9.12 to conclude that u is constant and, consequently, h is constant on Σ^n. Therefore, $\psi(\Sigma^n)$ must be contained in a slice $\{\tau\} \times M^{n+p-1}$. □

Considering once more the ambient space being a weighted product space of the form $\mathbb{R}^p \times M_\varphi^n$, we obtain our second codimension reduction result by applying recursively Theorem 9.10. More precisely, the main result reads as follows.

Corollary 9.10 *The only n-dimensional complete φ-minimal submanifolds having nonnegative Bakry-Émery-Ricci tensor and lying in a slab of a weighted product space $\mathbb{R}^p \times M_\varphi^n$ are the complete φ-minimal hypersurfaces immersed in M_φ^n.*

By Theorem 9.10 the following nonexistence result holds.

Corollary 9.11 *There are no complete φ-minimal submanifolds $\psi : \Sigma^n \to I \times_f M_\varphi^n$ having nonnegative Bakry-Émery-Ricci tensor and lying in a slab of a weighted warped product $I \times_f M_\varphi^n$ such that $(\log f)'' \geq 0$ and f' does not vanish on I.*

We recall that an important example of weighted manifold is the Gaussian space $\mathbb{G}^n$, which corresponds to the Euclidean space $\mathbb{R}^n$ endowed with the Gaussian probability measure

$$d\mu = (2\pi)^{-\frac{n}{2}} e^{-\frac{|x|^2}{2}} dx^2.$$

In the weighted product space $\mathbb{R} \times \mathbb{G}^n$ setting, Hieu and Nam extended the classical Bernstein's theorem [77] showing that the only weighted minimal graphs $\Sigma^n(u)$ of functions $u(x_2, \cdots, x_n) = x_1$ over $\mathbb{G}^n$ are the hyperplanes $x_1 = const.$; see [203, Theorem 4]. Taking into account Corollary 9.10, we can use [203, Theorem 4] to obtain a new Bernstein-type result.

In what follows a p-graph in $\mathbb{R}^p \times \mathbb{G}^n$ defined over $\mathbb{G}^n$ is a graph $u : \mathbb{G}^n \to \mathbb{R}^p$, with $(u(x), x) \in \mathbb{R}^p \times \mathbb{G}^n$.

Theorem 9.11 *The only complete φ-minimal bounded p-graphs in $\mathbb{R}^p \times \mathbb{G}^n$ defined over $\mathbb{G}^n$, having nonnegative Bakry-Émery-Ricci tensor, are the n-dimensional hyperplanes $\{q\} \times \mathbb{G}^n$ with $q \in \mathbb{R}^p$.*

Taking into account (5.2) and (9.60), by Theorem 9.11 we obtain the following

Corollary 9.12 *The only complete φ-minimal bounded p-graphs in $\mathbb{R}^p \times \mathbb{G}^n$ defined over $\mathbb{G}^n$, with the second fundamental form satisfying $|\alpha| \leq 1$, are the n-dimensional hyperplanes $\{q\} \times \mathbb{G}^n$ with $q \in \mathbb{R}^p$.*

In order to establish the last result of this chapter, define the space

$$\mathcal{L}_\varphi^k(\Sigma^n) = \left\{ u : \Sigma^n \to \mathbb{R} : \int_{\Sigma^n} |u|^k(x) e^{-\varphi(x)} d\Sigma^n < +\infty \right\},$$

and let us recall the next result that is a consequence of [255, Theorem 1.1].

Lemma 9.13 *Let u be a nonnegative smooth φ-subharmonic function on a complete Riemannian manifold Σ^n. If $u \in \mathcal{L}_\varphi^k(\Sigma^n)$, for some $k > 1$, then u is constant.*

By Lemmas 9.9 and 9.13 the following result holds.

Theorem 9.12 *Let $I \times_f M_\varphi^{n+p-1}$ be a weighted warped product and let $\psi : \Sigma^n \to I \times_f M_\varphi^{n+p-1}$ be a complete φ-minimal submanifold with $f'(h) \geq 0$. If $u = g(h) \in \mathcal{L}_\varphi^k(\Sigma^n)$,*

9.6 Submanifolds in a Weighted Riemannian Warped Product

for some $k > 1$, then $\psi(\Sigma^n)$ is contained in a slice $\{\tau\} \times M^{n+p-1}$, for some $\tau \in I$. Moreover, when $p = 2$, $\phi = \pi_M \circ \psi : \Sigma^n \to M^n$ is a φ-minimal hypersurface.

Remark 9.2 According to a result due to Wei and Wylie [294], all noncompact complete Riemannian manifolds Σ^n with nonnegative Bakry-Émery-Ricci tensor have at least a linear φ-volume growth, for some bounded weight function φ. For the sake of completeness, we recall here that a Riemannian manifold Σ^n has at least a linear φ-volume growth if for any $x \in \Sigma^n$, $\text{vol}_\varphi(B(x, R))$ has at least a linear growth in R, where $B(x, R)$ is the geodesic ball in Σ^n centered at x with radius R. Consequently, if in Theorem 9.12 we assume that Σ^n has nonnegative Bakry-Émery-Ricci tensor and $\varphi(h)$ is bounded, we conclude that Σ^n must be compact.

10 Submanifolds Immersed in a Killing Warped Product

10.1 Introduction

We recall that a *Killing vector field* defined on a Riemannian manifold is a smooth vector field whose flow constitutes a 1-parameter group of isometries of this manifold. It is well-known that Killing vector fields play an important role into submanifold theory. In this direction, Alías et al. [28] investigated complete minimal hypersurfaces immersed in Riemannian spaces of nonnegative Ricci curvature endowed with a globally defined Killing vector field K.

Assuming that the angle function defined by K and the Gauss map of the minimal hypersurface do not change sign, they obtained a nice extension of the classical Bernstein's Theorem [77]. Also working in this context, Dajczer et al. [134] established the concept of *Killing graph*, which is a graph constructed through the flow of such a Killing vector field K and whose domain is just a leaf of the orthogonal distribution $K^\perp$. In this setting, under appropriate assumptions involving domain data and the Ricci curvature of the ambient space, they solved the corresponding Dirichlet problem for prescribed mean curvature.

When the ambient space is a Riemannian product of the type $M^n \times \mathbb{R}$, whose base M^n is supposed to have nonnegative Ricci curvature and sectional curvature bounded from below, Rosenberg et al. [267] proved that any entire minimal graph defined over M^n with nonnegative height function must be a slice, extending a celebrated theorem due to Bombieri et al. [89] for entire minimal hypersurfaces in Euclidean space $\mathbb{R}^n$.

We also note that, taking into account some results of Chern [115] and Flanders [178], in [267] it was sufficient to suppose the constancy of the mean curvature. Afterwards, Dajczer and de Lira [131] extended Rosenberg-Schulze-Spruck's result [267], using a suitable version of the Omori-Yau's generalized maximum principle for the Laplacian (in the spirit of Pigola et al. [256]) to show that a constant mean curvature entire Killing graph contained into a *slab* (that is, a bounded region of the ambient space delimited by two slices) of a Killing warped product $M^n \times_\rho \mathbb{R}$ must be a totally geodesic slice, under certain restrictions on the curvature of the base M^n and on the warping function ρ. Next, in [132] the same authors also treated the case that the Killing graph lies inside a possible unbounded region of the ambient space $M^n \times_\rho \mathbb{R}$.

In this chapter, assuming suitable constraints on the warping function ρ and on the mean curvature vector field, we apply the Omori-Yau's generalized maximum principle in order to study the behavior of a support function naturally attached to a complete n-dimensional submanifold Σ^n immersed in a Killing warped product $M^{n+p} \times_\rho \mathbb{R}$ whose base M^{n+p} has sectional curvature bounded from below. When Σ^n has codimension 2, we conclude that there are no n-dimensional minimal compact submanifolds immersed in $M^{n+1} \times_\rho \mathbb{R}$ whose base is either $M^{n+1} = \mathbb{R}^{n+1}$ or $M^{n+1} = \mathbb{H}^{n+1}$. We also get that the only minimal topological 2-spheres immersed in $\mathbb{S}^3 \times_\rho \mathbb{R}$ are the totally geodesic ones in a slice $\mathbb{S}^3 \times \{t_0\}$.

Moreover, we obtain further results concerning the geometry of complete submanifolds immersed in a Killing warped product, under appropriate restrictions on the Bakry-Émery-Ricci tensor Ric_φ of these submanifolds with respect to the weight function $\varphi = \ln \rho^{-2}$. Afterwards, when $p = 0$, we establish sufficient conditions for the parabolicity of a complete two-sided hypersurface immersed into a Killing warped product $M^n \times_\rho \mathbb{R}$, whose base M^n has parabolic universal Riemannian covering. Besides, we apply a parabolicity criterion in order to obtain a rigidity result concerning hypersurfaces whose mean curvature is not supposed be constant. Parametric uniqueness results are applied to obtain suitable non-parametric ones, that is, to the case of entire Killing graphs.

Finally, we also study the uniqueness and nonexistence of mean curvature flow solitons (MCFS) with respect to a nowhere zero Killing vector field K globally defined in a Riemannian space, via suitable Liouville type results. To this scope, we consider that the ambient space is a Killing warped product $M^n \times_\rho \mathbb{R}$, where the base M^n is an arbitrarily fixed integral leaf of the distribution orthogonal to K and the warping function $\rho \in C^\infty(M)$ is given by $\rho = |K|$.

In particular, assuming that M^n is closed (that is, compact without boundary), we conclude that the only closed MCFS with respect to K are the totally geodesic slices. Furthermore, we prove Moser-Bernstein type results concerning entire Killing graphs constructed through the flow of K and which are complete MCFS with respect to it.

The results presented in this chapter are mainly based on the papers [56, 57] and [156, 183].

10.2 Recalling Some Basic Aspects of Submanifolds in a Killing Warped Product

Let $\overline{M}^{n+p+1}$ be an $(n + p + 1)$-dimensional Riemannian manifold endowed with a non-singular Killing vector field K. Suppose that the distribution $\mathcal{D}$ orthogonal to K is integrable. We denote by $\Psi : M^n \times \mathbb{I} \to \overline{M}^{n+1}$ the flow generated by K, where M^n is an arbitrarily fixed integral leaf of $\mathcal{D}$ labeled as $t = 0$, which we will suppose to be

10.2 Recalling Some Basic Aspects of Submanifolds in a Killing Warped...

connected, and $\mathbb{I}$ is the maximal interval of definition. Without lost of generality, in what follows we will also consider $\mathbb{I} = \mathbb{R}$. Hence, $\overline{M}^{n+p+1}$ is foliated by complete totally geodesic hypersurfaces.

We will assume that the ambient space $\overline{M}^{n+p+1}$ is a product manifold $M^{n+p} \times \mathbb{R}$ endowed with the warping metric

$$\langle \cdot, \cdot \rangle = \pi_M^* (\langle \cdot, \cdot \rangle_M) + (\rho \circ \pi_M)^2 \pi_{\mathbb{R}}^* \left(dt^2 \right), \qquad (10.1)$$

where π_M and $\pi_{\mathbb{R}}$ denote the canonical projections from $M^{n+p} \times \mathbb{R}$ onto each factor, $\langle \cdot, \cdot \rangle_M$ is the induced Riemannian metric on the base M^{n+p} and the warping function $\rho \in C^\infty$ is given by $\rho = |K| > 0$, where $|\cdot|$ denotes the norm of a vector field on $\overline{M}^{n+p+1}$. Such a Riemannian manifold $\overline{M}^{n+p+1}$ is said to be a *Killing warped product*.

Throughout this chapter, we will always deal with orientable submanifolds Σ^n immersed in $\overline{M}^{n+p+1} = M^{n+p} \times_\rho \mathbb{R}$ having codimension $p+1$, which means that the induced metric on Σ^n by the metric of $\overline{M}^{n+p+1}$ is positive definite and there exists an orthonormal frame $\{N_1, N_2, \ldots, N_{p+1}\}$ of the normal bundle $\mathfrak{X}^\perp(\Sigma^n)$. In this context, let $\overline{\nabla}, \widetilde{\nabla}$ and ∇ denote the Levi-Civita connections in $\overline{M}^{n+p+1}$, M^{n+p} and Σ^n, respectively. Then, as in [251], the curvature tensor R of the submanifold Σ^n is given by

$$R(X, Y)Z = \nabla_{[X,Y]} - [\nabla_X, \nabla_Y]Z,$$

where $[\cdot, \cdot]$ denotes the Lie bracket and $X, Y, Z \in \mathfrak{X}(\Sigma^n)$. A well-known fact is that the curvature tensor R of the submanifold Σ^n can be described in terms of the shape operators A_{N_i}, $1 \leq i \leq p+1$, and of the curvature tensor $\overline{R}$ of the ambient space $\overline{M}^{n+p+1} = M^{n+p} \times_\rho \mathbb{R}$ by the Gauss equation given by

$$R(X, Y)Z = (\overline{R}(X, Y)Z)^\top + \sum_{i=1}^{p+1} \left(\langle A_{N_i} X, Z \rangle A_{N_i} Y - \langle A_{N_i} Y, Z \rangle A_{N_i} X \right) \qquad (10.2)$$

for every tangent vector fields $X, Y, Z \in \mathfrak{X}(\Sigma^n)$, where $(\cdot)^\top$ denotes the tangential component of a vector field in $\mathfrak{X}(\overline{M}^{n+p+1})$ along Σ^n.

We will also consider some appropriate functions on a connected oriented submanifold $\varphi : \Sigma^n \to \overline{M}^{n+p+1}$ immersed in a Killing warped product $\overline{M}^{n+p+1} = M^{n+p} \times_\rho \mathbb{R}$, namely, the (vertical) height function $h = \pi_{\mathbb{R}} \circ \varphi$ and the angle functions $\Theta_i = \langle N_i, K \rangle$, where (as before) N_i, $i \in \{1, \ldots, p+1\}$, denote unit normal vector fields on Σ^n.

From the decomposition

$$K = K^\top + \sum_{i=1}^{p+1} \Theta_i N_i,$$

it is not difficult to see that

$$\nabla h = \frac{1}{\rho^2} K^\top \quad \text{and} \quad |\nabla h|^2 = \frac{\rho^2 - \sum_{i=1}^{p+1} \Theta_i^2}{\rho^4}. \tag{10.3}$$

In order establish our results concerning n-dimensional submanifolds immersed in a Killing warped product $M^{n+p} \times_\rho \mathbb{R}$, we observe that the mean curvature vector field $\mathbf{H}$ of such a submanifold Σ^n can be expressed in the following way

$$\mathbf{H} = \sum_{i=1}^{p} H_i N_i \tag{10.4}$$

for some smooth functions $H_1, H_2, \ldots, H_p$ defined on Σ^n, where we are taking $\{N_1, N_2, \ldots, N_{p+1}\}$ as an orthonormal frame of the normal bundle $\mathfrak{X}^\perp(\Sigma^n)$.

According to the notions and definitions of Chap. 5 for a smooth function $\varphi : \Sigma^n \to \mathbb{R}$, which is called a weight function on Σ^n, the φ-divergence operator on Σ^n is defined by

$$\operatorname{div}_\varphi(X) = e^\varphi \operatorname{div}(e^{-\varphi} X), \tag{10.5}$$

where X is a tangent vector field on Σ^n. Consequently, by (10.5), the drift Laplacian is given by

$$\Delta_\varphi u = \operatorname{div}_\varphi(\nabla u) = \Delta u - \langle \nabla u, \nabla \varphi \rangle, \tag{10.6}$$

where u is a smooth function on Σ^n. We will also refer to such an operator as been the φ-Laplacian of Σ^n.

Finally, we also recall that the Bakry-Émery-Ricci tensor is defined by

$$\operatorname{Ric}_\varphi = \operatorname{Ric} + \operatorname{Hess} \varphi, \tag{10.7}$$

see [59].

10.3 Rigidity Results for Submanifolds in a Killing Warped Product

This section is devoted to present our main nonexistence and rigidity results concerning submanifolds immersed in $M^{n+p} \times_\rho \mathbb{R}$. We will divide the main analysis into two cases.

10.3.1 The Compact Case

Considering the previous setting, we start applying the Divergence Theorem to obtain the following result.

Theorem 10.1 *Let Σ^n be a compact submanifold immersed in a Killing warped product $M^{n+p} \times_\rho \mathbb{R}$ such that $\langle \mathbf{H}, K \rangle$ does not change sign. Then Σ^n is a submanifold immersed in M^{n+p}, in particular $\mathbf{H}$ and K are orthogonal. Moreover, if $p = 0$ then M^{n+p} is compact.*

Proof Let Σ^n be a submanifold immersed in $M^{n+p} \times_\rho \mathbb{R}$. By (10.3) we have

$$\begin{aligned}
\Delta h &= \mathrm{div}\left(\rho^{-2} K^\top\right) \\
&= \langle \nabla \rho^{-2}, K^\top \rangle + \rho^{-2} \mathrm{div} K^\top \\
&= \langle \nabla \rho^{-2}, K^\top \rangle + \rho^{-2} \mathrm{div}\left(K - \sum_{i=1}^{p+1} \Theta_i N_i\right) \\
&= -\langle \rho^2 \nabla \rho^{-2}, \nabla h \rangle + n\rho^{-2} \sum_{i=1}^{p+1} \Theta_i H_i.
\end{aligned} \quad (10.8)$$

Hence, by (10.4) and (10.8) we get

$$\Delta h = -2\langle \nabla \ln \rho, \nabla h \rangle + \rho^{-2} \langle \mathbf{H}, K \rangle. \quad (10.9)$$

Taking $\varphi = \ln \rho^{-2}$, by (10.6) and (10.9) we obtain that

$$\Delta_\varphi h = n\rho^{-2} \langle \mathbf{H}, K \rangle. \quad (10.10)$$

Integrating (10.10) with respect to the measure $d\mu = \rho^2 d\Sigma^n$ and by using the compactness of Σ^n, we obtain that

$$\int_{\Sigma^n} \langle \mathbf{H}, K \rangle d\mu = 0,$$

thanks to the Divergence Theorem.

Since $\langle \mathbf{H}, K \rangle$ does not change sign, it must be identically zero. Consequently, we get that h is φ-harmonic in a compact manifold and, hence, it must be constant on Σ^n. Finally, when $p = 0$, we get that Σ^n is isometric to the base M^n and, in this case, M^n must be compact. □

The next result is a direct consequence of Theorem 10.1.

Corollary 10.1 *There are no n-dimensional compact submanifolds immersed in a Killing warped product $M^{n+p} \times_\rho \mathbb{R}$ with $\langle \mathbf{H}, K \rangle$ having strict sign.*

According to [36], an n-dimensional submanifold Σ^n immersed in a Killing warped product $M^{n+p} \times_\rho \mathbb{R}$ is said a *mean curvature flow soliton* (MCFS) with respect to K, if

$$\mathbf{H} = cK^\perp, \tag{10.11}$$

for some constant $c \in \mathbb{R}$, which is called the *soliton constant* of Σ^n; see Sect. 10.9 for more details concerning codimension one MCFS.

In this context, Theorem 10.1 reads as the following nonexistence result.

Corollary 10.2 *There does not exist an n-dimensional non-minimal compact MCFS immersed in $M^{n+p} \times_\rho \mathbb{R}$, with respect to K.*

Proof Indeed, supposing by contradiction that Σ^n is such a MCFS with respect to K, having soliton constant $c \neq 0$, by (10.11) we get that

$$\langle \mathbf{H}, K \rangle = \frac{|\mathbf{H}|^2}{c}.$$

Thus, $\langle \mathbf{H}, K \rangle$ does not change sign. Since Σ^n is assumed to be compact, by Theorem 10.1 we get that $\langle \mathbf{H}, K \rangle = 0$ and $K^\top = 0$. Then, we conclude that K vanishes identically on Σ^n, contradicting the non-singularity of K. □

When the codimension of Σ^n is 2, by Theorem 10.1 we get a characterization of totally geodesic hypersurfaces of M^{n+1}.

Theorem 10.2 *Let Σ^n be a compact submanifold immersed in a Killing warped product $M^{n+1} \times_\rho \mathbb{R}$ with parallel mean vector field such that $\langle \mathbf{H}, K \rangle$ does not change sign. Then, Σ^n is a constant mean curvature hypersurface immersed in M^{n+1}. In addition, if Ric_M is nonnegative and there exists a Killing vector field Y in M^{n+1} such that $\langle v, Y \rangle > 0$ for some orientation v of Σ^n in M^{n+1}, then Σ^n is a totally geodesic hypersurface of M^{n+1}.*

Proof The first part follows directly by Theorem 10.1. For the second one, we recall the following well-known formula

$$\Delta \langle v, Y \rangle = -(|A_v|^2 + \mathrm{Ric}_M(v, v))\langle v, Y \rangle,$$

see, for instance, Equation (24) of [97].

Therefore, by applying the Divergence Theorem to the above equation, we get that $A_v = 0$ and, hence, Σ^n must be totally geodesic. □

10.3 Rigidity Results for Submanifolds in a Killing Warped Product

From the first part of Theorem 10.2 we obtain the following codimension reduction, which allows us to get a nonexistence result when the base M^{n+1} is either $\mathbb{R}^{n+1}$ or $\mathbb{H}^{n+1}$, as well as a characterization of minimal topological 2-spheres immersed in $\mathbb{S}^3 \times_\rho \mathbb{R}$; see [18, 117, 164].

Corollary 10.3 *The only n-dimensional minimal compact submanifolds immersed in a Killing warped product $M^{n+1} \times_\rho \mathbb{R}$ are the minimal hypersurfaces immersed in M^{n+1}. In particular, there do not exist n-dimensional minimal compact submanifolds immersed in a Killing warped product $M^{n+1} \times_\rho \mathbb{R}$ with base either $M^{n+1} = \mathbb{R}^{n+1}$ or $M^{n+1} = \mathbb{H}^{n+1}$. Moreover, the only minimal topological 2-spheres immersed in $\mathbb{S}^3 \times_\rho \mathbb{R}$ are the totally geodesic ones in $\mathbb{S}^3$.*

10.3.2 The Complete Case

In order to apply the Omori-Yau's generalized maximum principle for complete submanifolds, we will need a reasonable set of sufficient conditions which guarantee that the Ricci curvature is bounded from below. To this goal, some preliminary computations are given below.

Let us consider an orientable submanifold $\varphi : \Sigma^n \to \overline{M}^{n+p+1}$ immersed in a Killing warped product $\overline{M}^{n+p+1} = M^{n+p} \times_\rho \mathbb{R}$. For any vector field U tangent to $\overline{M}^{n+p+1}$, we denote by U^* and $U^\perp$ the orthogonal projections of U onto TM and $T\mathbb{R}$, respectively. So, we can write

$$U = U^* + U^\perp. \tag{10.12}$$

Consequently, by (10.12) we get

$$U^\perp = \frac{\langle U, K \rangle}{\langle K, K \rangle} K = \frac{\langle U, K \rangle}{\rho^2} K. \tag{10.13}$$

Thus, for any vector fields U, V, W tangent to $\overline{M}^{n+p+1}$, by using (10.13) it is not difficult to verify that

$$\overline{R}(U, V)W = R_M(U^*, V^*)W^* - \frac{\langle V, K \rangle}{\rho^2} \overline{R}(K, U^*)W^*$$
$$+ \frac{\langle V, K \rangle \langle W, K \rangle}{\rho^4} \overline{R}(U^*, K)K + \frac{\langle U, K \rangle}{\rho^2} \overline{R}(K, V^*)W^*$$
$$- \frac{\langle U, K \rangle \langle W, K \rangle}{\rho^4} \overline{R}(V^*, K)K. \tag{10.14}$$

Then, by using [251, Lemma 7.34 and Proposition 7.42] into (10.14), we get

$$\overline{R}(U,V)W = R_M(U^*, V^*)W^* - \frac{\langle V, K\rangle \operatorname{Hess}_M \rho(U^*, W^*)}{\rho^3}K$$
$$+ \frac{\langle V, K\rangle\langle W, K\rangle}{\rho^3}D_{U^*}D\rho + \frac{\langle U, K\rangle \operatorname{Hess}_M \rho(V^*, W^*)}{\rho^3}K$$
$$- \frac{\langle U, K\rangle\langle W, K\rangle}{\rho^3}D_{V^*}D\rho, \tag{10.15}$$

where Hess_M stands for the Hessian on M^{n+p}. In particular, taking a local orthonormal frame $\{E_1, \ldots, E_n\}$ tangent to Σ^n and X a vector field tangent to Σ^n, we can consider $U = W = X$ and $V = E_i$ in (10.15) to obtain

$$\langle \overline{R}(X, E_i)X, E_i\rangle = \langle R_M(X^*, E_i^*)X^*, E_i^*\rangle$$
$$- \frac{\langle E_i, K\rangle^2}{\rho^3}\operatorname{Hess}_M \rho(X^*, X^*)$$
$$+ \frac{\langle E_i, K\rangle\langle X, K\rangle}{\rho^3}\operatorname{Hess}_M \rho(X^*, E_i^*)$$
$$+ \frac{\langle E_i, K\rangle\langle X, K\rangle}{\rho^3}\operatorname{Hess}_M \rho(X^*, E_i^*)$$
$$- \frac{\langle X, K\rangle^2}{\rho^3}\operatorname{Hess}_M(E_i^*, E_i^*). \tag{10.16}$$

Hence, by (10.16) it follows that

$$\langle \overline{R}(X, E_i)X, E_i\rangle = K_M(X^*, E_i^*)\left(\langle X^*, X^*\rangle\langle E_i^*, E_i^*\rangle - \langle X^*, E_i^*\rangle^2\right)$$
$$- \frac{1}{\rho}\operatorname{Hess}_M \rho(\widetilde{X}_i^* - \widetilde{E}_i^*, \widetilde{X}_i^* - \widetilde{E}_i^*), \tag{10.17}$$

where

$$\widetilde{X}_i^* = \frac{\langle E_i, K\rangle}{\rho}X^* \quad \text{and} \quad \widetilde{E}_i^* = \frac{\langle X, K\rangle}{\rho}E_i^*.$$

Therefore, by (10.17) we reach the following equation

$$\sum_{i=1}^n \langle \overline{R}(X, E_i)X, E_i\rangle = \sum_{i=1}^n K_M(X^*, E_i^*)\left(\langle X^*, X^*\rangle\langle E_i^*, E_i^*\rangle - \langle X^*, E_i^*\rangle^2\right)$$
$$- \sum_{i=1}^n \frac{1}{\rho}\operatorname{Hess}_M \rho(\widetilde{X}_i^* - \widetilde{E}_i^*, \widetilde{X}_i^* - \widetilde{E}_i^*). \tag{10.18}$$

10.3 Rigidity Results for Submanifolds in a Killing Warped Product

As a consequence of the computations given above, we obtain sufficient conditions which guarantee that the Ricci curvature of a submanifold is bounded from below.

Proposition 10.1 *Let $\overline{M}^{n+p+1} = M^{n+p} \times_\rho \mathbb{R}$ be a Killing warped product such that $\mathrm{Hess}_M \rho$ is bounded from above, ρ is bounded away from zero and whose base M^{n+p} has sectional curvature satisfying $K_M \geq -\kappa$, for some constant $\kappa \geq 0$. If Σ^n is a submanifold immersed in $\overline{M}^{n+p+1}$ with bounded second fundamental form, then its Ricci curvature is bounded from below.*

Proof From the Gauss equation (10.2), taking a local orthonormal frame $\{E_1, \ldots, E_n\}$ tangent to Σ^n, for all vector field X tangent to Σ^n we have that the Ricci curvature of Σ^n is given by

$$\mathrm{Ric}(X, X) = \sum_{i=1}^{n} \langle \overline{R}(X, E_i) X, E_i \rangle + n \langle A_\mathbf{H} X, X \rangle - \sum_{k=1}^{p+1} \langle A_{N_k}^2 X, X \rangle. \tag{10.19}$$

We observe that, for each $i = 1, \ldots, n$, (10.3) implies that

$$\langle X^*, X^* \rangle \langle E_i^*, E_i^* \rangle = \langle X - \langle X, \nabla h \rangle K, X - \langle X, \nabla h \rangle K \rangle$$
$$\langle E_i - \langle E_i, \nabla h \rangle K, E_i - \langle E_i, \nabla h \rangle K \rangle$$
$$= |X|^2 - \rho^2 |X|^2 \langle E_i, \nabla h \rangle^2 - \rho^2 \langle X, \nabla h \rangle^2$$
$$+ \rho^4 \langle X, \nabla h \rangle^2 \langle E_i, \nabla h \rangle^2 \tag{10.20}$$

and

$$\langle X^*, E_i^* \rangle^2 = \langle X - \langle X, \nabla h \rangle K, E_i - \langle E_i, \nabla h \rangle K \rangle^2$$
$$= \langle X, E_i \rangle^2 - 2\rho^2 \langle X, \nabla h \rangle \langle X, E_i \rangle \langle E_i, \nabla h \rangle$$
$$+ \rho^4 \langle X, \nabla h \rangle^2 \langle E_i, \nabla h \rangle^2. \tag{10.21}$$

Consequently, by (10.20) and (10.21) we get

$$\sum_{i=1}^{n} \langle X^*, X^* \rangle \langle E_i^*, E_i^* \rangle - \langle X^*, E_i^* \rangle^2 = (n-1)|X|^2 - \rho^2 |X|^2 |\nabla h|^2$$
$$- (n-2) \rho^2 \langle X, \nabla h \rangle^2$$
$$\leq (n-1)|X|^2. \tag{10.22}$$

Hence, taking into account our constraint on K_M, by (10.22) we obtain

$$\sum_{i=1}^{n} K_M(X^*, E_i^*)\left(\langle X^*, X^*\rangle\langle E_i^*, E_i^*\rangle - \langle X^*, E_i^*\rangle^2\right) \geq -\kappa(n-1)|X|^2. \tag{10.23}$$

Since we are assuming that $\mathrm{Hess}_M \rho$ is bounded from above, ρ is bounded away from zero and $K_M \geq -\kappa$, for some constant $\kappa \geq 0$, by (10.18), (10.19) and (10.23) it follows that

$$\mathrm{Ric}(X, X) \geq -\left(\kappa(n-1) - \beta + |A_\mathbf{H}| + |II|\right)|X|^2, \tag{10.24}$$

for some positive constant β, where II stands for the second fundamental form of Σ^n. Therefore, assuming that II is bounded, from the lower estimate (10.24) we conclude that the Ricci curvature of Σ^n is bounded from below. □

At this point, we quote once more the generalized maximum principle of Omori [250] and Yau [297]; see Sect. 1.6 for more details.

Lemma 10.1 *Let Σ^n be a n-dimensional complete Riemannian manifold whose Ricci curvature is bounded from below and let $u : \Sigma^n \to \mathbb{R}$ be a smooth function satisfying $\sup_\Sigma u < +\infty$. Then, there exists a sequence of points $\{p_k\} \subset \Sigma^n$ such that*

$$\lim_k u(p_k) = \sup_{\Sigma^n} u, \quad \lim_k |\nabla u(p_k)| = 0 \quad \text{and} \quad \limsup_k \Delta u(p_k) \leq 0.$$

Inspired by [131, Theorem 1], the following result holds.

Theorem 10.3 *Let $\overline{M}^{n+p+1} = M^{n+p} \times_\rho \mathbb{R}$ be a Killing warped product such that $\mathrm{Hess}_M \rho$ is bounded from above, ρ is bounded away from zero and whose base M^n has sectional curvature satisfying $K_M \geq -\kappa$, for some constant $\kappa \geq 0$. Let Σ^n be a complete submanifold immersed in $\overline{M}^{n+p+1}$ with second fundamental form II and $\nabla \ln \rho$ bounded. If Σ^n lies under (resp. above) a slice and $\langle \mathbf{H}, K \rangle \geq 0$ (resp. $\langle \mathbf{H}, K \rangle \leq 0$), then either ρ is unbounded or $\langle \mathbf{H}, K \rangle$ is not bounded away from zero. Assuming in addition that the Bakry-Émery-Ricci tensor of Σ^n with respect to the weight function $\varphi = \ln \rho^{-2}$ is nonnegative, if Σ^n lies in a slab of $\overline{M}^{n+p+1}$, ρ is bounded and $\langle \mathbf{H}, K \rangle$ is constant, then Σ^n must be a submanifold immersed in M^{n+p}.*

Proof Let us suppose by contradiction that ρ and $\langle \mathbf{H}, K \rangle^{-1}$ are bounded on Σ^n. Since the second fundamental form is bounded, by Proposition 10.1 we have that the Ricci curvature is bounded from below. Thus, taking into account (for instance) that Σ^n lies under a slice of $M^{n+1} \times_\rho \mathbb{R}$ and $\langle \mathbf{H}, K \rangle \geq 0$, by (10.9) and Lemma 10.1 we obtain a sequence of points $\{p_k\} \subset \Sigma^n$ such that

$$0 \geq \limsup_k \Delta h(p_k) = \limsup_k \left(\rho^{-2} \langle \mathbf{H}, K \rangle \right)(p_k) \geq 0.$$

Then, one has

$$\limsup_k \left(\langle \mathbf{H}, K \rangle \right)(p_k) = 0.$$

Therefore, $\langle \mathbf{H}, K \rangle^{-1}$ is unbounded, obtaining a contradiction.

Now, supposing in addition that ρ is bounded and $\langle \mathbf{H}, K \rangle$ is constant, we get that $\langle \mathbf{H}, K \rangle$ must be identically zero. So, by (10.10) we obtain that $\Delta_\varphi h = 0$, for $\varphi = \ln \rho^{-2}$.

Consequently, assuming that Σ^n lies in a slab of $\overline{M}^{n+p+1}$ and that Ric_φ is nonnegative, we can apply a Liouville-type Theorem due to Brighton [91] to conclude that h must be constant and, hence, Σ^n is a submanifold immersed in M^{n+p}. □

Remark 10.1 Related to the last part of Theorem 10.3, we point out that submanifolds immersed in $M^{n+p} \times_\rho \mathbb{R}$ with constant support function $\langle \mathbf{H}, K \rangle$ correspond to a natural extension of constant mean curvature hypersurfaces which make a constant angle with the tangent direction to $\mathbb{R}$. Such submanifolds had already been studied in reference [52] when the ambient space is a warped product of the type $\mathbb{R} \times_f M^{n+p}$, where $f : \mathbb{R} \to \mathbb{R}$ stands for the warping function.

10.4 Further Results for Submanifolds in a Killing Warped Product

An important and applicable class of examples relies in the case where the ambient space is a space form, or more generally, the warped product $M^{n+p} \times_\rho \mathbb{R}$ is an Einstein manifold furnished with the warped metric. Indeed, this structure of Einstein warped products has been extensively studied due to its connection with quasi-Einstein metrics (see, for instance, [102, 217] and references therein) having, on another hand, the interesting aspect of submanifolds in Einstein manifolds (see [219]).

A crucial observation in our context is that the assumption that $M^{n+p} \times_\rho \mathbb{R}$ is Einstein is equivalent that $\rho \in C^\infty(M^{n+p})$ is a solution to the following equation (see Corollary 9.107 of [80])

$$-(\Delta_g \rho)g + \mathrm{Hess}_g \rho - \rho \mathrm{Ric}_g = 0 \quad \text{in} \quad \mathrm{int}(M^{n+p}). \tag{10.25}$$

We observe that the left hand side of Eq. (10.25) is the formal L^2-adjoint of the linearization of the scalar curvature operator and this equation is closely related to the so-called static metrics (see [239]).

In our context, an operational advantage in dealing with these types of manifolds lies in the fact that the warping function ρ satisfies an equation that relates its Laplacian with its Hessian. From (10.25) it is not difficult to verify that we can improve Theorem 10.3 with a weaker hypothesis of subharmonicity of the warping function ρ, obtaining the following:

Theorem 10.4 *Let Σ^n be a complete submanifold immersed with bounded second fundamental form in an Einstein warped product $\overline{M}^{n+p+1} = M^{n+p} \times_\rho \mathbb{R}$ whose Laplacian of the warping function ρ is bounded from above and such that the sectional curvature of the base M^n is bounded from below. If Σ^n lies under (resp. above) a slice and $\langle \mathbf{H}, K \rangle \geq 0$ (resp. $\langle \mathbf{H}, K \rangle \leq 0$), then either ρ is unbounded or $\langle \mathbf{H}, K \rangle$ is not bounded away from zero.*

Next, it is worth to recall the weak Omori-Yau's generalized maximum principle for the drift Laplacian, whose proof can be found in [259].

Lemma 10.2 *Let Σ^n be a complete manifold whose Bakry-Émery-Ricci tensor with respect to a weight function φ is bounded from below and let $u : \Sigma^n \to \mathbb{R}$ be a smooth function satisfying $\sup_{\Sigma^n} u < +\infty$. Then, there exists a sequence of points $\{p_k\} \subset \Sigma^n$ such that*

$$\lim_k u(p_k) = \sup_{\Sigma^n} u \quad \text{and} \quad \limsup_k \Delta_\varphi u(p_k) \leq 0.$$

Taking into account Lemma 10.2 and (10.10), by arguing as in the proof of Theorem 10.3.

Theorem 10.5 *Let $\overline{M}^{n+p+1} = M^{n+p} \times_\rho \mathbb{R}$ be a Killing warped product such that ρ is bounded away from zero. Let Σ^n be a complete submanifold immersed in $\overline{M}^{n+p+1}$, whose Bakry-Émery-Ricci tensor with respect to the weight function $\varphi = \ln \rho^{-2}$ is bounded from below. If Σ^n lies under (resp. above) a slice and $\langle \mathbf{H}, K \rangle \geq 0$ (resp. $\langle \mathbf{H}, K \rangle \leq 0$), then either ρ is unbounded or $\langle \mathbf{H}, K \rangle$ is not bounded away from zero.*

In order to establish our next result, set

$$\mathcal{L}^k_\varphi(\Sigma^n) = \left\{ u : \Sigma^n \to \mathbb{R} : \int_{\Sigma^n} |u|^k d\mu < +\infty \right\},$$

where $d\mu = e^\varphi d\Sigma^n$ stands for the weighted measure related to the weight function $\varphi = \ln \rho^{-2}$. A special case, for $k = 1$, is the following.

10.4 Further Results for Submanifolds in a Killing Warped Product

Theorem 10.6 *Let Σ^n be a complete submanifold immersed in warped product $\overline{M}^{n+p+1} = M^{n+p} \times_\rho \mathbb{R}$ such that $\langle \mathbf{H}, K \rangle$ does not change sign. If $|\nabla h| \in \mathcal{L}_\varphi^1(\Sigma^n)$, then $\mathbf{H}$ must be orthogonal to K. Moreover, if the Bakry-Émery-Ricci tensor of Σ^n, Ric_φ, is nonnegative and it lies in a slab of $\overline{M}^{n+p+1}$, then Σ^n is submanifold immersed in M^{n+p}.*

Proof We argue as in the proof of [97, Proposition 2.1]. Since Σ^n is complete, there exists a sequence ϕ_j such that $\phi_j = 1$ in compact sets $C_j \subset \mathrm{supp}\phi_j \subset \Omega_j$ which are both exhaustion of Σ^n. Thus, $\phi_j \nearrow 1$ and $|\nabla \phi_j|_\infty \to 0$; see [284, Proposition 4.1] for $p = 1$. Thus, one has

$$\int_{\Sigma^n} \mathrm{div}_\varphi(\phi_j \nabla h) d\mu = \int_{\Sigma^n} \nabla \phi_j \cdot \nabla h d\mu + \int_{\Sigma^n} \phi_j \Delta_\varphi h d\mu.$$

By using the Divergence Theorem in the left hand side, it is equal to zero. Moreover, the Holder's inequality in the first term of the right hand side, we obtain that it goes to 0 as $j \to \infty$. By using the Monotone Convergence Theorem for the last term, we get

$$0 = \int_{\Sigma^n} \Delta_\varphi h d\mu.$$

Finally, since $\Delta_\varphi h \geq 0$, we obtain that $\Delta_\varphi h = 0$ and, consequently, $\langle \mathbf{H}, K \rangle = 0$ on Σ^n. Furthermore, assuming that $\mathrm{Ric}_\varphi \geq 0$, we can apply once more the Liouville-type theorem due to Brighton [91] to conclude that h must be constant and, hence, Σ^n is submanifold of M^{n+p}. □

Taking in account formula (10.10), the next consequence holds.

Corollary 10.4 *If Σ^n is a φ-parabolic submanifold immersed in a Killing warped product $\overline{M}^{n+p+1} = M^{n+p} \times_\rho \mathbb{R}$, lying in a upper (resp., lower) half space of $\overline{M}^{n+p}$ and such that $\langle \mathbf{H}, K \rangle$ is nonnegative (resp., nonpositive), then Σ^n is a submanifold immersed in M^{n+p} and, particularly, $\mathbf{H}$ and K are orthogonal.*

Proof By (10.10) we obtain that the height function h is φ-subharmonic since it is bounded from above and, therefore, it is constant. The conclusion immediately holds. □

Related to minimal φ-parabolic submanifolds into a Killing warped product, we also obtain the following result.

Corollary 10.5 *The only n-dimensional minimal φ-parabolic submanifolds immersed in a Killing warped product $\overline{M}^{n+p+1} = M^{n+p} \times_\rho \mathbb{R}$, lying in a vertical half space of $\overline{M}^{n+p+1}$, are those ones immersed in M^{n+p}.*

The following concept comes from the notion of probability on a manifold with a certain metric and a density not necessarily of finite integral, and it will be an important tool to get our next result.

Let us consider a real function $f : \Sigma^n \to \mathbb{R}$ and $1 \leq k < \infty$. Denoting $d\mu = e^\varphi d\Sigma^n$ and taking $\varphi = \ln \rho^{-2}$, the φ-weighted k-capacity of a compact subset $C \subset \Sigma^n$ is defined by

$$\operatorname{Cap}_{\varphi,k}(C) = \inf \left\{ \int_{\Sigma^n} |\nabla \phi|^k d\mu \, : \, \phi \equiv 1 \text{ in } C \text{ and } \phi \in C_0^1(\Sigma^n) \right\}.$$

It is natural to use the following definition of φ-weighted k-parabolicity: A Riemannian manifold Σ^n is said to be φ-weighted k-parabolic if the φ-weighted k-capacity of all compact subsets of Σ^n is zero.

The following result is mainly based on the above notion; see [284, Proposition 4.1] for additional comments and details.

Proposition 10.2 *A Riemannian manifold Σ^n is φ-weighted k-parabolic if, and only if, there exists a sequence of functions $\phi_j \in C_0^1(\Sigma^n)$ such that $0 \leq \phi_j \leq 1$, $\phi_j \nearrow 1$ uniformly on every compact subset of Σ^n and*

$$\int_{\Sigma^n} |\nabla \phi_j|^k d\mu \to 0.$$

Taking into account Proposition 10.2, as in the proof of Theorem 10.6, the following result holds.

Theorem 10.7 *Let Σ^n be a φ-weighted k-parabolic submanifold immersed in $\overline{M}^{n+p+1} = M^{n+p} \times_\rho \mathbb{R}$ such that $\langle \mathbf{H}, K \rangle$ does not change sign. If $|\nabla h| \in \mathcal{L}_\varphi^s(\Sigma^n)$, with $\frac{1}{k} + \frac{1}{s} = 1$, then $\mathbf{H}$ and K are orthogonal. Moreover, if $\operatorname{Ric}_\varphi$ is nonnegative and Σ^n is contained in a slab of $\overline{M}^{n+p+1} = M^{n+p} \times_\rho \mathbb{R}$, then Σ^n is a submanifold immersed in M^{n+p}.*

Finally, we end this section proving the next rigidity result.

Theorem 10.8 *Let Σ^n be a complete submanifold immersed in $M^{n+p} \times_\rho \mathbb{R}$, such that $\langle \mathbf{H}, K \rangle$ is nonnegative (resp. nonpositive) and $h \geq 0$ (resp. $h \leq 0$). If $h \in \mathcal{L}_\varphi^k(\Sigma^n)$ with $k > 1$, then Σ^n is a submanifold immersed in M^{n+p}. Moreover, if $\operatorname{Ric}_\varphi$ is nonnegative, ρ is bounded along Σ^n and $h > 0$ (resp. $h < 0$), then Σ^n is compact.*

Proof Under our constraints and taking into account that $\Delta_\varphi h = e^{-\varphi} \langle \mathbf{H}, K \rangle$, we obtain that h (respect. $-h$) is a nonnegative subharmonic function on Σ^n. Thus, since we are assuming that $h \in \mathcal{L}_\varphi^k(\Sigma^n)$ with $k > 1$, we can apply Theorem 1.1 of [255] to conclude that h is constant on Σ^n, that is, Σ^n is a submanifold immersed in M^{n+p}. Moreover, supposing that h has strict sign, we get that $\mathrm{vol}_\varphi(\Sigma)$ is finite. Hence, if in addition Ric_φ is nonnegative, ρ is bounded along Σ^n and $h > 0$ (respect. $h < 0$), we can apply Theorem 1.3 of [294] to conclude that Σ^n must be compact. □

10.5 Hypersurfaces in Killing Warped Products

In the remainder of this chapter, let us consider a connected two-sided hypersurface $\psi : \Sigma^n \to \overline{M}^{n+1}$ immersed into a Killing warped product $\overline{M}^{n+1} = M^n \times_\rho \mathbb{R}$, which means that there exists a globally defined unit normal vector field N on Σ^n. Let $\overline{\nabla}$ and ∇ denote the Levi-Civita connections in $\overline{M}^{n+1}$ and Σ^n, respectively. Then the Gauss and Weingarten formulas for the hypersurface $\psi : \Sigma^n \to \overline{M}^{n+1}$ are given by

$$\overline{\nabla}_X Y = \nabla_X Y + \langle AX, Y \rangle N, \tag{10.26}$$

and

$$AX = -\overline{\nabla}_X N \tag{10.27}$$

for every tangent vector fields $X, Y \in \mathfrak{X}(\Sigma^n)$. Here $A : \mathfrak{X}(\Sigma^n) \to \mathfrak{X}(\Sigma^n)$ stands for the shape operator (or Weingarten endomorphism) of Σ^n with respect to N. Moreover, the mean curvature is defined by

$$H = \frac{1}{n}\mathrm{trace}(A).$$

For our purposes, we will consider along Σ^n its height function $h = \pi_\mathbb{R} \circ \psi$ and its angle function $\Theta = \langle N, K \rangle$. By using the decomposition $K = K^\top + \Theta N$, where $(\ )^\top$ denotes the tangential component of a vector field in $\mathfrak{X}(\overline{M}^{n+1})$ along ψ, we obtain

$$\nabla h = \frac{1}{\rho^2} K^\top \quad \text{and} \quad |\nabla h|^2 = \frac{\rho^2 - \Theta^2}{\rho^4}. \tag{10.28}$$

To close this section, we recall that, for each $t \in \mathbb{R}$, the hypersurface $M^n \times \{t\}$ is called a slice of $\overline{M}^{n+1}$. We note that the slices of $\overline{M}^{n+1}$ are totally geodesic hypersurfaces and, by (10.28), a hypersurface is a slice if and only if $\rho^2 = \Theta^2$.

10.6 A Parabolicity Criterion for Two-Sided Hypersurfaces in a Killing Warped Product

Our aim in this section is to obtain an extension of the parabolicity criterion developed in [260] to the context of Killing warped products. We will assume that our hypersurfaces are connected. So, we get the following

Theorem 10.9 *Let $\overline{M}^{n+1} = M^n \times_\rho \mathbb{R}$ be a Killing warped product and let $\psi : \Sigma^n \to \overline{M}^{n+1}$ be a complete two-sided hypersurface such that the function Θ has strict sign and $\eta = \rho/\Theta$ is bounded. Then, M^n is complete. If in addition M^n has a parabolic universal Riemannian covering, then Σ^n parabolic.*

Proof The proof is based on two facts:

(i) Parabolicity is invariant under a quasi-isometry (cf. [194, 216]);
(ii) If the universal Riemannian covering $\widetilde{\Sigma}$ of Σ^n is parabolic, then Σ^n is also parabolic.

Since Θ has strict sign, for an appropriated choice of N we can suppose $\Theta > 0$. Now, denoting $\pi = \pi_M \circ \psi$, $\pi_* = d\pi$ and $h_* = dh$, for any tangent vector $v \in T\Sigma^n$, by using the Cauchy-Schwartz inequality it follows that

$$\langle v, v \rangle = \langle \pi_* v, \pi_* v \rangle_M + \rho^2 \langle h_* v, h_* v \rangle \leq \langle \pi_* v, \pi_* v \rangle_M + \rho^2 |\nabla h|^2 \langle v, v \rangle,$$

that is

$$(1 - \rho^2 |\nabla h|^2)\langle v, v \rangle \leq \langle \pi_* v, \pi_* v \rangle_M.$$

Now, the definition of the function η and (10.28) yields

$$\frac{1}{\eta^2} \langle v, v \rangle \leq \langle \pi_* v, \pi_* v \rangle_M.$$

Moreover, taking account our hypothesis, we have that

$$c^{-1} \langle v, v \rangle \leq \langle \pi_* v, \pi_* v \rangle_M, \qquad (10.29)$$

where $c = \sup_{\Sigma^n} \eta^2 \geq 1$.

Consequently, π is a local diffeomorphism and we can reason as in the proof of [163, Lemma 7.3.3] (see also [218, Lemma 8.8.1]) to conclude that π is a covering map and that M^n is complete.

10.6 A Parabolicity Criterion for Two-Sided Hypersurfaces in a Killing...

On the other hand, we see that

$$\langle v, v \rangle = \langle \pi_* v, \pi_* v \rangle_M + \rho^2 \langle h_* v, h_* v \rangle_\mathbb{R} \geq \langle \pi_* v, \pi_* v \rangle_{M^n}$$

Since $c \geq 1$, we obtain that

$$\langle \pi_* v, \pi_* v \rangle_{M^n} \leq c \langle v, v \rangle. \tag{10.30}$$

By (10.29) and (10.30) it follows that

$$c^{-1} \langle \pi_* v, \pi_* v \rangle \leq \langle v, v \rangle \leq c \langle \pi_* v, \pi_* v \rangle. \tag{10.31}$$

So, let $\widetilde{\Sigma}$ be the universal Riemannian covering of Σ^n with projection $\pi_\Sigma : \widetilde{\Sigma} \to \Sigma^n$. Then, the map $\pi_0 = \pi \circ \pi_{\Sigma^n} : \widetilde{\Sigma} \to M^n$ is a covering map. Now, if $\widetilde{M}$ is the universal Riemannian covering of M^n with projection $\widetilde{\pi} : \widetilde{M} \to M^n$, then there exists a diffeomorphism $\varphi : \widetilde{\Sigma} \to \widetilde{M}$ such that $\widetilde{\pi} \circ \varphi = \pi_0$. Moreover, φ is a quasi-isometry.

Indeed, if $v \in T\widetilde{\Sigma}$, by (10.31) we have that

$$\langle \varphi_* v, \varphi_* v \rangle_{\widetilde{M}} = \langle \widetilde{\pi}_*(\varphi_* v), \widetilde{\pi}_*(\varphi_* v) \rangle_M = \langle (\pi_0)_* v, (\pi_0)_* v \rangle_M$$
$$= \langle \pi_*((\pi_\Sigma)_* v), \pi_*((\pi_\Sigma)_* v) \rangle_M \leq c \langle (\pi_\Sigma)_* v, (\pi_\Sigma)_* v \rangle_{\Sigma^n} = c \langle v, v \rangle_{\widetilde{\Sigma}}.$$

Similarly, we obtain

$$\langle \varphi_* v, \varphi_* v \rangle_{\widetilde{M}} \geq c^{-1} \langle v, v \rangle_{\widetilde{\Sigma}}.$$

Therefore, since the universal Riemannian covering of M^n is parabolic, it follows that the universal Riemannian covering of Σ^n has the same property.

Hence, Σ^n must be also parabolic. □

Remark 10.2 We note that the function $\eta = \rho/\Theta$ defined in Theorem 10.9 is, in fact, given by $\eta = 1/\cos\theta$, where θ is the angle between N and K. In particular, the boundedness of η means that the vector fields N and K are away being orthogonal.

As a direct consequence of Theorem 10.9, we have the following corollary.

Corollary 10.6 Let $\overline{M}^{n+1} = M^n \times_\rho \mathbb{R}$ be a Killing warped product and let $\psi : \Sigma^n \to \overline{M}^{n+1}$ be a complete two-sided hypersurface such that the function Θ has strict sign and η is bounded. Then, M^n is complete. If in addition M^n is simply connected and parabolic, then Σ^n parabolic.

As well as

Corollary 10.7 *Let $\overline{M}^{n+1} = M^n \times \mathbb{R}$ be a Riemannian product and let $\psi : \Sigma^n \to \overline{M}^{n+1}$ be a complete two-sided hypersurface whose the function Θ is bounded away from zero. Then, M^n is complete. If in addition M^n has parabolic universal Riemannian covering, then Σ^n parabolic.*

10.7 Rigidity Results for Two-Sided Hypersurfaces in a Killing Warped Product

In this section, we will apply Theorem 10.9 in order to obtain a rigidity result for complete two-sided hypersurfaces immersed into a Killing warped product $\overline{M}^{n+1} = M^n \times_\rho \mathbb{R}$. It is worth pointing out that in our result the mean curvature H of the two-sided hypersurface is not supposed to be constant. In fact, we just assume that H does not change sign. The main result reads as follows.

Theorem 10.10 *Let $\overline{M}^{n+1} = M^n \times_\rho \mathbb{R}$ be a Killing warped product whose base M^n has parabolic universal Riemannian covering. Let $\psi : \Sigma^n \to \overline{M}^{n+1}$ be a complete two-sided hypersurface such that the function Θ has strict sign and η is bounded. Suppose that the height function h of Σ^n is bounded. If the mean curvature H and the function $\langle \nabla \rho, \nabla h \rangle$ have opposite signs, then Σ^n is a slice of $\overline{M}^{n+1}$.*

Proof Since $K = K^\top + \Theta N$, one has that

$$\overline{\nabla}_X K = \overline{\nabla}_X K^\top + X(\Theta) N + \Theta \overline{\nabla}_X N, \tag{10.32}$$

for every vector field X tangent to Σ^n.

Then, by using the Gauss and Weingarten formulas (10.26), (10.27) and (10.32), we get

$$\nabla_X K^\top = (\overline{\nabla}_X K)^\top + \Theta A X. \tag{10.33}$$

Consequently, by (10.28) and (10.33) we obtain

$$\begin{aligned}
\nabla_X \nabla h &= X\left(\frac{1}{\rho^2}\right) K^\top + \frac{1}{\rho^2} \nabla_X K^\top \\
&= -\frac{X(\rho^2)}{\rho^4} K^\top + \frac{1}{\rho^2} (\overline{\nabla}_X K)^\top + \frac{1}{\rho^2} \Theta A X \\
&= -\frac{2 X(\rho)}{\rho^3} K^\top + \frac{1}{\rho^2} (\overline{\nabla}_X K)^\top + \frac{1}{\rho^2} \Theta A X \\
&= -\frac{2 \langle \nabla \rho, X \rangle}{\rho^3} K^\top + \frac{1}{\rho^2} (\overline{\nabla}_X K)^\top + \frac{1}{\rho^2} \Theta A X. \tag{10.34}
\end{aligned}$$

10.7 Rigidity Results for Two-Sided Hypersurfaces in a Killing Warped...

Since K is a Killing vector field, taking a local orthonormal tangent frame $\{E_1, \ldots, E_n\}$ on Σ^n, by (10.34) we get

$$\Delta h = \sum_{i=1}^{n} \langle \nabla_{E_i} \nabla h, E_i \rangle$$

$$= -\sum_{i=1}^{n} \frac{2}{\rho^3} \langle \nabla \rho, E_i \rangle \langle K^\top, E_i \rangle + \sum_{i=1}^{n} \frac{1}{\rho^2} \langle (\overline{\nabla}_{E_i} K)^\top, E_i \rangle + \sum_{i=1}^{n} \frac{1}{\rho^2} \Theta \langle AE_i, E_i \rangle$$

$$= -\frac{2}{\rho^3} \langle \nabla \rho, K^\top \rangle + \sum_{i=1}^{n} \frac{1}{\rho^2} \langle \overline{\nabla}_{E_i} K, E_i \rangle + \frac{nH\Theta}{\rho^2}$$

$$= -\frac{2}{\rho} \langle \nabla \rho, \nabla h \rangle + \frac{nH\Theta}{\rho^2}. \tag{10.35}$$

Taking into account our hypothesis on H and on the function $\langle \nabla \rho, \nabla h \rangle$, by (10.35) we conclude that Δh does not change sign. Therefore, since Σ^n is parabolic thanks to Theorem 10.9, the bounded function h must be constant and, consequently, the hypersurface Σ^n is a slice of $\overline{M}^{n+1}$. □

Remark 10.3 Related to the assumption that the mean curvature H and the function $\langle \nabla \rho, \nabla h \rangle$ have opposite sings, despite of its technical nature, it can be regarded as a mild hypothesis in the sense that, in the case that ρ is constant, it is equivalent to ask that the mean curvature H has a sign.

From a classical result due to Ahlfors [2] and Blanc-Fiala-Huber [208], a complete Riemannian surface of nonnegative Gaussian curvature is parabolic. So, it is not difficult to see that we can reason as in the proof of Theorem 10.10 to get the following

Corollary 10.8 Let $\overline{M}^3 = M^2 \times_\rho \mathbb{R}$ be a Killing warped product whose base M^2 has nonnegative Gaussian curvature. Let $\psi : \Sigma^2 \to \overline{M}^3$ be a complete two-sided surface such that the function Θ has strict sign and η is bounded. Suppose that the height function h of Σ^2 is bounded. If the mean curvature H and the function $\langle \nabla \rho, \nabla h \rangle$ have opposite signs, then Σ^2 is a slice of $\overline{M}^3$.

Remark 10.4 Let M^2 be a complete Riemannian surface and let K_M be the Gaussian curvature of M^2. It is well-known that if

$$K_M \geq -\frac{1}{r^2 \log r},$$

where r, the distance of a reference point of M^2, is large enough, then M^2 must be parabolic [191, 240]. On the other hand, if the nonnegative part of K_M is integrable on M^2, then M^2 is also parabolic [229]. In this setting, since the assumption $K_M \geq 0$ is used to get the parabolicity of the Riemannian universal covering of M^2, it can be replaced in Corollary 10.8 by one of the two aforementioned assumptions.

The following consequence of Theorem 10.10 deals with minimal hypersurfaces, i.e. hypersurfaces whose mean curvature vanishes identically.

Corollary 10.9 *Let $\overline{M}^{n+1} = M^n \times_\rho \mathbb{R}$ be a Killing warped product whose base M^n has parabolic universal Riemannian covering. Let $\psi : \Sigma^n \to \overline{M}^{n+1}$ be a complete minimal two-sided hypersurface such that the function Θ has strict sign and η is bounded. Suppose that the height function h is bounded from below (resp. bounded from above). If the function $\langle \nabla \rho, \nabla h \rangle \geq 0$ (resp., $\langle \nabla \rho, \nabla h \rangle \leq 0$), then Σ^n is a slice of $\overline{M}^{n+1}$.*

10.8 Entire Killing Graphs in $M^n \times_\rho \mathbb{R}$

According to [134], we define the *entire Killing graph* $\Sigma(u)$ associated to a smooth function $u \in C^\infty(M^n)$ as the hypersurface given by

$$\Sigma(u) = \{\Psi(x, u(x)) : x \in M^n\} \subset M^n \times_\rho \mathbb{R},$$

where $\Psi : M^n \times \mathbb{I} \to \overline{M}^{n+1}$ stands for the flow generated by $K = \partial_t$.

The metric induced on M^n by the Riemannian metric (10.1) via $\Sigma(u)$ is given by

$$\langle \cdot, \cdot \rangle_u = \langle \cdot, \cdot \rangle_M + \rho^2 du^2. \qquad (10.36)$$

On the other hand, the function $g : M^n \times \mathbb{R} \to \mathbb{R}$ given by $g(x, t) = t - u(x)$ is such that $\Sigma(u) = \Psi(g^{-1}(0))$. Then, for any vector field X tangent to $M^n \times_\rho \mathbb{R}$, we have

$$X(g) = X^*(g) + \frac{1}{\rho^2} \langle X, \partial_t \rangle \partial_t(g) = \left\langle \frac{1}{\rho^2} \partial_t - Du, X \right\rangle,$$

where Du denotes the gradient of a function u with respect to the metric $\langle \cdot, \cdot \rangle_M$ of M^n and X^* is the orthogonal projection of X onto TM^n.

Thus, one has that

$$\overline{\nabla} g = \frac{1}{\rho^2} \partial_t - Du$$

10.8 Entire Killing Graphs in $M^n \times_\rho \mathbb{R}$

is a normal vector field on $g^{-1}(0)$ and, consequently

$$N_0 = \Psi_*(\overline{\nabla} g) = \frac{1}{\rho^2} K - \Psi_*(Du)$$

is a normal vector field on $\Sigma(u)$.

Since

$$|N_0| = \frac{(1+\rho^2|Du|_M^2)^{1/2}}{\rho},$$

it follows that (see [134, Eq. (1)])

$$N = \frac{N_0}{|N_0|} = \frac{1}{\rho(1+\rho^2|Du|_M^2)^{1/2}}(K - \rho^2 \Psi_*(Du))$$

gives an unit normal vector field on $\Sigma(u)$ such that the function Θ is given by

$$\Theta = \langle N, K \rangle = \frac{\rho}{(1+\rho^2|Du|_M^2)^{1/2}} > 0. \tag{10.37}$$

Moreover, the shape operator A of $\Sigma(u)$ with respect to N is given by

$$AX = \frac{\rho}{(1+\rho^2|Du|_M^2)^{1/2}} D_X Du - \frac{\rho^3 \langle D_X Du, Du \rangle}{(1+\rho^2|Du|_M^2)^{3/2}} Du$$

$$- \frac{\rho^2 \langle D\rho, X \rangle |Du|_{M^n}^2}{(1+\rho^2|Du|_M^2)^{3/2}} Du + \frac{\langle D\rho, X \rangle}{(1+\rho^2|Du|_M^2)^{1/2}} Du$$

$$+ \frac{\langle Du, X \rangle}{(1+\rho^2|Du|_M^2)^{1/2}} D\rho, \tag{10.38}$$

for any vector field X tangent to M^n. Moreover, by (10.38) it follows that the mean curvature H_u of an entire Killing graph $\Sigma(u)$ is given by

$$nH_u = \mathrm{Div}\left(\frac{\rho Du}{(1+\rho^2|Du|_M^2)^{1/2}}\right) + \frac{\langle Du, D\rho \rangle}{(1+\rho^2|Du|_M^2)^{1/2}}, \tag{10.39}$$

where Div stands for the divergence operator on M^n with respect to the metric $\langle \cdot, \cdot \rangle_M$. In particular, an entire Killing graph $\Sigma(u)$ is minimal if and only if the function $u \in C^\infty(M^n)$ satisfies the following elliptic partial differential equation of divergence form

$$\mathrm{Div}\left(\frac{\rho Du}{(1+\rho^2|Du|_M^2)^{1/2}}\right) + \frac{\langle Du, D\rho \rangle}{(1+\rho^2|Du|_M^2)^{1/2}} = 0.$$

In this context, as consequence of Theorem 10.10, we obtain the following result.

Theorem 10.11 *Let $\overline{M}^{n+1} = M^n \times_\rho \mathbb{R}$ be a Killing warped product whose base M^n is complete having parabolic universal Riemannian covering. Let $\Sigma(u)$ be an entire Killing graph of a bounded function $u \in C^\infty(M^n)$ such that $\rho^2 |Du|_M^2$ is bounded. If the mean curvature H_u and the function $\langle \nabla \rho, \nabla h \rangle$ have opposite signs, then u is constant on M^n.*

Proof By (10.37) and our hypothesis it follows that the function

$$\eta = \left(1 + \rho^2 |Du|_M^2\right)^{1/2}$$

is bounded on $\Sigma(u)$.

Now, let X be any vector field tangent to $\Sigma(u)$. By (10.36) we get

$$\langle X, X \rangle_u = \langle X^*, X^* \rangle_M + \rho^2 \langle Du, X^* \rangle_M^2 \geq \langle X^*, X^* \rangle_M.$$

This implies that

$$L_u(\gamma) \geq L_M(\gamma^*),$$

where $L_u(\gamma)$ stands for the length of a curve γ on $\Sigma(u)$ with respect to the induced metric (10.36) and $L_M(\gamma^*)$ denotes the length of the projection γ^* of γ onto M^n with respect to its metric $\langle \cdot, \cdot \rangle_M$.

Consequently, since projections onto M^n of divergent curves on $\Sigma(u)$ give divergent curves on M^n and as we are assume that the metric $\langle \cdot, \cdot \rangle_M$ is complete, we can apply Hopf-Rinow Theorem to conclude that the induced metric (10.36) is also complete. Therefore, we are in position to apply Theorem 10.10 to finish the proof. □

As consequence of Theorem 10.11 we also have the following

Corollary 10.10 *Let $\overline{M}^{n+1} = M^n \times_\rho \mathbb{R}$ be a Killing warped product whose base M^n is complete having parabolic universal Riemannian covering. Let $u \in C^\infty(M^n)$ be a bounded from below (resp., bounded from above) solution to the problem*

$$\begin{cases} \mathrm{Div}\left(\dfrac{\rho Du}{(1+\rho^2|Du|_M^2)^{1/2}}\right) + \dfrac{\langle Du, D\rho \rangle}{(1+\rho^2|Du|_M^2)^{1/2}} = 0, & \text{in } M^n \\ \rho^2 |Du|_M^2 < +\infty. \end{cases}$$

If the height function h of the entire Killing graph associate to u is such that $\langle \nabla \rho, \nabla h \rangle \geq 0$ (resp., $\langle \nabla \rho, \nabla h \rangle \leq 0$), then u is constant on M^n.

Remark 10.5 We notice that, as in Corollary 10.8, when $n = 2$ in Corollary 10.10 it is possible to assume that M^2 has nonnegative Gaussian curvature to arrive to the parabolicity of the Riemannian universal covering of M^2 and, therefore, the validity of the main result in the two dimensional case.

10.9 Mean Curvature Flow Solitons in Killing Warped Products

Let us consider an n-dimensional two-sided hypersurface $\psi : \Sigma^n \to \mathbb{R}^{n+1}$ whose position vector ψ evolves in the direction of the mean curvature vector **H**, then it gives rise to a solution to mean curvature flow

$$\Psi : [0, T) \times \Sigma^n \to \mathbb{R}^{n+1}$$

satisfying $\Psi(0, \cdot) = \psi(\cdot)$ and

$$\frac{\partial \Psi}{\partial t}(t, p) = \mathbf{H}(t, p),$$

where $\mathbf{H}(t, p)$ is the non-normalized mean curvature vector of the hypersurface $\Sigma_t^n = \Psi(t, \Sigma^n)$ at a point $\Psi(t, p)$.

This equation is called the *mean curvature flow equation*. The study of the mean curvature flow from the perspective of partial differential equations was started with Huisken [209] on the flow of convex hypersurfaces. One of the most interesting problems in the mean curvature flow is to understand its possible singularities, which are called *mean curvature flow solitons* (MCFS).

A key starting point for singularity analysis is Huisken's monotonicity formula [209] because the monotonicity implies that the flow is asymptotically self-similar near a given type I singularity and thus, is modeled by self-shrinking solutions of the flow.

More recently, Alías et al. [36] extended these investigations introducing the general definition of self-similar mean curvature flow in a Riemannian manifold $\overline{M}^{n+1}$ endowed with a conformal vector field K and establishing the corresponding notion of MCFS.

In particular, when $\overline{M}^{n+1}$ is a warped product of the type $I \times_f M^n$ and $K = f(t)\partial_t$, they applied weak maximum principles to guarantee that a complete n-dimensional MCFS is a slice of $\overline{M}^{n+1}$. In [123], Colombo et al. also studied some properties of MCFS in general Riemannian manifolds and in warped products, with emphasis on constant curvature and Schwarzschild type spaces. They also treated the case of MCFS which are entire graphs.

We also point out the recent work done by de Lira and Martín [162], where they investigated solitons invariant with respect to the flow generated by a complete parallel vector field in a Riemannian manifold. A special case occurs when the ambient manifold is the Riemannian product space $M^n \times \mathbb{R}$ and the complete parallel vector field is just the coordinate vector field ∂_t. When the metric of the base M^n is rotationally invariant and its sectional curvature is nonpositive, they characterized all the rotationally invariant MCFS.

In a similar way of the Euclidean context, the mean curvature flow

$$\Psi : \Sigma^n \times [0, T) \to \overline{M}^{n+1}$$

of a two-sided hypersurface $\psi : \Sigma^n \to \overline{M}^{n+1}$ in a $(n + 1)$-dimensional Riemannian manifold $\overline{M}^{n+1}$, satisfying $\Psi(0, \cdot) = \psi(\cdot)$, looks for solutions of the equation

$$\frac{\partial \Psi}{\partial t} = \mathbf{H}.$$

As before, $\mathbf{H}(\cdot, t)$ stands for the non-normalized mean curvature vector of $\Sigma_t^n = \Psi(\Sigma^n, t)$. So, according to [162, Definition 2], a two-sided hypersurface $\psi : \Sigma^n \to \overline{M}^{n+1}$ immersed in a Killing warped product $\overline{M}^{n+1} = M^n \times_\rho \mathbb{R}$ is called a *mean curvature flow soliton* (MCFS) with respect to K and with *soliton constant* $c \in \mathbb{R}$ if its (non-normalized) mean curvature function H satisfies

$$H = c \Theta, \tag{10.40}$$

see also [36, Definition 1.1], as well as [123, Definition 1.1].

In particular, we observe that each slice $M_t = M^n \times \{t\}$ of $\overline{M}^{n+1}$ is a MCFS with respect to K with soliton constant $c = 0$.

10.9.1 Uniqueness of MCFS in Killing Warped Products

Let us recall now an extension of the Hopf's Theorem on a complete Riemannian manifold (Σ^n, g) due to Yau [298]. Set

$$\mathcal{L}_g^p(\Sigma^n) = \left\{ u : \Sigma^n \to \mathbb{R} : \int_{\Sigma^n} |u|^p d_g \Sigma^n < +\infty \right\},$$

where $d_g \Sigma^n$ stands for the measure related to the Riemannian metric g.

The following result holds.

Lemma 10.3 *If u is a nonnegative smooth subharmonic function defined on a complete Riemannian manifold (Σ^n, g), such that $u \in \mathcal{L}_g^p(\Sigma)$ for some $p > 1$, then u must be constant.*

Now, we are in position to present our first uniqueness result concerning MCFS in a warped product $M^n \times_\rho \mathbb{R}$.

10.9 Mean Curvature Flow Solitons in Killing Warped Products

Theorem 10.12 *Let $\overline{M}^{n+1} = M^n \times_\rho \mathbb{R}$ be a Killing warped product with complete base M^n and let $\psi : \Sigma^n \to \overline{M}^{n+1}$ be a complete MCFS with respect to K and with soliton constant $c \geq 0$ (resp. $c \leq 0$). Suppose that $h \geq 0$ (resp. $h \leq 0$) and that ρ is bounded along Σ^n. If $h \in \mathcal{L}_g^p(\Sigma^n)$ for some $p > 1$, then Σ^n is a slice of $\overline{M}^{n+1}$.*

Proof It is well-known that, in local coordinates $(x_1, \ldots, x_n)$ of Σ^n, the Laplacian of its height function on a metric $\hat{g}$ is given by

$$\hat{\Delta} h = \frac{1}{\hat{G}} \sum_{k,l=1}^{n} \partial_k \left(\hat{g}^{kl} \hat{G} \partial_l(h) \right), \tag{10.41}$$

where

$$\hat{g}_{kl} = \hat{g}(\partial_k, \partial_l), \quad \hat{G} = \sqrt{\det(\hat{g}_{kl})} \quad \text{and} \quad (\hat{g}^{kl}) = (\hat{g}_{kl})^{-1}.$$

Taking the conformal metric

$$\hat{g} = \rho^{\frac{4}{n-2}} g,$$

where $g = \langle \cdot, \cdot \rangle$ stands for the induced metric of Σ^n via ψ, we have that

$$\hat{g}_{kl} = \rho^{\frac{4}{n-2}} g_{kl}, \quad \hat{g}^{kl} = \frac{1}{\rho^{\frac{4}{n-2}}} g^{kl}$$

and

$$\hat{G} = \sqrt{\det(\hat{g}_{kl})} = \sqrt{\rho^{\frac{4n}{n-2}} \det(g_{kl})} = \rho^{\frac{2n}{n-2}} G. \tag{10.42}$$

Thus, by (10.41) and (10.42) we obtain

$$\hat{\Delta} h = \frac{1}{\rho^{\frac{2n}{n-2}} G} \sum_{k,l=1}^{n} \partial_k \left(\frac{1}{\rho^{\frac{4}{n-2}}} g^{kl} \rho^{\frac{2n}{n-2}} G \partial_l(h) \right)$$

$$= \frac{\rho^{\frac{2n-4}{n-2}}}{\rho^{\frac{2n}{n-2}}} \sum_{k=1}^{n} \partial_k(\partial_k(h)) + \frac{1}{\rho^{\frac{2n}{n-2}}} \frac{2n-4}{n-2} \rho^{\frac{2n-4}{n-2}-1} \sum_{k=1}^{n} \partial_k(\rho) \partial_k(h)$$

$$= \frac{1}{\rho^{\frac{4}{n-2}}} \Delta h + \frac{2n-4}{(n-2)\rho^{\frac{n+2}{n-2}}} g(\nabla \rho, \nabla h). \tag{10.43}$$

By (10.35), (10.40) and (10.43), we get

$$\hat{\Delta}h = \frac{1}{\rho^{\frac{4}{n-2}}}\left(\frac{-2}{\rho}g(\nabla\rho, \nabla h) + \frac{c}{\rho^2}\Theta^2\right) + \frac{2n-4}{(n-2)\rho^{\frac{n+2}{n-2}}}g(\nabla\rho, \nabla h)$$

$$= \left(\frac{-4}{2\rho^{\frac{n+2}{n-2}}} + \frac{2n-4}{(n-2)\rho^{\frac{n+2}{n-2}}}\right)g(\nabla\rho, \nabla h) + \frac{c}{\rho^{\frac{2n}{n-2}}}\Theta^2$$

$$= \left(\frac{(-4n+8)\rho^{\frac{n+2}{n-2}} + (4n-8)\rho^{\frac{n+2}{n-2}}}{2(n-2)\rho^{\frac{2n+4}{n-2}}}\right)g(\nabla\rho, \nabla h) + \frac{c}{\rho^{\frac{2n}{n-2}}}\Theta^2. \quad (10.44)$$

Hence, by (10.44) one has

$$\hat{\Delta}h = \frac{c}{\rho^{\frac{2n}{n-2}}}\Theta^2. \quad (10.45)$$

Consequently, since we are assuming that $h \geq 0$ (resp. $h \leq 0$) and $c \geq 0$ (resp. $c \leq 0$), from (10.45) we conclude that h (resp. $-h$) is a subharmonic function with respect to the metric $\hat{g}$. On the other hand, since ρ is bounded along Σ^n, by (10.42) and $h \in \mathcal{L}_g^p(\Sigma)$ implies that $h \in \mathcal{L}_{\hat{g}}^p(\Sigma^n)$.

Therefore, we can apply Lemma 10.3 to guarantee that h is constant on Σ^n, that is, Σ^n must be a slice of $\overline{M}^{n+1}$. □

By Theorem 10.12 we get a rigidity result related to closed (compact without boundary) MCFS.

Corollary 10.11 *The only closed MCFS with respect to the Killing vector field K of a Killing warped product $M^n \times_\rho \mathbb{R}$ whose base M^n is closed, are the totally geodesic slices.*

Next, we quote the next auxiliary result due to Yau [298].

Lemma 10.4 *All complete noncompact Riemannian manifold with nonnegative Ricci curvature has at least linear volume growth.*

Theorem 10.12 in addition to Lemma 10.4 lead us to get the following nonexistence result.

Theorem 10.13 *Let $\overline{M}^{n+1} = M^n \times_\rho \mathbb{R}$ be a Killing warped product whose base M^n is complete noncompact with nonnegative Ricci curvature, and having bounded warping function ρ. There is no complete MCFS with respect to K, with soliton constant $c \geq 0$ (resp. $c \leq 0$) and positive (resp. negative) height function satisfying $h \in \mathcal{L}_g^p(\Sigma^n)$ for some $p > 1$.*

10.9 Mean Curvature Flow Solitons in Killing Warped Products

Proof Let us suppose the existence of such a MCFS, namely $\psi : \Sigma^n \to \overline{M}^{n+1}$. By Theorem 10.12, we get that Σ^n is a slice of $\overline{M}^{n+1}$. Consequently, $|h|$ must be equal to a positive constant α and, since we are assuming that $h \in \mathcal{L}_g^p(\Sigma^n)$, we obtain

$$\mathrm{vol}_{g_M}(M^n) = \mathrm{vol}_g(\Sigma^n) = \frac{1}{\alpha^p} \int_{\Sigma^n} |h|^p d_g \Sigma^n < +\infty. \tag{10.46}$$

On the other hand, taking into account that M^n is complete noncompact manifold with nonnegative Ricci curvature, Lemma 10.4 assures that M^n has at least linear volume growth, against (10.46). □

According to [79, Definition 1], we say that a smooth Riemannian manifold (Σ^n, g) satisfies the $\mathcal{L}_g^1$-*Liouville property*, when every nonnegative superharmonic function $u \in \mathcal{L}_g^1(\Sigma^n)$ must be constant.

Bessa et al. [79, Corollary 3] ensure that a stochastically complete manifold (and, in particular, a parabolic manifold) always satisfies the $\mathcal{L}_g^1$-Liouville property. However, in [79, Section 2] the authors constructed examples of stochastically incomplete (and, in particular, nonparabolic) manifolds satisfying the $\mathcal{L}_g^1$-Liouville property.

It is not difficult to verify that arguing as in the proof of Theorem 10.12 we obtain the following uniqueness result concerning MCFS satisfying the $\mathcal{L}_g^1$-Liouville property.

Theorem 10.14 *Let $\overline{M}^{n+1} = M^n \times_\rho \mathbb{R}$ be a Killing warped product with base M^n and let $\psi : \Sigma^n \to \overline{M}^{n+1}$ be a MCFS with respect to K and with soliton constant $c \geq 0$ (resp. $c \leq 0$). Suppose that $h \geq 0$ (resp. $h \leq 0$) and that ρ is bounded along Σ^n. If Σ^n satisfies the $\mathcal{L}_g^1$-Liouville property and $h \in \mathcal{L}_g^1(\Sigma^n)$, then Σ^n is contained into a slice of $\overline{M}^{n+1}$.*

The next result is due to Yau [298].

Lemma 10.5 *Let u be a smooth function defined on a complete Riemannian manifold (Σ^n, g), such that Δu does not change sign on Σ^n. If $|\nabla u| \in \mathcal{L}_g^1(\Sigma^n)$, then Δu vanishes identically on Σ^n.*

In our next uniqueness result, we will suppose that the MCFS Σ^n lies in *slab* of $\overline{M}^{n+1}$, which means that Σ^n is contained in a bounded region of the type

$$M^n \times [t_1, t_2] = \{(p, t) \in M^n \times_\rho \mathbb{R} : t_1 \leq t \leq t_2 \text{ and } p \in M^n\}.$$

Theorem 10.15 *Let $\overline{M}^{n+1} = M^n \times_\rho \mathbb{R}$ be a Killing warped product with complete base M^n and let $\psi : \Sigma^n \to \overline{M}^{n+1}$ be a MCFS with respect to K and with soliton constant*

c, lying in a slab of $\overline{M}^{n+1}$. Suppose in addition that ρ is bounded along Σ^n. If $|\nabla h| \in \mathcal{L}_g^1(\Sigma^n)$, then Σ^n is a slice of $\overline{M}^{n+1}$.

Proof Considering the conformal metric

$$\hat{g} = \rho^{\frac{4}{n-2}} g,$$

since

$$\hat{g}^{kl} = \frac{1}{\rho^{\frac{4}{n-2}}} g^{kl},$$

we get

$$\widehat{\nabla} h = \sum_{k,l=1}^{n} \hat{g}^{kl} \partial_l(h) \partial_k = \frac{1}{\rho^{\frac{4}{n-2}}} \nabla h. \qquad (10.47)$$

Thus, assuming that ρ is bounded along Σ^n, by (10.42) and (10.47) we obtain

$$\int_{\Sigma^n} |\widehat{\nabla} h|_{\hat{g}} d_{\hat{g}} \Sigma^n = \int_{\Sigma^n} \rho^{\frac{2(n-1)}{n-2}} |\nabla h| d_g \Sigma^n \leq \left(\sup_{\Sigma^n} \rho^{\frac{2(n-1)}{n-2}} \right) \int_{\Sigma^n} |\nabla h| d_g \Sigma^n. \qquad (10.48)$$

Consequently, since we are supposing that

$$|\nabla h| \in \mathcal{L}_g^1(\Sigma^n),$$

by (10.48) we have that

$$|\widehat{\nabla} h|_{\hat{g}} \in \mathcal{L}_{\hat{g}}^1(\Sigma^n).$$

So, taking into account (10.45), we can apply Lemma 10.5 to conclude that $\hat{\Delta} h$ vanishes identically on Σ^n.

On the other hand, we have that

$$|\widehat{\nabla} h^2|_{\hat{g}} = 2|h| |\widehat{\nabla} h|_{\hat{g}}.$$

Thus, assuming that Σ^n lies in a slab of $\overline{M}^{n+1}$, we also get that

$$|\widehat{\nabla} h^2|_{\hat{g}} \in \mathcal{L}_{\hat{g}}^1(\Sigma^n).$$

10.9 Mean Curvature Flow Solitons in Killing Warped Products

Moreover, we have that

$$\hat{\Delta}h^2 = 2h\hat{\Delta}h + 2|\widehat{\nabla}h|_{\hat{g}}^2 = 2|\widehat{\nabla}h|_{\hat{g}}^2 \geq 0. \tag{10.49}$$

Hence, we can apply once more Lemma 10.5 to infer that $\hat{\Delta}h^2 = 0$ on Σ^n and, thanks to (10.49), we conclude that $|\widehat{\nabla}h|_{\hat{g}}$ is identically zero on Σ^n.

Therefore, Σ^n must be a slice of $\overline{M}^{n+1}$. □

Let (Σ^n, g) be a complete noncompact Riemannian manifold and let us denote by $d(\cdot, o) : \Sigma^n \to [0, +\infty)$ be the Riemannian distance of (Σ^n, g), measured from a fixed point $o \in \Sigma^n$. According to [35], we say that a smooth function $\zeta \in C^\infty(\Sigma^n)$ *converges to zero at infinity*, when it satisfies the following condition

$$\lim_{d(x,o) \to +\infty} \zeta(x) = 0. \tag{10.50}$$

Keeping in mind this concept, the following lemma corresponds to item (*a*) of [35, Theorem 2.2].

Lemma 10.6 *Let (Σ^n, g) be a complete noncompact Riemannian manifold and let $X \in \mathfrak{X}(\Sigma)$ be a vector field on Σ^n. Assume that there exists a nonnegative, non-identically vanishing function $\zeta \in C^\infty(\Sigma^n)$ which converges to zero at infinity and such that $g(\nabla\zeta, X) \geq 0$. If $\mathrm{div}_g X \geq 0$ on Σ^n, then $g(\nabla\zeta, X) \equiv 0$ on Σ^n.*

Given a complete noncompact Riemannian immersion $\psi : \Sigma^n \to M^n \times_\rho \mathbb{R}$ and $t_* \in \mathbb{R}$, we say that a function ζ defined on Σ^n *converges from below (above) to t_* at infinity* when $\zeta \leq t_*$ ($\zeta \geq t_*$) and the function $\tilde{\zeta} = \zeta - t_*$ converges to zero at infinity.

We are in position now to present our last uniqueness result concerning MCFS in a Killing warped product.

Theorem 10.16 *Let $\overline{M}^{n+1} = M^n \times_\rho \mathbb{R}$ be a Killing warped product with complete noncompact base M^n. The only complete noncompact MCFS $\psi : \Sigma^n \to \overline{M}^{n+1}$ with respect to K and with soliton constant $c \geq 0$ (resp., $c \leq 0$), such that ρ is bounded on Σ^n and h converges from above (resp., below) to t_* at infinity, is the slice $M_{t_*} = M^n \times \{t_*\}$.*

Proof Let $\psi : \Sigma^n \to \overline{M}^{n+1}$ be such a mean curvature flow soliton and let us consider on Σ^n the metric

$$\hat{g} = \rho^{\frac{4}{n-2}} g,$$

which is conformal to its induced metric g. Since h converges from above (resp., below) to t_* and $c \geq 0$ (resp., $c \leq 0$), by choosing the smooth function $\zeta = h - t_*$ (resp., $\zeta = t_* - h$)

and the vector field $X = \hat{\nabla}\zeta$, by (10.45) we get that

$$\mathrm{div}_{\hat{g}} X \geq 0. \tag{10.51}$$

Moreover, we have

$$\hat{g}(\hat{\nabla}\zeta, X) = |\hat{\nabla}\zeta|_{\hat{g}}^2 \geq 0. \tag{10.52}$$

In addition, since h converges to t_* at infinity, we have that ζ is a nonnegative non-identically vanishing function which converges to zero (also related to the metric $\hat{g}$, since ρ is bounded on Σ^n).

Thus, by (10.51) and (10.52) we can apply Lemma 10.6 to get that $\hat{g}(\hat{\nabla}\zeta, X)$ is identically zero on Σ^n. Hence, thanks to (10.52) we conclude that $\hat{\nabla} h$ vanishes identically on Σ^n, which means that h is constant and (since it converges to t_* at infinity) Σ^n must be the slice M_{t_*}. □

10.9.2 Moser-Bernstein Type Results for MCFS

The study of the rigidity of entire minimal or, more generally, constant mean curvature graphs in a Riemannian space is a classical and fruitful theme into the theory of geometric analysis and it was started with Bernstein's Theorem [78] (amended by Hopf in [207]), which asserts that the only entire minimal graphs in $\mathbb{R}^3$ are the planes. Later on, Simons [278] proved a result that, together with some theorems of De Giorgi [135] and Fleming [179], yield a proof of the extension of the Bernstein's theorem to $\mathbb{R}^n$, for $n = 7$. However, Bombieri et al. [88] astonishingly showed that Bernstein's Theorem does not hold for $n = 8$.

Consequently, it turns an interesting research topic in geometric analysis has been the possible extension of Bernstein's result to either higher dimension or another ambient space. A very notable contribution in this direction was made by Moser [246], who showed that the hyperplanes are the only entire minimal graphs of $\mathbb{R}^n$ whose gradient of the corresponding function has bounded norm.

In this subsection, we will apply the theorems of the previous one in order to establish new Moser-Bernstein type results concerning entire Killing graphs constructed over the base M^n of a warped product $M^n \times_\rho \mathbb{R}$, which are MCFS.

To this aim, we observe that (10.39) and (10.40) assure that an entire Killing graph $\Sigma(u)$ is a MCFS with respect to K and with soliton constant c if and only if u is a solution of the following elliptic non-linear partial differential equation

$$\mathrm{Div}_M \left(\frac{\rho Du}{(1 + \rho^2 |Du|_M^2)^{1/2}} \right) = \frac{1}{(1 + \rho^2 |Du|_M^2)^{1/2}} \left(c\rho - g(Du, D\rho) \right). \tag{10.53}$$

10.9 Mean Curvature Flow Solitons in Killing Warped Products

We are in position now to state and prove the following Moser-Bernstein type result.

Theorem 10.17 *Let $\overline{M}^{n+1} = M^n \times_\rho \mathbb{R}$ be a Killing warped product with complete base M^n and bounded warping function ρ. Let $\Sigma(u)$ be an entire Killing graph determined by a smooth function $u \in C^\infty(M^n)$, which is nonnegative (resp., nonpositive) solution of equation (10.53) with $c \geq 0$ (resp., $c \leq 0$). Suppose in addition that $|Du|_M$ is bounded on M^n. If $u \in L^p_{g_M}(M^n)$ for some $p > 1$, then $u \equiv t_0$ for some nonnegative (resp., nonpositive) $t_0 \in \mathbb{R}$.*

Proof As in the proof of Theorem 10.11, we have that the induced metric $g_u = \langle \cdot, \cdot \rangle_u$ given in (10.36) is complete. Moreover, by (10.36) it follows that

$$d_g \Sigma^n = \sqrt{|G|} d_{g_M} M^n,$$

where $d_{g_M} M^n$ and $d_g \Sigma^n$ stand for the volume elements of (M^n, g_M) and $(\Sigma(u), g_u)$, respectively, and $G = \det(g_{ij})$ with

$$g_{ij} = g_u(E_i, E_j) = \rho^2 E_i E_j + \delta_{ij}, \tag{10.54}$$

for each $i, j \in \{1, \ldots, n\}$. Here, $\{E_1, \ldots, E_n\}$ denotes a local orthonormal frame with respect to the metric g_{M^n}.

Clearly, one has

$$\det(g_{ij}) = \sum_{\sigma \in S_n} (\text{sign}\,\sigma) g_{1\sigma(1)} g_{2\sigma(2)} \cdots g_{n\sigma(n)}, \tag{10.55}$$

where S_n is the set of bijective functions

$$\sigma : \{1, \ldots, n\} \to \{1, \ldots, n\}$$

and sign σ is the sign of the permutation σ.

By (10.54) and (10.55), we get

$$\det(g_{ij}) = \sum_\sigma (\text{sign}\,\sigma) \Pi_{l=1}^n (\rho^2 E_l E_{\sigma(l)} + \delta_{l\sigma(l)}). \tag{10.56}$$

Now, since (10.56) holds, a straightforward computation ensures that

$$\det(g_{ij}) = \sum_{\sigma,k} (\text{sign}\,\sigma) \rho^2 E_{i_1} E_{\sigma(i_1)} \rho^2 E_{i_2} E_{\sigma(i_2)} \cdots$$

$$\cdots \rho^2 E_{i_k} E_{\sigma(i_k)} \delta_{i_{k+1}\sigma(i_{k+1})} \cdots \delta_{i_n \sigma(i_n)}$$

$$+ \sum_\sigma (\text{sign}\sigma)\rho^{2n} E_1 E_{\sigma(1)} \cdot \ldots \cdot E_n E_{\sigma(n)}$$

$$+ \sum_\sigma (\text{sign}(\sigma))\delta_{1\sigma(1)} \cdot \ldots \cdot \delta_{n\sigma(n)}.$$

On the other hand, we notice that

$$\sum_\sigma (\text{sign}\sigma)\rho^{2n} E_1 E_{\sigma(1)} \cdot \ldots \cdot E_n E_{\sigma(n)} = 0$$

and

$$\sum_\sigma (\text{sign}\sigma)\delta_{1\sigma(1)} \cdot \ldots \cdot \delta_{n\sigma(n)} = 1.$$

Thus, it follows that

$$\det(g_{ij}) = \sum_{\sigma,k} (\text{sign } \sigma)\rho^2 E_{i_1} E_{\sigma(i_1)} \rho^2 E_{i_2} E_{\sigma(i_2)} \cdot \ldots$$

$$\ldots \cdot \rho^2 E_{i_k} E_{\sigma(i_k)} \delta_{i_{k+1}\sigma(i_{k+1})} \cdot \ldots \cdot \delta_{i_n\sigma(i_n)}$$

$$= \sum_\sigma (\text{sign } \sigma)\rho^2 E_{i_1} E_{\sigma(i_1)} \delta_{i_2\sigma(i_2)} \cdot \ldots \cdot \delta_{i_n\sigma(i_n)}$$

$$+ \sum_{\sigma,k\geq 2} (\text{sign } \sigma)\rho^{2k} E_{i_1} E_{\sigma(i_1)} E_{i_2} E_{\sigma(i_2)} \cdot \ldots$$

$$\ldots \cdot E_{i_k} E_{\sigma(i_k)} \delta_{i_{k+1}\sigma(i_{k+1})} \cdot \ldots \cdot \delta_{i_n\sigma(i_n)}.$$

If σ is different from the identity, then

$$\rho^2 E_{i_1} E_{\sigma(i_1)} \delta_{i_2\sigma(i_2)} \cdot \ldots \cdot \delta_{i_n\sigma(i_n)} = 0.$$

Otherwise, we have

$$\sum_\sigma (\text{sign } \sigma)\rho^2 E_{i_1} E_{\sigma(i_1)} \delta_{i_2\sigma(i_2)} \cdot \ldots \cdot \delta_{i_n\sigma(i_n)} = \rho^2 \left(E_1^2 + E_2^2 + \cdots + E_n^2\right).$$

Now, considering

$$\sum_{\sigma,k\geq 2} (\text{sign } \sigma)\rho^{2k} E_{i_1} E_{\sigma(i_1)} E_{i_2} E_{\sigma(i_2)} \cdot \ldots \cdot E_{i_k} E_{\sigma(i_k)} \delta_{i_{k+1}\sigma(i_{k+1})} \cdot \ldots \cdot \delta_{i_n\sigma(i_n)},$$

10.9 Mean Curvature Flow Solitons in Killing Warped Products

for each fixed $k \geq 2$ and each fixed index $i_1, i_2, \ldots, i_n$, we have

$$\sum_{\sigma, k \geq 2} (\text{sign } \sigma) \rho^{2k} E_{i_1}^2 E_{i_2}^2 \cdot \ldots \cdot E_{i_k}^2 = 0, \tag{10.57}$$

where sign $\sigma = 1$ if the permutation in σ is even and sign $\sigma = -1$ if the permutation in σ is odd.

Thus, one has

$$\sum_{\sigma, k \geq 2} (\text{sign } \sigma) \rho^{2k} E_{i_1} E_{\sigma(i_1)} E_{i_2} E_{\sigma(i_2)} \cdot \ldots \cdot E_{i_k} E_{\sigma(i_k)} \delta_{i_{k+1}\sigma(i_{k+1})} \cdot \ldots \cdot \delta_{i_n\sigma(i_n)} = 0.$$

Then, by (10.9.2) and (10.57), we obtain

$$|G| = 1 + \rho^2 |Du|_M^2.$$

Consequently, the following relation holds

$$d_g \Sigma^n = \left(1 + \rho^2 |Du|_M^2\right)^{1/2} d_{g_M} M^n. \tag{10.58}$$

Hence, since $u \in \mathcal{L}_{g_M}^p(M^n)$ for some $p > 1$ and ρ and $|Du|_M$ are bounded, relation (10.58) guarantees that $h \in \mathcal{L}_g^p(\Sigma^n)$ for some $p > 1$. Therefore, since the metric (10.36) is complete, the result follows by applying Theorem 10.12. $\square$

The theorem stated below ensures a nonexistence result for the non-linear PDE given in (10.53).

Theorem 10.18 *Let $\overline{M}^{n+1} = M^n \times_\rho \mathbb{R}$ be a Killing warped product with complete noncompact base M^n having nonnegative Ricci curvature and with bounded warping function ρ. There is no entire Killing graph $\Sigma(u)$ determined by a positive (resp., negative) smooth function $u \in C^\infty(M^n)$, which is solution of equation (10.53) with $c \geq 0$ (resp., $c \leq 0$), such that $|Du|_M$ is bounded on M^n and $u \in \mathcal{L}_{g_M}^p(M^n)$ for some $p > 1$.*

Proof Let us suppose the existence of such an entire Killing graph $\Sigma(u)$, determined by a positive (resp. negative) smooth function $u \in C^\infty(M^n)$. By Theorem 10.17, we get that $u \equiv t_0$ for some positive (resp., negative) $t_0 \in \mathbb{R}$. Since $u \in \mathcal{L}_{g_M}^p(M^n)$ for some $p > 1$, we obtain

$$\text{vol}_{g_M}(M^n) = \frac{1}{|t_0|^p} \int_{M^n} |u|^p d_{g_M} M^n < +\infty. \tag{10.59}$$

Now, taking into account that M^n is complete noncompact with nonnegative Ricci curvature, Lemma 10.4 assures that M^n has at least linear volume growth, against (10.59). □

Taking into account relation (10.58), it is easily seen that the following nonparametric version of Theorem 10.14 holds.

Theorem 10.19 *Let $\overline{M}^{n+1} = M^n \times_\rho \mathbb{R}$ be a Killing warped product with base M^n satisfying the $\mathcal{L}^1_{g_{M^n}}$-Liouville property and with bounded warping function ρ. Let $\Sigma(u)$ be an entire Killing graph determined by a smooth function $u \in C^\infty(M^n)$, which is nonnegative (resp., nonpositive) solution of equation (10.53) with $c \geq 0$ (resp., $c \leq 0$). Suppose in addition that $|Du|_M$ is bounded on M^n. If $u \in \mathcal{L}^1_{g_M}(M)$, then $u \equiv t_0$ for some nonnegative (resp., nonpositive) $t_0 \in \mathbb{R}$.*

Finally, by Theorem 10.15, the following Moser-Bernstein type result holds.

Theorem 10.20 *Let $\overline{M}^{n+1} = M^n \times_\rho \mathbb{R}$ be a Killing warped product with complete base M^n and with bounded warping function ρ. Let $\Sigma(u)$ be an entire Killing graph determined by a bounded smooth function $u \in C^\infty(M^n)$, which is solution of equation (10.53). If $|Du|_M \in \mathcal{L}^1_{g_M}(M^n)$, then $u \equiv t_0$ for some $t_0 \in \mathbb{R}$.*

Proof As in the proof of Theorem 10.11, we get that the entire Killing graph $\Sigma(u)$ is complete with respect to the induced metric (10.36).

On the other hand, since

$$N^* = N - N^\perp = \frac{\rho \Psi_*(Du)}{(1 + \rho^2 |Du|_M^2)^{1/2}},$$

we have that

$$|N^*|_M^2 = \frac{\rho^2 |Du|_{M^n}^2}{1 + \rho^2 |Du|_M^2}. \tag{10.60}$$

Thus, by (10.60) we get

$$|\nabla h|^2 = \frac{1}{\rho^2}|N^*|_M^2 = \frac{|Du|_M^2}{1 + \rho^2 |Du|_M^2}. \tag{10.61}$$

Hence, by (10.58) and (10.61) we have

$$|\nabla h| d_g \Sigma^n = |Du|_M d_{g_M} M^n. \tag{10.62}$$

Finally, since $|Du|_M \in \mathcal{L}^1_{g_M}(M^n)$, relation (10.62) assures that $|\nabla h| \in \mathcal{L}^1_g(\Sigma^n)$. Therefore, we can apply Theorem 10.15 to conclude the proof. □

Weakly Trapped Submanifolds Immersed in a Generalized Robertson-Walker Spacetime

11.1 Introduction

The singularity theorems of Hawking and Penrose [200, 201] assert the presence of incomplete geodesics in the time evolution of Cauchy data with physically reasonable matter containing a trapped surface. Studying the structure of these singularities and understanding whether generically they entail a loss of predictability power of the theory have become central issues in classical General Relativity.

More precisely, the weak cosmic censorship conjecture asserts, roughly speaking, that in the asymptotically flat case the singularity will be generically hidden from infinity by an event horizon and that a black hole will form. Since this conjecture aims precisely at showing that a black hole forms, any sensible approach to its proof should not make strong a priori assumptions on the global structure of the spacetime. It is therefore necessary to replace to concept of black hole, which requires full knowledge of the future evolution of a spacetime, with a quasi-local concept that captures its main features and that can be used as a tool to show the existence of a black hole.

In this setting, a marginally trapped surface in a spacetime represents an extreme gravitational situation. More precisely, under suitable circumstances, the occurrence of a marginally trapped surface signals the presence of a black hole [119, 200]. For this and other reasons marginally trapped surfaces have played a fundamental role in quasi-local descriptions of black holes; see, for instance, [58].

From a mathematical point of view, marginally trapped surfaces are the natural generalization of minimal surfaces to a Lorentzian setting [42] and they arose in a more purely mathematical context in the work of Schoen and Yau [276] concerning the existence of solutions of Jang's equation, in connection with their proof of the positive mass theorem. The mathematical theory of marginally trapped surfaces has been greatly developed in recent years. We refer the reader to the survey article [43], which describes many of these developments.

More recently, Alías et al. [34] obtained several rigidity results for codimension two marginally trapped submanifolds immersed in a generalized Robertson-Walker (GRW) spacetime $-I \times_f M^{n+1}$, which guarantee that, under appropriate geometric constraints, these submanifolds are contained in slices of the form $\{t\} \times M^{n+1}$, with $t \in I \subset \mathbb{R}$. They also derived some interesting nonexistence results for *weakly trapped submanifolds* (that is, the mean curvature vector field is supposed to be causal) in $-I \times_f M^{n+1}$.

In particular, they gave applications to some cases of physical relevance such as the Einstein-de Sitter spacetime and certain open regions of de Sitter spacetime, including the so called steady state spacetime. We also refer to the recent work [70] where the authors obtained a nice classification concerning marginally outer trapped submanifolds of codimension two in the de Sitter space.

In this chapter, we also deal with codimension two weakly trapped submanifolds immersed in a GRW spacetime. Under suitable hypothesis, we are able to prove codimension reduction results for these spacelike submanifolds in the sense that they must be immersed into a slice of the ambient GRW spacetime.

To this scope, we use three main core concepts: the generalized maximum principle of Omori and Yau, stochastic completeness and another appropriate maximum principle at infinity due to Yau. We also present a construction of a nontrivial example of weakly trapped submanifold in the static GRW spacetime $-\mathbb{R} \times \mathbb{H}^2 \times \mathbb{R}$, illustrating the importance of our results.

The results presented in this chapter can be found in the papers [128]. They are strongly motivated by the approach recently developed in [34].

11.2 Spacelike Submanifolds in a GRW Spacetime

In this section we will introduce some basic facts and notations that will appear throughout this chapter. Let M^{n+1} be a connected $(n+1)$-dimensional oriented Riemannian manifold, $I \subset \mathbb{R}$ an open interval and $f : I \to \mathbb{R}$ a positive smooth function.

In the product manifold $\overline{M}^{n+2} = I \times M^{n+1}$ denote by π_I and π_M the projections onto the factors I and M^{n+1}, respectively.

The class of Lorentzian manifolds which will be of our concern here is the one obtained by furnishing $\overline{M}^{n+2}$ with the Lorentzian metric

$$\langle v, w \rangle_p = -\langle (\pi_I)_* v, (\pi_I)_* w \rangle_{\pi_I(p)} + (f \circ \pi_I)(p)^2 \langle (\pi_M)_* v, (\pi_M)_* w \rangle_{\pi_M(p)},$$

for all $p \in \overline{M}^{n+2}$ and $v, w \in T_p \overline{M}$.

In such a case, we write $\overline{M}^{n+2} = -I \times_f M^{n+1}$ and, according to the terminology established in [24], we recall that $\overline{M}^{n+2}$ is called a *generalized Robertson-Walker* (GRW)

11.2 Spacelike Submanifolds in a GRW Spacetime

spacetime. When the Riemannian fiber M^{n+1} has constant sectional curvature, $\overline{M}^{n+2}$ has been known in the mathematical literature as a Robertson-Walker (RW) spacetime, an allusion to the fact that, for $n = 2$, it is an exact solution of Einstein's equations for a perfect fluid (for more details, see [251, Chapter 12]).

We will choose on $\overline{M}^{n+2}$ the time-orientation given by the globally defined timelike unit vector field

$$\partial_t = \frac{\partial}{\partial t}\bigg|_{(t,x)}, \quad (t,x) \in -I \times_f M^{n+1}.$$

In what follows we deal with a codimension two connected spacelike submanifold $\psi : \Sigma^n \to \overline{M}^{n+2}$, which means that the induced metric on Σ^n by the metric of $\overline{M}^{n+2}$ is positive definite and there exists an orthonormal frame $\{N, v\}$ of the normal bundle $\mathfrak{X}^\perp(\Sigma^n)$ constituted by a future (past) timelike normal unit vector field N, that is, $\langle N, \partial_t \rangle \leq -1$ ($\langle N, \partial_t \rangle \geq 1$) on Σ^n, and a spacelike normal unit vector field v.

In this framework, the Gauss and Weingarten formulas for $\psi : \Sigma^n \to \overline{M}^{n+2}$ are given by

$$\overline{\nabla}_X Y = \nabla_X Y - \langle A_N X, Y \rangle N + \langle A_v X, Y \rangle v \tag{11.1}$$

and

$$A_\xi X = -\overline{\nabla}_X \xi^\top, \tag{11.2}$$

for $\xi \in \{N, v\}$ and tangent vector fields $X, Y \in \mathfrak{X}(\Sigma^n)$, where the operator $A_\xi : \mathfrak{X}(\Sigma^n) \to \mathfrak{X}(\Sigma^n)$ stands for the Weingarten operator of the spacelike submanifold Σ^n with respect to ξ.

Similarly to the context of spacelike hypersurfaces immersed in a GRW spacetime, we get the height by projection into the open interval $I \subset \mathbb{R}$. More precisely, given $\phi : \Sigma^n \to \overline{M}^{n+2}$ we get $h : \Sigma^n \to I$ such that $h = \pi_I \circ \phi$. However, the support function depends on the normal that we select. To this scope we consider ξ a normal vector field on Σ^n and define the support function as $\langle \xi, \partial_t \rangle$, where we require that ξ denotes a future-pointing normal map of Σ^n when it is causal.

Let us denote by $\overline{\nabla}$ and ∇ the gradients with respect to the metrics of $-I \times_f M^{n+1}$ and Σ^n, respectively. Then, a simple computation shows that the gradient of π_I on $-I \times_f M^{n+1}$ is given by

$$\overline{\nabla}\pi_I = -\langle \overline{\nabla}\pi_I, \partial_t \rangle \partial_t = -\partial_t. \tag{11.3}$$

So, by (11.3) we deduce that the gradient of h on Σ^n is

$$\nabla h = (\overline{\nabla}\pi_I)^\top = -\partial_t^\top = -\partial_t - \langle N, \partial_t \rangle N + \langle \nu, \partial_t \rangle \nu. \tag{11.4}$$

Alternatively, we can rewrite (11.4) as follows

$$\nabla h = (\overline{\nabla}\pi_I)^\top = -\partial_t^\top = -\partial_t + \sum_{i=1}^{2} \epsilon_i \langle \xi_i, \partial_t \rangle \xi_i, \tag{11.5}$$

where $N = \xi_1$, $\nu = \xi_2$, $\epsilon_i = \langle \xi_i, \xi_i \rangle$ and $(\cdot)^\top$ denotes the tangential component of a vector field in $\mathfrak{X}(\overline{M}^{n+2})$ along Σ^n. Thus, by (11.5) we get

$$|\nabla h|^2 = \langle N, \partial_t \rangle^2 - (1 + \langle \nu, \partial_t \rangle^2), \tag{11.6}$$

where $|\cdot|$ denotes the norm of a vector field on Σ^n.

Consequently, denoting $\Theta_1 = \langle N, \partial_t \rangle$ and $\Theta_2 = \langle \nu, \partial_t \rangle$, by (11.6) we derive

$$|\nabla h|^2 = \Theta_1^2 - (1 + \Theta_2^2). \tag{11.7}$$

11.3 Weakly Trapped Submanifolds in GRW Spacetimes

In recent decades, interest has increased considerably in the study of spacelike submanifolds immersed in a Lorentzian manifold. Within this field, trapped submanifolds appear as an important particular case. We recall that a spacelike submanifold of a spacetime is said to be trapped if its mean curvature vector is timelike. The concept of a trapped submanifold was first introduced by Penrose in his seminal work [253] in order to study singularities of spacetime. In General Relativity, a trapped surface is a two-dimensional embedded spatial surface such that the product of the traces of its two future-directed null second fundamental forms is everywhere positive; see, for instance, references [223, 271].

The limiting case of the trapped submanifolds are the marginally trapped submanifolds, which are defined as being spacelike submanifolds whose mean curvature vector field is lightlike. We note that marginally trapped submanifolds can also be regarded as the Lorentzian dual of minimal submanifolds, when compared with the Riemannian context. For a thorough discussion concerning marginally trapped submanifolds, we indicate to the reader the references [39–41, 108, 110].

In this chapter, our purpose is study the geometry of certain codimension two spacelike submanifolds immersed in a general Robertson-Walker (GRW) spacetime. We recall that a GRW spacetime is a Lorentzian warped product $-I \times_f M^{n+1}$, with 1-dimensional negative base I, an $(n + 1)$-dimensional Riemannian fiber M^{n+1} and smooth positive warping function $f \in C^\infty(I)$. For every $t \in I$, the slice $M_t = \{t\} \times M$ is an embedded

11.4 Codimension Reduction Results in GRW Spacetimes

spacelike hypersurface of $-I \times_f M^{n+1}$ and $t \in I \mapsto M_t$ determines a foliation of $-I \times_f M^{n+1}$ by totally umbilical spacelike hypersurfaces with constant mean curvature $(\log f)'(t)$ with respect to the orientation given by the globally defined timelike unit vector field ∂_t. In [180], Flores, Haesen and Ortega showed that hypersurfaces with constant mean curvature in M^{n+1} produce codimension two marginally trapped submanifolds of $-I \times_f M^{n+1}$ when contained in a slice M_t at an appropriate $t \in I$. More recently, Alías et al. [34] obtained rigidity results for codimension two marginally trapped submanifolds in $-I \times_f M^{n+1}$ which guarantee that, under appropriate hypothesis, the only such submanifolds are of this form. They also derived some interesting nonexistence results for *weakly trapped submanifolds* (that is, the mean curvature vector field is supposed to be causal) immersed in $-I \times_f M^{n+1}$.

In the next section, we deal with n-dimensional spacelike submanifolds and, in particular, weakly trapped submanifolds immersed in a GRW spacetime $-I \times_f M^{n+1}$. Under suitable hypotheses, we will prove that such a spacelike submanifold must be immersed into a slice of $-I \times_f M^{n+1}$. So, these results can be regarded as codimension reduction results for spacelike submanifolds in GRW spacetimes.

11.4 Codimension Reduction Results in GRW Spacetimes

In this section we present some rigidity results concerning codimension two spacelike submanifolds immersed in a GRW spacetime $-I \times_f M^{n+1}$. More precisely, we establish sufficient conditions such that a spacelike submanifold Σ^n is contained in the Riemannian fiber M^{n+1}. To this aim, we observe that the mean curvature vector field $\mathbf{H}$ of Σ^n, which is defined by

$$\mathbf{H} = \frac{1}{n} \sum_{i=1}^{n} II(E_i, E_i),$$

for some orthonormal frame $\{E_i\}_{i=1}^n$ on Σ^n, where II stands for the second fundamental form of Σ^n, can be expressed in the following way

$$\mathbf{H} = -H_N N + H_\nu \nu, \tag{11.8}$$

for some smooth functions H_N and H_ν defined on Σ^n.

11.4.1 Codimension Reduction Via Certain Integrability Constraints

In order to establish our result, we will need some auxiliary lemmas. The first one gives us suitable formulas for the Laplacian of the height function of a spacelike submanifold in a GRW spacetime.

Lemma 11.1 *Let Σ^n be a spacelike submanifold immersed in $-I \times_f M^{n+1}$. Then*

(i) $\Delta h = -(\log f)'(h)[n + |\nabla h|^2] - n\langle \mathbf{H}, \partial_t \rangle$;
(ii) $\Delta \sigma(h) = -n(f'(h) + f(h)\langle \mathbf{H}, \partial_t \rangle)$, *where*

$$\sigma(t) = \int_0^t f(s)ds.$$

Proof We recall that the vector field $(f \circ \pi_I)\partial_t$ is metrically equivalent to a closed 1-form. Thus, we have that

$$\overline{\nabla}_X (f \circ \pi_I)\partial_t = (f' \circ \pi_I)X, \tag{11.9}$$

for all $X \in T(I \times_f M^{n+1})$.

After a direct computation, by (11.9) we conclude that

$$\begin{aligned} f(h)\mathrm{div}\partial_t &= f'(h)[n - \langle \nabla h, \partial_t \rangle] \\ &= f'(h)\left(n - \langle \nabla h, \partial_t^\top - \Theta_1 N + \Theta_2 \nu \rangle\right) \\ &= f'(h)\left(n + |\nabla h|^2\right). \end{aligned} \tag{11.10}$$

Then, by (11.10) we get

$$\begin{aligned} \Delta h &= \mathrm{div}\left(-\partial_t - \Theta_1 N + \Theta_2 \nu\right) \\ &= -(\log f)'(h)\left(n + |\nabla h|^2\right) + n\Theta_1 H_N - n\Theta_2 H_\nu \\ &= -(\log f)'(h)\left(n + |\nabla h|^2\right) - n\langle \mathbf{H}, \partial_t \rangle, \end{aligned} \tag{11.11}$$

and item (i) is proved.

Moreover, we also have that

$$\Delta \sigma(h) = f'(h)|\nabla h|^2 + f(h)\Delta h. \tag{11.12}$$

Therefore, by (11.11) and (11.12) also the validity of item (ii) is showed. □

Our second auxiliary lemma is an extension of Hopf's theorem on a complete Riemannian manifold due to Yau in [298]. As customary, $\mathcal{L}_g^1(\Sigma^n)$ denotes here the space of Lebesgue integrable functions on Σ^n.

11.4 Codimension Reduction Results in GRW Spacetimes

Lemma 11.2 *Let u be a smooth function defined on a complete Riemannian manifold Σ^n, such that Δu does not change sign on Σ^n. If $|\nabla u| \in \mathcal{L}_g^1(\Sigma^n)$, then Δu vanishes identically on Σ^n.*

In this setting, a slab

$$[t_1, t_2] \times M^{n+1} = \left\{ (t, q) \in -I \times_f M^{n+1} : t_1 \leq t \leq t_2 \right\}$$

will be called a *timelike bounded region*.

Theorem 11.1 *Let Σ^n be a complete spacelike submanifold which lies in a timelike bounded region of $-I \times_f M^{n+1}$ and such that f is monotone non-increasing (non-decreasing). If $\langle \mathbf{H}, \partial_t \rangle$ is non-positive (nonnegative) on Σ^n and $|\nabla h| \in \mathcal{L}_g^1(\Sigma^n)$, then Σ^n is contained in a slice $\{t\} \times M^{n+1}$ with $f'(t) = 0$.*

Proof Taking into account that f is monotone non-increasing (non-decreasing) and $\langle \mathbf{H}, \partial_t \rangle$ is non-positive (nonnegative) on Σ^n, by item (i) of Lemma 11.1 we obtain that Δh does not change sign on Σ^n. Since $|\nabla h| \in \mathcal{L}_g^1(\Sigma^n)$, by Lemma 11.2 we get that $\Delta h = 0$ on Σ^n.

On the other hand, since Σ^n lies in a timelike bounded region of $-I \times_f M^{n+1}$, we have $|\nabla h^2| \in \mathcal{L}_g^1(\Sigma^n)$. Now, we notice that

$$\Delta h^2 = 2h \Delta h + 2|\nabla h|^2 = 2|\nabla h|^2 \geq 0.$$

Thus, we can apply Lemma 11.2 to obtain that $\Delta h^2 = 0$ and, hence, $|\nabla h| = 0$ on Σ^n.

Therefore, h is constant on Σ^n, which means that it is contained in a slice $\{t\} \times M^{n+1}$. Moreover, by item (i) of Lemma 11.1 we also conclude that $f'(t) = 0$. □

According to [34], we recall that a spacelike submanifold is said *weakly future (past) trapped submanifold* when its mean vector field is causal and future-pointing (past-pointing) everywhere. Thus, we observe that when Σ^n is a weakly future (past) trapped submanifold the hypothesis that the support function $\langle \mathbf{H}, \partial_t \rangle$ be non-positive (nonnegative) is automatically satisfied.

Consequently, by Theorem 11.1 we obtain the following nonexistence property.

Corollary 11.1 *There is no a weakly future (past) trapped submanifold Σ^n which lies in a timelike bounded region of $-I \times_f M^{n+1}$, with f being monotone non-increasing (non-decreasing), and such that $|\nabla h| \in \mathcal{L}_g^1(\Sigma^n)$.*

We say that the spacelike submanifold Σ^n is ν-*minimal* if the component of its mean curvature in the spacelike direction H_ν, defined in (11.8), vanishes identically on Σ^n. In this setting, we have the following

Theorem 11.2 *Let Σ^n be a complete spacelike submanifold which lies in a timelike bounded region of $-I \times_f M^{n+1}$. Suppose that Σ^n is ν-minimal and*

$$H_N \geq \sup_{\Sigma^n}(\log f)'(h) > 0, \tag{11.13}$$

with respect to the past-pointing orientation N of Σ^n. If $|\nabla h| \in \mathcal{L}^1_g(\Sigma^n)$, then Σ^n is contained in a slice $\{t\} \times M^{n+1}$.

Proof Since $\langle N, \partial_t \rangle \geq 1$, by Lemma 11.1 we get

$$\Delta \sigma(h) = -nf(h)\left((\log f)'(h) - \langle N, \partial_t \rangle H_N\right)$$

$$\geq nf(h)\left(H_N - (\log f)'(h)\right)$$

$$\geq nf(h)\left(H_N - \sup_{\Sigma^n}(\log f)'(h)\right). \tag{11.14}$$

Thus, taking into account our hypothesis (11.13), by (11.14) we have that $\Delta \sigma(h) \geq 0$ on Σ^n.

On the other hand, since Σ^n lies in a timelike bounded region of $-I \times_f M^{n+1}$, there exists a positive constant C such that

$$|\nabla \sigma(h)| = f(h)|\nabla h| \leq C|\nabla h|.$$

Consequently, since $|\nabla h| \in \mathcal{L}^1_g(\Sigma^n)$ it follows that $|\nabla \sigma(h)| \in \mathcal{L}^1_g(\Sigma^n)$.

So, we can apply Lemma 11.2 obtaining that $\Delta \sigma(h) = 0$ on Σ^n. Thus, thanks to (11.14), we get

$$(\log f)'(h) = \langle N, \partial_t \rangle H_N.$$

Then, we have that

$$H_N = \sup_{\Sigma^n}(\log f)'(h) \geq (\log f)'(h) = \langle N, \partial_t \rangle H_N.$$

Therefore, $\langle N, \partial_t \rangle = 1$ on Σ^n and, hence, we conclude that Σ^n is contained in a slice $\{t\} \times M^{n+1}$. □

From the proof of Theorem 11.2 it is easily seen that the following result holds.

11.4 Codimension Reduction Results in GRW Spacetimes

Corollary 11.2 *Let Σ^n be a complete spacelike submanifold which lies in a timelike bounded region of $-I \times_f M^{n+1}$. Suppose that Σ^n is ν-minimal and H_N is constant. Assuming that*

$$H_N \geq \sup_{\Sigma^n}(\log f)'(h) \geq 0$$

and $|\nabla h| \in \mathcal{L}_g^1(\Sigma^n)$, one has that Σ^n is either a minimal submanifold or it is contained in a slice $\{t\} \times M^{n+1}$.

11.4.2 Codimension Reduction Via Omori-Yau's Generalized Maximum Principle

According to [19], we suppose that the GRW spacetime $-I \times_f M^{n+1}$ obeys the so-called *strong null convergence condition*

$$K_M \geq \sup_I(f^2(\log f)''), \tag{11.15}$$

where K_M stands for the sectional curvature of the Riemannian base M^{n+1}.

In this setting, we obtain the following lower estimate for the Ricci curvature of a codimension two spacelike submanifold in a GRW spacetime.

Lemma 11.3 *Let Σ^n be a spacelike submanifold immersed in a GRW spacetime $-I \times_f M^{n+1}$, which obeys the strong null convergence condition (11.15). Then, for all $X \in \mathfrak{X}(\Sigma^n)$, the Ricci curvature of Σ^n satisfies the following inequality*

$$\mathrm{Ric}(X, X) \geq (n-1)\frac{f''}{f}|X|^2 + \left|AX - \frac{nH_N}{2}X\right|^2 - \left|SX - \frac{nH_\nu}{2}X\right|^2 + \epsilon\frac{n^2|\mathbf{H}|^2}{4}|X|^2,$$

where $\mathbf{H}$ is the mean curvature vector field of Σ^n,

$$\epsilon = \mathrm{sign}\langle \mathbf{H}, \mathbf{H}\rangle, \quad |\mathbf{H}|^2 = |\langle \mathbf{H}, \mathbf{H}\rangle|$$

and the functions H_ν and H_N are defined in (11.8).

Proof For simplicity, we will assume that $\mathbf{H}$ is causal. By (11.1) and (11.2) we can deduce that the curvature tensor R of the spacelike submanifold Σ^n is given in terms of the

Weingarten operator A and the curvature tensor $\overline{R}$ of the GRW spacetime $-I \times_f M^{n+1}$ by the Gauss equation

$$\langle R(X,Y)Z, W\rangle = \langle \overline{R}(X,Y)Z, W\rangle + \langle II(X,Z), II(Y,W)\rangle$$
$$- \langle II(X,W), II(Y,Z)\rangle \tag{11.16}$$

for every tangent vector fields $X, Y, Z, W \in \mathfrak{X}(\Sigma^n)$.

Thus, considering $X \in \mathfrak{X}(\Sigma^n)$ and a local orthonormal frame $\{E_1, \ldots, E_n\}$ of $\mathfrak{X}(\Sigma^n)$, it follows by the Gauss equation (11.16) that the Ricci curvature tensor of Σ^n is given by

$$\mathrm{Ric}(X,X) = \sum_{i=1}^n \langle \overline{R}(X,E_i)X, E_i\rangle + n\langle II(X,X), \mathbf{H}\rangle - \sum |II(X,E_i)|^2$$

$$= \sum_{i=1}^n \langle \overline{R}(X,E_i)X, E_i\rangle + n\langle -\langle AX, X\rangle N + \langle SX, X\rangle v, \mathbf{H}\rangle$$

$$+ \sum \langle AX, E_i\rangle^2 - \sum \langle SX, E_i\rangle^2, \tag{11.17}$$

where $II(X,Y) = -\langle AX, Y\rangle N + \langle SX, Y\rangle v$. Thus, by (11.17) we obtain

$$\mathrm{Ric}(X,X) = \sum_{i=1}^n \langle \overline{R}(X,E_i)X, E_i\rangle - n\langle AX, X\rangle H_N$$

$$+ n\langle SX, X\rangle H_v + |AX|^2 - |SX|^2, \tag{11.18}$$

where $\mathbf{H} = -H_N N + H_v v$.

Consequently, by (11.18) we get

$$\mathrm{Ric}(X,X) = \sum_{i=1}^n \langle \overline{R}(X,E_i)X, E_i\rangle + \left|AX - \frac{nH_N}{2}X\right|^2$$

$$- \left|SX - \frac{nH_v}{2}X\right|^2 + \frac{n^2}{4}(-H_N^2 + H_v^2)|X|^2$$

$$= \sum_{i=1}^n \langle \overline{R}(X,E_i)X, E_i\rangle + \left|AX - \frac{nH_N}{2}X\right|^2$$

$$- \left|SX - \frac{nH_v}{2}X\right|^2 + \frac{n^2}{4}\langle \mathbf{H}, \mathbf{H}\rangle|X|^2. \tag{11.19}$$

11.4 Codimension Reduction Results in GRW Spacetimes

Moreover, since the GRW spacetime $-I \times_f M^{n+1}$ obeys the strong null convergence condition (11.15), by [47, formula (4.9)] we have that

$$\sum_i \langle \overline{R}(X, E_i)X, E_i \rangle \geq (n-1) \frac{f''}{f} |X|^2, \qquad (11.20)$$

see also [95, Section 3].

Therefore, by (11.20) and (11.19) the conclusion follows. □

Next, we present a generalization of the classical Omori and Yau maximum principle [250, 297]. This result has been used by many authors in the last few decades. Among others, we mention here Pigola et al. [256] and Borbély [90]; see also [33] as a general reference on the generalized maximum principle of Omori and Yau.

Lemma 11.4 *Let Σ^n be an n-dimensional complete Riemannian manifold such that*

$$\text{Ric}(\nabla r, \nabla r) \geq -G(r),$$

for a radial function $G \in C^1([0, +\infty))$ obeying the following properties

$$G(0) > 0, \quad G' \geq 0 \quad \text{and} \quad G^{-1/2} \notin L^1[0, +\infty),$$

where r is the distance function on Σ^n to a fixed reference point $o \in \Sigma^n$. If $u \in C^2(\Sigma^n)$ with $\sup_{\Sigma^n} u < +\infty$, then there exists a sequence of points $\{p_k\} \subset \Sigma^n$ satisfying

$$\lim_k u(p_k) = \sup_{\Sigma^n} u, \quad \lim_k |\nabla u(p_k)| = 0 \quad \text{and} \quad \lim_k \Delta u(p_k) \leq 0.$$

We are in position now to present the following result.

Theorem 11.3 *Let Σ^n be a complete weakly trapped submanifold which lies in a bounded timelike region of a GRW spacetime $-I \times_f M^{n+1}$ obeying the strong null convergence condition (11.15). Suppose that the spatial Weingarten operator of Σ^n satisfies $|S|^2 \leq G(r)$, for some radial function $G(r)$ satisfying the properties in Lemma 11.4. If $|\mathbf{H}| = \sqrt{-\langle \mathbf{H}, \mathbf{H} \rangle}$ is bounded and*

$$|\nabla h| \leq C \inf_{\Sigma^n} \left(|\mathbf{H}| - (\log f)'(h) \right) \qquad (11.21)$$

for some positive constant C, then Σ^n is contained in a slice $\{t\} \times M^{n+1}$.

Proof Let us suppose that Σ^n is a weakly future trapped submanifold. Since **H** is causal and future-pointing, we have that

$$\langle \mathbf{H}, \partial_t \rangle \leq -|\mathbf{H}|.$$

Thus, by item (i) of Lemma 11.1 we get that

$$\Delta h \geq n(|\mathbf{H}| - (\log f)'(h)) - (\log f)'(h)|\nabla h|^2. \tag{11.22}$$

Moreover, since Σ^n lies in a bounded timelike region of $-I \times_f M^{n+1}$, taking into account the main assumptions on Σ^n and **H**, we can apply Lemmas 11.3 and 11.4 obtaining a sequence $\{p_k\} \subset \Sigma^n$ such that

$$\lim_k h(p_k) = \sup_{\Sigma^n} h, \quad \lim_k |\nabla h(p_k)| = 0 \quad \text{and} \quad \limsup_k \Delta h(p_k) \leq 0.$$

Hence, by (11.22) one has

$$0 \geq \limsup_k \Delta h(p_k) \geq n \lim_k (|\mathbf{H}| - (\log f)'(h))(p_k) \geq 0.$$

Therefore, we infer

$$\lim_k (|\mathbf{H}| - (\log f)'(h))(p_k) = 0.$$

Taking into account (11.21), the conclusion follows.

Finally, the case of a weakly past trapped submanifold Σ^n is similar. $\square$

Remark 11.1 When the ambient space is a static GRW spacetime $-I \times M^{n+1}$ whose Riemannian fiber M^{n+1} has nonnegative sectional curvature, by Lemma 11.3 we see that the assumptions on $|S|$ and $|\mathbf{H}|$ are enough to guarantee that the Ricci curvature of Σ^n satisfies the radial control asked in Lemma 11.4. Consequently, in Theorem 11.3 we can relax the hypothesis that Σ^n is contained in a timelike bounded region by supposing that Σ^n is either under a slice, when it is weakly future trapped, or over a slice, when it is weakly past trapped.

We recall now that a Riemannian manifold Σ^n is said to be *stochastically complete* if, for some (and, hence, for any) $(x, t) \in \Sigma^n \times (0, +\infty)$, the heat kernel $p(x, y, t)$ of the Laplace-Beltrami operator Δ satisfies the conservation property

$$\int_\Sigma p(x, y, t) d\mu(y) = 1. \tag{11.23}$$

11.4 Codimension Reduction Results in GRW Spacetimes

From the probabilistic viewpoint, the stochastic completeness is the property of a stochastic process to have infinite life time. For the Brownian motion on a manifold, the conservation property (11.23) means that the total probability of the particle to be found in the state space is constantly equal to one; for more details, see references [173, 193, 194, 280].

On the other hand, Pigola, Rigoli and Setti showed that stochastic completeness turns out to be equivalent to the validity of a weak form of the Omori-Yau maximum principle; see [254, Theorem 1.1] or [256, Theorem 3.1].

More precisely, one has

Lemma 11.5 *A Riemannian manifold Σ^n is stochastically complete if and only if for every $g \in C^2(\Sigma^n)$ satisfying $\sup_{\Sigma^n} g < +\infty$, there exists a sequence of points $\{p_k\} \subset \Sigma^n$ such that*

$$\lim_{k \to \infty} g(p_k) = \sup_{\Sigma^n} g \quad \text{and} \quad \limsup_{k \to \infty} \Delta g(p_k) \leq 0.$$

In the next result, we apply Lemma 11.5 in order to infer the rigidity of stochastically complete weakly trapped submanifolds in a GRW spacetime.

Theorem 11.4 *Let Σ^n be a stochastically complete weakly future (past) trapped submanifold which lies in a timelike bounded region of a GRW spacetime $-I \times_f M^{n+1}$, with $f' \leq 0$ ($f' \geq 0$) on Σ^n. If $|\mathbf{H}|$ is bounded and (11.21) is satisfied, then Σ^n is a complete minimal hypersurface of a slice $\{t_0\} \times_f M^{n+1}$, with $f'(t_0) = 0$.*

Proof The main goal of this proof is to use the stochastic completeness of the trapped submanifold Σ^n that is suppose to be contained in a timelike bounded region of $-I \times_f M^{n+1}$. We will prove that Σ^n it must be contained in a totally geodesic slice $\{t_0\} \times M^{n+1}$ and $|\mathbf{H}| = 0$. To this aim, let us assume, for instance, that Σ^n is a weakly future trapped submanifold. Hence, by item (ii) of Lemma 11.1 and Lemma 11.5, there exists a sequence of points $\{p_k\} \subset \Sigma^n$ such that

$$0 \geq \limsup_k \Delta \sigma(h)(p_k) \geq nC \lim_k (|\mathbf{H}| - (\log f)'(h))(p_k) \geq 0,$$

where $C = \inf_{\Sigma^n} f > 0$. Consequently, taking into account that $|\mathbf{H}|$ is bounded and $f' \leq 0$, we get $(\log f)'(t_0) = 0$, for some $t_0 \in I$, as well as $|\mathbf{H}| = 0$. Therefore, by (11.21), we conclude that Σ^n is a complete minimal hypersurface of a slice $\{t_0\} \times_f M^{n+1}$, with $f'(t_0) = 0$. □

Remark 11.2 Related to Theorems 11.3 and 11.4, we observe that the hypothesis that $|\mathbf{H}|$ be bounded is weaker than to assume that $\mathbf{H}$ be parallel (that is, $\mathbf{H}$ is parallel as a section of the normal bundle) and that it is automatically satisfied when Σ^n is marginally trapped.

11.5 Constructing a Nontrivial Weakly Trapped Submanifold

We end this chapter constructing a nontrivial example of weakly trapped submanifold in the static GRW spacetime $-\mathbb{R} \times \mathbb{H}^2 \times \mathbb{R}$, which is not contained in a slice of the ambient space. This simple prototype gives a simple motivation of the interest about the results presented in the previous section.

Let us consider the smooth functions $u_a : \mathbb{H}^2 \to \mathbb{R}$ given by $u_a(x, y) = a \ln y$ and the corresponding entire graph

$$\Sigma(a, b) = \{(a \ln y, x, y, b \ln y) : y > 0\} \subset -\mathbb{R} \times \mathbb{H}^2 \times \mathbb{R}.$$

We have that $Du_a(x, y) = (0, ay)$ and, hence, $|Du_a(x, y)|^2 = |a|^2$. If we take $0 < |a| < 1$, we have that $\Sigma(a, b)$ will be a complete spacelike surface in $-\mathbb{R} \times \mathbb{H}^2 \times \mathbb{R}$.

Moreover, the normal vectors are given by

$$N = \frac{\partial_t + Du_a}{\sqrt{1 - a^2}} \quad \text{and} \quad \bar{v} = \frac{\partial_s - Du_b}{\sqrt{1 + b^2}}.$$

Indeed, for all $v \in T\Sigma(a, b)$ we have that

$$\langle N, v \rangle = \frac{1}{\sqrt{1 - a^2}} \{v(\pi_{-\mathbb{R}}|_{\Sigma(a,b)}) + v(u_a)\} = \frac{1}{\sqrt{1 - a^2}} \{-av(\ln y) + av(\ln y)\} = 0.$$

and

$$\langle \bar{v}, v \rangle = \frac{1}{\sqrt{1 - b^2}} \{v(\pi_\mathbb{R}|_{\Sigma(a,b)}) - v(u_b)\} = \frac{1}{\sqrt{1 - b^2}} \{bv(\ln y) - bv(\ln y)\} = 0.$$

We also notice that N and $\bar{v}$ represent the unit normal vectors of the graphs of u_a in $-\mathbb{R} \times \mathbb{H}^2$ and of u_b in $\mathbb{H}^2 \times \mathbb{R}$, respectively. Despite these unit vectors being perpendicular to $\Sigma(a, b)$, they are not orthogonal to each other. Nonetheless, we can use the Gram-Schmidt process to obtain a spacelike normal v.

More precisely, taking the vector

$$\tilde{v} = \bar{v} + \langle N, \bar{v} \rangle N,$$

the spatial normal vector v will be given by

$$v = \frac{\tilde{v}}{|\tilde{v}|}.$$

11.5 Constructing a Nontrivial Weakly Trapped Submanifold

Now, let us observe that $u_a = au_1$. Thus, a direct computation yields

$$\tilde{v} = \frac{\partial_s - Du_b}{\sqrt{1+b^2}} + \left\langle \frac{\partial_t + Du_a}{\sqrt{1-a^2}}, \frac{\partial_s - Du_b}{\sqrt{1+b^2}} \right\rangle \frac{\partial_t + Du_a}{\sqrt{1-a^2}}$$

$$= \frac{\partial_s - Du_b}{\sqrt{1+b^2}} - \frac{ab}{\sqrt{1+b^2}} \frac{\partial_t + Du_a}{1-a^2}$$

$$= \frac{1}{\sqrt{1+b^2}} \left(\partial_s - \frac{ab}{1-a^2} \partial_t - b\frac{Du}{1-a^2} \right),$$

where we set $u = u_1$.

We also observe that ∂_t, ∂_s and Du are unit and orthogonal vectors, therefore

$$|\tilde{v}|^2 = \frac{1-a^2+b^2}{(1+b^2)(1-a^2)}$$

In the special case $a = b$, we get

$$|\tilde{v}|^2 = \frac{1}{1-a^4}$$

Moreover, if $a = 0$ this vector is unit.

By Eq. (11.6) the height function h satisfies

$$|\nabla h|^2 = \frac{a^2}{1-a^2} + \frac{1}{|\tilde{v}|^2} \langle \tilde{v}, \partial_t \rangle^2.$$

Consequently,

$$|\nabla h|^2 = \frac{a^2}{1-a^2} + \frac{a^2 b^2}{(1-a^2)(1-a^2+b^2)}. \tag{11.24}$$

Clearly, if $a = b$ we obtain

$$|\nabla h|^2 = \frac{a^4 + a^2}{1-a^2}.$$

Moreover, if $b = 0$ we get

$$|\nabla h|^2 = \frac{a^2}{1-a^2},$$

which is the same formula valid for a surface in $-\mathbb{R} \times \mathbb{H}^2$.

So, if $a = 0$ we get $\nabla h \equiv 0$, since it is contained in the slice $\{0\} \times \mathbb{H}^2 \times \mathbb{R}$. Furthermore, the timelike mean curvature H_N of $\Sigma(a,b)$ is given by

$$2H_N = \mathrm{Div}\left(\frac{aDu}{\sqrt{1 - a^2|Du|^2}}\right) = \mathrm{Div}\left(\frac{aDu}{\sqrt{1 - a^2}}\right),$$

where Div is the divergence on $\mathbb{H}^2$. So, using that $\mathrm{Div} = \mathrm{Div}_0 - \frac{2}{y} dy$, where Div_0 denotes the usual divergence on $\mathbb{R}^2$, we have

$$\Delta u = -1,$$

Then

$$H_N = -\frac{a}{2\sqrt{1 - a^2}}. \tag{11.25}$$

Moreover, for the spatial mean curvature we have that $2H_\nu = \mathrm{Div}\left(\frac{\tilde{\nu}}{|\tilde{\nu}|}\right)$, that is

$$2H_\nu = \frac{1}{|\tilde{\nu}|} \frac{b}{\sqrt{1 + b^2(1 - a^2)}}.$$

Therefore, one has

$$H_\nu = \frac{1}{2} \frac{b}{\sqrt{(1 - a^2 + b^2)(1 - a^2)}}. \tag{11.26}$$

Consequently, by (11.8), (11.25) and (11.26) we have that $|\mathbf{H}|$ is bounded. Hence, taking into account (11.24), hypothesis (11.21) is satisfied.

Furthermore, we claim that $\Sigma(a,b)$ is stochastically complete. Indeed, the induced metric on $\Sigma(a,b)$ has the form

$$-dt^2 + \frac{1}{y^2}(dx^2 + dy^2) + ds^2|_{\Sigma(a,b)} = -\frac{a^2}{y^2}dy^2 + \frac{1}{y^2}(dx^2 + dy^2) + \frac{b^2}{y^2}dy^2$$

$$= (1 + b^2 - a^2)\left(\frac{dx^2}{(\sqrt{1 + b^2 - a^2}\, y)^2} + \frac{dy^2}{y^2}\right)$$

$$= (1 + b^2 - a^2)\left(\frac{dx^2}{\bar{y}^2} + \frac{d\bar{y}^2}{\bar{y}^2}\right),$$

where $\bar{y} = \sqrt{1 + b^2 - a^2}\, y$.

11.5 Constructing a Nontrivial Weakly Trapped Submanifold

Hence, $\Sigma(a,b)$ is homothetic to $\mathbb{H}^2$ for $a^2 \neq b^2$ and isometric when $a^2 = b^2$. Consequently, since $\mathbb{H}^2$ is stochastically complete, it follows that $\Sigma(a,b)$ must be also stochastically complete.

Therefore, in Theorem 11.4, the assumption that the trapped submanifold Σ^n must be contained in a timelike bounded region of the ambient space cannot be dropped.

Studying the Geometry of Weakly Trapped Submanifolds in Standard Static Spacetimes

12.1 Introduction

Our purpose in this chapter is to study weakly trapped submanifolds of codimension two immersed in a standard static spacetime.

In this setting, we apply some generalized maximum principles in order to investigate the geometry of these trapped submanifolds. For instance, we establish sufficient conditions to guarantee that such a spacelike submanifold must be a hypersurface of the Riemannian base of the ambient spacetime.

As a consequence, we prove that there are no n-dimensional compact (without boundary) trapped submanifolds immersed in an $(n + 2)$-dimensional standard static spacetime. This nonexistence result was originally obtained for stationary spacetimes by Mars and Senovilla [235].

Furthermore, we investigate parabolic weakly trapped submanifolds immersed in a standard static spacetime.

The results presented in this chapter are mainly based on the paper [182].

12.2 Spacelike Submanifolds in Standard Static Spacetimes

Let $\overline{M}^{n+2}$ be a $(n + 2)$-dimensional Lorentzian manifold endowed with a timelike Killing vector field K. Suppose that the distribution $\mathcal{D}$ orthogonal to K is of constant rank and integrable. We denote by $\Psi : M^{n+1} \times \mathbb{I} \to \overline{M}^{n+2}$ the flow generated by K, where M^{n+1} is an arbitrarily fixed spacelike integral leaf of $\mathcal{D}$ labeled as $t = 0$, which will be assumed connected, and $\mathbb{I}$ is the maximal interval of definition. In the sequel, we consider $\mathbb{I} = \mathbb{R}$.

In this setting, $\overline{M}^{n+2}$ can be regard as a *standard static spacetime* $M^{n+1} \times_\rho \mathbb{R}_1$, that is, the product manifold $M^{n+1} \times \mathbb{R}$ endowed with the warping metric

$$\langle \cdot, \cdot \rangle = \pi_M^* \left(\langle \cdot, \cdot \rangle_M \right) - (\rho \circ \pi_M)^2 \pi_\mathbb{R}^* \left(dt^2 \right), \tag{12.1}$$

where π_M and $\pi_\mathbb{R}$ denote the canonical projections from $M^{n+1} \times \mathbb{R}$ onto each factor, $\langle \cdot, \cdot \rangle_M$ is the induced Riemannian metric on the base M^{n+1}, $\mathbb{R}_1$ is the manifold $\mathbb{R}$ endowed with the metric $-dt^2$ and the warping function

$$\rho = |K| = \sqrt{-\langle K, K \rangle} \in C^\infty,$$

where $|\cdot|$ denotes the norm of a timelike vector field on $\overline{M}^{n+2}$.

Along this chapter, we will consider a connected and oriented spacelike submanifold $\psi : \Sigma^n \to \overline{M}^{n+2}$ immersed in a standard static spacetime $\overline{M}^{n+2} = M^{n+1} \times_\rho \mathbb{R}_1$, that is, the metric induced on Σ^n via ψ is a Riemannian metric. As usual, we also denote by $\langle \cdot, \cdot \rangle$ the metric on Σ^n induced via ψ. Since K is a globally defined timelike vector field on $\overline{M}^{n+2}$, it follows that there exists a unitary timelike normal vector field N globally defined on Σ^n which is in the same time-orientation of K (one can define N as been the unitary direction of K minus its projection on Σ^n) and, as we are assuming that Σ^n is oriented, a spacelike normal vector field ν globally defined on Σ^n, such that $\{N, \nu\}$ constitutes an orthonormal frame for the normal bundle of Σ^n. By using the inverse Cauchy-Schwarz inequality, we get

$$\langle N, K \rangle \leq -\rho < 0 \tag{12.2}$$

on Σ^n. The normal vector field N is said to be the *future-pointing Gauss map* of Σ^n. Throughout this chapter, N will always denote the future-pointing Gauss map of a spacelike submanifold $\psi : \Sigma^n \to \overline{M}^{n+2}$.

It is well-known that the curvature tensor R of the submanifold Σ^n can be described in terms of the second fundamental form II and of the curvature tensor $\overline{R}$ of the ambient spacetime $\overline{M}^{n+2} = M^{n+1} \times_\rho \mathbb{R}_1$ by the Gauss equation as follows

$$\langle R(X, Y)Z, W \rangle = \langle \overline{R}(X, Y)Z, W \rangle + \langle II(X, Z), II(Y, W) \rangle$$
$$- \langle II(X, W), II(Y, Z) \rangle, \tag{12.3}$$

for every tangent vector fields $X, Y, Z, W \in \mathfrak{X}(\Sigma^n)$; see, for instance, [251].

In this previous setting, we will consider some natural smooth functions defined on a connected spacelike submanifold $\psi : \Sigma^n \to \overline{M}^{n+2}$ immersed in a standard static spacetime $\overline{M}^{n+2} = M^{n+1} \times_\rho \mathbb{R}_1$, namely, the (vertical) height function $h = \pi_\mathbb{R} \circ \psi$, the

12.3 Some Auxiliary Results

angle function $\Theta_1 = \langle N, K \rangle$, where we recall that N denotes the future-pointing Gauss map of Σ^n, and the angle function $\Theta_2 = \langle \nu, K \rangle$.

From the decomposition $K = K^\top - \Theta_1 N + \Theta_2 \nu$, where $(\cdot)^\top$ denotes the tangential component of a vector field in $\mathfrak{X}(\overline{M}^{n+2})$ along Σ^n, we obtain:

$$\nabla h = -\frac{1}{\rho^2} K^\top \quad \text{and} \quad \langle \nabla h, \nabla h \rangle = \frac{\Theta_1^2 - \Theta_2^2 - \rho^2}{\rho^4}. \tag{12.4}$$

We also point out that the mean curvature vector field $\mathbf{H}$ along Σ^n is defined by

$$\mathbf{H} = -H_N N + H_\nu \nu,$$

where H_N and H_ν denote the mean curvature functions in relation to the future-pointing Gauss map N and the spacelike vector field ν, respectively.

Remark 12.1 Since the slices are totally geodesic spacelike hypersurfaces, we have that the second fundamental form of a submanifold $\psi : \Sigma^n \to \{t_0\} \times M^{n+1} \hookrightarrow \overline{M}^{n+2}$ is the lifting, via π_M, of the second fundamental form of the hypersurface $\tilde{\psi} = \pi_M \circ \psi : \Sigma^n \to M^{n+1}$. Moreover, $N = K/|K|$ and ν give an orthonormal frame of the normal fiber bundle of ψ.

12.3 Some Auxiliary Results

In order to prove in the next section our main results concerning spacelike submanifolds immersed in a standard static spacetime, we will need some key lemmas. The first one gives a suitable formula for the Laplacian of the height function of such a spacelike submanifold.

Lemma 12.1 *Let Σ^n be a spacelike submanifold immersed in a standard static spacetime $M^{n+1} \times_\rho \mathbb{R}_1$. Then,*

$$\Delta h = -2\langle \nabla \ln \rho, \nabla h \rangle - \rho^{-2} \langle \mathbf{H}, \partial_t \rangle. \tag{12.5}$$

Proof By direct computations, one has

$$\Delta h = -\mathrm{div}\left(\rho^{-2} K^\top\right) = -\langle \nabla \rho^{-2}, K^\top \rangle - \rho^{-2} \mathrm{div} K^\top$$
$$= -\langle \nabla \rho^{-2}, K^\top \rangle - \rho^{-2} \mathrm{div}(K + \Theta_1 N - \Theta_2 \nu)$$
$$= \langle \rho^2 \nabla \rho^{-2}, \nabla h \rangle - n\rho^{-2}(-\Theta_1 H_N + \Theta_2 H_\nu)$$
$$= -\langle 2\nabla \ln \rho, \nabla h \rangle - n\rho^{-2} \langle \mathbf{H}, \partial_t \rangle.$$

□

The next key lemma gives sufficient conditions to guarantee that the Ricci curvature of a spacelike submanifold Σ^n immersed in a standard static spacetime $M^{n+1} \times_\rho \mathbb{R}_1$ is bounded from below.

For this, we will denote by A and S the components of the second fundamental form of Σ^n with respect to N and v, respectively, that is

$$II(X, Y) = \langle SX, Y\rangle v + \langle AX, Y\rangle N. \tag{12.6}$$

Lemma 12.2 *Let $\overline{M}^{n+2} = M^{n+1} \times_\rho \mathbb{R}_1$ be a standard static spacetime whose Riemannian base M^{n+1} has nonnegative sectional curvature K_M and convex warping function ρ. Let $\psi : \Sigma^n \to \overline{M}^{n+2}$ be a spacelike submanifold. Then, one has*

$$\mathrm{Ric}(X, X) \geq \left|AX + \frac{nH_N}{2}X\right|^2 - \left|SX - \frac{nH_v}{2}X\right|^2 + \epsilon \frac{n^2|\mathbf{H}|^2}{4}|X|^2,$$

where ϵ stands for the sign of $\langle \mathbf{H}, \mathbf{H}\rangle$.

Proof For vector fields U, V, W tangent to $\overline{M}^{n+2}$, we can write

$$U = U^* + U^\perp,$$

where U^* and $U^\perp$ are the orthogonal projections of U onto TM and $T\mathbb{R}_1$, respectively. Thus,

$$U^\perp = \frac{\langle U, K\rangle}{\langle K, K\rangle}K = -\frac{\langle U, K\rangle}{\rho^2}K.$$

On the other hand, with a straightforward computation it is easily seen that

$$\overline{R}(U, V)W = R_M(U^*, V^*)W^* + \frac{\langle V, K\rangle}{\rho^2}\overline{R}(K, U^*)W^* + \frac{\langle V, K\rangle\langle W, K\rangle}{\rho^4}\overline{R}(U^*, K)K$$

$$-\frac{\langle U, K\rangle}{\rho^2}\overline{R}(K, V^*)W^* - \frac{\langle U, K\rangle\langle W, K\rangle}{\rho^4}\overline{R}(V^*, K)K.$$

Then, by O'Neill [251, Lemma 7.34 and Proposition 7.42] we get

$$\overline{R}(U, V)W = R_M(U^*, V^*)W^* + \frac{\langle V, K\rangle \mathrm{Hess}_M \rho(U^*, W^*)}{\rho^3}K$$

$$+ \frac{\langle V, K\rangle\langle W, K\rangle\langle K, K\rangle}{\rho^5}\overline{\nabla}_{U^*}\overline{\nabla}(\rho \circ \pi_M)$$

$$- \frac{\langle U, K\rangle \mathrm{Hess}_M \rho(V^*, W^*)}{\rho^3}K - \frac{\langle U, K\rangle\langle W, K\rangle\langle K, K\rangle}{\rho^5}\overline{\nabla}_{V^*}\overline{\nabla}(\rho \circ \pi_M)$$

12.3 Some Auxiliary Results

$$= R_M(U^*, V^*)W^* + \frac{\langle V, K \rangle \text{Hess}_M \rho(U^*, W^*)}{\rho^3} K$$

$$- \frac{\langle V, K \rangle \langle W, K \rangle}{\rho^3} D_{U^*} D\rho$$

$$- \frac{\langle U, K \rangle \text{Hess}_M \rho(V^*, W^*)}{\rho^3} K + \frac{\langle U, K \rangle \langle W, K \rangle}{\rho^3} D_{V^*} D\rho,$$

where Hess_M is the Hessian on M^{n+1}. In particular, taking a local orthonormal frame $\{E_1, \ldots, E_n\}$ tangent to Σ^n and X a vector field tangent to Σ^n, we can take $U = W = X$ and $V = E_i$ in the last equation to obtain

$$\overline{R}(X, E_i)X = R_M(X^*, E_i^*)X^* + \frac{\langle E_i, K \rangle \text{Hess}_M \rho(X^*, X^*)}{\rho^3} K$$

$$- \frac{\langle E_i, K \rangle \langle X, K \rangle}{\rho^3} D_{X^*} D\rho$$

$$- \frac{\langle X, K \rangle \text{Hess}_M \rho(E_i^*, X^*)}{\rho^3} K + \frac{\langle X, K \rangle^2}{\rho^3} D_{E_i^*} D\rho.$$

Hence, the last equation ensures that

$$\langle \overline{R}(X, E_i)X, E_i \rangle = \langle R_M(X^*, E_i^*)X^*, E_i \rangle + \frac{\langle E_i, K \rangle^2}{\rho^3} \text{Hess}_M \rho(X^*, X^*)$$

$$- \frac{\langle E_i, K \rangle \langle X, K \rangle}{\rho^3} \langle D_{X^*} D\rho, E_i \rangle$$

$$- \frac{\langle E_i, K \rangle \langle X, K \rangle}{\rho^3} \text{Hess}_M \rho(E_i^*, X^*) + \frac{\langle X, K \rangle^2}{\rho^3} \langle D_{E_i^*} D\rho, E_i \rangle$$

$$= \langle R_M(X^*, E_i^*)X^*, E_i^* \rangle + \frac{\langle E_i, K \rangle^2}{\rho^3} \text{Hess}_M \rho(X^*, X^*)$$

$$- \frac{\langle E_i, K \rangle \langle X, K \rangle}{\rho^3} \text{Hess}_M \rho(X^*, E_i^*)$$

$$- \frac{\langle E_i, K \rangle \langle X, K \rangle}{\rho^3} \text{Hess}_M \rho(X^*, E_i^*)$$

$$+ \frac{\langle X, K \rangle^2}{\rho^3} \text{Hess}_M \rho(E_i^*, E_i^*).$$

Then

$$\langle \overline{R}(X, E_i)X, E_i\rangle = K_M(X^*, E_i^*)\left(\langle X^*, X^*\rangle\langle E_i^*, E_i^*\rangle - \langle X^*, E_i^*\rangle^2\right)$$

$$+ \frac{\langle E_i, K\rangle^2}{\rho^3}\text{Hess}_M\rho(X^*, X^*)$$

$$- 2\frac{\langle E_i, K\rangle\langle X, K\rangle}{\rho^3}\text{Hess}_M\rho(X^*, E_i^*)$$

$$+ \frac{\langle X, K\rangle^2}{\rho^3}\text{Hess}_M\rho(E_i^*, E_i^*)$$

$$= K_M(X^*, E_i^*)\left(\langle X^*, X^*\rangle\langle E_i^*, E_i^*\rangle - \langle X^*, E_i^*\rangle^2\right)$$

$$+ \frac{1}{\rho}\text{Hess}_M\rho(\widetilde{X}_i^*, \widetilde{X}_i^*)$$

$$- \frac{2}{\rho}\text{Hess}_M\rho(\widetilde{X}_i^*, \widetilde{E}_i^*) + \frac{1}{\rho}\text{Hess}_M\rho(\widetilde{E}_i^*, \widetilde{E}_i^*),$$

where

$$\widetilde{X}_i^* = \frac{\langle E_i, K\rangle}{\rho}X^* \quad \text{and} \quad \widetilde{E}_i^* = \frac{\langle X, K\rangle}{\rho}E_i^*.$$

Hence, we deduce that

$$\langle \overline{R}(X, E_i)X, E_i\rangle = K_M(X^*, E_i^*)\left(\langle X^*, X^*\rangle\langle E_i^*, E_i^*\rangle - \langle X^*, E_i^*\rangle^2\right)$$

$$+ \frac{1}{\rho}\text{Hess}_M\rho(\widetilde{X}_i^* - \widetilde{E}_i^*, \widetilde{X}_i^* - \widetilde{E}_i^*). \qquad (12.7)$$

Thus, we obtain that

$$\sum_{i=1}^n \langle \overline{R}(X, E_i)X, E_i\rangle = \sum_{i=1}^n K_M(X^*, E_i^*)\left(\langle X^*, X^*\rangle\langle E_i^*, E_i^*\rangle - \langle X^*, E_i^*\rangle^2\right)$$

$$+ \sum_{i=1}^n \frac{1}{\rho}\text{Hess}_M\rho(\widetilde{X}_i^* - \widetilde{E}_i^*, \widetilde{X}_i^* - \widetilde{E}_i^*). \qquad (12.8)$$

12.4 Weakly Trapped Submanifolds in Standard Static Spacetimes

On the other hand, taking a local orthonormal frame $\{E_1, \ldots, E_n\}$ tangent to Σ^n and by using the Gauss equation (12.3) in addition to (12.6), we have that the Ricci curvature Ric of Σ^n is given by

$$\mathrm{Ric}(X, X) = \sum_{i=1}^{n} \langle \overline{R}(X, E_i)X, E_i \rangle + \left| AX + \frac{nH_N}{2} X \right|^2$$

$$- \left| SX - \frac{nH_v}{2} X \right|^2 + \epsilon \frac{n^2 |\mathbf{H}|^2}{4} |X|^2. \tag{12.9}$$

Therefore, since the sectional curvature K_M is nonnegative and the warping function ρ is convex, by (12.8) and (12.9) it follows that

$$\mathrm{Ric}(X, X) \geq \left| AX + \frac{nH_N}{2} X \right|^2 - \left| SX - \frac{nH_v}{2} X \right|^2 + \epsilon \frac{n^2 |\mathbf{H}|^2}{4} |X|^2 \tag{12.10}$$

for every tangent vector field $X \in \mathfrak{X}(\Sigma^n)$. □

12.4 Weakly Trapped Submanifolds in Standard Static Spacetimes

This section is devoted to present our first results concerning the geometry of weakly trapped submanifolds immersed with codimension two in a standard static spacetime. For this reason, we will recall in the next paragraph the concept of weakly trapped submanifold (with a slight change from the one given in [34]).

A *future (past) trapped submanifold* is a submanifold such that the mean curvature vector **H** is timelike and it is future (past) pointing.

A *marginally future (past) trapped submanifold* is a submanifold with **H** null, such that it is future (past) pointing. A *weakly future (past) trapped submanifold* is a submanifold with **H** causal or zero, such that it is future (past) pointing when it is causal.

Finally, we recall that a *minimal submanifold* is a submanifold such that the mean curvature vector **H** is identically zero.

In what follows, according to the nomenclature of [6], we say that a spacelike submanifold Σ^n immersed in a standard static spacetime $\overline{M}^{n+2} = M^{n+1} \times_\rho \mathbb{R}_1$ is *bounded away from the future infinity* of $\overline{M}^{n+2}$ if there exists $\bar{t} \in \mathbb{R}$ such that

$$\Sigma^n \subset \left\{ (x, t) \in \overline{M}^{n+2} : t \leq \bar{t} \right\}.$$

Analogously, we say that Σ^n is *bounded away from the past infinity* of $\overline{M}^{n+2}$ if there exists $\underline{t} \in \mathbb{R}$ such that

$$\Sigma^n \subset \left\{ (x,t) \in \overline{M}^{n+2} : t \geq \underline{t} \right\}.$$

Finally, Σ^n is said *bounded away from the infinity* of $\overline{M}^{n+2}$ if it is both bounded away from the past and the future infinity of $\overline{M}^{n+2}$. In other words, Σ^n is bounded away from the infinity if there exist $\underline{t} < \overline{t}$ such that Σ^n is contained in the slab bounded by the slices $M^{n+1} \times \{\underline{t}\}$ and $M^{n+1} \times \{\overline{t}\}$.

In this setting, we obtain the following result

Theorem 12.1 *Let Σ^n be a complete weakly future (past) trapped submanifold immersed with bounded second fundamental form in a standard static spacetime $\overline{M}^{n+2} = M^{n+1} \times_\rho \mathbb{R}_1$, whose Riemannian base M^{n+1} has nonnegative sectional curvature and such that the warping function ρ is convex on M^{n+1} outside a compact set. If Σ^n is bounded away from the future (past) infinity of $\overline{M}^{n+2}$ and $\nabla \log \rho$ is bounded then either the inverse support function $\langle \mathbf{H}, \partial_t \rangle^{-1}$ or the warping function ρ is unbounded.*

Proof Let Σ^n be the weakly future trapped submanifold. Suppose by contradiction that ρ and $\langle \mathbf{H}, \partial_t \rangle^{-1}$ are bounded on Σ. Since the second fundamental form is bounded, by Lemma 12.2 we have that the Ricci curvature is bounded from below.

Thus, taking into account that Σ^n is bounded away from the future infinity of $M^{n+1} \times_\rho \mathbb{R}_1$, by Lemma 12.1 jointly with the generalized maximum principle of Omori [250] and Yau [297] we obtain a sequence of points $\{p_k\} \subset \Sigma^n$ such that

$$0 \geq \limsup_k \Delta h(p_k) = \limsup_k \left(-\rho^{-2} \langle \mathbf{H}, \partial_t \rangle \right)(p_k) \geq 0. \tag{12.11}$$

Hence, one has

$$\limsup_k (\langle \mathbf{H}, \partial_t \rangle)(p_k) = 0.$$

Therefore $\langle \mathbf{H}, \partial_t \rangle^{-1}$ is unbounded, but it is a contradiction. □

By Theorem 12.1 we obtain the following property.

Corollary 12.1 *Let Σ^n be a complete future (past) trapped submanifold immersed with bounded second fundamental form and with mean curvature bounded away from zero, that is, $|\mathbf{H}| \geq \alpha > 0$ for some positive constant α, in a standard static spacetime $\overline{M}^{n+2} = M^{n+1} \times_\rho \mathbb{R}_1$, whose Riemannian base M^{n+1} has nonnegative sectional curvature and such* that the warping function ρ is convex on M^{n+1} outside a compact set. If Σ^n is bounded away from the future (past) infinity of $\overline{M}^{n+2}$ and $\nabla \log \rho$ is bounded, then the warping function ρ is unbounded.

Proof The result follows by the proof of Theorem 12.1 just observing that our assumptions on Σ^n allow us to conclude that

$$-\rho^{-2}\langle \mathbf{H}, \partial_t \rangle \geq \rho^{-1}|\mathbf{H}| \geq \rho^{-1}\alpha.$$

The proof is complete. □

As a byproduct of the strategy used along the proof of Theorem 12.1 it is possible to prove the nonexistence of compact (without boundary) two codimensional trapped submanifolds immersed in a standard static spacetime.

Corollary 12.2 *There are no n-dimensional compact (without boundary) trapped submanifolds immersed in a standard static spacetime $\overline{M}^{n+2} = M^{n+1} \times_\rho \mathbb{R}_1$.*

Proof Let us observe that for any compact submanifold immersed in a standard static spacetime $\overline{M}^{n+2} = M^{n+1} \times_\rho \mathbb{R}_1$ the strong maximum principle holds. Therefore, the conclusion follows arguing as in the proof of Theorem 12.1 by using Lemma 12.1. □

Such a nonexistence result was originally obtained for stationary spacetimes by Mars and Senovilla in [235].

12.5 Einstein Standard Static Spacetimes

Now we investigate a particular case where the above mentioned standard static spacetime $M^{n+1} \times_\rho \mathbb{R}_1$ is an Einstein manifold. An $(n+2)$-dimensional vacuum spacetime with cosmological constant Λ is a Lorentzian manifold (L, h) satisfying the Einstein equation $Ric = \Lambda h$.

The vacuum is said to be static when $L = M^{n+1} \times_\rho \mathbb{R}_1$ and $h = -\rho^2 dt^2 + g$, where (M^{n+1}, g) is an $(n+1)$-dimensional connected Riemannian manifold, that we will take to be orientable. Furthermore, a complete and connected Riemannian manifold (M^{n+1}, g) with boundary ∂M (possibly empty) is said to be *static* if there exists a non-trivial solution $\rho \in C^\infty(M^{n+1})$ of

$$-(\Delta_g \rho)g + \nabla^2_g \rho - \rho Ric_g = 0 \quad \text{in} \quad \text{int}(M^{n+1}). \tag{12.12}$$

The left hand side of Eq. (12.12) is the formal L^2-adjoint of the linearization of the scalar curvature operator (see [239]). We call the function ρ a static potential of M^{n+1}. On the other hand, a solution of (12.12) in a manifold allows us to construct a spacetime satisfying the vacuum Einstein equations (with cosmological constant), whose properties, physically interpreted, justify the name static; see [124–126].

In our case, an operational advantage in dealing with these types of manifolds lies in the fact that the warping function ρ satisfies an equation involving the Laplace and the Hessian operators. Taking into account (12.12), it is not difficult to verify that we can improve the last results with a weaker hypothesis of subharmonicity of the warping function ρ, obtaining the following result

Theorem 12.2 *Let Σ^n be a complete weakly future (past) trapped submanifold immersed with bounded second fundamental form in an Einstein standard static spacetime $\overline{M}^{n+2} = M^{n+1} \times_\rho \mathbb{R}_1$, with nonnegative Ricci curvature, whose Riemannian base M^{n+1} has nonnegative sectional curvature and such that the warping function ρ is subharmonic in M^{n+1}. If Σ^n is bounded away from the future (past) infinity of $\overline{M}^{n+2}$ and $\nabla \log \rho$ is bounded, then either the inverse support function $\langle \mathbf{H}, \partial_t \rangle^{-1}$ or the warping function ρ is unbounded.*

12.6 Further Results via a Suitable Density Function

In order to prove the next results, we denote by $\varphi = \log \rho^2$ the density function on $\overline{M}^{n+2} = M^{n+1} \times_\rho \mathbb{R}_1$. To this aim, let us define the weighted Laplacian as follows

$$\Delta_\varphi u = e^{-\varphi} \mathrm{div}(e^\varphi \nabla u),$$

where u is a smooth function in $\overline{M}^{n+2}$. Moreover, set

$$\mathcal{L}_\varphi^p(\Sigma) = \left\{ u : \Sigma^n \to \mathbb{R} : \int_{\Sigma^n} |u|^p(x) d\mu < +\infty \right\},$$

where $d\mu = e^{\varphi(x)} d\Sigma^n$ stands for the weighted measure defined on Σ^n.

In this setting, arguing as in the proof of [97, Proposition 2.1], we obtain the following result.

Theorem 12.3 *Let Σ^n be a complete weakly trapped submanifold immersed in a standard static spacetime $\overline{M}^{n+2} = M^{n+1} \times_\rho \mathbb{R}_1$ and such that $|\nabla h| \in \mathcal{L}_\varphi^1(\Sigma^n)$. Then, Σ^n is a minimal spacelike submanifold. If in addition Σ^n is bounded away from the infinity of $\overline{M}^{n+2}$, then Σ^n is contained in a slice $M^{n+1} \times \{t_0\}$.*

12.6 Further Results via a Suitable Density Function

Proof Since Σ^n is complete, there exists a sequence ϕ_j such that $\phi_j = 1$ in compact sets $K_j \subset \text{supp}\phi_j \subset \Omega_j$ which are both exhaustion of Σ^n. Thus, $\phi_j \nearrow 1$ and $|\nabla \phi_j|_\infty \to 0$; see [284, Proposition 4.1] for $p = 1$). Thus, one has

$$\int_{\Sigma^n} \text{div}_\varphi(\phi_j \nabla h) d\mu = \int_{\Sigma^n} \nabla \phi_j \cdot \nabla h d\mu + \int_{\Sigma^n} \phi_j \Delta_\varphi h d\mu,$$

where $d\mu = e^\varphi d\Sigma^n$.

Consequently, by using the Divergence Theorem, the left hand side is zero. By Holder's inequality in the first term of the right hand side, we obtain that it goes to 0 as $j \to +\infty$. Finally, by using the Monotone Convergence Theorem for the last term, we get

$$0 = \int_{\Sigma^n} \Delta_\varphi h d\mu.$$

Hence, since $\Delta_\varphi h \geq 0$, we obtain that $\Delta_\varphi h = 0$. Therefore, since **H** is casual or zero, we must have $\mathbf{H} \equiv 0$.

Now, let us observe that

$$\Delta_\varphi h^2 = 2h \Delta_\varphi h + 2|\nabla h|^2 \geq 0.$$

Since h is bounded, we obtain $|\nabla h^2| \in \mathcal{L}_\varphi^1(\Sigma^n)$. Thus, we can apply once more the previous argument to guarantee that $\Delta_\varphi h^2 = 0$ and, hence, $|\nabla h|^2 = 0$ on Σ^n, which implies that h is constant. □

In the spirit of Theorem 12.3, we get the next result.

Corollary 12.3 *The only compact weakly trapped submanifolds Σ^n immersed in a standard static spacetime $M^{n+1} \times_\rho \mathbb{R}_1$ are the minimal hypersurfaces of the slices $M^{n+1} \times \{t_0\}$. If in addition Ric_M is nonnegative on M^{n+1} and there is a Killing vector field Y on M^{n+1} such that the angle between ν and Y is nonnegative, then Σ^n is a totally geodesic hypersurface of a slice $M^{n+1} \times \{t_0\}$.*

Proof The first statement directly follows by Theorem 12.3. Now, let us recall the following well-known formula

$$\Delta \langle \nu, Y \rangle = -(|A_\nu|^2 + \text{Ric}_M(\nu, \nu))\langle \nu, Y \rangle,$$

see, for instance, the paper [97, Eq. (24)].

Therefore, by applying the Divergence Theorem to the above equation we get that $A_\nu = 0$. In conclusion, Σ^n must be totally geodesic. □

Arguing as in the proof of Theorem 12.3, the following property holds.

Corollary 12.4 *The only φ-parabolic weakly future (past) trapped submanifold Σ^n immersed in a standard static spacetime $\overline{M}^{n+2} = M^{n+1} \times_\rho \mathbb{R}_1$, bounded away from the future (past) infinity of $\overline{M}^{n+2}$, are the minimal hypersurfaces of the slices $M^{n+1} \times \{t_0\}$.*

Proof As in the proof of Theorem 12.3, we have that the height function h is φ-subharmonic and bounded from above. Therefore, the function h is constant. Standard arguments used above ensure the result. □

The following concept is crucial in order to prove the next results. To this scope, let us consider a real function $f : \Sigma^n \to \mathbb{R}$ and $1 \leq p < \infty$. We define the φ-weighted p-capacity of a compact subset $K \subset \Sigma^n$ as follows

$$\mathrm{Cap}_{\varphi,p}(K) = \inf\left\{\int_{\Sigma^n} |\nabla \phi|^p d\mu \,:\, \phi \equiv 1 \text{ in } K \quad \text{and} \quad \phi \in C_0^1(\Sigma^n)\right\},$$

where $d\mu = e^\varphi d\Sigma^n$.

We also define the notion of φ-weighted p-parabolicity:
A Riemannian manifold Σ^n is said to be φ-weighted p-parabolic if the φ-weighted p-capacity of all compact subsets of Σ^n is zero.

We are in position now to prove the following carachterization result; see [284, Proposition 4.1].

Proposition 12.1 *A Riemannian manifold Σ^n is φ-weighted p-parabolic if and only if there exists a sequence of functions $\phi_j \in C_0^1(\Sigma^n)$ such that $0 \leq \phi_j \leq 1$, $\phi_j \nearrow 1$ uniformly on every compact subset of Σ^n and*

$$\int_{\Sigma^n} |\nabla \phi_j|^p d\mu \to 0.$$

Taking into account Proposition 12.1 and by arguing as in the proof of Theorem 12.3, the next result holds.

Theorem 12.4 *The only n-dimensional weakly trapped φ-weighted p-parabolic submanifolds immersed in a standard static spacetime $\overline{M}^{n+2} = M^{n+1} \times_\rho \mathbb{R}_1$ such that $|\nabla h| \in \mathcal{L}_\varphi^q(\Sigma^n)$, with $\dfrac{1}{p} + \dfrac{1}{q} = 1$, are the minimal spacelike submanifolds.*

Finally, let us recall that the Bakry-Émery Ricci tensor is defined by

$$\mathrm{Ric}_\varphi = \mathrm{Ric} + \mathrm{Hess}\varphi,$$

12.6 Further Results via a Suitable Density Function

see [59] for additional comments and remarks.

In this setting, we obtain the following

Theorem 12.5 *Let Σ^n be a complete weakly future (past) trapped submanifold immersed in the standard static spacetime $M^{n+1} \times_\rho \mathbb{R}_1$, such that $h \geq 0 (h \leq 0)$. If $h \in \mathcal{L}_\varphi^p(\Sigma^n)$ with $p > 1$, then Σ^n is a hypersurface contained in a slice $M^{n+1} \times \{t_0\}$. Moreover, if the Bakry-Émery Ricci tensor of Σ^n is nonnegative, ρ is bounded along Σ^n and $h > 0 (h < 0)$ then Σ^n is compact.*

Proof Since $\Delta_\varphi h = -e^{-\varphi} \langle \mathbf{H}, \partial_t \rangle$, thanks to our assumptions we obtain that h (or $-h$) is a nonnegative subharmonic function on Σ^n. Thus, since we are assuming that $h \in \mathcal{L}_\varphi^p(\Sigma)$ with $p > 1$, we can apply [255, Theorem 1.1] to conclude that h is constant on Σ^n, that is, Σ^n is a hypersurface contained in a slice $M^{n+1} \times \{t_0\}$. Moreover, supposing that h has strict sign, we get that $\text{Vol}(\Sigma)$ is finite. Hence, if in addition the Bakry-Émery Ricci tensor of Σ^n is nonnegative and ρ is bounded along Σ^n, we can apply [294, Theorem 1.3] to conclude that Σ^n must be compact. □

Part IV

Stability of Riemannian Immersions in Semi-Riemannian Warped Products

A Notion of Stability to Closed Hypersurfaces in the Hyperbolic Space

13.1 Introduction

The notion of stability concerning hypersurfaces of constant mean curvature in Riemannian ambient spaces was first studied by Barbosa and do Carmo in [62], and Barbosa, do Carmo and Eschenburg in [64], where they proved that spheres are the only stable critical points of the area functional for volume-preserving variations.

In [16], Alencar, do Carmo and Colares extended to hypersurfaces with constant scalar curvature the above stability result on constant mean curvature.

In the context of higher order mean curvatures H_r, with $r \in \{1, \ldots, n\}$, first Alencar, do Carmo and Rosenberg in [17] and shortly after Barbosa and Colares [60] studied closed hypersurfaces with constant $(r+1)$-th mean curvature H_{r+1} immersed in a space form and established the concept of r-stability.

They showed that such hypersurfaces are r-stable if and only if they are geodesic spheres.

Futhermore, Velásquez et al. [287] studied the notion of (r, s)-stability concerning closed hypersurfaces with higher order mean curvatures linearly related in a space form, showing that, if the hypersurface Σ^n is contained either in an open hemisphere of the Euclidean sphere $\mathbb{S}^{n+1}$ or in the hyperbolic space $\mathbb{H}^{n+1}$, then Σ^n is (r, s)-stable if and only if Σ^n is a geodesic sphere.

Successively, Velásquez et al. [129], through the development of a different technique, managed to complete this characterization of (r, s)-stable hypersurfaces in the Euclidean space $\mathbb{R}^{n+1}$.

More recently, Velásquez, de Lima and de Sousa in [148] established the notion of strong stability (that is, stability with respect to not necessarily volume-preserving variations) related to closed linear Weingarten hypersurfaces satisfying

$$na_0 H + n(n+1)a_1 H_2 = \text{const.},$$

where a_0 and a_1 are nonnegative constants (with at least one non zero), immersed in the hyperbolic space $\mathbb{H}^{n+1}$.

In this setting, initially they showed that geodesic spheres are strongly stable and, afterwards, under a suitable restriction on the mean and scalar curvatures, they proved that if a closed linear Weingarten hypersurface into $\mathbb{H}^{n+1}$ is strongly stable and its image of its Gauss mapping is contained in the chronological future (or past) of an equator of the de Sitter space, then it must be a geodesic sphere.

In this chapter, we consider an appropriated warped product model of the hyperbolic space $\mathbb{H}^{n+1}$ to study the notion of strong (r, k, a, b)-stability related to closed hypersurfaces immersed in $\mathbb{H}^{n+1}$, where r and k are entire numbers satisfying the inequality $0 \leq k < r \leq n-2$ and a and b are real numbers (at least one nonzero).

In this setting, considering some appropriate restrictions on the constants a and b, we show that geodesic spheres of $\mathbb{H}^{n+1}$ are strongly (r, k, a, b)-stable. Afterwards, under a suitable restriction on the higher order mean curvatures H_{r+1} and H_{k+1}, we prove that if a closed hypersurface into the hyperbolic space $\mathbb{H}^{n+1}$ is strongly (r, k, a, b)-stable, then it must be a geodesic sphere, provided that the image of its Gauss mapping is contained in the chronological future (or past) of an equator of the de Sitter space $\mathbb{S}_1^{n+1}$.

The results of this chapter are mainly based on the paper [288].

13.2 A Warped Product Model for the Hyperbolic Space

Let $\mathbb{L}^{n+2}$ denote the $(n+2)$-dimensional Lorentz-Minkowski space ($n \geq 2$), that is, the real vector space $\mathbb{R}^{n+2}$ endowed with the Lorentz metric

$$\langle v, w \rangle = \sum_{i=1}^{n+1} v_i w_i - v_{n+2} w_{n+2},$$

for all $v, w \in \mathbb{R}^{n+2}$. We recall that the $(n+1)$-dimensional hyperbolic space $\mathbb{H}^{n+1}$ can be regarded as the following hyperquadric

$$\mathbb{H}^{n+1} = \left\{ p \in \mathbb{L}^{n+2} : \langle p, p \rangle = -1 \text{ and } p_{n+2} \geq 1 \right\},$$

which is a spacelike hypersurface of $\mathbb{L}^{n+2}$. In other words, the induced metric via the inclusion $\iota : \mathbb{H}^{n+1} \hookrightarrow \mathbb{L}^{n+2}$ is a Riemannian metric on $\mathbb{H}^{n+1}$. Indeed, this is the (complete) metric of $\mathbb{H}^{n+1}$ of constant sectional curvature -1. In this setting, we will denote by $\overline{\nabla}$ the Levi-Civita connection of $\mathbb{H}^{n+1}$.

13.2 A Warped Product Model for the Hyperbolic Space

According to Montiel in [242], any geodesic sphere of $\mathbb{H}^{n+1}$ is given by

$$L^n_\varrho = \left\{ p \in \mathbb{H}^{n+1} : \langle p, \mathbf{a} \rangle = \varrho \right\}, \tag{13.1}$$

where $\mathbf{a} \in \mathbb{L}^{n+2}$ is a unit timelike vector (that is, $\langle \mathbf{a}, \mathbf{a} \rangle = -1$) and ϱ is any real number such that $\varrho^2 > 1$. In this case, L^n_ϱ is isometric to n-dimensional Euclidean sphere $\mathbb{S}^n(\sqrt{\varrho^2 - 1})$. When one fixes such a vector $\mathbf{a}$ and moves $\varrho \in (-\infty, -1) \cup (1, +\infty)$ one obtains a complete foliation of $\mathbb{H}^{n+1} \setminus \{\mathbf{n}\}$ by means of geodesic spheres, where $\mathbf{n} \in \mathbb{L}^{n+2}$ denotes the north pole of $\mathbb{H}^{n+1}$.

More precisely, we have that

$$V(p) = \mathbf{a} + \langle p, \mathbf{a} \rangle p, \quad p \in \mathbb{H}^{n+1}, \tag{13.2}$$

is a conformal, and closed vector field (in the sense that its dual 1-form is closed) globally defined in the hyperbolic space $\mathbb{H}^{n+1}$; see [242, Proposition 1].

Arguing as in [46, Section 3] we have that

$$N_\varrho(p) = -\frac{1}{\sqrt{\varrho^2 - 1}}(\mathbf{a} + \varrho p), \quad p \in L^n_\varrho, \tag{13.3}$$

is a unit normal vector field on L^n_ϱ, with shape operator

$$A_\varrho = \frac{\varrho}{\sqrt{\varrho^2 - 1}} \, \mathrm{Id},$$

where Id denotes the identity operator on $\mathfrak{X}(L^n_\varrho)$, and constant mean curvature H_ϱ satisfying the relation

$$H^2_\varrho = \frac{\varrho^2}{\varrho^2 - 1} \in (1, +\infty).$$

In order to prove our main result in this chapter, we note that, according to [242, Example 4.3], the hyperbolic space $\mathbb{H}^{n+1}$ without the north pole $\mathbf{n}$ can be identified with the following warped product

$$\mathbb{H}^{n+1} \setminus \{\mathbf{n}\} \simeq (0, +\infty) \times_{\sinh \tau} \mathbb{S}^n, \quad \tau \in (0, +\infty), \tag{13.4}$$

($\simeq$ means *isometric to*) where $\mathbb{S}^n$ stands for the Euclidean unit sphere. More precisely, if $d\tau^2$ and $d\sigma^2$ denote the metrics of $(0, +\infty)$ and $\mathbb{S}^n$, respectively, one has that

$$\langle \cdot, \cdot \rangle = (\pi_1)^* \left(d\tau^2 \right) + (\sinh \tau)^2 (\pi_{\mathbb{S}^n})^* \left(d\sigma^2 \right),$$

is the tensor metric of $(0, +\infty) \times_{\sinh \tau} \mathbb{S}^n$, where π_1 and $\pi_{\mathbb{S}^n}$ denote the projections onto the $(0, +\infty)$ and $\mathbb{S}^n$ factors, respectively. We notice that, in this warped product model, the *slices*

$$\Sigma_\tau^n = \{\tau\} \times \mathbb{S}^n, \quad \tau \in (0, +\infty),$$

are, exactly, the geodesic spheres L_ϱ^n given in (13.1).

Moreover, if we orient such slices by the unit normal vector field $-\dfrac{\partial}{\partial \tau}$, then the r-th mean curvature H_r^τ of Σ_τ^n is constant. More precisely

$$H_r^\tau = (\coth \tau)^r, \quad r \in \{1, \ldots, n\}. \tag{13.5}$$

Finally, we observe that

$$W = (\sinh \tau)\frac{\partial}{\partial \tau} \tag{13.6}$$

is a closed conformal vector field, which corresponds to V given in (13.2), with

$$\overline{\nabla}_Y W = (\cosh \tau) Y$$

for any smooth vector field Y defined in $(0, +\infty) \times_{\sinh \tau} \mathbb{S}^n$.

13.3 The Notion of Strong (r, k, a, b)-Stability

Let $\psi : \Sigma^n \to \mathbb{H}^{n+1}$ be a closed hypersurface, namely, a isometric immersion from a closed, n-dimensional orientable Riemannian manifold Σ^n into the hyperbolic space $\mathbb{H}^{n+1}$. Here, closed means compact without boundary. Let us denote by ∇ and $d\Sigma^n$, respectively the Levi-Civita connection and the volume element of Σ^n. Since Σ^n is orientable, one can choose a globally defined unit normal vector field N on Σ^n. Finally, let A be the shape operator with respect to N.

A variation of $\psi : \Sigma^n \to \mathbb{H}^{n+1}$ is a smooth mapping $X : (-\epsilon, \epsilon) \times \Sigma^n \to \mathbb{H}^{n+1}$ satisfying the following condition:

For every $t \in (-\epsilon, \epsilon)$, the map $X_t : \Sigma^n \hookrightarrow \mathbb{H}^{n+1}$ given by $X_t(p) = X(t, p)$ is an immersion such that $X_0 = \psi$.

In the sequel, we denote by $d\Sigma_t$ the volume element of the metric induced on Σ^n by X_t and by N_t the unit normal vector field along X_t.

The variational field associated to the variation $X : (-\epsilon, \epsilon) \times \Sigma^n \to \mathbb{H}^{n+1}$ is $\dfrac{\partial X}{\partial t}\bigg|_{t=0}$.

If we set

13.3 The Notion of Strong (r, k, a, b)-Stability

$$f = \left\langle \frac{\partial X}{\partial t}, N_t \right\rangle, \qquad (13.7)$$

then

$$\frac{\partial X}{\partial t} = f N_t + \left(\frac{\partial X}{\partial t}\right)^\top,$$

where $(\cdot)^\top$ stands for tangential component.

The balance of volume of the variation $X : (-\epsilon, \epsilon) \times \Sigma^n \to \mathbb{H}^{n+1}$ is the functional

$$\begin{aligned}
\mathcal{V} : (-\epsilon, \epsilon) &\to \mathbb{R} \\
t &\mapsto \mathcal{V}(t) = \int_{\Sigma^n \times [0,t]} X^*(d\mathbb{H}^{n+1}),
\end{aligned}$$

where $d\mathbb{H}^{n+1}$ denotes the volume element of $\mathbb{H}^{n+1}$.

The following crucial lemma can be found in [60].

Lemma 13.1 *If $X : (-\epsilon, \epsilon) \times \Sigma^n \to \mathbb{H}^{n+1}$ is a variation of a closed hypersurface $\psi : \Sigma^n \to \mathbb{H}^{n+1}$, then*

$$\mathcal{V}'(t) = \int_{\Sigma^n} f \, d\Sigma_t,$$

where f is the function defined in (13.7).

According to [60], we define the r-area functional associated to the variation $X : (-\epsilon, \epsilon) \times \Sigma^n \to \mathbb{H}^{n+1}$ by

$$\begin{aligned}
\mathcal{A}_r : (-\epsilon, \epsilon) &\to \mathbb{R} \\
t &\mapsto \mathcal{A}_r(t) = \int_{\Sigma^n} F_r(S_1, S_2, \ldots, S_r) d\Sigma_t,
\end{aligned}$$

where $S_r = S_r(t)$ and F_r is recursively defined by setting $F_0 = 1$, $F_1 = S_1$ and

$$F_r = S_r - \frac{(n-r+1)}{r-1} F_{r-2},$$

for $2 \le r \le n-1$.

We notice that if $r = 0$, the functional $\mathcal{A}_0$ is the classical area functional. The following result follows by Barbosa and Colares [60, Proposition 4.1].

Lemma 13.2 *If $X : (-\epsilon, \epsilon) \times \Sigma^n \to \mathbb{H}^{n+1}$ is a variation of $\psi : \Sigma^n \to \mathbb{H}^{n+1}$, then*

$$\frac{\partial H_{r+1}}{\partial t} = \frac{r+1}{b_r}\left\{L_r f + \left(-\mathrm{tr}(P_r) + \mathrm{tr}(A^2 P_r)\right) f\right\} + \left\langle \left(\frac{\partial X}{\partial t}\right)^\top, \nabla H_{r+1}\right\rangle, \quad (13.8)$$

where $b_r = (r+1)\binom{n}{r+1}$ and f is the function defined in (13.7).

The previous lemma allows us to obtain the first variation of the r-area functional; see, for example, [287, Lemma 3.4].

Lemma 13.3 *If $X : (-\epsilon, \epsilon) \times \Sigma^n \to \mathbb{H}^{n+1}$ is a variation of a closed hypersurface $\psi : \Sigma^n \to \mathbb{H}^{n+1}$, then*

$$\mathcal{A}'_r(t) = -b_r \int_{\Sigma^n} H_{r+1} f\, d\Sigma_t,$$

where $b_r = (r+1)\binom{n}{r+1}$ and f is the function defined in (13.7).

Let r and k be entire numbers, satisfying the inequality $0 \leq k < r \leq n-2$, and consider real numbers a and b (with at least one non zero). In order to characterize hypersurfaces whose elementary functions satisfy a certain constant quotient, we define

$$\begin{aligned} \mathcal{C}_{k,a,b} : (-\epsilon, \epsilon) &\to \mathbb{R} \\ t &\mapsto \mathcal{C}_{k,a,b}(t) = a\mathcal{A}_k(t) + b\mathcal{V}(t), \end{aligned}$$

and we say that the variation $X : (-\epsilon, \epsilon) \times \Sigma^n \to \mathbb{H}^{n+1}$ of $\psi : \Sigma^n \to \mathbb{H}^{n+1}$ preserve the linear combination $\mathcal{C}_{k,a,b}$ if $\mathcal{C}_{k,a,b}(t) = \mathcal{C}_{k,a,b}(0)$ for all $t \in (-\epsilon, \epsilon)$.

Now, we consider the variational problem of minimizing the r-area functional $\mathcal{A}_r$ for any variation that preserves the functional $\mathcal{C}_{k,a,b}$. The Jacobi functional associated to the problem is given by

$$\begin{aligned} \mathcal{J}_{r,k,a,b} : (-\epsilon, \epsilon) &\to \mathbb{R} \\ t &\mapsto \mathcal{J}_{r,k,a,b}(t) = \mathcal{A}_r(t) + \rho \mathcal{C}_{k,a,b}(t), \end{aligned}$$

where ρ is a constant to be determined. As an immediate consequence of Lemmas 13.1 and 13.3 we get

$$\mathcal{J}'_{r,k,a,b}(t) = -\int_{\Sigma^n} \{b_r H_{r+1} + \rho\, (ab_k H_{k+1} - b)\}\, f\, d\Sigma_t,$$

where f is the function defined in (13.7). To choose ρ, let

13.3 The Notion of Strong (r, k, a, b)-Stability

$$\overline{\mathcal{H}} = \frac{1}{\mathcal{A}_0(0)} \int_{\Sigma^n} \left\{ \frac{b_r H_{r+1}(0)}{a\, b_k H_{k+1}(0) - b} \right\} d\Sigma^n,$$

be the mean of the function $b_r H_{r+1}(0)/\{a b_k H_{k+1}(0) - b\}$ along Σ^n, where $H_j(0)$ stands for the j-th mean curvature of the immersion $X_0 = \psi$.

We emphasize that, if $b_r H_{r+1}(0)/\{a b_k H_{k+1}(0) - b\}$ is constant, one has

$$\overline{\mathcal{H}} = \frac{b_r H_{r+1}(0)}{a b_k H_{k+1}(0) - b} = \frac{b_r H_{r+1}}{a b_k H_{k+1} - b}. \tag{13.9}$$

This notation will be used here without additional comments. Therefore, if we choose $\rho = -\overline{\mathcal{H}}$, we obtain

$$\mathcal{J}'_{r,k,a,b}(t) = -\int_{\Sigma^n} \left\{ b_r H_{r+1} - \overline{\mathcal{H}}(a b_k H_{k+1} - b) \right\} f\, d\Sigma_t, \tag{13.10}$$

where f is the function defined in (13.7).

Now, arguing as in [62, Proposition 2.7], by (13.10) we have the following result; see also [287, Proposition 3.6].

Proposition 13.1 *Let r and k be entire numbers satisfying the inequality $0 \leq k < r \leq n - 2$, and let $\psi : \Sigma^n \to \mathbb{H}^{n+1}$ be a closed hypersurface. The following facts are equivalent:*

(a) $\psi : \Sigma^n \to \mathbb{H}^{n+1}$ *have higher order mean curvatures H_{k+1} and H_{r+1} verifying*

$$\frac{b_r H_{r+1}}{a b_k H_{k+1} - b} = \mathrm{const.},$$

with $a b_k H_{k+1} - b \neq 0$ on Σ^n, where a and b are real numbers (with at least one non zero) and

$$b_j = (j+1) \binom{n}{j+1}$$

for $j \in \{k, r\}$;

(b) *For any variation $X : (-\epsilon, \epsilon) \times \Sigma^n \to \mathbb{H}^{n+1}$ of $\psi : \Sigma^n \to \mathbb{H}^{n+1}$ that preserves the functional $C_{k,a,b}$, we have $\mathcal{A}'_r(0) = 0$;*

(c) *For any variation $X : (-\epsilon, \epsilon) \times \Sigma^n \to \mathbb{H}^{n+1}$ of $\psi : \Sigma^n \to \mathbb{H}^{n+1}$, we have $\mathcal{J}'_{r,k,a,b}(0) = 0$.*

Motivated by de Lima et al. [148], the main goal in this chapter is to detect closed hypersurfaces $\psi : \Sigma^n \to \mathbb{H}^{n+1}$ that minimize the Jacobi functional $\mathcal{J}_{r,k,a,b}$ for every variation $X : (-\epsilon, \epsilon) \times \Sigma^n \to \mathbb{H}^{n+1}$ of $\psi : \Sigma^n \to \mathbb{H}^{n+1}$.

Next, Proposition 13.1 shows that the critical points of $\mathcal{J}_{r,k,a,b}$ are hypersurfaces $\psi : \Sigma^n \to \mathbb{H}^{n+1}$ such that

$$\frac{b_r H_{r+1}}{a b_k H_{k+1} - b} = \text{const.},$$

with $a b_k H_{k+1} - b \neq 0$ on Σ^n. So, for such a hypersurface, we aim to compute the second variation of $\mathcal{J}_{r,k,a,b}$.

The following notion of stability will be crucial in the sequel.

Definition 13.1 Let r and s be entire numbers satisfying the inequality $0 \leq k < r \leq n-2$, and let $\psi : \Sigma^n \to \mathbb{H}^{n+1}$ be a closed hypersurface whose higher order mean curvatures H_{k+1} and H_{r+1} satisfy

$$\frac{b_r H_{r+1}}{a b_k H_{k+1} - b} = \text{const.},$$

with $a b_k H_{k+1} - b \neq 0$ on Σ^n, where a and b are real numbers (with at least one non zero) and

$$b_j = (j+1) \binom{n}{j+1}$$

for $j \in \{k, r\}$. We say that $\psi : \Sigma^n \to \mathbb{H}^{n+1}$ is *strongly (r, k, a, b)-stable* if $\mathcal{J}''_{r,k,a,b}(0)(f) \geq 0$ for every $f \in C^\infty(\Sigma^n)$.

The formula for the second variation of $\mathcal{J}_{r,k,a,b}$ is a straightforward consequence of Lemmas 13.2 and 13.3.

More precisely, one has the next result.

Proposition 13.2 Let r and k be entire numbers satisfying the inequality $0 \leq k < r \leq n-2$, and let $\psi : \Sigma^n \to \mathbb{H}^{n+1}$ be a closed hypersurface whose higher order mean curvatures H_{k+1} and H_{r+1} satisfy

$$\frac{b_r H_{r+1}}{a b_k H_{k+1} - b} = \text{const.},$$

with $a b_k H_{k+1} - b \neq 0$ on Σ^n, where a and b are real numbers (with at least one non zero) and

$$b_j = (j+1) \binom{n}{j+1}$$

13.3 The Notion of Strong (r, k, a, b)-Stability

for $j \in \{k, r\}$. If $X : (-\epsilon, \epsilon) \times \Sigma^n \to \mathbb{H}^{n+1}$ is a variation of $\psi : \Sigma^n \to \mathbb{H}^{n+1}$, then $\mathcal{J}''_{r,k,a,b}(0)$ is given by

$$\mathcal{J}''_{r,k,a,b}(0)(f) = -(r+1) \int_{\Sigma^n} \left\{ \widetilde{L}_{r,k,a,b}(f) + \left(-\operatorname{tr}(P_r) + \operatorname{tr}(A^2 P_r) \right. \right. \tag{13.11}$$
$$\left. \left. - \Lambda_{r,k,a,b} \left(-\operatorname{tr}(P_k) + \operatorname{tr}(A^2 P_k) \right) \right) f \right\} f d\Sigma^n,$$

for any $f \in C^\infty(\Sigma^n)$, where $\widetilde{L}_{r,k,a,b}$ is the differential operator

$$\begin{aligned}\widetilde{L}_{r,k,a,b} : C^\infty(\Sigma^n) &\to C^\infty(\Sigma^n) \\ f &\mapsto \widetilde{L}_{r,k,a,b}(f) = L_r(f) - \Lambda_{r,k,a,b} L_k(f),\end{aligned} \tag{13.12}$$

and

$$\Lambda_{r,k,a,b} = \frac{a(k+1) b_r H_{r+1}}{a(r+1) b_k H_{k+1} - (r+1) b}. \tag{13.13}$$

Proof By (13.8), (13.9) and (13.10), we obtain

$$\mathcal{J}''_{r,k,a,b}(0) = \frac{\partial}{\partial t} \left(-\int_\Sigma \left\{ b_r H_{r+1} - \overline{\mathcal{H}}(ab_k H_{k+1} - b) \right\} f d\Sigma_t \right) \bigg|_{t=0}$$

$$= -\int_{\Sigma^n} \left(b_r \frac{\partial H_{r+1}}{\partial t} \bigg|_{t=0} - \overline{\mathcal{H}} a b_k \frac{\partial H_{k+1}}{\partial t} \bigg|_{t=0} \right) f dM$$

$$- \int_{\Sigma^n} \underbrace{\left(b_r H_{r+1} - \overline{\mathcal{H}}(ab_k H_{k+1} - b) \right)}_{0} \frac{\partial}{\partial t}(f dM_t) \bigg|_{t=0},$$

$$= -(r+1) \int_{\Sigma^n} \left\{ \left(L_r - \Lambda_{r,k,a,b} L_k \right)(f) + \left(-\operatorname{tr}(P_r) + \operatorname{tr}(A^2 P_r) \right. \right.$$
$$\left. \left. - \Lambda_{r,k,a,b} \left(-\operatorname{tr}(P_k) + \operatorname{tr}(A^2 P_k) \right) \right) f \right\} f d\Sigma^n$$

$$- \int_{\Sigma^n} \left\langle \left(\frac{\partial X}{\partial t} \right)^T \nabla \underbrace{\left(b_r H_{r+1} - \overline{\mathcal{H}}(a b_k H_{k+1} - b) \right)}_{0} \right\rangle f d\Sigma^n.$$

To finish the proof, we observe that the above expression depends only on the hypersurface $\psi : \Sigma^n \to \mathbb{H}^{n+1}$ and on the function $f \in C^\infty(\Sigma^n)$. $\square$

13.4 Strongly (r, k, a, b)-Stable Hypersurfaces in $\mathbb{H}^{n+1}$

A similar result to the statements found in [60, Proposition 5.1] and [287, Proposition 4.1] is established below.

Proposition 13.3 *If $\Lambda_{r,k,a,b}$ is nonpositive, then the geodesic spheres of $\mathbb{H}^{n+1}$ are strongly (r, k, a, b)-stable.*

Proof Let $\psi : \Sigma^n \to \mathbb{H}^{n+1}$ be a geodesic sphere of $\mathbb{H}^{n+1}$. Since $\psi : \Sigma^n \to \mathbb{H}^{n+1}$ is totally umbilical then its principal curvatures are all equal to a certain constant κ. By choosing a suitably normal vector we may assume that $\kappa > 0$. Thus, we have

$$S_j = \binom{n}{j} \kappa^j, \quad H_j = \kappa^j, \quad S_j(A_i) = \binom{n-1}{j} \kappa^j.$$

Moreover, if $e_1, \ldots, e_n$ are the principal directions of Σ^n, then, by (1.6), one has

$$L_j(f) = \sum_{i=1}^{n} \langle \text{Hess}(f)(e_i), P_j(e_i) \rangle = \binom{n-1}{j} \kappa^j \Delta f.$$

for any $j \in \{0, \ldots, n\}$ and every $f \in C^\infty(\Sigma^n)$.

Next, for entire numbers r and k, satisfying the inequality $0 \leq k < r \leq n-2$, and real numbers a and b (with at least one non zero) such that $a(k+1)\binom{n}{k+1}\kappa^{k+1} \neq b$, we have

$$\frac{b_r H_{r+1}}{ab_k H_{k+1} - b} = \frac{b_r \kappa^{r+1}}{ab_k \kappa^{k+1} - b} = \text{const.},$$

where $b_j = (j+1)\binom{n}{j+1}$ for $j \in \{k, r\}$.

Then, by Lemma 1.2 and (13.11) we obtain

$$\mathcal{J}''_{r,k,a,b}(0)(f) = -(r+1) \int_{\Sigma^n} \left\{ \Gamma^{n-1}_{r,k,a,b} \Delta f + \left(-(n-r)S_r + S_1 S_{r+1} \right. \right.$$
$$\left. -(r+2)S_{r+2} \right) f - \Lambda_{r,k,a,b} \left(-(n-k)S_k + S_1 S_{k+1} \right.$$
$$\left. \left. -(k+2)S_{k+2} \right) f \right\} f d\Sigma^n,$$

where

$$\Gamma^{n-1}_{r,k,a,b} = \binom{n-1}{r} \kappa^r - \Lambda_{r,k,a,b} \binom{n-1}{k} \kappa^k \tag{13.14}$$

13.4 Strongly (r, k, a, b)-Stable Hypersurfaces in $\mathbb{H}^{n+1}$

and $\Lambda_{r,k,a,b}$ is defined in (13.13).

Thus, one has

$$\mathcal{J}''_{r,k,a,b}(0)(f) = -(r+1) \int_{\Sigma^n} \left\{ \Gamma^{n-1}_{r,k,a,b} \Delta f + \left(-(n-r)\binom{n}{r} \kappa^r \right. \right.$$

$$+ n \binom{n}{r+1} \kappa^{r+2} - (r+2)\binom{n}{r+2} \kappa^{r+2} \right) f$$

$$- \Lambda_{r,k,a,b} \left(-(n-k)\binom{n}{k} \kappa^k + n \binom{n}{k+1} \kappa^{k+2} \right.$$

$$\left. \left. - (k+2)\binom{n}{k+2} \kappa^{k+2} \right) f \right\} f \, d\Sigma^n.$$

So that

$$\mathcal{J}''_{r,k,a,b}(0)(f) = -(r+1) \int_{\Sigma^n} \left\{ \Gamma^{n-1}_{r,k,a,b} \Delta f \right.$$

$$- \left((n-r)\binom{n}{r} \kappa^r - \Lambda_{r,k,a,b}(n-k)\binom{n}{k} \kappa^k \right) f$$

$$+ \kappa^{r+2} \left(n \binom{n}{r+1} - (r+2)\binom{n}{r+2} \right) f$$

$$\left. - \kappa^{k+2} \Lambda_{r,k,a,b} \left(n \binom{n}{k+1} - (k+2)\binom{n}{k+2} \right) f \right\} f \, d\Sigma^n,$$

that is

$$\mathcal{J}''_{r,k,a,b}(0)(f) = -(r+1) \int_{\Sigma^n} \left\{ \Gamma^{n-1}_{r,k,a,b} \Delta f - n \Gamma^{n-1}_{r,k,a,b} f + n \Gamma^{n-1}_{r,k,a,b} \kappa^2 f \right\} f \, d\Sigma^n$$

$$= (r+1) \Gamma^{n-1}_{r,k,a,b} \int_{\Sigma^n} \left\{ -f \Delta f - n(-1+\kappa^2) f^2 \right\} d\Sigma^n, \quad (13.15)$$

for every $f \in C^\infty(\Sigma^n)$.

Hence, if λ_1 denote the first eigenvalue of the Laplacian of Σ^n, by (13.14) and (13.15), thanks to the assumption of function $\Lambda_{r,k,a,b}$, we get

$$\mathcal{J}''_{r,k,a,b}(0)(f) \geq (r+1) \Gamma^{n-1}_{r,k,a,b} \int_{\Sigma^n} \left\{ \lambda_1 - n(-1+\kappa^2) \right\} f^2 d\Sigma^n = 0,$$

for any $f \in C^\infty(\Sigma^n)$, where the last equality was obtained by observing that Σ^n is isometric to a n-dimensional Euclidean sphere with constant sectional curvature equal to $\kappa^2 - 1$; hence $\lambda_1 = n(\kappa^2 - 1)$. Therefore, we conclude that $\psi : \Sigma^n \to \mathbb{H}^{n+1}$ is strongly (r, k, a, b)-stable. $\square$

The next result, whose proof is a consequence of a suitable formula due to Barros and Sousa [66], will be crucial in the sequel.

Lemma 13.4 *Let r and k be entire numbers satisfying the inequality $0 \leq k < r \leq n-2$, and let $\psi : \Sigma^n \to (0, +\infty) \times_{\sinh \tau} \mathbb{S}^n$ be a hypersurface whose higher order mean curvatures H_{k+1} and H_{r+1} satisfy*

$$\frac{b_r H_{r+1}}{ab_k H_{k+1} - b} = \text{const.},$$

with $ab_k H_{k+1} - b \neq 0$ on Σ^n, where a and b are real numbers (with at least one non zero) and

$$b_j = (j+1)\binom{n}{j+1}$$

for $j \in \{k, r\}$. If N is the unit normal vector field defined globally on Σ^n and $\eta = \langle (\sinh \tau)\frac{\partial}{\partial \tau}, N \rangle$, then

$$\widetilde{L}_{r,k,a,b}(\eta) = -\left\{\left(-\mathrm{tr}(P_r) + \mathrm{tr}(A^2 P_r)\right) - \Lambda_{r,k,a,b}\left(-\mathrm{tr}(P_k) + \mathrm{tr}(A^2 P_k)\right)\right\}\eta$$
$$- \{b_r H_r - \Lambda_{r,k,a,b}\, b_k H_k\}\left\langle \frac{\partial}{\partial \tau}, N \right\rangle \sinh \tau$$
$$- \{b_r H_{r+1} - \Lambda_{r,k,a,b}\, b_k H_{k+1}\} \cosh \tau. \qquad (13.16)$$

where $\widetilde{L}_{r,k,a,b}$ is the differential operator defined in (13.12) and $\Lambda_{r,k,a,b}$ is defined in (13.13).

Proof By Barros and Sousa [66, Theorem 2], one has

$$L_j(\eta) = -\left\{\mathrm{tr}(A^2 P_j) - \mathrm{tr}(P_j)\right\}\eta$$
$$- b_j H_j N(\cosh \tau) - b_j H_{j+1} \cosh \tau - \frac{b_j}{j+1}\left\langle (\sinh \tau)\frac{\partial}{\partial \tau}, \nabla H_{j+1} \right\rangle$$

for $j \in \{k, r\}$. Thus, it follows that

$$\widetilde{L}_{r,k,a,b}(\eta) = L_r(\eta) - \Lambda_{r,k,a,b} L_k(\eta)$$
$$= -\left\{-\mathrm{tr}(P_r) + \mathrm{tr}(A^2 P_r)\right\}\eta$$
$$- b_r H_r N(\cosh \tau) - b_r H_{r+1} \cosh \tau - \frac{b_r}{r+1}\left\langle (\sinh \tau)\frac{\partial}{\partial \tau}, \nabla H_{r+1} \right\rangle$$

13.4 Strongly (r, k, a, b)-Stable Hypersurfaces in $\mathbb{H}^{n+1}$

$$-\Lambda_{r,k,a,b}\left(-\left\{-\text{tr}(P_k) + \text{tr}(A^2 P_k)\right\}\eta - b_k H_k N(\cosh\tau)\right.$$

$$-b_k H_{k+1}\cosh\tau - \frac{b_k}{k+1}\left\langle(\sinh\tau)\frac{\partial}{\partial\tau}, \nabla H_{k+1}\right\rangle\right)$$

$$= -\left\{\left(-\text{tr}(P_r) + \text{tr}(A^2 P_r)\right) - \Lambda_{r,k,a,b}\left(-\text{tr}(P_k) + \text{tr}(A^2 P_k)\right)\right\}\eta$$

$$- \left\{b_r H_r - \Lambda_{r,k,a,b}\, b_k H_k\right\} N(\cosh\tau)$$

$$- \left\{b_r H_{r+1} - \Lambda_{r,k,a,b}\, b_k H_{k+1}\right\}\cosh\tau$$

$$-\underbrace{\left\langle(\sinh\tau)\frac{\partial}{\partial\tau}, \nabla\left(-\frac{b_r}{r+1}H_{r+1} + \Lambda_{r,k,a,b}\frac{b_k}{k+1}H_{k+1}\right)\right\rangle}_{0}. \qquad (13.17)$$

Now, observing that

$$\overline{\nabla}\cosh\tau = \left\langle\overline{\nabla}\cosh\tau, \frac{\partial}{\partial\tau}\right\rangle\frac{\partial}{\partial\tau} = (\cosh\tau)'\frac{\partial}{\partial\tau} = (\sinh\tau)\frac{\partial}{\partial\tau},$$

we have

$$N(\cosh\tau) = \langle\overline{\nabla}\cosh\tau, N\rangle = \left\langle\frac{\partial}{\partial\tau}, N\right\rangle\sinh\tau. \qquad (13.18)$$

Finally, by (13.18) and (13.17) we obtain (13.16). $\square$

Now, we recall that the $(n+1)$-dimensional de Sitter space $\mathbb{S}_1^{n+1}$ is defined as the following hyperquadric of $\mathbb{L}^{n+2}$ given by

$$\mathbb{S}_1^{n+1} = \left\{p \in \mathbb{L}^{n+2} : \langle p, p\rangle = 1\right\}.$$

The metric induced by $\langle\cdot,\cdot\rangle$ makes $\mathbb{S}_1^{n+1}$ a Lorentz manifold with constant sectional curvature equal to one.

Now, taking again a unit timelike vector $\mathbf{a} \in \mathbb{L}^{n+2}$, we have the vector field

$$K(p) = \mathbf{a} - \langle p, \mathbf{a}\rangle p, \quad p \in \mathbb{S}_1^{n+1},$$

is a conformal and closed timelike vector field globally defined in $\mathbb{S}_1^{n+1}$.

By Montiel [243, Proposition 1], we see that such a vector field K foliates $\mathbb{S}_1^{n+1}$ by means of totally umbilical round spheres, which are described as the following level sets

$$\mathcal{L}_\varepsilon^n = \left\{p \in \mathbb{S}_1^{n+1} : \langle p, \mathbf{a}\rangle = \varepsilon\right\}, \quad \varepsilon \in \mathbb{R}.$$

In particular, the level set

$$\mathcal{L}_0^n = \left\{ p \in \mathbb{S}_1^{n+1} : \langle p, \mathbf{a} \rangle = 0 \right\}$$

defines a round sphere of radius one which is a totally geodesic hypersurface in $\mathbb{S}_1^{n+1}$.

According to [13], we will refer to that sphere as the *equator* of $\mathbb{S}_1^{n+1}$ determined by $\mathbf{a}$. This equator divides $\mathbb{S}_1^{n+1}$ into two connected components, the *chronological future* which is given by

$$\left\{ p \in \mathbb{S}_1^{n+1} : \langle \mathbf{a}, p \rangle < 0 \right\}, \tag{13.19}$$

and the *chronological past*, given by

$$\left\{ p \in \mathbb{S}_1^{n+1} : \langle \mathbf{a}, p \rangle > 0 \right\}.$$

On the other hand, we observe that the unit normal vector field N of a hypersurface $\psi : \Sigma^n \to \mathbb{H}^{n+1}$ can be regarded as a map $N : \Sigma^n \to \mathbb{S}_1^{n+1}$, called *Gauss mapping* of immersion ψ. In this setting, the image $N(\Sigma^n)$ will be called the *Gauss image* of $\psi : \Sigma^n \to \mathbb{H}^{n+1}$.

Remark 13.1 By fixing a unit timelike vector $\mathbf{a} \in \mathbb{L}^{n+2}$ and considering in $\mathbb{H}^{n+1}$ as well as in $\mathbb{S}_1^{n+1}$ the foliations previously described, we have that the Gauss mapping $N_\varrho : L_\varrho^n \to \mathbb{S}_1^{n+1}$ of a geodesic sphere L_ϱ^n of the hyperbolic space $\mathbb{H}^{n+1}$ is given by (13.3). Consequently, we have that $N_\varrho(L_\varrho^n) \subset \mathcal{L}_\varepsilon^n$ for $\varepsilon = -\sqrt{\varrho^2 - 1} < 0$. Therefore, we conclude that the Gauss image of a geodesic sphere of $\mathbb{H}^{n+1}$ is contained in the chronological future (or past) of the equator of $\mathbb{S}_1^{n+1}$ determined by $\mathbf{a}$.

We are in position now to state and prove our main result.

Theorem 13.1 *Let r and k be entire numbers satisfying the inequality $0 \leq k < r \leq n-2$, let a and b be real numbers, with $b \neq 0$, and let $\psi : \Sigma^n \to \mathbb{H}^{n+1}$ be a strongly (r, k, a, b)-stable closed hypersurface. Suppose that $\Lambda_{r,k,a,b}$ is nonpositive and the higher order mean curvatures H_{k+1} and H_{r+1} of $\psi : \Sigma^n \to \mathbb{H}^{n+1}$ satisfy*

$$H_{j+1} \geq H_j, \quad j \in \{k, r\}. \tag{13.20}$$

If the Gauss image of $\psi : \Sigma^n \to \mathbb{H}^{n+1}$ is contained in the chronological future (or past) of an equator of $\mathbb{S}_1^{n+1}$ then $\psi(\Sigma^n)$ is a geodesic sphere of $\mathbb{H}^{n+1}$.

Proof We first claim that $H_j > 0$ everywhere on Σ^n, for all $j \in \{0, \ldots, r+1\}$. Indeed, as $\psi : \Sigma^n \to \mathbb{H}^{n+1}$, we may assume that the orientation N of $\psi : \Sigma^n \to \mathbb{H}^{n+1}$ is considered

13.4 Strongly (r, k, a, b)-Stable Hypersurfaces in $\mathbb{H}^{n+1}$

such that its principal curvatures are positive at a point $p_0 \in \Sigma^n$. Moreover, the strong (r, k, a, b)-stability of $\psi : \Sigma^n \to \mathbb{H}^{n+1}$ implies that the quotient $b_r H_{r+1}/\{a b_k H_{k+1} - b\}$ is constant, with $a b_k H_{k+1} - b \neq 0$ on Σ^n. Set

$$\beta = \frac{(r+1) S_{r+1}}{a b_k H_{k+1} - b}(p_0) \equiv \frac{b_r H_{r+1}}{a b_k H_{k+1} - b}. \tag{13.21}$$

If $a b_k H_{k+1} - b < 0$ then $\beta < 0$, so, by (13.21), $H_{r+1} = b_r^{-1} \beta \{a b_k H_{k+1} - b\} > 0$ on Σ^n. On the other hand, if $a b_k H_{k+1} - b > 0$, we have $\beta > 0$, so, again by (13.21), $H_{r+1} = b_r^{-1} \beta \{a b_k H_{k+1} - b\} > 0$ on Σ^n. Anyway $H_{r+1} > 0$ on Σ^n. A classical inequalities due to Gärding [190] ensures that our assertion follows.

Now, without loss of generality, let us suppose that the Gauss image $N(\Sigma^n)$ of the hypersurface $\psi : \Sigma^n \to \mathbb{H}^{n+1}$ is contained in the chronological future of the equator of $\mathbb{S}_1^{n+1}$ determined by a unit timelike vector $\mathbf{a} \in \mathbb{L}^{n+2}$. For such vector $\mathbf{a}$, let us consider the warped product given in (13.4), which identifies the hyperbolic space $\mathbb{H}^{n+1}$ (minus a point) with $(0, +\infty) \times_{\sinh \tau} \mathbb{S}^n$.

In this setting, we consider the *normal angle* θ of $\psi : \Sigma^n \to \mathbb{H}^{n+1}$, which is the smooth function $\theta : \Sigma^n \to [0, \pi]$ given by

$$\cos \theta(p) = -\left\langle \Phi_* N, \frac{\partial}{\partial \tau}\right\rangle_{(\Phi \circ \psi)(p)}, \tag{13.22}$$

where Φ stands for an isometry between the hyperquadric $\mathbb{S}_1^{n+1}$ and $(0, +\infty) \times_{\sinh \tau} \mathbb{S}^n$. By (13.22) and 13.6, we have

$$\cos \theta(p) = -\left\langle \Phi_* N((\Phi \circ \psi)(p)), \frac{W((\Phi \circ \psi)(p))}{|W((\Phi \circ \psi)(p))|}\right\rangle$$

$$= -\frac{1}{|\Phi_*^{-1} W(\psi(p))|} \left\langle N(\psi(p)), \Phi_*^{-1} W(\psi(p))\right\rangle \tag{13.23}$$

for any $p \in \Sigma^n$.

Since $\Phi_*^{-1} W = V$, bt (13.23) we get

$$\cos \theta(p) = -\frac{1}{|V(\psi(p))|} \left\langle N(\psi(p)), \mathbf{a} + \langle \mathbf{a}, \psi(p)\rangle \psi(p)\right\rangle$$

$$= -\frac{1}{|V(\psi(p))|} \left\langle N(x(p)), \mathbf{a}\right\rangle, \tag{13.24}$$

where V is the closed conformal vector field given by (13.2).

Hence, since the Gauss image $N(\Sigma^n)$ is contained in the chronological future of $\mathbb{S}_1^{n+1}$ determined by $\mathbf{a}$, by (13.19) and (13.24)) we conclude that

$$0 < \cos \theta \leq 1. \tag{13.25}$$

On the other hand, since $\psi : \Sigma^n \to \mathbb{H}^{n+1}$ is strongly (r, k, a, b)-stable, by Definition 13.1 and Eq. (13.11) we obtain

$$0 \leq \mathcal{J}''_{r,k,a,b}(0)(f) = -(r+1) \int_{\Sigma^n} \left\{ \tilde{L}_{r,k,a,b}(f) + \left(-\mathrm{tr}(P_r) + \mathrm{tr}(A^2 P_r) \right. \right.$$
$$\left. \left. - \Lambda_{r,k,a,b} \left(-\mathrm{tr}(P_k) + \mathrm{tr}(A^2 P_k) \right) \right) f \right\} f d\Sigma^n,$$

for every $f \in C^\infty(\Sigma^n)$, where $\tilde{L}_{r,k,a,b}$ is the differential operator defined in (13.12) and $\Lambda_{r,k,a,b}$ is defined in (13.13). In particular, taking

$$f = \eta \circ \Phi^{-1} = \left\langle (\sinh \tau) \frac{\partial}{\partial \tau}, \Phi_* N \right\rangle = -\sinh \tau \cos \theta \tag{13.26}$$

and considering $N = \Phi_* N$ and $H_j = H_j \circ \Phi^{-1}$ for $j \in \{k, k+1, r, r+1\}$, by Lemma 13.4 we obtain

$$0 \leq (r+1) \int_{\Phi(\psi(\Sigma^n))} \left\{ \left(b_r H_r - \Lambda_{r,k,a,b} b_k H_k \right) \sinh \tau \cos \theta \right.$$
$$\left. - \left(b_r H_{r+1} - \Lambda_{r,k,a,b} b_k H_{k+1} \right) \cosh \tau \right\} \sinh \tau \cos \theta \, d\Phi(\Sigma)$$
$$\leq (r+1) \int_{\Phi(\psi(\Sigma^n))} \left\{ \left(b_r H_r - \Lambda_{r,k,a,b} b_k H_k \right) \cos \theta \right. \tag{13.27}$$
$$\left. - \left(b_r H_{r+1} - \Lambda_{r,k,a,b} b_k H_{k+1} \right) \right\} \cosh \tau \sinh \tau \cos \theta \, d\Phi(\Sigma),$$

where

$$b_j = (n-j) \binom{n}{j},$$

with $j \in \{k, r\}$. Now, since $\Lambda_{r,k,a,b} \leq 0$, by (13.20) we also have that

$$b_k (H_{k+1} - H_k) \Lambda_{r,k,a,b} - b_r (H_{r+1} - H_r) \leq 0.$$

Equivalently,

$$b_r H_{r+1} - \Lambda_{r,k,a,b} b_k H_{k+1} \geq b_r H_r - \Lambda_{r,k,a,b} b_k H_k. \tag{13.28}$$

By (13.28) and (13.27) we have

$$0 \leq (r+1) \int_{\Phi(\psi(\Sigma^n))} (b_r H_r - \Lambda_{r,k,a,b} b_k H_k)(\cos \theta - 1) \cosh \tau \sinh \tau \cos \theta d\Phi(\Sigma) \leq 0,$$

13.4 Strongly (r, k, a, b)-Stable Hypersurfaces in $\mathbb{H}^{n+1}$

thanks to (13.25) and bearing in mind that $\Lambda_{r,k,a,b} \leq 0$, $H_r > 0$, $H_k > 0$ and $\tau > 0$. Hence, $\cos\theta = 1$ and, consequently, there exists $\tau_0 \in (0, +\infty)$ such that $(\Phi \circ \psi)(\Sigma^n) = \{\tau_0\} \times \mathbb{S}^n$. □

Remark 13.2 We would like to point out that, taking into account that the higher order mean curvatures H_j^τ, $j \in \{0, \ldots, n\}$, of each slice $\Sigma_\tau^n = \{\tau\} \times \mathbb{S}^n$ verify

$$H_{r+1}^\tau > H_r^\tau \geq H_{k+1}^\tau > H_k^\tau > 1$$

(as defined in (13.5)) for any entire numbers r and k satisfying the inequality $0 \leq k < r \leq n-2$, our restriction on the values of the higher order mean curvatures H_{k+1} and H_{r+1} in Theorem 13.1 constitutes a mild hypothesis in the sense that, in the light of Proposition 13.3 and Remark 13.1, it is natural to detect geodesic spheres of $\mathbb{H}^{n+1}$.

Finally, let us recall that a hypersurface $\psi : \Sigma^n \to \mathbb{H}^{n+1}$ is *linear Weingarten* when the mean curvature H and normalized scalar curvature R satisfy

$$\delta_0 H + \delta_1 R = \delta_2,$$

for some constants $\delta_0, \delta_1, \delta_2 \in \mathbb{R}$.

Now, the Gauss equation of $\psi : \Sigma^n \to \mathbb{H}^{n+1}$ yields $R = -1 + H^2$. Hence, taking $r = 1$ and $k = 0$ in Theorem 13.1, the following result holds.

Corollary 13.1 *Let a and b be real numbers, with $b \neq 0$, and let $\psi : \Sigma^n \to \mathbb{H}^{n+1}$ be a strongly $(1, 0, a, b)$-stable closed linear Weingarten hypersurface such that*

$$n(n-1)(R+1) - naH\delta = -b\delta,$$

where $\delta \in \mathbb{R} \setminus \{0\}$. Suppose that $\Lambda_{1,0,a,b}$ is nonpositive and $1 \leq H \leq R+1$. If the Gauss image of $\psi : \Sigma^n \to \mathbb{H}^{n+1}$ is contained in the chronological future (or past) of an equator of $\mathbb{S}_1^{n+1}$ then Σ^n is a sphere and ψ is its inclusion as a geodesic sphere of the hyperbolic space $\mathbb{H}^{n+1}$.

Stable Closed Spacelike Hypersurfaces in the de Sitter Space

14.1 Introduction

It is well-known that immersions with constant mean curvature in real space forms are solutions for the variational problem of minimizing the area functional while keeping null balance of volume. A local solution for this variational problem is said to be stable. This concept was introduced and studied by Barbosa and do Carmo [62], and Barbosa et al. [64], where they proved that spheres are the only stable critical points of the area functional for volume-preserving variations.

In the Lorentzian setting, Barbosa and Oliker [63] obtained an analogous result, proving that constant mean curvature space-like hypersurfaces in Lorentz manifolds are also critical points of the area functional for any variation that keep the volume constant. They also computed the second variation formula and showed, for the de Sitter space $\mathbb{S}_1^{n+1}$, that spheres maximize the area functional for volume-preserving variations.

Barros et al. [67] studied the problem of strong stability (that is, stability with respect to not necessarily volume-preserving variations) for spacelike hypersurfaces with constant mean curvature in a generalized Robertson Walker (GRW) spacetime of constant sectional curvature, giving a characterization of maximal spacelike hypersurfaces and spacelike slices.

In [93] Camargo, Caminha, da Silva and de Lima obtained an extension of the main result of [67] concerning spacelike hypersurfaces with constant r-th mean curvature into a GRW spacetime, giving a characterization of r-maximal spacelike hypersurfaces and spacelike slices. In particular, they treated the case in which the ambient GRW spacetime is the de Sitter space $\mathbb{S}_1^{n+1}$.

Later on, Zhang [300] considered the stability of closed linear Weingarten spacelike hypersurfaces in $\mathbb{S}_1^{n+1}$. This concept of stability arises from considering the variational problem of minimizing a suitable linear combination of the second area for volume-preserving variations. More precisely, he proved that a closed orientable hypersurface $\psi : \Sigma^n \to \mathbb{S}_1^{n+1}$ with positive mean curvature H and second mean curvature H_2 satisfying

$$(n-1)H_2 + aH = b$$

for some constants $a > 0$ and $b \in \mathbb{R}$, is stable (in the sense defined along the cited paper) if and only if it is totally umbilical.

In this chapter, motivated by these works, we introduce a new concept of stability for closed spacelike hypersurface immersed in the de Sitter space $\mathbb{S}_1^{n+1}$. Namely, we define the notion of strong (r, s, a, b)-stability concerning closed spacelike hypersurfaces $\psi : \Sigma^n \to \mathbb{S}_1^{n+1}$ immersed into $\mathbb{S}_1^{n+1}$, which comes from the study of the variational problem of maximizing a certain Jacobi functional.

Under a suitable constraint on a constant that appears in the computation of the second variation of this functional, we use an appropriated warped product model of $\mathbb{S}_1^{n+1}$ in order to prove that a closed spacelike hypersurface $\psi : \Sigma^n \to \mathbb{S}_1^{n+1}$ contained in a chronological future (or past) of $\mathbb{S}_1^{n+1}$, with positive $(s+1)$-th curvature and such that the mean curvature H satisfies the inequality $H \le 1$, must be a totally umbilical round sphere.

The results presented in this chapter are mainly based on the paper [289].

14.2 A Warped Product Model for the de Sitter Space

As in Sect. 13.4, let us denote by

$$\mathbb{S}_1^{n+1} = \left\{ p \in \mathbb{L}^{n+2} : \langle p, p \rangle = 1 \right\},$$

the $(n+1)$-dimensional de Sitter space. We have that $\mathbb{S}_1^{n+1}$ can be described as the spacetime

$$\mathbb{S}_1^{n+1} = -\mathbb{R} \times_{\cosh t} \mathbb{S}^n, \quad t \in \mathbb{R},$$

where $\mathbb{S}^n$ is the n-dimensional Euclidean unitary sphere. We observe that there are many possible choices for the unitary time-like vector $\mathbf{a} \in \mathbb{L}^{n+2}$ and, hence, a lot of ways to describe $\mathbb{S}_1^{n+1}$ as a warped product; see, for instance, [243].

We notice that in this model the totally umbilical round spheres

$$\mathcal{L}_\varepsilon^n = \left\{ p \in \mathbb{S}_1^{n+1} : \langle p, \mathbf{a} \rangle = \varepsilon \right\}, \quad \varepsilon \in \mathbb{R}.$$

of $\mathbb{S}_1^{n+1}$ are the slices

$$M_{t_0}^n = \{t_0\} \times \mathbb{S}^n, \quad t_0 \in \mathbb{R}.$$

When oriented by the unitary normal vector field ∂_t, $M_{t_0}^n$ has constant k-th mean curvature

$$H_k^{t_0} = (\tanh t_0)^k.$$

Moreover, the equator of $\mathbb{S}_1^{n+1}$ is the slice

$$M_0^n = \{0\} \times \mathbb{S}^n;$$

consequently, $(\cosh t)' = \sinh t$ vanishes only on this slice.

Finally, the vector field

$$K = \cosh t \, \partial_t$$

is conformal Killing, timelike and closed in $\mathbb{S}_1^{n+1}$, with associated conformal factor $\psi = \sinh t$.

14.3 Description of the Variational Problem

Let $\psi : \Sigma^n \to \mathbb{S}_1^{n+1}$ be an isometric immersion of a closed spacelike hypersurface M^n. In accordance with last chapter, here, closed also means compact without boundary. In that case, since $\mathbb{S}_1^{n+1}$ is time-orientable, Σ^n is orientable. Consequently, one can choose a globally defined unitary normal timelike vector field N on Σ^n having the same time orientation of $\mathbb{S}_1^{n+1}$. In such a case we say that N is the future-pointing Gauss map of $\psi : \Sigma^n \to \mathbb{S}_1^{n+1}$.

A variation of $\psi : \Sigma^n \to \mathbb{S}_1^{n+1}$ is a smooth map $X : \Sigma^n \times (-\epsilon, \epsilon) \to \mathbb{S}_1^{n+1}$ such that, for each $t \in (-\epsilon, \epsilon)$, the map $X_t : M^n \to \mathbb{S}_1^{n+1}$ given by $X_t(p) = X(p, t)$ is a spacelike immersion such that $X_0 = \psi$. Here and in the sequel $d\Sigma_t$ denotes the volume element of the metric induced on Σ^n by X_t and N_t is the future-pointing Gauss map along X_t.

The variational field associated to the variation $X : \Sigma^n \times (-\epsilon, \epsilon) \to \mathbb{S}_1^{n+1}$ is the vector field $\left.\dfrac{\partial X}{\partial t}\right|_{t=0}$. If we set

$$f = -\left\langle \frac{\partial X}{\partial t}, N_t \right\rangle, \qquad (14.1)$$

then

$$\frac{\partial X}{\partial t} = f N_t + \left(\frac{\partial X}{\partial t}\right)^\top,$$

where $(\cdot)^\top$ stands for tangential component.

The balance of volume of X is the functional

$$\mathcal{V} : (-\epsilon, \epsilon) \to \mathbb{R}$$
$$t \mapsto \mathcal{V}(t) = \int_{\Sigma^n \times [0,t]} X^*(dV),$$

where dV is the volume element of $\mathbb{S}_1^{n+1}$.

The following classical result can be found, for instance, in [296].

Lemma 14.1 *Let $\psi : \Sigma^n \to \mathbb{S}_1^{n+1}$ a closed spacelike hypersurface. If $X : \Sigma^n \times (-\epsilon, \epsilon) \to \mathbb{S}_1^{n+1}$ is a variation of $\psi : \Sigma^n \to \mathbb{S}_1^{n+1}$, then*

$$\mathcal{V}'(t) = \int_{\Sigma^n} f \, d\Sigma_t,$$

where f is the function defined in (14.1).

Define now the k-area functional associated to the variation $X : \Sigma^n \times (-\epsilon, \epsilon) \to \mathbb{S}_1^{n+1}$ as the functional given by

$$\mathcal{A}_k : (-\epsilon, \epsilon) \to \mathbb{R}$$
$$t \mapsto \mathcal{A}_k(t) = \int_{\Sigma^n} F_k(S_1, \ldots, S_k) d\Sigma_t,$$

where $S_k = S_k(t)$ and F_k is recursively defined by setting $F_0 = 1$, $F_1 = -S_1$ and

$$F_k = (-1)^r S_k + \frac{n-k+1}{k-1} F_{k-2},$$

for every $2 \leq k \leq n-1$.

The following result follows by Camargo et al. [93, Lemma 2.2].

Lemma 14.2 *Let $\psi : \Sigma^n \to \mathbb{S}_1^{n+1}$ a spacelike hypersurface. If $X : \Sigma^n \times (-\epsilon, \epsilon) \to \mathbb{S}^{n+1}$ is a variation of $\psi : \Sigma^n \to \mathbb{S}_1^{n+1}$, then*

$$\frac{\partial H_{k+1}}{\partial t} = \frac{k+1}{b_k} \left\{ L_k(f) + \mathrm{tr}(P_k)f - \mathrm{tr}(A^2 P_k)f \right\} + \left\langle \left(\frac{\partial X}{\partial t}\right)^\top, \nabla H_{k+1} \right\rangle, \quad (14.2)$$

where $b_k = (k+1)\binom{n}{k+1}$ and f is the function defined in (14.1).

The previous lemma allows us to compute the first variation of the r-area functional; see [93, Proposition 2.3].

14.3 Description of the Variational Problem

Lemma 14.3 *Let* $\psi : \Sigma^n \to \mathbb{S}_1^{n+1}$ *a closed spacelike hypersurface. If* $X : \Sigma^n \times (-\epsilon, \epsilon) \to \mathbb{S}_1^{n+1}$ *is a variation of* $\psi : \Sigma^n \to \mathbb{S}_1^{n+1}$, *then*

$$\mathcal{A}_k'(t) = \int_{\Sigma^n} (b_k H_{k+1} + c_k) f \, d\Sigma_t, \tag{14.3}$$

where f is the function defined in (14.1), $c_0 = 0$, $c_1 = 1$ *and*

$$c_k = -\frac{(n-k+1)}{k-1} c_{k-2},$$

if $k \geq 2$.

Let r and s be entire numbers satisfying the inequalities $0 \leq r < s \leq n-2$, and consider real numbers a and b (at least one nonzero). We define the following functional

$$\begin{aligned} C_{r,a,b} : (-\epsilon, \epsilon) &\to \mathbb{R} \\ t &\mapsto C_{r,a,b}(t) = a\mathcal{A}_r(t) + b\mathcal{V}(t), \end{aligned} \tag{14.4}$$

and we say that the variation $X : \Sigma^n \times (-\epsilon, \epsilon) \to \mathbb{S}_1^{n+1}$ preserves the linear combination $C_{r,a,b}$ if $C_{r,a,b}(t) = C_{r,a,b}(0)$ for every $t \in (-\epsilon, \epsilon)$.

Now, we consider the variational problem of maximizing the s-area functional $\mathcal{A}_s$ for any variation which preserves the functional $C_{r,a,b}$. The Jacobi functional associated to this problem is given by

$$\begin{aligned} \mathcal{J}_{r,s,a,b} : (-\epsilon, \epsilon) &\to \mathbb{R} \\ t &\mapsto \mathcal{J}_{r,s,a,b}(t) = \mathcal{A}_s(t) + \gamma \, C_{r,a,b}(t), \end{aligned}$$

where γ is a constant to be chosen later. As an immediate consequence of Lemmas 14.1 and 14.3, we get

$$\mathcal{J}'_{r,s,a,b}(t) = \int_{\Sigma^n} \{b_s H_{s+1} + c_s + \gamma \left(a(b_r H_{r+1} + c_r) + b\right)\} f \, d\Sigma_t,$$

where f is the function defined in (14.1). In order to make a convenient choice of γ, let us suppose that $a(b_r H_{r+1} + c_r) + b$ never vanishes on Σ^n, and let

$$\overline{\mathcal{H}} = \frac{1}{\mathcal{A}_0(0)} \int_{\Sigma^n} \frac{b_s H_{s+1}(0) + c_s}{a(b_r H_{r+1}(0) + c_r) + b} d\Sigma^n$$

be the integral mean of the function $(b_s H_{s+1}(0) + c_s)/[a(b_r H_{r+1}(0) + c_r) + b]$ over Σ^n. When the function $(b_s H_{s+1}(0) + c_s)/[a(b_r H_{r+1}(0) + c_r) + b]$ is constant, one has

$$\overline{\mathcal{H}} = \frac{b_s H_{s+1}(0) + c_s}{a(b_r H_{r+1}(0) + c_r) + b} = \frac{b_s H_{s+1} + c_s}{a(b_r H_{r+1} + c_r) + b}, \tag{14.5}$$

This notation will be used in what follows without further comments. Thus, if we choose $\gamma = -\overline{\mathcal{H}}$, we arrive at

$$\mathcal{J}'_{r,s,a,b}(t) = \int_{\Sigma^n} \left\{ b_s H_{s+1} + c_s - \overline{\mathcal{H}}(a(b_r H_{r+1} + c_r) + b) \right\} f d\Sigma_t, \tag{14.6}$$

where f is the function defined in (14.1).

Let r and s be two entire numbers satisfying the inequalities $0 \le r < s \le n-2$ and a and b two real numbers, at least one nonzero. Let also $\psi : \Sigma^n \to \mathbb{S}_1^{n+1}$ be a closed hypersurface such that the function $a(b_r H_{r+1} + c_r) + b$ never vanishes on Σ^n.

Arguing as in [62, Proposition 2.7], by (14.6), the next result holds.

Proposition 14.1 *The following statements are equivalent:*

(a) $\psi : \Sigma^n \to \mathbb{S}_1^{n+1}$ *have higher order mean curvatures* H_{r+1} *and* H_{s+1} *verifying*

$$\frac{b_s H_{s+1} + c_s}{a(b_r H_{r+1} + c_r) + b} = \text{const.},$$

where

$$b_j = (j+1) \binom{n}{j+1}$$

for $j \in \{r, s\}$ *and the constant* c_r *is defined recursively by* $c_0 = 0$, $c_1 = 1$ *and*

$$c_i = -\frac{(n-i+1)}{i-1} c_{i-2},$$

if $i \ge 2$.
(b) *For any variation* $X : \Sigma^n \times (-\epsilon, \epsilon) \to \mathbb{S}_1^{n+1}$ *of* $\psi : \Sigma^n \to \mathbb{S}_1^{n+1}$ *that preserves the functional* $\mathcal{C}_{r,a,b}$, *we have* $\mathcal{A}'_s(0) = 0$.
(c) *For any variation* $X : \Sigma^n \times (-\epsilon, \epsilon) \to \mathbb{S}_1^{n+1}$ *of* $\psi : \Sigma^n \to \mathbb{S}_1^{n+1}$, *we have* $\mathcal{J}'_{r,s,a,b}(0) = 0$.

Motivated by Barros et al. [67] and da Silva et al. [129], the main goal now is to detect closed spacelike hypersurfaces $\psi : \Sigma^n \to \mathbb{S}_1^{n+1}$ which maximize the Jacobi functional $\mathcal{J}_{r,s,a,b}$ for all the variations $X : \Sigma^n \times (-\epsilon, \epsilon) \to \mathbb{S}_1^{n+1}$ of $\psi : \Sigma^n \to \mathbb{S}_1^{n+1}$. Proposition 14.1 shows that the critical points of $\mathcal{J}_{r,s,a,b}$ are closed spacelike hypersurfaces $\psi : \Sigma^n \to \mathbb{S}_1^{n+1}$ whose higher order mean curvatures H_{r+1} and H_{s+1} verify

14.3 Description of the Variational Problem

$$\frac{b_s H_{s+1}}{a(b_r H_{r+1} + c_r) + b} = \text{const.},$$

with $a(b_r H_{r+1} + c_r) + b \neq 0$ on Σ^n.

The following notion of stability can be given.

Definition 14.1 Let $\psi : \Sigma^n \to \mathbb{S}_1^{n+1}$ be a closed spacelike hypersurface whose higher order mean curvatures H_{r+1} and H_{s+1} satisfy

$$\frac{b_s H_{s+1} + c_s}{a(b_r H_{r+1} + c_r) + b} = \text{const.},$$

with $a(b_r H_{r+1} + c_r) + b \neq 0$ on Σ^n, where $0 \leq r < s \leq n - 2$ are natural numbers and a and b are real numbers (at least one nonzero) and the constant c_r is defined recursively by $c_0 = 0$, $c_1 = 1$ and

$$c_i = -\frac{(n - i + 1)}{i - 1} c_{i-2}$$

if $i \geq 2$. We say that $\psi : \Sigma^n \to \mathbb{S}_1^{n+1}$ is strongly (r, s, a, b)-stable provided that $\mathcal{J}''_{r,s,a,b}(0)(f) \leq 0$ for every $f \in C^\infty(\Sigma^n)$.

Now, we want to compute the second variation of $\mathcal{J}_{r,s,a,b}$ for the above class of hypersurfaces. As a direct consequence of Lemmas 14.2 and 14.3, we get the following result.

Proposition 14.2 Let $\psi : \Sigma^n \to \mathbb{S}_1^{n+1}$ a closed spacelike hypersurface whose higher order mean curvatures H_{s+1} and H_{r+1} satisfy

$$\frac{b_r H_{r+1} + c_r}{a(b_s H_{s+1} + c_s) + b} = \text{const.},$$

with $a(b_s H_{s+1} + c_s) + b \neq 0$ on Σ^n, where $0 \leq r < s \leq n - 2$ are natural numbers and a and b are real numbers (at least one nonzero) and the constant c_r is defined recursively by $c_0 = 0$, $c_1 = 1$ and

$$c_i = -\frac{(n - i + 1)}{i - 1} c_{i-2}$$

if $i \geq 2$. If $X : \Sigma^n \times (-\epsilon, \epsilon) \to \mathbb{S}_1^{n+1}$ is a variation of $\psi : \Sigma^n \to \mathbb{S}_1^{n+1}$, then $\mathcal{J}''_{r,s,a,b}$ is given by

$$\mathcal{J}''_{r,s,a,b}(0)(f) = (s+1) \int_{\Sigma^n} \left\{ \mathcal{L}_{r,s,a,b}(f) + \left(\text{tr}(P_s) - \text{tr}(A^2 P_s) \right) - \Upsilon_{r,s,a,b} \left(\text{tr}(P_r) - \text{tr}(A^2 P_r) \right) \right\} f \, d\Sigma^n, \quad (14.7)$$

for all $f \in C^\infty(M)$, where $\mathfrak{L}_{r,s,a,b}$ is the differential operator

$$\mathfrak{L}_{r,s,a,b} : C^\infty(\Sigma^n) \to C^\infty(\Sigma^n)$$
$$f \mapsto \mathfrak{L}_{r,s,a,b}(f) = L_s(f) - \Upsilon_{r,s,a,b} L_r(f), \qquad (14.8)$$

and

$$\Upsilon_{r,s,a,b} = \frac{a(r+1)(b_s H_{s+1} + c_s)}{(s+1)(a(b_r H_{r+1} + c_r) + b)}. \qquad (14.9)$$

Proof By (14.6), (14.5), and (14.2), we obtain

$$\mathcal{J}''_{r,s,a,b}(0) = \frac{\partial}{\partial t}\left(\int_{\Sigma^n} \left\{ b_s H_{s+1} + c_s - \overline{\mathcal{H}}(a(b_r H_{r+1} + c_r) + b) \right\} f d\Sigma_t \right)\bigg|_{t=0}$$

$$= \int_{\Sigma^n} \left(b_s \frac{\partial H_{s+1}}{\partial t}\bigg|_{t=0} - \overline{\mathcal{H}} a b_r \frac{\partial H_{r+1}}{\partial t}\bigg|_{t=0} \right) f d\Sigma$$

$$+ \int_{\Sigma^n} \underbrace{\left(b_s H_{s+1} + c_s - \overline{\mathcal{H}}(a(b_r H_{r+1} + c_r) + b) \right)}_{=0} \frac{\partial}{\partial t}(f d\Sigma_t)\bigg|_{t=0}$$

$$= (r+1)\int_{\Sigma^n} \left\{ \left(L_s - \Upsilon_{r,s,a,b} L_r\right)(f) + \left(\mathrm{tra}(P_s) - \mathrm{tr}(A^2 P_s)\right.\right.$$
$$\left.\left. - \Upsilon_{r,s,a,b}\left(\mathrm{tr}(P_r) - \mathrm{tr}(A^2 P_r)\right)\right) f \right\} f d\Sigma$$

$$+ \int_{\Sigma^n} \left\{ \left(\frac{\partial X}{\partial t}\right)^T, \underbrace{\nabla\left(b_s H_{s+1} - \overline{\mathcal{H}}(a(b_r H_{r+1} + c_r) + b)\right)}_{=0} \right\} f d\Sigma^n.$$

To finish the proof, we observe that the above expression depends only on the hypersurface $\psi : \Sigma^n \to \mathbb{S}_1^{n+1}$ and on the function $f \in C^\infty(\Sigma^n)$. □

14.4 Strongly (r, s, a, b)-Stable Hypersurfaces in $\mathbb{S}_1^{n+1}$

We start this section by showing that the totally umbilical round spheres of $\mathbb{S}_1^{n+1}$ are strongly (r, s, a, b)-stable.

Proposition 14.3 *If the function $\Upsilon_{r,s,a,b}$ is nonpositive, then the totally umbilical spheres of $\mathbb{S}_1^{n+1}$ are strongly (r, s, a, b)-stable.*

14.4 Strongly (r, s, a, b)-Stable Hypersurfaces in $\mathbb{S}_1^{n+1}$

Proof Let $\psi : \Sigma^n \to \mathbb{S}_1^{n+1}$ be a totally umbilical sphere. Since $\psi : \Sigma^n \to \mathbb{S}_1^{n+1}$ is totally umbilical, then its principal curvatures are all equal to a certain constant κ. By choosing the orientation, we may assume that $\kappa < 0$. Thus, for $i \in \{0, \ldots, n\}$ we have

$$S_i = \binom{n}{i}\kappa^i, \quad H_i = (-\kappa)^i > 0, \quad \text{and} \quad S_i(A_j) = \binom{n-1}{i}\kappa^i.$$

Moreover, if $e_1, \ldots, e_n$ are the principal directions of $\psi : \Sigma^n \to \mathbb{S}_1^{n+1}$, by (1.6), one has

$$L_i(f) = \sum_{j=1}^{n} \langle \text{Hess}(f)(e_j), P_i(e_j) \rangle = \binom{n-1}{i}(-\kappa)^i \Delta f,$$

for every $f \in C^\infty(\Sigma^n)$.

Next, for entire numbers r and s satisfying the inequality $0 \leq r < s \leq n - 2$, and real numbers a and b (with at least one nonzero) such that

$$a\left((r+1)\binom{n}{r+1}(-\kappa)^{r+1} + c_r\right) + b \neq 0,$$

we get

$$\frac{b_s H_{s+1} + c_s}{a(b_r H_{r+1} + c_r) + b} = \frac{b_s(-\kappa)^{s+1} + c_s}{a(b_r(-\kappa)^{r+1} + c_r) + b} = \text{const.},$$

where $b_i = (i+1)\binom{n}{i+1}$ for $i \in \{r, s\}$. Here, the constant c_r is defined recursively by $c_0 = 0$, $c_1 = 1$ and $c_i = -\dfrac{(n-i+1)}{i-1} c_{i-2}$ if $i \geq 2$.

Then, by Lemma 1.2 and (14.7) we obtain

$$\mathcal{J}''_{r,s,a,b}(0)(f) = (s+1)\int_{\Sigma^n} \left\{ \Gamma_{r,s,a,b} \Delta f + \left(b_s H_s - n\frac{b_s}{s+1} H_1 H_{s+1}\right.\right.$$
$$+ b_{s+1} H_{s+2}\right) f - \Upsilon_{r,k,a,b}\left(b_r H_r - n\frac{b_r}{r+1} H_1 H_{r+1}\right.$$
$$\left.\left. + b_{r+1} H_{r+2}\right) f \right\} f \, d\Sigma^n,$$

where

$$\Gamma_{r,s,a,b} = \binom{n-1}{s}(-\kappa)^s - \Upsilon_{r,s,a,b}\binom{n-1}{r}(-\kappa)^r. \tag{14.10}$$

Thus, it follows that

$$\mathcal{J}''_{r,k,a,b}(0)(f) = (s+1)\int_{\Sigma^n}\left\{\Gamma_{r,s,a,b}\,\Delta f + \left(b_s(-\kappa)^s\right.\right.$$

$$\left.-\frac{nb_s}{s+1}(-\kappa)^{s+2} + b_{s+1}(-\kappa)^{r+2}\right)f$$

$$\left.-\Upsilon_{r,s,a,b}\left(b_r(-\kappa)^k - \frac{nb_r}{r+1}(-\kappa)^{k+2} + b_{r+1}(-\kappa)^{k+2}\right)f\right\}fd\Sigma^n$$

$$= (s+1)\int_{\Sigma^n}\left\{\Gamma_{r,s,a,b}\,\Delta f\right.$$

$$+\left((s+1)\binom{n}{s+1}(-\kappa)^s - \Upsilon_{r,s,a,b}(r+1)\binom{n}{r+1}(-\kappa)^r\right)f$$

$$+(-\kappa)^{s+2}\left(-n\binom{n}{s+1} + (s+2)\binom{n}{s+2}\right)f$$

$$\left.-(-\kappa)^{r+2}\Upsilon_{r,s,a,b}\left(-n\binom{n}{r+1} + (r+2)\binom{n}{r+2}\right)f\right\}fd\Sigma^n$$

$$= (s+1)\int_{\Sigma^n}\left\{\Gamma_{r,s,a,b}\,\Delta f + n\,\Gamma_{r,s,a,b}\,f - n\Gamma_{r,s,a,b}\kappa^2 f\right\}fd\Sigma^n$$

$$= (s+1)\Gamma_{r,s,a,b}\int_{\Sigma^n}\left\{f\Delta f + n(1-\kappa^2)f^2\right\}d\Sigma^n, \qquad (14.11)$$

for any $f \in C^\infty(\Sigma^n)$.

Hence, if λ_1 denotes the first eigenvalue of the Laplacian of Σ^n, the assumption on $\Upsilon_{r,k,a,b}$, by (14.10) and (14.11) yield

$$\mathcal{J}''_{r,s,a,b}(0)(f) \leq (s+1)\Gamma_{r,s,a,b}\int_{\Sigma^n}\left\{-\lambda_1 + n(1-\kappa^2)\right\}f^2 d\Sigma^n \leq 0,$$

for every $f \in C^\infty(\Sigma^n)$, where the last inequality was obtained by observing that Σ^n is isometric to a n-dimensional Euclidean sphere with constant sectional curvature equal to $\kappa^2 + 1$. Consequently, one has $\lambda_1 = n(\kappa^2 + 1)$. Therefore, the totally umbilical sphere $\psi : \Sigma^n \to \mathbb{S}^{n+1}_1$ is strongly (r, s, a, b)-stable. □

Now, we show the following preparatory result.

Lemma 14.4 *Let* $\psi : \Sigma^n \to -\mathbb{R} \times_{\cosh t} \mathbb{S}^n$ *be a spacelike hypersurface whose higher order mean curvatures* H_{r+1} *and* H_{s+1} *satisfy*

$$\frac{b_s H_{s+1} + c_s}{a(b_r H_{r+1} + c_r) + b} = \text{const.}, \qquad (14.12)$$

14.4 Strongly (r, s, a, b)-Stable Hypersurfaces in $\mathbb{S}_1^{n+1}$

with $a(b_r H_{r+1} + c_r) + b \neq 0$ on Σ^n, where $0 \leq r < s \leq n - 2$ are natural numbers and a and b are real numbers (at least one nonzero) and the constant c_r is defined recursively by $c_0 = 0$, $c_1 = 1$ and $c_i = -\dfrac{(n - i + 1)}{i - 1} c_{i-2}$ for $i \geq 2$. If N stands for the Gauss map of $\psi : \Sigma^n \to -\mathbb{R} \times_{\cosh t} \mathbb{S}^n$ and $K = \cosh t\, \partial_t$, the differential operator $\mathfrak{L}_{r,s,a,b}$ given in (14.8) acting on $\langle N, K \rangle$ is given by

$$\mathfrak{L}_{r,s,a,b}(\langle N, K \rangle) = \left(\mathrm{tr}(A^2 P_s) - \mathrm{tr}(P_s)\right) \langle N, K \rangle$$

$$- \Upsilon_{r,s,a,b} \left(\mathrm{tr}(A^2 P_r) - \mathrm{tr}(P_r)\right) \langle N, K \rangle$$

$$+ \left(b_s H_s - \Upsilon_{r,s,a,b} b_r H_r\right) \langle N, K \rangle$$

$$+ \left(b_s H_{s+1} - \Upsilon_{r,s,a,b} b_r H_{r+1}\right) \sinh t. \qquad (14.13)$$

Proof By Alías and Colares [19, Lemma 8.1], we get

$$L_k(\langle N, K \rangle) = \left(\mathrm{tr}(A^2 P_k) - \mathrm{tr}(P_k)\right) \langle N, K \rangle - b_k H_k N(\sinh t)$$

$$+ b_k H_{k+1} \sinh t + \frac{b_k}{k+1} \langle \sinh t\, \partial_t, \nabla H_{k+1} \rangle,$$

for $k \in \{r, s\}$; see also [94, Lemma 3.1]. Next, by (14.8), we have

$$\mathfrak{L}_{r,s,a,b}(\langle N, K \rangle) = \left(\mathrm{tr}(A^2 P_s) - \mathrm{tr}(P_s)\right) \langle N, K \rangle$$

$$- \Upsilon_{r,s,a,b} \left(\mathrm{tr}(A^2 P_r) - \mathrm{tr}(P_r)\right) \langle N, K \rangle$$

$$+ \left(b_s H_{s+1} - \Upsilon_{r,s,a,b} b_r H_{r+1}\right) \sinh t$$

$$\left(-b_s H_s + \Upsilon_{r,s,a,b} b_r H_r\right) N(\sinh t)$$

$$+ \left\langle \sinh t\, \partial_t, \nabla \left(\frac{b_s}{s+1} H_{s+1} - \Upsilon_{r,s,a,b} \frac{b_r}{r+1} H_{r+1}\right)\right\rangle \qquad (14.14)$$

Now, by (14.12) and (14.9), a simple computation shows that

$$\nabla \left(\frac{b_s}{s+1} H_{s+1} - \Lambda_{r,s,a,b} \frac{b_r}{r+1} H_{r+1}\right) = 0. \qquad (14.15)$$

Finally, we observe that

$$\overline{\nabla} \sinh t = -\langle \overline{\nabla} \sinh t, \partial_t \rangle = -(\sinh t)' \partial_t = -\cosh t\, \partial_t,$$

so that

$$N(\sinh t) = \langle N, \overline{\nabla} \sinh t \rangle = -\langle N, K \rangle. \tag{14.16}$$

By (14.15), (14.16) and (14.14), we obtain (14.13). □

The main result of this chapter reads as follows.

Theorem 14.1 *Let $\psi : \Sigma^n \to \mathbb{S}_1^{n+1}$ be a compact strongly (r, s, a, b)-stable spacelike hypersurface that is contained in a chronological future (or past) of $\mathbb{S}_1^{n+1}$. Suppose that $\Upsilon_{r,s,a,b}$ is nonpositive. If $H_{s+1} > 0$ and $H \leq 1$, then $\psi(\Sigma^n)$ is a totally umbilical round sphere.*

Proof Without loss of generality, we can assume that the closed spacelike hypersurface $\psi : \Sigma^n \to \mathbb{S}_1^{n+1} \hookrightarrow \mathbb{L}^{n+2}$ is contained in the chronological future of the equator determined by a unitary timelike vector $\mathbf{a} \in \mathbb{L}^{n+2}$. With this choice of $\mathbf{a}$, we use the warped product model $-\mathbb{R} \times_{\cosh t} \mathbb{S}^n$ for the de Sitter space $\mathbb{S}_1^{n+1}$.

By hypothesis, we have

$$0 \geq \mathcal{J}''_{r,s,a,b}(0)(f) = (s+1) \int_{\Sigma^n} \left\{ \mathcal{L}_{r,s,a,b}(f) + \left(\mathrm{tr}(P_s) - \mathrm{tr}(A^2 P_s) \right.\right.$$
$$\left.\left. - \Upsilon_{r,s,a,b} \left(\mathrm{tr}(P_r) + \mathrm{tr}(A^2 P_r) \right) \right) f \right\} f \, d\Sigma^n,$$

for every $f \in C^\infty(\Sigma^n)$. In particular, if we take $f = \langle \cosh t \, \partial_t, N \rangle$, by Lemma 14.4 we get

$$0 \geq \int_{\Sigma^n} \left\{ \left(b_s H_s - \Upsilon_{r,s,a,b} \, b_r H_r \right) \cosh \theta \, \cosh t \right.$$
$$\left. + \left(b_s H_{s+1} - \Upsilon_{r,s,a,b} \, b_r H_{r+1} \right) \sinh t \right\} \cosh \theta \, \cosh t \, d\Sigma^n, \tag{14.17}$$

where θ is the hyperbolic angle between N and ∂_t. In other words, θ is the only nonnegative real number such that $\cosh \theta = -\langle N, \partial_t \rangle$.

Now, since $\psi : \Sigma^n \to \mathbb{S}_1^{n+1}$ is closed and contained in the chronological future of $\mathbb{S}_1^{n+1}$, arguin as in the proof of [13, Theorem 1] we can prove the existence of a elliptic point in Σ^n. Thus, since $H_{s+1} > 0$ on Σ^n, by Colares and de Lima [136, Lemma 2.1] (see also [96, Proposition 2.3]) we conclude that the following hold:

(i) Each H_k is positive on Σ^n, $k \in \{1, \ldots, s+1\}$;
(ii) $H_{k-1} \geq H_k^{(k-1)/k}$ and $H \geq H_k^{1/k}$, with $k \in \{1, \ldots, s+1\}$. If $k \geq 2$, then in the above inequalities, the equality holds only at umbilical points of $\psi : \Sigma^n \to \mathbb{S}_1^{n+1}$.

Therefore, taking into account that by assumption $H \leq 1$ on Σ^n, we have

14.4 Strongly (r, s, a, b)-Stable Hypersurfaces in $\mathbb{S}_1^{n+1}$

$$1 \geq H \geq H_{k+1}^{1/(k+1)}, \quad k \in \{1, \ldots, s+1\}.$$

Hence $H_{k+1} \leq 1$ on Σ^n for every $k \in \{1, \ldots, s+1\}$. Consequently, on Σ^n we get

$$H_k \geq H_{k+1}^{1/(k+1)} \geq H_{k+1}, \quad k \in \{1, \ldots, s+1\}.$$

On the other hand, $\sinh y \leq \cosh y$ and $\cosh y \geq 1$, for every $y \in \mathbb{R}$. Hence, by (14.17), we have

$$0 \geq \int_{\Sigma^n} (\cosh\theta - 1) \left(b_s H_s - \Upsilon_{r,s,a,b} \, b_r H_r\right) \cosh^2 t \, \cosh\theta \, d\Sigma^n \geq 0.$$

Then $\cosh\theta \equiv 1$ on Σ^n. Consequently, there exists $t_0 \in \mathbb{R} \setminus \{0\}$ such that $\psi(\Sigma^n) = \{t_0\} \times \mathbb{S}^n$. □

Remark 14.1 We point out that Barbosa and Oliker in [63] introduce a definition of stability proving that a complete, connected and spacelike hypersurface $\psi : \Sigma^n \to \mathbb{S}_1^{n+1}$, with constat mean curvature H, is stable (according to their definition) provided that (i) Σ^n is compact; (ii) $H^2 \geq 1$ or (iii) $H^2 < 4(n-1)/n^2$. It is worth noting that, when $a = 0$, our stability condition in Definition 14.1 becomes $H_k =$ constant for some $k \in \{2, \ldots, n\}$. Thus, our Theorem 14.1 can be viewed as a generalization of [63, Theorem 4.7].

Remark 14.2 According to [241, Theorem 4], a closed spacelike hypersurface of $\mathbb{S}_1^{n+1}$, with constant mean curvature, is umbilical and coincides, up to a rigid motion of $\mathbb{S}_1^{n+1}$, with an n-dimensional sphere described in Example 1 (c) of the same paper. The square of the mean curvature, H^2, of such sphere assumes all possible values in the interval $[0, 1)$. Taking into account this facts, we would like to point out that our assumption $H \leq 1$ in Theorem 14.1 is a mild hypothesis, since we are interested in detecting totally umbilical round spheres of $\mathbb{S}_1^{n+1}$.

15 Stability in Certain Semi-Riemannian Manifolds with Density

15.1 Introduction

The study of variational questions associated to the area functional in weighted semi-Riemannian manifolds has been a focus of attention in the last years. In this direction, Rosales et al. [265] investigated the isoperimetric problem for Euclidean space endowed with a continuous density, showing that, for a radial log-convex density, balls about the origin are isoperimetric regions.

Afterwards, Cañete and Rosales [100] studied smooth Euclidean solid cones endowed with a smooth homogeneous weighted function. In this context, they proved that the unique compact, orientable, second order minima of the weighted area under variations preserving the weighted volume and with free boundary in the boundary of the cone are intersections with the cone of round spheres centered at the vertex.

In [211] Impera and Rimoldi established stability properties concerning φ-minimal hypersurfaces (that is, with identically zero φ-mean curvature) isometrically immersed in a weighted manifold endowed with a weight function φ and having nonnegative Bakry-Émery Ricci curvature, under certain volume growth conditions. Meanwhile, Castro and Rosales [103] obtained variational characterizations of critical points and second order minima of the weighted area with or without a volume constraint in weighted Riemannian manifolds with boundary.

Furthermore, in this setting, Batista et al. [69] showed some general inequalities involving the weighted mean curvature of compact submanifolds immersed in weighted Riemannian manifolds. As application, they obtained an isoperimetric inequality for such submanifolds. Moreover, they also proved an extrinsic upper bound to the first nonzero eigenvalue of the drift Laplacian on closed submanifolds of weighted Riemannian manifolds. Afterwards, McGonagle and Ross [237] showed that the hyperplane is the only stable, smooth solution to the isoperimetric problem in the $(n+1)$-dimensional Gaussian space $\mathbb{G}^{n+1}$.

This chapter is devoted to the study of stability in certain weighted semi-Riemannian. Firstly, we deal with conformal Killing graphs in a Riemannian manifold $\overline{M}_\varphi^{n+1}$ endowed with a weight function φ and having a closed conformal Killing vector field V with conformal factor ϕ_V, that is graphs constructed through the flow generated by V and which are defined over an integral leaf of the foliation $V^\perp$ orthogonal to V.

In this context, we obtain some stability results for φ-minimal conformal Killing graphs of $\overline{M}_\varphi^{n+1}$ according to the behavior of ϕ_V. Moreover, related to conformal Killing graphs immersed in $\overline{M}_\varphi^{n+1}$ with constant φ-mean curvature, we also study the strong stability.

Afterwards, we establish the notions of φ-stability and strong φ-stability concerning closed spacelike hypersurfaces immersed with constant φ-mean curvature in a conformally stationary spacetime endowed with a conformal timelike vector field V and a weighted function φ. When V is closed, with the aid of the φ-Laplacian of a suitable support function, we characterize φ-stable closed spacelike hypersurfaces through of the analysis of the first eigenvalue of the Jacobi operator associated to the corresponding variational problem.

Furthermore, we obtain sufficient conditions which assure that a strongly φ-stable closed spacelike hypersurface must be either φ-maximal or isometric to a leaf orthogonal to V.

Finally, we study a notion of stability for closed spacelike hypersurface immersed in a weighted standard static spacetime $M_\varphi^n \times_\rho \mathbb{R}_1$ with constant weighted mean curvature H_φ, via the first eigenvalue of the drift Laplacian Δ_φ.

15.2 Stability of Conformal Killing Graphs in Foliated Riemannian Spaces with Density

In this section, we investigate the stability of conformal Killing graphs in foliated Riemannian spaces with density. The results that will be presented are mainly based on the paper [292].

15.2.1 Conformal Killing Graphs

Let $(\overline{M}^{n+1}, \overline{g}, \overline{\nabla}, d\overline{\mu})$ be a *weighted oriented Riemannian manifold*, that is, a oriented Riemannian manifold $\overline{M}^{n+1}$ with metric tensor $\overline{g}$, Levi-Civita connection $\overline{\nabla}$, and endowed with a weighted volume form $d\overline{\mu} = e^{-\varphi} d\overline{M}^{n+1}$, where φ is a real-valued smooth function on $\overline{M}^{n+1}$, which is called *weight function*, and $d\overline{M}^{n+1}$ is the volume element induced by the metric $\overline{g}$. For simplicity of notation, we will denote $(\overline{M}^{n+1}, \overline{g}, \overline{\nabla}, d\overline{\mu})$ by $\overline{M}_\varphi^{n+1}$.

15.2 Stability of Conformal Killing Graphs in Foliated Riemannian...

In this setting, we recall that the *Bakry-Émery-Ricci tensor* $\overline{\text{Ric}}_\varphi$ of $\overline{M}_\varphi^{n+1}$ is defined by

$$\overline{\text{Ric}}_\varphi = \overline{\text{Ric}} + \overline{\text{Hess}}\,\varphi, \tag{15.1}$$

where $\overline{\text{Ric}}$ and $\overline{\text{Hess}}$ are respectively the standard Ricci tensor and the Hessian in the product $\overline{M}_\varphi^{n+1}$.

Along this section, we will consider hypersurfaces $\psi : \Sigma^n \to \overline{M}_\varphi^{n+1}$, namely, isometric immersions from a connected, n-dimensional oriented Riemannian manifold Σ^n into $\overline{M}_\varphi^{n+1}$, and ∇ and $g = \overline{g}|_{\Sigma^n}$ will denote the Levi-Civita connection of Σ^n and the induced metric on Σ^n, respectively. Let N be the unit normal vector field, called the Gauss map of $\psi : \Sigma^n \to \overline{M}_\varphi^{n+1}$, globally defined on Σ^n.

In this setting, let $A : \mathfrak{X}(\Sigma^n) \to \mathfrak{X}(\Sigma^n)$ be the shape operator of $\psi : \Sigma^n \to \overline{M}_\varphi^{n+1}$ with respect to N. So, at each $p \in \Sigma^n$, A restricts to a self-adjoint linear map $A_p : T_p \Sigma^n \to T_p \Sigma^n$ which is defined by $A_p(v) = -\overline{\nabla}_v N$ for every $v \in T_p \Sigma^n$. The φ-mean curvature of $\psi : \Sigma^n \to \overline{M}_\varphi^{n+1}$ is the function H_φ defined by

$$n H_\varphi = n H + \overline{g}(\overline{\nabla}\varphi, N), \tag{15.2}$$

where $H = \text{tr}(A)/n$ denotes the classical mean curvature of $\psi : \Sigma^n \to \overline{M}_\varphi^{n+1}$ with respect to N. The φ-divergence on Σ^n for any $X \in \mathfrak{X}(\Sigma^n)$ is defined by

$$\text{div}_\varphi X = \text{div}(X) - g(\nabla\varphi, X), \tag{15.3}$$

where $\text{div}(X) = \text{tr}\{Y \mapsto \nabla_Y X\}$ denotes the divergence relative to Σ^n.

A direct calculation assures us that

$$\text{div}_\varphi(\xi X) = \xi \text{div}_\varphi X + g(\nabla\xi, X) \tag{15.4}$$

for every $X \in \mathfrak{X}(\Sigma^n)$ and any $\xi \in C^\infty(\Sigma^n)$. We define the φ-Laplacian (or drift Laplacian) relative to Σ^n as follows

$$\Delta_\varphi(\xi) = \text{div}_\varphi(\nabla\xi) = \Delta\xi - g(\nabla\varphi, \nabla\xi), \tag{15.5}$$

for all $\xi \in C^\infty(\Sigma^n)$, where Δ is the standard Laplacian relative to Σ^n. By (15.4) and (15.5) we can easily obtain the expression

$$\Delta_\varphi(\xi\eta) = \xi\Delta_\varphi(\eta) + \eta\Delta_\varphi(\xi) + 2g(\nabla\xi, \nabla\eta), \tag{15.6}$$

which is valid for any pair of functions $\xi, \eta \in C^\infty(\Sigma^n)$.

In what follows, let us consider an $(n+1)$-dimensional weighted Riemannian manifold $\overline{M}_\varphi^{n+1}$ endowed with a conformal Killing vector field V whose orthogonal distribution $\mathcal{D}$ is integrable. Thus, there exists a smooth real-valued function ϕ_V defined on $\overline{M}_\varphi^{n+1}$ such that

$$\mathcal{L}_V \overline{g} = 2\phi_V \overline{g}, \qquad (15.7)$$

where $\mathcal{L}$ stands for the Lie derivative of the metric $\overline{g}$ of $\overline{M}_\varphi^{n+1}$. The function ϕ_V is called the conformal factor of V.

In this setting, we denote by $\Phi : \mathbb{I} \times \mathbb{M}^n \to \overline{M}_\varphi^{n+1}$ the flow generated by V, where $\mathbb{I} = (-\infty, a)$ is an interval with $a > 0$ and $\mathbb{M}^n$ is an arbitrarily fixed integral leaf of $\mathcal{D}$ labeled as $t = 0$ and which we will suppose to be connected and complete. It may happen that $a = +\infty$, i.e., the vector field V is complete. Since $\Phi_t = \Phi(t, \cdot)$ is a conformal map for any fixed $t \in \mathbb{R}$, there exists a positive function $\lambda \in C^\infty(\mathbb{I} \times \mathbb{M}^n)$ such that $\lambda(0, u) = 1$ and $\Phi_t^* \overline{g}(u) = \lambda^2(t, u) \overline{g}(u)$, for any $u \in \mathbb{M}^n$.

Throughout this section, we restrict ourselves to the case where the function λ depends only on the variable t, that is, $\lambda \in C^\infty(\mathbb{I})$. Geometrically, as it was already observed in [130], this hypothesis allows us to relate the induced metrics in distinct leaves of the foliation orthogonal to V, which we will denote by $V^\perp$.

In the sequel, for $X, Y \in \mathfrak{X}(\overline{M}^{n+1})$ we will write $\langle X, Y \rangle = \overline{g}(X, Y)$. By (15.7) we easily deduce the conformal Killing equation

$$\langle \overline{\nabla}_X V, Y \rangle + \langle X, \overline{\nabla}_Y V \rangle = 2\phi_V \langle X, Y \rangle,$$

for any $X, Y \in \mathfrak{X}(\overline{M}^{n+1})$.

An interesting particular case of a conformal Killing vector field V is that in which

$$\overline{\nabla}_X V = \phi_V X \qquad (15.8)$$

for all $X \in \mathfrak{X}(\overline{M}^{n+1})$. In this case we say that V is closed, an allusion to the fact that its dual 1-form is closed. Yet more particularly, a closed and conformal Killing vector field V is said to be parallel if its conformal factor ϕ_V vanishes identically, and homothetic if ϕ_V is constant.

Let $\mathbb{M}_t^n = \Phi_t(\mathbb{M}^n)$ be a leaf of $V^\perp$ furnished with the induced metric. By (15.8) we get

$$\overline{\nabla}\langle V, V \rangle = 2\phi_V V. \qquad (15.9)$$

Consequently, $|V|^2$ is constant on the leaves of $V^\perp$. Moreover, computing covariant derivative in (15.9), we have

$$\left(\overline{\mathrm{Hess}}\langle V, V \rangle\right)(X, Y) = 2X(\phi_V)\langle V, Y \rangle + 2\phi_V^2 \langle X, Y \rangle.$$

15.2 Stability of Conformal Killing Graphs in Foliated Riemannian...

Now, since both $\overline{\text{Hess}}$ and the metric $\langle \cdot, \cdot \rangle$ are symmetric tensors, we get

$$X(\phi_V)\langle V, Y \rangle = Y(\phi_V)\langle V, X \rangle$$

for every $X, Y \in \mathfrak{X}(\overline{M})$. Now, taking $Y = V$ we have

$$\overline{\nabla}\phi_V = \frac{V(\phi_V)}{|V|^2} V = \nu(\phi_V)\nu, \qquad (15.10)$$

where $\nu = \frac{V}{|V|}$ and, hence, ϕ_V is also constant on the leaves of $V^\perp$.

Furthermore, with a straightforward computation, we verify that the shape operator $A_t : \mathfrak{X}(\mathbb{M}^n_t) \to \mathfrak{X}(\mathbb{M}^n_t)$ of a leaf $\mathbb{M}^n_t \in V^\perp$ with respect to ν is given by

$$A_t(X) = \overline{\nabla}_X \nu = \frac{\phi_V}{|V|} X,$$

for any $X \in \mathfrak{X}(\mathbb{M}^n_t)$. Hence, the leaves $\mathbb{M}^n_t$ are totally umbilical hypersurfaces with constant mean curvature $\mathcal{H} = \mathcal{H}(t)$ with respect to ν given by

$$\mathcal{H} = \frac{\phi_V}{|V|}. \qquad (15.11)$$

Under the additional condition that the weight function φ of $\overline{M}^{n+1}_\varphi$ does not depend on the parameter of the flow associated to unit vector field ν, which means that $\langle \overline{\nabla}\varphi, \nu \rangle = 0$ on $\overline{M}^{n+1}_\varphi$, we obtain by (15.2) and (15.11) that the φ-mean curvature of a leaf $\mathbb{M}^n_t \in V^\top$ is given by

$$\mathcal{H}_\varphi = \frac{\phi_V}{|V|}. \qquad (15.12)$$

At the end of this first section, our purpose will be to give a description of conformal Killing graphs immersed in a weighted Riemannian manifold $\overline{M}^{n+1}_\varphi$ endowed with a closed conformal Killing vector field V.

In this sense, following the ideas developed in [130], given a domain Ω in $\mathbb{M}^n = \mathbb{M}^n_0$, we define the *conformal Killing graph* $\Sigma(z)$ of a smooth function z on $\overline{\Omega}$ as the hypersurface of $\overline{M}^{n+1}_\varphi$ given by

$$\Sigma(z) = \{ \Phi(z(u), u) : u \in \overline{\Omega} \},$$

where Φ is the flow generated by V. When $\Omega = \mathbb{M}^n$, $\Sigma(z)$ is said to be *entire*.

If we assign coordinates $x_0 = t, x_1, \ldots, x_n$ to points in $\overline{M}^{n+1}_\varphi$ of the form $\overline{u} = \Phi(t, u)$, where $x_1, \ldots, x_n$ are local coordinates in $\mathbb{M}^n$, then the corresponding coordinate vector fields are

$$\partial_0|_{\bar{u}} = V(t) \quad \text{and} \quad \partial_i|_{\bar{u}} = \Phi_{t*}\partial_i|_u, \quad \text{for every } i \in \{1,\ldots,n\}.$$

Thus, the conformal Killing graph $\Sigma(z)$ is parameterized in terms of local coordinates by $z(x_1,\ldots,x_n), x_1,\ldots,x_n$ and the tangent space to $\Sigma(z)$ is spanned by the vectors

$$\frac{\partial z}{\partial x_i}\partial_0|_{\Phi(z(u),u)} + \partial_i|_{\Phi(z(u),u)}, \quad \text{for all } i \in \{1,\ldots,n\}. \tag{15.13}$$

Hence, by (15.13) it is easily seen that the metric induced on $\Sigma(z)$ is given by

$$\lambda^2(z(u))\left(\frac{1}{\gamma}dz^2 + d\sigma^2\right),$$

where

$$\gamma = \frac{1}{|V(0)|^2}$$

and $d\sigma^2$ stands for the metric of the leaf $\mathbb{M}^n$.

Moreover, denoting by Dz the gradient of the function z with respect to the metric $d\sigma^2$, a straightforward computation ensures that

$$N = \frac{1}{\lambda(z(u))\sqrt{\gamma+|Dz(u)|^2}}\left(\Phi_{z(u)*}Dz(u) - \gamma\,\partial_0|_{\Phi(z(u),u)}\right), \tag{15.14}$$

gives an orientation on $\Sigma(z)$ such that $\langle N, V \rangle < 0$.

15.2.2 An Auxiliary Lemma

A particular class of Riemannian manifolds endowed by a closed conformal Killing vector field is the so-called warped product of the type $I \times_\varphi M^n$, where $I \subset \mathbb{R}$ is an open interval with the metric dt^2, M^n is an n-dimensional Riemannian manifold and $f : I \to \mathbb{R}$ is positive and smooth. A warped product $I \times_\varphi M^n$ endowed with a weight function φ will be called a weighted warped product and it will be also denoted by

$$\left(I \times_\varphi M^n\right)_\varphi. \tag{15.15}$$

In this setting, if π_I is the canonical projection onto I, then the vector field $V = (\phi \circ \pi_I)\partial_t$ is a conformal Killing and closed, with conformal factor $\phi_V = f' \circ \pi_I$, where the prime denotes differentiation with respect to $t \in I$.

15.2 Stability of Conformal Killing Graphs in Foliated Riemannian...

Moreover, for $t_0 \in I$, the leaf $M_{t_0}^n = \{t_0\} \times M^n$ (also called slice) is totally umbilical, with constant mean curvature

$$\mathcal{H}(t_0) = \frac{\phi'(t_0)}{\phi(t_0)},$$

with respect to $-\partial_t$; see [242].

Conversely, let $\overline{M}_\varphi^{n+1}$ be a weighted Riemannian manifold endowed with a closed conformal Killing vector field V. If $p \in \overline{M}_\varphi^{n+1}$ and $\mathbb{M}_p^n$ is the leaf of $V^\perp$ passing through p, then we can find a neighborhood $\mathcal{U}_p$ of p in $\mathbb{M}_p^n$ and an open interval $I \subset \mathbb{R}$ containing 0 such that the flow Φ of V is defined on $\mathcal{U}_p$ for every $t \in I$.

Besides, if V is complete, arguing as in [242, Section 3], one can prove that

$$\begin{aligned} \mathbb{R} \times \mathbb{M}_p^n &\longrightarrow \overline{M}_\varphi^{n+1} \\ (t, u) &\mapsto \Phi(t, u) \end{aligned} \tag{15.16}$$

is a global parametrization on $\overline{M}_\varphi^{n+1}$, so that $\overline{M}_\varphi^{n+1}$ is isometric to the weighted warped product

$$\left(\mathbb{R} \times_\varphi \mathbb{M}_p^n\right)_\varphi, \tag{15.17}$$

where $f(t) = |V(\Phi(t, u))|$, $t \in \mathbb{R}$ and $u \in \mathbb{M}_p^n$ is an arbitrary point.

In the sequel we assume that the weight function φ of $\overline{M}_\varphi^{n+1}$ does not depend on the parameter of the flow associated with the unit vector field $\nu = -V/|V|$, that is, $\langle \overline{\nabla}\varphi, \nu \rangle = 0$. This condition has already been used in (15.12) computing the φ-mean curvature of the leaves of $V^\perp$. In particular, when the ambient space is a warped product of the type $I \times_\varphi M^n$, we will explicit this condition simply writing

$$I \times_\varphi M_\varphi^n. \tag{15.18}$$

In this case, by (15.12) we get that the φ-mean curvature of the slice $\{t\} \times M^n$ is given by

$$\mathcal{H}_\varphi(t) = \frac{\phi'(t)}{\phi(t)} \tag{15.19}$$

with respect to the orientation given by $-\partial_t$.

In this scenario, we will consider the support function η_V on a conformal Killing graph $\Sigma(z)$ immersed in $\overline{M}_\varphi^{n+1}$, which is defined by

$$\eta_V : \Sigma(z) \to \mathbb{R}$$
$$p \mapsto \eta_V(p) = \langle V(p), N(p) \rangle, \qquad (15.20)$$

where N is the Gauss map of $\Sigma(z)$ given in (15.14). We have that η_V is negative and

$$\nabla \eta_V = -A(V^\top), \qquad (15.21)$$

where A is the shape operator of $\Sigma(z)$ with respect to N and $V^\top$ is the projection of the vector field V on the tangent bundle of $\Sigma(z)$.

In the next lemma, we present a suitable formula for the drift Laplacian of η_V.

Lemma 15.1 *Let $\overline{M}_\varphi^{n+1}$ be a weighted Riemannian manifold endowed with a closed conformal Killing vector field V having conformal factor ϕ_V and such that the weight function φ does not depend on the parameter of the flow associated to $v = -V/|V|$. If $\psi : \Sigma(z) \to \overline{M}_\varphi^{n+1}$ is a conformal Killing graph in $\overline{M}_\varphi^{n+1}$, with Gauss map N given in (15.14), and η_V is the smooth function on $\Sigma(z)$ defined in (15.20) then*

$$\Delta_\varphi(\eta_V) = -\left\{\overline{\mathrm{Ric}}_\varphi(N,N) + |A|^2\right\}\eta_V - nV^\top(H_\varphi) - n\left\{\phi_V H_\varphi + N(\phi_V)\right\}, \qquad (15.22)$$

where A and H_φ are the shape operator and the φ-mean curvature of $\psi : \Sigma(z) \to \overline{M}_\varphi^{n+1}$ with respect to N, respectively, and $\overline{\mathrm{Ric}}_\varphi$ denotes the Bakry-Émery-Ricci tensor of $\overline{M}_\varphi^{n+1}$.

Proof According to our previous digression, we have that (up to isometry) $\overline{M}_\varphi^{n+1}$ can be regarded locally as a weighted warped product of the type (15.17). In this setting, we have that $V = f\partial_t$, $\phi_V = f'$, $v = -\partial_t$, $|V| = f$, and, consequently, $\langle \overline{\nabla}\varphi, \partial_t \rangle = 0$.

By (15.2) we get

$$n\langle \partial_t, \nabla H \rangle = n\langle \partial_t^\top, \nabla H \rangle = n\langle \partial_t^\top, \nabla H_\varphi \rangle - \partial_t^\top \langle \overline{\nabla}\varphi, N \rangle, \qquad (15.23)$$

where $\partial_t^\top = \partial_t - \langle N, \partial_t \rangle N$ is the projection of ∂_t on the tangent bundle of $\Sigma(z)$.

On the other hand

$$\partial_t^\top \langle \overline{\nabla}\varphi, N \rangle = \langle \overline{\nabla}_{\partial_t^\top} \overline{\nabla}\varphi, N \rangle + \langle \overline{\nabla}\varphi, \overline{\nabla}_{\partial_t^\top} N \rangle$$
$$= \langle \overline{\nabla}_{\partial_t - \langle N, \partial_t \rangle N} \overline{\nabla}\varphi, N \rangle - \langle \overline{\nabla}\varphi, A(\partial_t^\top) \rangle$$
$$= \langle \overline{\nabla}_{\partial_t} \overline{\nabla}\varphi, N \rangle - \langle N, \partial_t \rangle \overline{\mathrm{Hess}}\,\varphi(N, N) - \langle \overline{\nabla}\varphi, A(\partial_t^\top) \rangle. \qquad (15.24)$$

Taking into account that $\langle \overline{\nabla}\varphi, \partial_t \rangle = 0$ and denoting by $\widetilde{\nabla}$ the Levi-Civita connection on $\mathbb{M}_p^n$, we have $\overline{\nabla}\varphi = f^{-2}\widetilde{\nabla}\varphi$. Then,

15.2 Stability of Conformal Killing Graphs in Foliated Riemannian...

$$\langle \overline{\nabla}_{\partial_t} \overline{\nabla} \varphi, N \rangle = \langle \overline{\nabla}_{\partial_t} (f^{-2} \widetilde{\nabla} \varphi), N \rangle$$
$$= \langle -2f^{-3} f' \widetilde{\nabla} \varphi + f^{-2} \overline{\nabla}_{\partial_t} \widetilde{\nabla} \varphi, N \rangle. \quad (15.25)$$

Hence, applying [251, Proposition 7.35], by (15.25) we get

$$\langle \overline{\nabla}_{\partial_t} \overline{\nabla} \varphi, N \rangle = \langle -2f^{-3} f' \widetilde{\nabla} \varphi + f^{-2} f^{-1} f' \widetilde{\nabla} \varphi, N \rangle$$
$$= -f' f^{-3} \langle \widetilde{\nabla} \varphi, N \rangle = -f' f^{-1} \langle \overline{\nabla} \varphi, N \rangle. \quad (15.26)$$

Now, by (15.26) and (15.24) we get that

$$\partial_t^\top \langle \overline{\nabla} \varphi, N \rangle = -\langle \overline{\nabla} \varphi, N \rangle f^{-1} f' - \langle N, \partial_t \rangle \overline{\text{Hess}} \, \varphi(N, N) - \langle \overline{\nabla} \varphi, A(\partial_t^\top) \rangle. \quad (15.27)$$

By (15.23) and (15.27) we conclude that

$$-nf \langle \partial_t, \nabla H \rangle = -nf \langle \partial_t^\top, \nabla H_\varphi \rangle - f' \langle \overline{\nabla} f, N \rangle$$
$$- f \langle N, \partial_t \rangle \overline{\text{Hess}} \, \varphi(N, N) - f \langle \overline{\nabla} \varphi, A(\partial_t^\top) \rangle. \quad (15.28)$$

On the other hand, from [98, Proposition 2.1] we have that

$$\Delta \langle N, f \partial_t \rangle = -n \langle f \partial_t, \nabla H \rangle - n \left\{ f' H + N(f') \right\}$$
$$- \langle N, f \partial_t \rangle \left\{ \overline{\text{Ric}}(N, N) + |A|^2 \right\}. \quad (15.29)$$

So, by (15.28), (15.29) and (15.1) we obtain

$$\Delta \langle N, f \partial_t \rangle = -n \langle f \partial_t, \nabla H_\varphi \rangle - \langle N, f \partial_t \rangle \left\{ \overline{\text{Ric}}_\varphi(N, N) + |A|^2 \right\}$$
$$- n \left\{ f' H_\varphi + N(f') \right\} - \langle \overline{\nabla} \varphi, A(f \partial_t^\top) \rangle. \quad (15.30)$$

Moreover, by (15.21) we have

$$\nabla \langle N, f \partial_t \rangle = -A(f \partial_t^\top). \quad (15.31)$$

By (15.30), (15.31) and (15.5) the conclusion is attained. □

Remark 15.1 We observe that the following result is a consequence of a Cheeger-Gromoll splitting type theorem due to G. Wei and W. Wylie (cf. [294, Theorem 6.1]; see also [176, Theorem 1.1]):

Let $\overline{M}_\varphi^{n+1}$ be a weighted Riemannian manifold that contains a line. If the Bakry-Émery-Ricci tensor of $\overline{M}_\varphi^{n+1}$ is nonnegative and the weight function φ is bounded, then φ must be constant along the line.

Consequently, in any weighted Riemannian manifold $\overline{M}_\varphi^{n+1}$ endowed with a complete closed conformal Killing vector field V, having nonnegative Bakry-Émery-Ricci tensor and with bounded weight function φ, we have that φ does not depend on the parameter of the flow associated with the unit vector field $\nu = -V/|V|$, that is, $\langle \overline{\nabla}\varphi, \nu \rangle = 0$. In this case, it is easily seen the condition adopted in Lemma 15.1 on the weight function φ are naturally verified.

15.2.3 Stability of φ-Minimal Conformal Killing Graphs

Let $\overline{M}_\varphi^{n+1}$ be a weighted Riemannian manifold, with weight function φ and endowed with a closed conformal Killing vector field V, and let $\psi : \Sigma(z) \to \overline{M}_\varphi^{n+1}$ be a conformal Killing graph with Gauss map N as in (15.14). In this setting, we denote by $d\Sigma(z)$ the volume element with respect to the metric induced by $\psi : \Sigma(z) \to \overline{M}_\varphi^{n+1}$ and we mean by $C_0^\infty(\Sigma(z))$ the set of functions of class C^∞ compactly supported on $\Sigma(z)$.

It is well-known that, given a function $\varphi \in C_0^\infty(\Sigma(z))$ there exists a normal variation with compact support an fixed boundary

$$\psi_s : \Sigma(z) \to \overline{M}_\varphi^{n+1}, \quad \text{for } s \in (-\epsilon, \epsilon), \tag{15.32}$$

of $\psi : \Sigma(z) \to \overline{M}_\varphi^{n+1}$, that is

(i) $\psi_s = \text{Id}$ outside a compact subset of $\Sigma(z)$;
(ii) For any $s \in (-\epsilon, \epsilon)$, the map $\psi_s : \Sigma(z) \to \overline{M}_\varphi^{n+1}$ is an immersion such that $\psi_0(p) = \psi(p)$ for every $p \in \Sigma(z)$;
(iii) $\psi_s(p) = p$ for every $p \in \partial\Sigma(z)$.

Moreover, associated with $\psi_s : \Sigma(z) \to \overline{M}_\varphi^{n+1}$ we have that the variational normal field is φN and the first variation of the weighted area functional

$$\mathcal{A}_\varphi : (-\epsilon, \epsilon) \to \mathbb{R}$$
$$s \mapsto \mathcal{A}_\varphi(s) = \text{Area}_\varphi\big(\psi_s(\Sigma(z))\big) = \int_{\Sigma(z)} d\mu_s, \tag{15.33}$$

where $d\mu_s = e^{-\varphi} d\Sigma(z)_s$ and $d\Sigma(z)_s$ denotes the volume element of $\Sigma(z)$ with respect to the metric induced by $\psi_s : \Sigma(z) \to \overline{M}_\varphi^{n+1}$, is given by

15.2 Stability of Conformal Killing Graphs in Foliated Riemannian...

$$\delta_\xi (\mathcal{A}_\varphi) = \frac{d\mathcal{A}}{ds}(0) = n \int_{\Sigma(z)} \xi H_\varphi d\mu, \qquad (15.34)$$

see, for instance, [103, Lemma 3.2].

As a consequence, $\psi : \Sigma(z) \to \overline{M}_\varphi^{n+1}$ is a φ-minimal if and only if $\delta_\xi (\mathcal{A}_\varphi) = 0$ for every smooth function $\xi \in C_0^\infty(\Sigma(z))$. In other words, φ-minimal conformal Killing graphs in $\overline{M}_\varphi^{n+1}$ are characterized as critical points of $\mathcal{A}_\varphi$.

The stability operator of this variational problem is given by the second variation formula for the φ-area, which in our case is written as follows (see [103, Proposition 3.5] for $H_\varphi = 0$)

$$\delta_\xi^2 (\mathcal{A}_\varphi) = \frac{d^2 \mathcal{A}}{ds^2}(0) = -\int_{\Sigma(z)} \xi L_\varphi(\xi) d\mu \qquad (15.35)$$

with

$$L_\varphi = \Delta_\varphi + |A|^2 + \overline{\mathrm{Ric}}_\varphi(N, N),$$

where Δ_φ is the drift Laplacian operator on $\Sigma(z)$, N and $|A|$ are the Gauss map and the length of the shape operator A of $\psi : \Sigma(z) \to \overline{M}_\varphi^{n+1}$, respectively, and $\overline{\mathrm{Ric}}_\varphi$ is the Bakry-Émery-Ricci tensor of $\overline{M}_\varphi^{n+1}$.

For φ-minimal conformal Killing graphs in $\overline{M}_\varphi^{n+1}$, the following notion of stability will be crucial in the sequel.

Definition 15.1 Let $\overline{M}_\varphi^{n+1}$ be a weighted Riemannian manifold, with weight function φ and endowed with a closed conformal Killing vector field V, and let $\psi : \Sigma(z) \to \overline{M}_\varphi^{n+1}$ be a φ-minimal conformal Killing graph. We say that $\psi : \Sigma(z) \to \overline{M}_\varphi^{n+1}$ is L_φ-stable if $\delta_\xi^2 (\mathcal{A}_\varphi) \geq 0$ for every $\xi \in C_0^\infty(\Sigma(z))$.

In order to proof our main theorem in this section, we will need to use the following auxiliary result, which gives a sufficient condition for a φ-minimal hypersurfaces to be L_φ-stable.

Lemma 15.2 Let $\overline{M}_\varphi^{n+1}$ be a weighted Riemannian manifold, with weight function φ and endowed with a closed conformal Killing vector field V, and let $\psi : \Sigma(z) \to \overline{M}_\varphi^{n+1}$ be a φ-minimal conformal Killing graph. If there exists a positive smooth function $u \in C^\infty(\Sigma(z))$ such that $L_\varphi(u) \leq 0$, then $\psi : \Sigma(z) \to \overline{M}_\varphi^{n+1}$ is L_φ-stable.

Proof Let us assume that there exists such a function u and take $\xi \in C_0^\infty(\Sigma(z))$. Then, we can choose $\varrho \in C_0^\infty(\Sigma(z))$ satisfying $\xi = \varrho u$. Hence, from (15.6) and (15.35) we have

$$\delta_\xi^2(\mathcal{A}_\varphi) = -\int_{\Sigma(z)} \xi L_\varphi(\xi) d\mu = -\int_{\Sigma(z)} \varrho\, u L_\varphi(\varrho\, u) d\mu$$

$$= -\int_{\Sigma(z)} \left(\varrho^2 u L_\varphi(u) + \varrho\, u^2 \Delta_\varphi(\varrho) + 2\varrho\, u \langle \nabla\varrho, \nabla u \rangle \right) d\mu$$

$$\geq -\int_{\Sigma(z)} \left(\varrho\, u^2 \Delta(\varrho) + 2\varrho\, u \langle \nabla\varrho, \nabla u \rangle - \varrho\, u^2 \langle \nabla\varrho, \nabla\varphi \rangle \right) d\mu$$

$$= -\int_{\Sigma(z)} \left(\varrho\, u^2 \Delta(\varrho) + \frac{1}{2} \langle \nabla\varrho^2, \nabla u^2 \rangle - \varrho\, u^2 \langle \nabla\varrho, \nabla\varphi \rangle \right) d\mu. \quad (15.36)$$

On the other hand, we can see that

$$\text{div}(u^2 \nabla\varrho^2) = \langle \nabla u^2, \nabla\varrho^2 \rangle + u^2 \Delta(\varrho^2) = \langle \nabla u^2, \nabla\varrho^2 \rangle + 2\varrho\, u^2 \Delta(\varrho) + 2u^2 |\nabla\varrho|^2.$$

Therefore, by the weighted version of Divergence Theorem (see [100, Lemma 2.2]), the above equation and (15.36) yield

$$\delta_\xi^2(\mathcal{A}_\varphi) \geq -\int_{\Sigma(z)} \left(\frac{1}{2} \text{div}(u^2 \nabla\varrho^2) - u^2 |\nabla\varrho|^2 - \varrho\, u^2 \langle \nabla\varrho, \nabla\varphi \rangle \right) d\mu$$

$$= -\int_{\Sigma(z)} \left(\frac{1}{2} \text{div}_\varphi(u^2 \nabla\varrho^2) - u^2 |\nabla\varrho|^2 \right) d\mu$$

$$= \int_{\Sigma(z)} u^2 |\nabla\varrho|^2 d\mu \geq 0.$$

Hence, $\psi : \Sigma(z) \to \overline{M}_\varphi^{n+1}$ is L_φ-stable. $\square$

Now, analyzing the behavior of the conformal factor ϕ_V along a conformal Killing graph $\psi : \Sigma(z) \to \overline{M}_\varphi^{n+1}$, we will state and prove our main result concerning L_φ-stability. In what follows, $t \in \mathbb{R}$ denotes the parameter of the flow associated with the unit vector field $\nu = -V/|V|$.

Theorem 15.1 *Let $\overline{M}_\varphi^{n+1}$ be a weighted Riemannian manifold nonnegative Bakry-Émery-Ricci tensor, endowed with a complete closed conformal Killing vector field V having conformal factor ϕ_V and whose weight function φ is bounded, and let $\psi : \Sigma(z) \to \overline{M}_\varphi^{n+1}$ be a φ-minimal conformal Killing graph. The following facts hold:*

(a) If

$$\frac{\partial \phi_V}{\partial t} \leq 0 \text{ on } \Sigma(z)$$

15.2 Stability of Conformal Killing Graphs in Foliated Riemannian... 403

then $\psi : \Sigma(z) \to \overline{M}_\varphi^{n+1}$ is L_φ-stable.

(b) If $\Sigma(z)$ is compact and

$$\frac{\partial \phi_V}{\partial t} \geq 0 \text{ on } \Sigma(z)$$

then $\psi : \Sigma(z) \to \overline{M}_\varphi^{n+1}$ is L_φ-stable if and only if ϕ_V is constant on $\Sigma(z)$.

(c) If $\Sigma(z)$ is compact and

$$\frac{\partial \phi_V}{\partial t} > 0 \text{ on } \Sigma(z)$$

then $\psi : \Sigma(z) \to \overline{M}_\varphi^{n+1}$ cannot be L_φ-stable.

Proof The weight function φ does not depend on the parameter of the flow associated with v; see Remark 15.1. By (15.10) we observe that

$$N(\phi_V) = \langle N, \overline{\nabla}\phi_V \rangle = -\frac{v(\phi_V)}{|V|}\eta_V = -\frac{1}{|V|}\frac{\partial \phi_V}{\partial t}\eta_V, \qquad (15.37)$$

where η_V is the negative support function defined in (15.20).

Let us consider on $\Sigma(z)$ the smooth positive function $u = -\eta_V$. Then, by (15.22) and (15.37) we obtain

$$L_\varphi(u) = \frac{n}{|V|}\frac{\partial \phi_V}{\partial t} u. \qquad (15.38)$$

Hence, a direct application of Lemma 15.2 ensures that item (*a*) holds.

We show now item (*b*). In this case $C_0^\infty(\Sigma(z)) = C^\infty(\Sigma(z))$. So, if $\varphi : \Sigma(z) \to \overline{M}_\varphi^{n+1}$ is L_φ-stable, by Definition 15.1 and Eq. (15.38) we get

$$0 \leq \delta_u^2(\mathcal{A}_\varphi) = -\int_{\Sigma(z)} u L_\varphi(u) d\mu = -n \int_{\Sigma(z)} \frac{u^2}{|V|}\frac{\partial \phi_V}{\partial t} d\mu \leq 0. \qquad (15.39)$$

Consequently

$$\frac{\partial \phi_V}{\partial t} = 0 \text{ on } \Sigma(z).$$

The converse is also true thanks to item (*a*).

Finally, let us prove (*c*). Arguing by contradiction, assume that $\varphi : \Sigma(z) \to \overline{M}_\varphi^{n+1}$ L_φ-stable. On the other hand, by (15.39), it follows that

$$0 \le -n \int_{\Sigma(z)} \frac{u^2}{|V|} \frac{\partial \phi_V}{\partial t} d\mu < 0,$$

which is absurd. The proof is now complete. □

When the ambient space is a weighted warped product of the type (15.18), we can apply Theorem 15.1 to obtain the following result.

Corollary 15.1 *Let $\varphi : \Sigma(z) \to \mathbb{R} \times_\varphi M_\varphi^n$ be a φ-minimal conformal Killing graph.*

(a) *If the warping function f satisfies $f'' \le 0$ on $\Sigma(z)$ then $\varphi : \Sigma(z) \to \mathbb{R} \times_\varphi M_\varphi^n$ is L_φ-stable.*
(b) *If $\Sigma(z)$ compact and the warping function f satisfies $f'' \ge 0$ on $\Sigma(z)$ then $\varphi : \Sigma(z) \to \mathbb{R} \times_\varphi M_\varphi^n$ is L_φ-stable if and only if $f = at + b$ on $\Sigma(z)$, for some $a, b \in \mathbb{R}$.*
(c) *If $\Sigma(z)$ compact and the warping function f satisfies $f'' > 0$ on $\Sigma(z)$ then $\varphi : \Sigma(z) \to \mathbb{R} \times_\varphi M_\varphi^n$ cannot be L_φ-stable.*

15.2.4 Stability of Constant φ-Mean Curvature Conformal Killing Graphs

Let $\overline{M}_\varphi^{n+1}$ be a weighted Riemannian manifold, with weight function φ and endowed with closed conformal Killing vector field V. Moreover, let $\psi : \Sigma(z) \to \overline{M}_\varphi^{n+1}$ be a closed (i.e. compact and without boundary) conformal Killing graph with Gauss map N defined in (15.14).

Set

$$\mathcal{G} = \left\{ \xi \in C^\infty(\Sigma(z)) : \int_{\Sigma(z)} \xi d\mu = 0 \right\},$$

given by the smooth functions on $\Sigma(z)$ with weighted integral mean equal to zero, where $d\mu = e^{-f} d\Sigma(z)$ and $d\Sigma(z)$ is the volume element with respect to the metric induced by $\psi : \Sigma(z) \to \overline{M}_\varphi^{n+1}$.

According to [64, Lemma 2.1 and Lemma 2.2] (see also [103, Lemma 3.2]), every smooth function $\varphi \in \mathcal{G}$ induces a normal variation (namely, a smooth function of the form given in (15.32) and such that item (ii) holds) of $x : \Sigma(z) \to \overline{M}_\varphi^{n+1}$, with variational normal field ξN and whose first variation of the weighted area functional $\mathcal{A}_\varphi : (-\epsilon, \epsilon) \to \mathbb{R}$, namely $\delta_\xi (\mathcal{A}_\varphi)$, defined as in (15.33), is given by (15.34).

As a consequence of (15.34), any closed conformal Killing graph $\psi : \Sigma(z) \to \overline{M}_\varphi^{n+1}$ with constant φ-mean curvature H_φ is a critical point of $\mathcal{A}_\varphi$ restricted to all functions ξ belonging to $\mathcal{G}$. Geometrically, this condition means that the variations under consideration preserve a certain weighted volume function; see [103, Section 3] for more details. For

15.2 Stability of Conformal Killing Graphs in Foliated Riemannian...

these critical points, [103, Proposition 3.5] (see also [64, Proposition 2.5]) asserts that the stability of the corresponding variational problem is given by the second variation

$$\delta_\xi^2 (\mathcal{A}_\varphi) = -\int_{\Sigma(z)} \left\{ \Delta_\varphi(\xi) + \left(|A|^2 + \overline{\mathrm{Ric}}_\varphi(N,N) \right) \xi \right\} \xi d\mu, \tag{15.40}$$

where Δ_φ is the drift Laplacian operator on $\Sigma(z)$, as well as N and $|A|$ are respectively the Gauss map and the length of the shape operator A of $\psi : \Sigma(z) \to \overline{M}_\varphi^{n+1}$. Finally, $\overline{\mathrm{Ric}}_\varphi$ is the Bakry-Émery-Ricci tensor of $\overline{M}_\varphi^{n+1}$.

By (15.40), we notice that $\delta_\xi^2 (\mathcal{A}_\varphi)$ depends only on $\xi \in C^\infty(\Sigma(z))$.

Definition 15.2 Let $\overline{M}_\varphi^{n+1}$ be a weighted Riemannian manifold, with weight function φ and endowed with a closed conformal Killing vector field V. Moreover, let $\psi : \Sigma(z) \to \overline{M}_\varphi^{n+1}$ be a closed conformal Killing graph with constant φ-mean curvature H_φ. We say that $\psi : \Sigma(z) \to \overline{M}_\varphi^{n+1}$ is *strongly φ-stable* when $\delta_\xi^2(\mathcal{A}_\varphi) \geq 0$ for every $\xi \in C^\infty(\Sigma(z))$.

We are in position now to state and prove the following rigidity result for strongly φ-stable conformal Killing graphs $\psi : \Sigma(z) \to \overline{M}_\varphi^{n+1}$.

Theorem 15.2 *Let $\overline{M}_\varphi^{n+1}$ be a weighted Riemannian manifold nonnegative Bakry-Émery-Ricci tensor, endowed with a complete closed conformal Killing vector field V having conformal factor ϕ_V and whose weight function φ is bounded. Moreover, let $\psi : \Sigma(z) \to \overline{M}_\varphi^{n+1}$ be a strongly φ-stable closed conformal Killing graph. Suppose that*

$$\frac{\partial \phi_V}{\partial t} \geq \max\{\phi_V H_\varphi, 0\}, \tag{15.41}$$

where $t \in \mathbb{R}$ is the parameter of the flow associated with the unit vector field $\nu = -V/|V|$. If the set where $\phi_V = 0$ has empty interior in $\Sigma(z)$, then $\psi : \Sigma(z) \to \overline{M}_\varphi^{n+1}$ is either φ-minimal or isometric to a leaf of the foliation $V^\perp$.

Proof As observed in Remark 15.1, we have that φ does not depend on $t \in \mathbb{R}$. Now, let us consider in $\overline{M}_\varphi^{n+1}$ the global parametrization (15.16). Since $\psi : \Sigma(z) \to \overline{M}_\varphi^{n+1}$ is strongly φ-stable, by Definition 15.5 and (15.40) it follows that

$$-\int_{\Sigma(z)} \left\{ \Delta_\varphi(\xi) + \left(\overline{\mathrm{Ric}}_\varphi(N,N) + |A|^2 \right) \xi \right\} \xi d\mu \geq 0, \tag{15.42}$$

for every $\xi \in C^\infty(\Sigma(z))$.

In particular, since H_φ is constant on $\Sigma(z)$, taking the negative function η_V defined in (15.20) by (15.22) we have

$$\Delta_\varphi(\eta_V) + \left(\overline{\mathrm{Ric}}_\varphi(N,N) + |A|^2\right)\eta_V = -n\left\{\phi_V H_\varphi + N(\phi_V)\right\}.$$

Thus, by (15.42) it follows that

$$\int_{\Sigma(z)} \left\{\phi_V H_\varphi + N(\phi_V)\right\} \eta_V d\mu \geq 0. \tag{15.43}$$

On the other hand, by (15.10) one has

$$N(\phi_V) = \langle N, \overline{\nabla}\phi_V \rangle = \nu(\phi_V)\langle N, \nu \rangle = -\frac{\partial \phi_V}{\partial t}\cos\theta,$$

where θ is the angle between N and $-\nu$. The above equation in addition to (15.43), immediately yields

$$\int_{\Sigma(z)} \left(\phi_V H_\varphi - \frac{\partial \phi_V}{\partial t}\cos\theta\right) |V|\cos\theta d\mu \geq 0.$$

By (15.41), one has

$$0 \leq \int_{\Sigma(z)} \left\{\phi_V H_\varphi - \frac{\partial \phi_V}{\partial t}\cos\theta\right\} |V|\cos\theta d\mu$$
$$\leq \int_{\Sigma(z)} (1 - \cos\theta) \frac{\partial \phi_V}{\partial t} |V|\cos\theta, d\mu \leq 0.$$

We can write

$$(1 - \cos\theta)\frac{\partial \phi_V}{\partial t} = 0 \quad \text{and} \quad \frac{\partial \phi_V}{\partial t} = -\phi_V H_\varphi$$

on $\Sigma(z)$.

Now, since H_φ is constant on $\Sigma(z)$, $\psi : \Sigma(z) \to \overline{M}_\varphi^{n+1}$ is either φ-minimal or $H_\varphi \neq 0$ on $\Sigma(z)$. If the last case occurs, since the zero set of ϕ_V has empty interior in $\Sigma(z)$, taking into account the above expression of $\frac{\partial \phi_V}{\partial t}$, one has that $\frac{\partial \phi_V}{\partial t} \neq 0$ on a dense subset of $\Sigma(z)$.

Hence, $\cos\theta = 1$ on this set. By continuity, $\cos\theta = 1$ on $\Sigma(z)$. Therefore, in this case, $\psi : \Sigma(z) \to \overline{M}_\varphi^{n+1}$ must be a leaf of the foliation $V^\perp$. □

We close our section observing that, when the ambient space is a weighted warped product of the type (15.18), we can apply Theorem 15.2 to obtain the following result.

Corollary 15.2 *Let* $\psi : \Sigma(z) \to \mathbb{R} \times_\varphi M_\varphi^n$ *be a strongly φ-stable closed conformal Killing graph. Suppose that the warped function f satisfies*

$$f'' \geq \max\{f' H_\varphi, 0\}.$$

If the set where $f' = 0$ has empty interior in $\Sigma(z)$, then $\psi : \Sigma(z) \to \mathbb{R} \times_\varphi M_\varphi^n$ is either φ-minimal or isometric to the slice $\{t_0\} \times M^n$, for some $t_0 \in \mathbb{R}$.

15.3 Stability of Spacelike Hypersurfaces in Weighted Spacetimes

This section is dedicated to the study of the stability of spacelike hypersurfaces immersed in certain weighted spacetimes. The results that will be presented can be founded in reference [152].

15.3.1 Background

This section is devoted to recall some basic facts concerning spacelike hypersurfaces immersed in a weighted Lorentzian space.

Let $(\overline{M}^{n+1}, \langle \cdot, \cdot \rangle, \overline{\nabla}, d\overline{\mu})$ be a weighted time-oriented Lorentzian manifold, that is, a time-oriented Lorentzian manifold $\overline{M}^{n+1}$ with metric tensor $\langle \cdot, \cdot \rangle$, Levi-Civita conection $\overline{\nabla}$ and endowed with a weighted volume form $d\overline{\mu} = e^{-\varphi} d\overline{M}^{n+1}$, where φ is a real-valued smooth function on $\overline{M}^{n+1}$, called weighted function, and $d\overline{M}^{n+1}$ is the volume element induced by the metric $\langle \cdot, \cdot \rangle$. In order to shorten our notation, we denote $(\overline{M}^{n+1}, \langle \cdot, \cdot \rangle, \overline{\nabla}, d\overline{\mu})$ simply by $\overline{M}_\varphi^{n+1}$. We mean by $C^\infty(\overline{M}^{n+1})$ the ring of real functions of class C^∞ on $\overline{M}^{n+1}$ and by $\mathfrak{X}(\overline{M}^{n+1})$ the $C^\infty(\overline{M}^{n+1})$-module of vector fields of class C^∞ on $\overline{M}^{n+1}$. For $\overline{M}_\varphi^{n+1}$, the Bakry-Émery-Ricci tensor $\overline{\text{Ric}}_\varphi$ is defined by

$$\overline{\text{Ric}}_\varphi = \overline{\text{Ric}} + \overline{\text{Hess}}\,\varphi, \tag{15.44}$$

where $\overline{\text{Ric}}$ and $\overline{\text{Hess}}$ are the Ricci tensor and the Hessian operator in $\overline{M}_\varphi^{n+1}$, respectively.

In this context, we consider spacelike hypersurfaces $\psi : \Sigma^n \to \overline{M}_\varphi^{n+1}$, namely, isometric immersions from a connected, n-dimensional orientable Riemannian manifold Σ^n into $\overline{M}_\varphi^{n+1}$. We let ∇ denote the Levi-Civita connection of Σ^n.

Since we are supposing that $\overline{M}_\varphi^{n+1}$ is time-orientable and $\psi : \Sigma^n \to \overline{M}_\varphi^{n+1}$ is a spacelike hypersurface, then Σ^n is orientable and one can choose a globally defined unit normal vector field N on Σ^n having the same time-orientation of $\overline{M}_\varphi^{n+1}$, which is called the future-pointing Gauss map of $\psi : \Sigma^n \to \overline{M}_\varphi^{n+1}$. Furthermore, let $A : \mathfrak{X}(\Sigma^n) \to \mathfrak{X}(\Sigma^n)$

be the shape operator of $\psi : \Sigma^n \to \overline{M}_\varphi^{n+1}$ with respect to N, so that at each $p \in \Sigma^n$, A restricts to a self-adjoint linear map

$$A_p : T_p\Sigma^n \to T_p\Sigma^n$$
$$v \mapsto A_p v = -\overline{\nabla}_v N.$$

The φ-mean curvature of $\psi : \Sigma^n \to \overline{M}_\varphi^{n+1}$ is the function H_φ given by

$$nH_\varphi = nH - \langle \overline{\nabla}\varphi, N \rangle, \qquad (15.45)$$

where $H = -\mathrm{tr}(A)/n$ denotes the standard mean curvature of $\psi : \Sigma^n \to \overline{M}_\varphi^{n+1}$ with respect to its future-pointing Gauss map N. The φ-divergence on Σ^n is defined by

$$\mathrm{div}_\varphi : \mathfrak{X}(\Sigma^n) \to C^\infty(\Sigma^n)$$
$$X \mapsto \mathrm{div}_\varphi X = \mathrm{div} X - \langle \nabla\varphi, X \rangle, \qquad (15.46)$$

where div denotes the standard divergence of Σ^n.

A direct computation assures that

$$\mathrm{div}_\varphi(\xi X) = \xi \, \mathrm{div}_\varphi(X) + \langle \nabla\xi, X \rangle$$

for every $X \in \mathfrak{X}(\Sigma^n)$ and any $\xi \in C^\infty(\Sigma^n)$.

Finally, let us define the φ-Laplacian on Σ^n by

$$\Delta_\varphi : C^\infty(\Sigma^n) \to C^\infty(\Sigma^n)$$
$$\xi \mapsto \Delta_\varphi(\xi) = \mathrm{div}_\varphi \nabla\xi = \Delta(\xi) - \langle \nabla\varphi, \nabla\xi \rangle \qquad (15.47)$$

where Δ is the standard Laplacian of Σ^n.

15.3.2 Description of the Variational Problem

Now, let us consider immersions $\psi : \Sigma^n \to \overline{M}_\varphi^{n+1}$ of compact spacelike hypersurfaces Σ^n with boundary $\partial\Sigma^n$ (possibly empty). A variation of $\psi : \Sigma^n \to \overline{M}_\varphi^{n+1}$ is a smooth mapping

$$X : \Sigma^n \times (-\epsilon, \epsilon) \to \overline{M}_\varphi^{n+1}$$

satisfying the following two conditions:

15.3 Stability of Spacelike Hypersurfaces in Weighted Spacetimes

1. For any $t \in (-\epsilon, \epsilon)$, the map $X_t : \Sigma^n \to \overline{M}_\varphi^{n+1}$ given by $X_t(p) = X(t, p)$ is a spacelike immersion such that $X_0 = \psi$.
2. $X_t|_{\partial \Sigma^n} = \psi|_{\partial \Sigma^n}$, for every $t \in (-\epsilon, \epsilon)$.

In the sequel, let us denote by dM_t the volume element of the metric induced on Σ^n by X_t and let N_t be the unit normal vector field along X_t. Moreover, we also consider in Σ^n the weighted volume form given by $d\mu_t = e^{-\varphi} dM_t$. When $t = 0$ all these mathematical objects coincide with ones defined in Σ^n, respectively.

Set

$$u_t = -\left\langle \frac{\partial X}{\partial t}, N_t \right\rangle, \tag{15.48}$$

then

$$\left.\frac{\partial X}{\partial t}\right|_{t=0} = u_0 N + \left(\left.\frac{\partial X}{\partial t}\right|_{t=0}\right)^\top,$$

where $(\cdot)^\top$ stands for tangential component.

The weighted volume of the variation X is the functional

$$\begin{aligned} \mathcal{V}_\varphi : (-\epsilon, \epsilon) &\to \mathbb{R} \\ t &\mapsto \mathcal{V}_\varphi(t) = \int_{\Sigma^n \times [0,t]} X^*(d\overline{\mu}). \end{aligned}$$

We say that X is volume-preserving if $\mathcal{V}_\varphi(t) = \mathcal{V}_\varphi(0)$, for every $t \in (-\epsilon, \epsilon)$.

The following result is well-known and, in the context of weighted manifolds, can be found in [103] or [265].

Lemma 15.3 Let $\overline{M}_\varphi^{n+1}$ be a weighted time-oriented Lorentzian manifold and let $\psi : \Sigma^n \to \overline{M}_\varphi^{n+1}$ be a closed spacelike hypersurface. If $X : \Sigma^n \times (-\epsilon, \epsilon) \to \overline{M}_\varphi^{n+1}$ is a variation of $\psi : \Sigma^n \to \overline{M}_\varphi^{n+1}$, then

$$\frac{d\mathcal{V}_\varphi}{dt} = \int_{\Sigma^n} u_t d\mu_t,$$

where u_t is given in (15.48). In particular, X is volume-preserving if and only if

$$\int_{\Sigma^n} u_t d\mu_t = 0$$

for every $t \in (-\epsilon, \epsilon)$.

Remark 15.2 It is easy to see that [64, Lemma 2.2] still holds for weighted time-oriented Lorentzian manifolds. More preceisely, if $u \in C^\infty(\Sigma^n)$ is such that $u|_{\partial \Sigma} = 0$ and

$$\int_{\Sigma^n} u d\mu = 0,$$

then there exists a volume-preserving variation of $\psi : \Sigma^n \to \overline{M}_\varphi^{n+1}$ whose variational field is uN.

The weighted area functional associated to the variation X is given by

$$\begin{aligned} \mathcal{A}_\varphi : (-\epsilon, \epsilon) &\to \mathbb{R} \\ t &\mapsto \mathcal{A}_\varphi(t) = \int_{\Sigma^n} d\mu_t. \end{aligned}$$

Arguing as in the proof of [103, Lemma 3.2], the following result holds.

Lemma 15.4 *Let $\overline{M}_\varphi^{n+1}$ be a weighted time-oriented Lorentzian manifold and $\psi : \Sigma^n \to \overline{M}_\varphi^{n+1}$ a closed spacelike hypersurface. If $X : \Sigma^n \times (-\epsilon, \epsilon) \to \overline{M}_\varphi^{n+1}$ is a variation of $\psi : \Sigma^n \to \overline{M}_\varphi^{n+1}$, then*

$$\mathcal{A}_\varphi'(t) = n \int_{\Sigma^n} \left(H_\varphi\right)_t u_t d\mu_t, \tag{15.49}$$

where u_t is given in (15.48) and $\left(H_\varphi\right)_t = H_\varphi(t, \cdot)$ denotes the φ-mean curvature of $\psi : \Sigma^n \to \overline{M}_\varphi^{n+1}$ with respect to the metric induced by X_t.

In order to characterize hypersurfaces with constant φ-mean curvature, we consider the variational problem of maximizing the functional $\mathcal{A}_\varphi$ for any variation $X : \Sigma^n \times (-\epsilon, \epsilon) \to \overline{M}_\varphi^{n+1}$ of $\psi : \Sigma^n \to \overline{M}_\varphi^{n+1}$ that preserve the weighted volume $\mathcal{V}_\varphi$. The Jacobi functional associated to this problem is given by

$$\begin{aligned} \mathcal{J}_\varphi : (-\epsilon, \epsilon) &\to \mathbb{R} \\ t &\mapsto \mathcal{J}_\varphi(t) = \mathcal{A}_\varphi(t) - \varrho \mathcal{V}_\varphi(t), \end{aligned} \tag{15.50}$$

where ϱ is a constant to be determined. As an immediate consequence of Lemmas 15.3 and 15.4 we get

$$\mathcal{J}_\varphi'(t) = \int_{\Sigma^n} \left\{n \left(H_\varphi\right)_t - \varrho\right\} u_t d\mu_t. \tag{15.51}$$

In order to make an appropriated choice of ϱ, let

15.3 Stability of Spacelike Hypersurfaces in Weighted Spacetimes

$$\overline{\mathcal{H}}_\varphi = \frac{1}{\mathcal{A}_\varphi(0)} \int_{\Sigma^n} (H_\varphi)_0 \, d\mu$$

be an integral mean of the φ-mean curvature of $\psi : \Sigma^n \to \overline{M}_\varphi^{n+1}$. If $(H_\varphi)_0$ is constant, we have

$$\overline{\mathcal{H}}_\varphi = (H_\varphi)_0 = H_\varphi. \tag{15.52}$$

This notation will be used in the sequel. Therefore, if we choose $\varrho = n\overline{\mathcal{H}}_\varphi$, by (15.51) we have

$$\mathcal{J}'_\varphi(t) = n \int_{\Sigma^n} \left\{ (H_\varphi)_t - \overline{\mathcal{H}}_\varphi \right\} u_t \, d\mu_t. \tag{15.53}$$

Arguing as in the proof of [62, Proposition 2.7], by (15.53), we get

Proposition 15.1 *Let $\overline{M}_\varphi^{n+1}$ be a weighted time-oriented Lorentzian manifold and let $\psi : \Sigma^n \to \overline{M}_\varphi^{n+1}$ be a closed spacelike hypersurface. The following statements are equivalent:*

(a) $\psi : \Sigma^n \to \overline{M}_\varphi^{n+1}$ *has constant φ-mean curvature H_φ;*
(b) *For any variation $X : \Sigma^n \times (-\epsilon, \epsilon) \to \overline{M}_\varphi^{n+1}$ of $\psi : \Sigma^n \to \overline{M}_\varphi^{n+1}$ that preserves the volume, we have that $\mathcal{J}'_\varphi(0) = 0$;*
(c) *For any variation $X : \Sigma^n \times (-\epsilon, \epsilon) \to \overline{M}_\varphi^{n+1}$ of $\psi : \Sigma^n \to \overline{M}_\varphi^{n+1}$, we have that $\mathcal{J}'_\varphi(0) = 0$.*

In particular, Proposition 15.1 guarantees that a spacelike hypersurface $\psi : \Sigma^n \to \overline{M}_\varphi^{n+1}$ is a critical point of the variational problem described above if and only if its φ-mean curvature H_φ is constant.

Definition 15.3 Let $\overline{M}_\varphi^{n+1}$ be a weighted time-oriented Lorentzian manifold and let $\psi : \Sigma^n \to \overline{M}_\varphi^{n+1}$ be a closed spacelike hypersurface having constant φ-mean curvature H_φ. We say that $\psi : \Sigma^n \to \overline{M}_\varphi^{n+1}$ is φ-*stable* if $\mathcal{A}''_\varphi(0) \leq 0$, for all volume-preserving variation $X : \Sigma^n \times (-\epsilon, \epsilon) \to \overline{M}_\varphi^{n+1}$ of $\psi : \Sigma^n \to \overline{M}_\varphi^{n+1}$.

Remark 15.3 Let $\psi : \Sigma^n \to \overline{M}_\varphi^{n+1}$ be a closed spacelike hypersurface as described in Definition 15.3. We consider the set

$$\mathcal{G} = \left\{ u \in C^\infty(\Sigma^n) : \int_{\Sigma^n} u \, d\mu = 0 \right\}. \tag{15.54}$$

As in [62], we can establish the following criterion of φ-stability: $\psi : \Sigma^n \to \overline{M}_\varphi^{n+1}$ is φ-stable if and only if $\mathcal{J}_\varphi''(0)(u) \leq 0$, for any $u \in \mathcal{G}$.

Now, let us consider closed spacelike hyperpurfaces $\psi : \Sigma^n \to \overline{M}_\varphi^{n+1}$ which maximize the functional Jacobi $\mathcal{J}_\varphi$ for any variation $X : \Sigma^n \times (-\epsilon, \epsilon) \to \overline{M}_\varphi^{n+1}$ of $\psi : \Sigma^n \to \overline{M}_\varphi^{n+1}$. By Proposition 15.1 one has that $\psi : \Sigma^n \to \overline{M}_\varphi^{n+1}$ is a critical point of $\mathcal{J}_\varphi$ if and only if its φ-mean curvature H_φ is constant.

Definition 15.4 Let $\overline{M}_\varphi^{n+1}$ be a weighted time-oriented Lorentzian manifold and let $\psi : \Sigma^n \to \overline{M}_\varphi^{n+1}$ be a closed spacelike hypersurface whose φ-mean curvature H_φ is constant. We say that $\psi : \Sigma^n \to \overline{M}_\varphi^{n+1}$ is *strongly φ-stable* if $\mathcal{J}_\varphi''(0)(u) \leq 0$, for any $u \in C^\infty(\Sigma^n)$.

The sought formula for the second variation $\mathcal{J}_\varphi''$ of the Jacobi functional $\mathcal{J}_\varphi$ is given in the following

Proposition 15.2 *Let $\overline{M}_\varphi^{n+1}$ be a weighted time-oriented Lorentzian manifold and let $\psi : \Sigma^n \to \overline{M}_\varphi^{n+1}$ be a closed spacelike hypersurface whose φ-mean curvature H_φ is constant. If $X : \Sigma^n \times (-\epsilon, \epsilon) \to \overline{M}_\varphi^{n+1}$ is a variation of $\psi : \Sigma^n \to \overline{M}_\varphi^{n+1}$, then $\mathcal{J}_\varphi''(0)$ is given by*

$$\mathcal{J}_\varphi''(0)(u) = \int_{\Sigma^n} \left\{ \Delta_\varphi(u) - \left\{ \overline{\mathrm{Ric}}_\varphi(N, N) + |A|^2 \right\} u \right\} u d\mu, \tag{15.55}$$

for all $u \in C^\infty(\Sigma^n)$.

Proof Since H_φ is constant, by (15.53) and (15.52) we have that

$$\mathcal{J}_\varphi''(0) = \int_{\Sigma^n} n \left(\frac{\partial H_\varphi}{\partial t} \bigg|_{t=0} \right) u_0 d\mu + \int_{\Sigma^n} n \underbrace{(H_\varphi - \overline{\mathcal{H}}_\varphi)}_{=0} \frac{\partial}{\partial t} (u_t d\mu_t) \bigg|_{t=0}.$$

On the other hand, arguing as in the proof of equation (3.5) of [103], we obtain

$$n \frac{\partial H_\varphi}{\partial t} \bigg|_{t=0} = \Delta_\varphi(u_0) - \left\{ \overline{\mathrm{Ric}}_\varphi(N, N) + |A^2| \right\} u_0.$$

Hence,

$$\mathcal{J}_\varphi''(0) = \int_{\Sigma^n} \left\{ \Delta_\varphi(u_0) - \left\{ \overline{\mathrm{Ric}}_\varphi(N, N) + |A|^2 \right\} u_0 \right\} u_0 d\mu.$$

Finally, we observe that the above expression depends only on the hypersurface $\psi : \Sigma^n \to \overline{M}_\varphi^{n+1}$ and on the function $u_0 \in C^\infty(\Sigma^n)$. □

15.3.3 The φ-Laplacian of a Support Function

Proceeding with the context of the previous section, let $\overline{M}_\varphi^{n+1}$ be a weighted Lorentzian manifold. As in the Riemannian setting of the previous section, a vector field V on $\overline{M}_\varphi^{n+1}$ is said to be conformal if

$$\mathcal{L}_V \langle \cdot, \cdot \rangle = 2\phi \langle \cdot, \cdot \rangle$$

for some function $\phi \in C^\infty(\overline{M}^{n+1})$, where $\mathcal{L}$ stands for the Lie derivative of the Lorentzian metric of $\overline{M}_\varphi^{n+1}$.

The function ϕ is called the conformal factor of V. So, extending the terminology established in [26], a weighted Lorentzian manifold $\overline{M}_\varphi^{n+1}$ endowed with a globally defined timelike conformal vector field will be called a weighted conformally stationary spacetime.

Since $\mathcal{L}_V(X) = [V, X]$ for any $X \in \mathfrak{X}(\overline{M}^{n+1})$, it follows by the tensorial character of $\mathcal{L}_V$ that $V \in \mathfrak{X}(\overline{M})$ is conformal if and only if

$$\langle \overline{\nabla}_X V, Y \rangle + \langle X, \overline{\nabla}_Y V \rangle = 2\phi_V \langle X, Y \rangle, \tag{15.56}$$

for all $X, Y \in \mathfrak{X}(\overline{M})$. In particular, V is a Killing vector field if and only if $\phi_V \equiv 0$.

Let us suppose, in addition, that the conformal timelike vector field V is closed, that is,

$$\overline{\nabla}_X V = \phi_V X \tag{15.57}$$

for every $X \in \mathfrak{X}(\overline{M}^{n+1})$.

Assuming that V has no singularities on an open set $\mathcal{U} \subset \overline{M}_\varphi^{n+1}$, the distribution $V^\perp$ on $\mathcal{U}$ of vector fields orthogonal to V is integrable, if for every $X, Y \in V^\perp$, one has

$$\langle [X, Y], V \rangle = \langle \overline{\nabla}_X Y - \overline{\nabla}_Y X, V \rangle = -\langle Y, \overline{\nabla}_X V \rangle + \langle X, \overline{\nabla}_Y V \rangle = 0.$$

Let Ξ^n be a leaf of $V^\perp$ furnished with the induced metric. By (15.57) we get

$$\overline{\nabla} \langle V, V \rangle = 2\phi_V V, \tag{15.58}$$

so that $\langle V, V \rangle$ is constant on connected leaves of $V^\perp$.

Computing the covariant derivative in (15.58), we have

$$\overline{\text{Hess}} \langle V, V \rangle (X, Y) = 2X(\phi_V) \langle V, Y \rangle + 2\phi_V^2 \langle X, Y \rangle.$$

Since both $\overline{\text{Hess}}$ and the metric $\langle \cdot, \cdot \rangle$ are symmetric tensors, we get

$$X(\phi_V)\langle V, Y\rangle = Y(\phi_V)\langle V, X\rangle$$

for every $X, Y \in \mathfrak{X}(\overline{M}^{n+1})$. Taking $Y = V$ we have

$$\overline{\nabla}\phi_V = \frac{V(\phi_V)}{\langle V, V\rangle} V = -\nu_V(\phi_V)\nu_V, \quad (15.59)$$

where $\nu_V = \dfrac{V}{\sqrt{-\langle V, V\rangle}}$. Hence, ϕ_V is also constant on the connected leaves of $V^\perp$. If Ξ^n is such a leaf and A_Ξ denotes its shape operator with respect to ν_V, we get

$$A_\Xi(X) = -\overline{\nabla}_X \nu_V = \phi_V X$$

for any $X \in \mathfrak{X}(\Xi^n)$ and, hence, Ξ^n is a totally umbilical spacelike hypersurface with constant mean curvature $\mathcal{H}$ given by

$$\mathcal{H} = \frac{\phi_V}{\sqrt{-\langle V, V\rangle}}. \quad (15.60)$$

Under the additional hypothesis that the weighted function φ of $\overline{M}_\varphi^{n+1}$ does not depend on the parameter of the flow associated to unit timelike vector field ν_V, which means that $\langle \overline{\nabla}\varphi, \nu_V\rangle = 0$ on $\overline{M}_\varphi^{n+1}$, by (15.45) and (15.60) we obtain that the φ-mean curvature of a leaf Ξ^n of $V^\top$ with respect to ν_V is given by

$$\mathcal{H}_\varphi = \frac{\phi_V}{\sqrt{-\langle V, V\rangle}}. \quad (15.61)$$

According to the terminology established in [24], a particular class of conformally stationary spacetimes is constituted by the so-called generalized Robertson-Walker (GRW) spacetimes, namely, Lorentzian warped product spaces of the type $-I \times_\varphi M^n$, where $I \subset \mathbb{R}$ is an open interval with the metric $-dt^2$, M^n is an n-dimensional Riemannian manifold and $f : I \to \mathbb{R}$ is positive and smooth.

A GRW spacetime $-I \times_\varphi M^n$ endowed with a weighted function φ is called a weighted GRW spacetime and it will be denoted by

$$\left(-I \times_\varphi M^n\right)_\varphi.$$

For such a space, if π_I is the canonical projection onto I, then the vector field

$$V = (f \circ \pi_I)\partial_s \quad (15.62)$$

15.3 Stability of Spacelike Hypersurfaces in Weighted Spacetimes

is a conformal, timelike and closed, with conformal factor $\phi_V = f' \circ \pi_I$, where f' denotes first derivative with respect to the parameter s of f. Moreover, it follows by [243, Proposition 1] that each spacelike leaf $M_s^n = \{s\} \times M^n$ is totally umbilical and, supposing that φ does not depend on the parameter s, the φ-mean curvature of M_s^n with respect to ∂_s is equal to $\dfrac{f'(s)}{f(s)}$.

Conversely, let $\overline{M}_\varphi^{n+1}$ be a weighted conformally stationary Lorentzian manifold endowed with closed conformal vector field V. If $p \in \overline{M}_\varphi^{n+1}$ and Ξ_p^n is the leaf of $V^\perp$ passing through p, then we can find a neighborhood $\mathcal{U}_p$ of p in Ξ_p^n and an open interval $I \subset \mathbb{R}$ containing 0 such that the flow $\mathcal{F}$ of V is defined on $\mathcal{U}_p$ for every $s \in I$. Besides, if $\overline{M}_\varphi^{n+1}$ is timelike geodesically complete, which means that any timelike geodesic of $\overline{M}_\varphi^{n+1}$ is defined for all values of the parameter $s \in \mathbb{R}$, by [243, Section 3] we have that

$$\begin{aligned} \mathbb{R} \times \Xi_p^n &\to \overline{M}_\varphi^{n+1} \\ (s, q) &\mapsto \mathcal{F}(s, q) \end{aligned} \quad (15.63)$$

is a global parametrization on $\overline{M}_\varphi^{n+1}$, such that $\overline{M}_\varphi^{n+1}$ is isometric to the weighted GRW spacetime

$$\left(-\mathbb{R} \times_\varphi \Xi_p^n\right)_\varphi, \quad (15.64)$$

where

$$f(s) = \sqrt{-\langle V(\mathcal{F}(s, q)), V(\mathcal{F}(s, q))\rangle},$$

$s \in \mathbb{R}$, and $q \in \Xi_p^n$ is an arbitrary point.

On the other hand, we observe that by Case [101, Theorem 1.2] it follows that if $\overline{M}_\varphi^{n+1}$ is a weighted timelike geodesic complete conformally stationary spacetime, with closed conformal timelike vector field V, endowed with a weighted function $\varphi \in C^\infty(\overline{M}^{n+1})$ which is bounded and such that its Bakry-Émery-Ricci tensor $\overline{\mathrm{Ric}}_\varphi$ satisfies $\overline{\mathrm{Ric}}_\varphi(X, X) \geq 0$ for all the timelike vector fieds X, then φ must be constant along the timelike line contained em $\overline{M}_\varphi^{n+1}$, given via isometry by (15.63).

So, motivated by this result, along this section we will consider weighted conformally stationary spacetimes $\overline{M}_\varphi^{n+1}$ endowed with a closed conformal timelike vector field V and whose weighted function φ does not depend on the parameter of the flow associated with the unit timelike vector field

$$\nu_V = \frac{V}{\sqrt{-\langle V, V \rangle}},$$

that is, $\langle \overline{\nabla}\varphi, v_V \rangle = 0$. This condition has already been used in (15.61) computing the φ-mean curvature of the leaves $V^\top$.

In particular, when the ambient space is a weighted GRW spacetime, we will explicit this condition simply writing

$$-I \times_\varphi M_\varphi^n. \tag{15.65}$$

In the scenario described above, if $\psi : \Sigma^n \to \overline{M}_\varphi^{n+1}$ is a spacelike hypersurface, then the smooth function

$$\begin{aligned} \eta_V : \Sigma^n &\to \mathbb{R} \\ p &\mapsto \eta_V(p) = \langle V(p), N(p) \rangle \end{aligned} \tag{15.66}$$

is negative and, after some standard calculations, we get

$$\nabla \eta_V = -A(V^\top), \tag{15.67}$$

where A is the shape operator of $\psi : \Sigma^n \to \overline{M}_\varphi^{n+1}$.

The following proposition gives a suitable formula for the φ-Laplacian of the support function η, which will be crucial to obtain our criteria of φ-stability along the next section.

Proposition 15.3 *Moreover, let $\overline{M}_\varphi^{n+1}$ be a weighted conformally stationary spacetime with a closed conformal timelike vector field V and whose weighted function φ does not depend on the parameter of the flow associated to v_V. Let $\psi : \Sigma^n \to \overline{M}_\varphi^{n+1}$ be a spacelike hypersurface with future-pointing Gauss map N and support function $\eta_V = \langle V, N \rangle$. Then*

$$\Delta_\varphi(\eta_V) = \left\{ \overline{\mathrm{Ric}}_\varphi(N, N) + |A|^2 \right\} \eta_V + nV^\top(H_\varphi) + n \left\{ \phi_V H_\varphi - N(\phi_V) \right\}, \tag{15.68}$$

where ϕ_V is the conformal factor of V, $\overline{\mathrm{Ric}}_\varphi$ denotes the Bakry-Émery-Ricci tensor of $\overline{M}_\varphi^{n+1}$, A and H_φ are the shape operator and the φ-mean curvature of $\psi : \Sigma^n \to \overline{M}_\varphi^{n+1}$ with respect to N, respectively, and ∇H_φ stands for the gradient of H_φ in the induced metric on Σ^n.

Proof By (15.64) and (15.65) we have that (up to isometry) $\overline{M}_\varphi^{n+1}$ can be locally viewed as a weighted GRW spacetime of the form

$$-\mathbb{R} \times_\varphi \left(\Xi_p^n \right)_\varphi.$$

In this setting, by (15.62) we have that $V = f \partial_t$, $\phi_V = f'$, $v_V = \partial_t$, $\sqrt{-\langle V, V \rangle} = f$ and, consequently, $\langle \overline{\nabla}\varphi, \partial_t \rangle = 0$.

15.3 Stability of Spacelike Hypersurfaces in Weighted Spacetimes

Note that, by (15.45) we get

$$n\langle \partial_t, \nabla H\rangle = n\langle \partial_t^\top, \nabla H\rangle = n\langle \partial_t^\top, \nabla H_\varphi\rangle + \partial_t^\top\langle \overline{\nabla}\varphi, N\rangle. \tag{15.69}$$

On the other hand

$$\begin{aligned}\partial_t^\top\langle \overline{\nabla}\varphi, N\rangle &= \langle \overline{\nabla}_{\partial_t^\top}\overline{\nabla}\varphi, N\rangle + \langle \overline{\nabla}\varphi, \overline{\nabla}_{\partial_t^\top} N\rangle \\ &= \langle \overline{\nabla}_{\partial_t+\Theta N}\overline{\nabla}\varphi, N\rangle - \langle \overline{\nabla}\varphi, A(\partial_t^\top)\rangle \\ &= \langle \overline{\nabla}_{\partial_t}\overline{\nabla}\varphi, N\rangle + \Theta\, \overline{\mathrm{Hess}}_\varphi(N, N) - \langle \overline{\nabla}\varphi, A(\partial_t^\top)\rangle, \end{aligned} \tag{15.70}$$

where $\partial_t^\top = \partial_t + \Theta N$ and $\Theta = \langle N, \partial_t\rangle$.

Now, taking into account that $\langle \overline{\nabla}\varphi, \partial_t\rangle = 0$ and denoting by $\widetilde{\nabla}$ the Levi-Civita connection on Ξ_p^n, we have $\overline{\nabla}\varphi = f^{-2}\widetilde{\nabla}\varphi$. Then, one has

$$\begin{aligned}\langle \overline{\nabla}_{\partial_t}\overline{\nabla}\varphi, N\rangle &= \langle \overline{\nabla}_{\partial_t}(f^{-2}\widetilde{\nabla}\varphi), N\rangle \\ &= \langle -2f^{-3}f'\widetilde{\nabla}\varphi + f^{-2}\overline{\nabla}_{\partial_t}\widetilde{\nabla}\varphi, N\rangle.\end{aligned} \tag{15.71}$$

Hence, on account of [251, Proposition 7.35], by (15.71) we have

$$\begin{aligned}\langle \overline{\nabla}_{\partial_t}\overline{\nabla}\varphi, N\rangle &= \langle -2f^{-3}f'\widetilde{\nabla}\varphi + f^{-2}f^{-1}f'\widetilde{\nabla}\varphi, N\rangle \\ &= -f'f^{-3}\langle \widetilde{\nabla}\varphi, N\rangle \\ &= -f'f^{-1}\langle \overline{\nabla}\varphi, N\rangle.\end{aligned} \tag{15.72}$$

By (15.72) and (15.70) we get that

$$\partial_t^\top\langle \overline{\nabla}\varphi, N\rangle = -\langle \overline{\nabla}\varphi, N\rangle f^{-1}f' + \Theta\,\overline{\mathrm{Hess}}_\varphi(N, N) - \langle \overline{\nabla}\varphi, A(\partial_t^\top)\rangle. \tag{15.73}$$

Thus, by (15.69) and (15.73) we conclude that

$$\begin{aligned}nf\langle \partial_t, \nabla H\rangle &= nf\langle \partial_t^\top, \nabla H_\varphi\rangle - f'\langle \overline{\nabla}\varphi, N\rangle \\ &\quad + f\Theta\,\overline{\mathrm{Hess}}_\varphi(N, N) - f\langle \overline{\nabla}\varphi, A(\partial_t^\top)\rangle.\end{aligned} \tag{15.74}$$

On the other hand, by Barros et al. [67, Proposition 3.1] it follows that

$$\Delta\langle N, f\partial_t\rangle = n\langle f\partial_t, \nabla H\rangle + f\Theta\left\{\overline{\mathrm{Ric}}(N, N) + |A|^2\right\} + n\left\{f'H - N(f')\right\}. \tag{15.75}$$

So, by (15.44), (15.74) and (15.75) we get

$$\Delta \langle N, f\partial_t \rangle = n\langle f\partial_t, \nabla H_\varphi \rangle + f\Theta \left\{ \overline{\mathrm{Ric}}_\varphi(N, N) + |A|^2 \right\}$$
$$+ n \left\{ f' H_\varphi - N(f') \right\} - f \langle \overline{\nabla}\varphi, A(\partial_t^\top) \rangle. \tag{15.76}$$

Moreover, by (15.67) one has

$$\nabla \langle N, f\partial_t \rangle = -f A(\partial_t^\top). \tag{15.77}$$

Therefore, by (15.76), (15.77) and (15.47), we have (15.68). □

In particular, by Proposition 15.3 we obtain the following

Corollary 15.3 *Let $\overline{M}_\varphi^{n+1}$ be a weighted conformally stationary spacetime with a Killing timelike vector field W and whose weighted function φ does not depend on the parameter of the flow associated to unit vector field v_W. Moreover, let $\psi : \Sigma^n \to \overline{M}_\varphi^{n+1}$ be a spacelike hypersurface with future-pointing Gauss map N and support function $\eta_W = \langle W, N \rangle$. Then*

$$\Delta_\varphi(\eta_W) = \left\{ \overline{\mathrm{Ric}}_\varphi(N, N) + |A|^2 \right\} \eta_W + n W^\top (H_\varphi).$$

In addition, if $\psi : \Sigma^n \to \overline{M}_\varphi^{n+1}$ is closed and H_φ and $\lambda = \overline{\mathrm{Ric}}_\varphi(N, N) + |A|^2$ are constants on Σ^n, then λ is an eigenvalue of the operator Δ_φ on Σ^n with eigenfunction η_W.

15.3.4 Stability Results in Weighted Spacetimes

We present now our first φ-stability criterion concerning closed spacelike hypersurfaces immersed in a weighted conformally stationary spacetime.

Theorem 15.3 *Let $\overline{M}_\varphi^{n+1}$ be a weighted conformally stationary spacetime with a closed conformal timelike vector field V and whose weighted function φ does not depend on the parameter of the flow associated to v_V. Suppose that $\overline{M}_\varphi^{n+1}$ is also equipped with a Killing timelike vector field W and that φ does not depend on the parameter of the flow associated to unit vector field v_W. Let $\psi : \Sigma^n \to \overline{M}_\varphi^{n+1}$ be a closed spacelike hypersurface with constant φ-mean curvature H_φ and such that $\lambda = \overline{\mathrm{Ric}}_\varphi(N, N) + |A|^2$ is constant. Then $\psi : \Sigma^n \to \overline{M}_\varphi^{n+1}$ is φ-stable if and only if λ is the first eigenvalue of φ-Laplacian Δ_φ on Σ^n.*

Proof Since that λ is constant and W is a Killing timelike vector field on $\overline{M}_\varphi^{n+1}$, Corollary 15.3 guarantees that λ is in the spectrum of Δ_φ.

15.3 Stability of Spacelike Hypersurfaces in Weighted Spacetimes

Let λ_1 be the first eigenvalue of Δ_φ on Σ^n. If $\lambda = \lambda_1$, then the variational characterization of λ_1 (see, for instance, [69, Section 1]) gives

$$\lambda = \min_{u \in \mathcal{G} \setminus \{0\}} \frac{-\int_{\Sigma^n} u \Delta_\varphi(u) \, d\mu}{\int_{\Sigma^n} u^2 \, d\mu},$$

where $\mathcal{G}$ is defined in (15.54). It follows that, for any $u \in \mathcal{G}$,

$$\mathcal{J}''_\varphi(0)(u) = \int_{\Sigma^n} \left\{ u \Delta_\varphi(u) + \lambda u^2 \right\} d\mu \leq (-\lambda + \lambda) \int_{\Sigma^n} u^2 \, d\mu = 0,$$

and, according to Remark 15.3, $\psi : \Sigma^n \to \overline{M}_\varphi^{n+1}$ is φ-stable.

Now, suppose that $\psi : \Sigma^n \to \overline{M}_\varphi^{n+1}$ is φ-stable, so that $\mathcal{J}''_\varphi(0)(u) \leq 0$ for every $u \in \mathcal{G}$. Let u be an eigenfunction associated to the first eigenvalue λ_1 of Δ_φ.

By Remark 15.2 we obtain that there exists a volume-preserving variation of $\psi : \Sigma^n \to \overline{M}_\varphi^{n+1}$ whose variational field is uN. Consequently, by (15.55) we get

$$0 \geq \mathcal{J}''_\varphi(0)(u) = (-\lambda_1 + \lambda) \int_{\Sigma^n} u^2 \, d\mu.$$

Therefore, since $\lambda_1 \leq \lambda$, we must have $\lambda_1 = \lambda$. □

The following rigidity result holds.

Theorem 15.4 *Let $\overline{M}_\varphi^{n+1}$ be a timelike geodesically complete weighted conformally stationary spacetime endowed with a closed conformal timelike vector field V and whose weighted function φ does not depend on the parameter $s \in \mathbb{R}$ of the flow associated to v_V. Moreover, let $\psi : \Sigma^n \to \overline{M}_\varphi^{n+1}$ be a strongly φ-stable closed spacelike hypersurface. Suppose that the conformal factor ϕ_V of V satisfies the condition*

$$\frac{\partial \phi_V}{\partial s} \geq \max \{\phi_V H_\varphi, 0\}. \tag{15.78}$$

If the set where $\phi_V = 0$ has empty interior in Σ^n, then $\psi : \Sigma^n \to \overline{M}_\varphi^{n+1}$ is either φ-maximal or isometric to a leaf of the foliation $V^\perp$.

Proof Let us consider in $\overline{M}_\varphi^{n+1}$ the global parametrization (15.63). Since $\psi : \Sigma^n \to \overline{M}_\varphi^{n+1}$ is strongly φ-stable, by (15.55) it follows that

$$\int_{\Sigma^n} \left\{ \Delta_\varphi(u) - \{\overline{\mathrm{Ric}}_\varphi(N,N) + |A|^2\} u \right\} u \, d\mu \leq 0, \tag{15.79}$$

for any $u \in C^\infty(\Sigma^n)$.

Since H_φ is constant on Σ^n, making use of the smooth function η_V defined in (15.66), by (15.68) we have

$$\Delta_\varphi(\eta_V) - \{\overline{\mathrm{Ric}}_\varphi(N,N) + |A|^2\} \eta_V = n\{\phi_V H_\varphi - N(\phi_V)\}.$$

Thus, by (15.79) we have that

$$\int_{\Sigma^n} \{\phi_V H_\varphi - N(\phi_V)\} \eta_V d\mu \leq 0. \tag{15.80}$$

On the other hand, by (15.59) it follows that

$$N(\phi_V) = \langle N, \overline{\nabla}\phi_V \rangle = -v_V(\phi_V)\langle N, v_V \rangle = \frac{\partial \phi_V}{\partial s} \cosh \theta,$$

where θ is the hyperbolic angle between N and v. Substituting the above expression into (15.80), we have

$$\int_{\Sigma^n} \left(\frac{\partial \phi_V}{\partial s} \cosh \theta - \phi_V H_\varphi \right) \sqrt{-\langle V, V \rangle} \cosh \theta \, d\mu \leq 0.$$

Now, by (15.78) we obtain

$$0 \geq \int_{\Sigma^n} \left\{ \frac{\partial \phi_V}{\partial s} \cosh \theta - \phi_V H_\varphi \right\} \sqrt{-\langle V, V \rangle} \cosh \theta \, d\mu$$

$$\geq \int_{\Sigma^n} (\cosh \theta - 1) \frac{\partial \phi_V}{\partial s} \sqrt{-\langle V, V \rangle} \cosh \theta \, d\mu \geq 0.$$

Hence, we can write

$$\frac{\partial \phi_V}{\partial s} (\cosh \theta - 1) = 0 \quad \text{and} \quad \frac{\partial \phi_V}{\partial s} = \phi_V H_\varphi$$

on Σ^n.

Now, since H_φ is constant on $\Sigma(z)$, $\psi: \Sigma(z) \to \overline{M}_\varphi^{n+1}$ is either φ-minimal or $H_\varphi \neq 0$ on $\Sigma(z)$. If the last case occurs, since the zero set of ϕ_V has empty interior in $\Sigma(z)$, taking into account the above expression of $\frac{\partial \phi_V}{\partial t}$, one has that $\frac{\partial \phi_V}{\partial t} \neq 0$ on a dense subset of $\Sigma(z)$.

Hence, $\cos\theta = 1$ on this set. By continuity, $\cos\theta = 1$ on $\Sigma(z)$. Therefore, in this case, $\psi : \Sigma(z) \to \overline{M}_\varphi^{n+1}$ must be a leaf of the foliation $V^\perp$. □

We close this section observing that, when the ambient space is a weighted GRW spacetime, we can apply Theorem 15.4 obtaining the following extension of [67, Theorem 1.1].

Corollary 15.4 *Let* $\psi : \Sigma^n \to -\mathbb{R} \times_\varphi M_\varphi^n$ *be a closed, strongly φ-stable spacelike hypersurface. Suppose that*

$$f'' \geq \max\{f'H_\varphi, 0\}.$$

If the set where $f' = 0$ has empty interior in Σ^n, then $\psi : \Sigma^n \to -\mathbb{R} \times_\varphi M_\varphi^n$ is either φ-maximal or isometric to a slice $\{s_0\} \times M^n$, for some $s_0 \in \mathbb{R}$.

15.4 A Notion of φ-Stability in Weighted Standard Static Spacetimes

In this section we will study a notion of stability for closed spacelike hypersurface immersed in a weighted standard static spacetime with constant weighted mean curvature, via the first eigenvalue of the drift Laplacian. The results that will be presented are mainly based on Section 5 of [159].

Let $\psi : \Sigma^n \to M_\varphi^n \times_\rho \mathbb{R}_1$ be a compact spacelike hypersurface with boundary $\partial \Sigma^n$ (possibly empty). A *variation* of Ψ is as a smooth mapping

$$\begin{aligned} \Psi : (-\epsilon, \epsilon) \times \Sigma^n &\to M_\varphi^n \times_\rho \mathbb{R}_1 \\ (s, p) &\mapsto \Psi(s, p) \end{aligned}$$

such that the following conditions hold:

(i) For any $s \in (-\epsilon, \epsilon)$, the map $\Psi_s : \Sigma^n \to M_\varphi^n \times_\rho \mathbb{R}_1$ given by $\Psi_s(p) = \Psi(s, p)$ is a Riemannian immersion such that $\Psi_0 = \psi$;
(ii) $\Psi_s|_{\partial \Sigma^n} = \psi|_{\partial \Sigma^n}$, for all $s \in (-\epsilon, \epsilon)$.

In the sequel, let $d\Sigma_s$ be the volume element of the warping metric (4.1) induced on $\Sigma_s^n = \Psi_s(\Sigma^n)$ and let N_s be the future-pointing Gauss map along Σ_s^n. Moreover, we also consider in Σ_s^n the weighted volume form given by $d\sigma_s = e^{-\varphi}d\Sigma_s$. If $s = 0$ the classical notions on Σ^n are recovered. Moreover for any open subset Ω of $M_\varphi^n \times_\rho \mathbb{R}_1$ with compact closure, $\mathrm{Vol}_\varphi(\Omega)$ and $\mathrm{Area}_\varphi(\Omega)$ will denote the *weighted volume* and *weighted area* of Ω, respectively.

The *variational field* associated to the variation Ψ is the smooth vector field

$$\left.\frac{\partial \Psi}{\partial s}\right|_{s=0}.$$

If

$$u_s = -\left\langle \frac{\partial \Psi}{\partial s}, N_s \right\rangle, \tag{15.81}$$

we get

$$\left.\frac{\partial \Psi}{\partial s}\right|_{s=0} = u_0 N + \left(\left.\frac{\partial \Psi}{\partial s}\right|_{s=0}\right)^T.$$

The *balance of weighted volume* and the *weighted area functional* associated to the variation Ψ are respectively the functionals

$$\begin{aligned} \mathcal{V}_\varphi : (-\epsilon, \epsilon) &\to \mathbb{R} \\ s &\mapsto \mathcal{V}_\varphi(s) = \mathrm{Vol}_\varphi\left(\Psi\left([0,s] \times \Sigma^n\right)\right) = \int_{[0,s] \times \Sigma^n} \Psi^*(d\overline{\sigma}) \end{aligned}$$

and

$$\begin{aligned} \mathcal{A}_\varphi : (-\epsilon, \epsilon) &\to \mathbb{R} \\ s &\mapsto \mathcal{A}_\varphi(s) = \mathrm{Area}_\varphi\left(\Sigma_s^n\right) = \int_{\Sigma_s^n} d\sigma_s, \end{aligned}$$

where $d\overline{v}$ is the volume element on induced by the warping metric (4.1).

We say that the variation Ψ is *weighted volume-preserving* of Σ^n if $\mathcal{V}_\varphi(s) = \mathcal{V}_\varphi(0) = 0$, for all $s \in (-\epsilon, \epsilon)$.

The following result is a consequence in [152, Lemma 1 and Lemma 2].

Lemma 15.5 *Let* $\Psi : (-\epsilon, \epsilon) \times \Sigma^n \to M_\varphi^n \times_\rho \mathbb{R}_1$ *be a variation of the closed (i.e. compact and without boundary) spacelike hypersurface* $\psi : \Sigma^n \to M_\varphi^n \times_\rho \mathbb{R}_1$. *If u_s is the smooth function given in (15.81), then*

$$\frac{d}{ds}\mathcal{V}_\varphi(s) = \int_{\Sigma_s^n} u_s \, d\sigma_s \quad \text{and} \quad \frac{d}{ds}\mathcal{A}_\varphi(s) = n \int_{\Sigma_s^n} (H_\varphi)_s u_s \, d\sigma_s$$

where $(H_\varphi)_s = H_\varphi(s, \cdot)$ *denotes the φ-mean curvature of Σ_s^n. In particular, Ψ is weighted volume-preserving of Σ^n if and only if*

15.4 A Notion of φ-Stability in Weighted Standard Static Spacetimes

$$\int_{\Sigma_s^n} u_s \, d\sigma_s = 0$$

for all $s \in (-\epsilon, \epsilon)$.

Remark 15.4 Applying the same topological arguments used along the proof of [24, Proposition 3.2], we conclude that a closed spacelike hypersurface Σ^n immersed in a standard static spacetime $M^n \times_\rho \mathbb{R}_1$ exist if and only if the Riemannian base M^n is also compact. On the other hand, it is not difficult to verify that [64, Lemma 2.2] still remains valid for weighted standard static spacetimes. More precisely, given a closed spacelike hypersurface $\psi : \Sigma^n \to M_\varphi^n \times_\rho \mathbb{R}_1$, if $u \in C^\infty(\Sigma^n)$ is such that

$$\int_{\Sigma^n} u \, d\sigma = 0,$$

then there exists a weighted volume-preserving variation $\Psi : (-\epsilon, \epsilon) \times \Sigma^n \to M_\varphi^n \times_\rho \mathbb{R}_1$ of $\psi : \Sigma^n \to M_\varphi^n \times_\rho \mathbb{R}_1$ whose variational field is

$$\left.\frac{\partial \Psi}{\partial s}\right|_{s=0} = uN.$$

In order to characterize closed spacelike hypersurfaces $\psi : \Sigma^n \to M_\varphi^n \times_\rho \mathbb{R}_1$ with constant φ-mean curvature, we consider the variational problem of maximizing the weighted area functional $\mathcal{A}_\varphi$ for any variation $\Psi : (-\epsilon, \epsilon) \times \Sigma^n \to M_\varphi^n \times_\rho \mathbb{R}_1$ of $\psi : \Sigma^n \to M_\varphi^n \times_\rho \mathbb{R}_1$ that keeps the balance of weighted volume $\mathcal{V}_\varphi$ equal to zero.

To this aim, let us consider the weighted Jacobi functional

$$\begin{aligned} \mathcal{J}_\varphi : (-\epsilon, \epsilon) &\to \mathbb{R} \\ s &\mapsto \mathcal{J}_\varphi(s) = \mathcal{A}_\varphi(s) - \lambda \mathcal{V}_\varphi(s), \end{aligned} \tag{15.82}$$

where λ is a constant to be determined. As an immediate consequence of Lemma 15.5 we get that the first variation of $\mathcal{J}_\varphi$ takes the following form

$$\frac{d}{ds} \mathcal{J}_\varphi(s) = \int_{\Sigma_s^n} \{n (H_\varphi)_s - \lambda\} u_s \, d\sigma_s, \tag{15.83}$$

where u_s is the smooth function given in (15.81).

Furthermore, let

$$\overline{\mathcal{H}}_\varphi = \frac{1}{\text{Area}_\varphi(\Sigma^n)} \int_{\Sigma^n} H_\varphi \, d\sigma \tag{15.84}$$

be an integral mean of the φ-mean curvature H_φ on Σ^n. If H_φ is constant, we have

$$\overline{\mathcal{H}}_\varphi = H_\varphi. \tag{15.85}$$

Therefore, if we take $\lambda = n\overline{\mathcal{H}}_\varphi$, by (15.83) we have

$$\frac{d}{ds}\mathcal{J}_\varphi(s) = n\int_{\Sigma^n_s}\left\{(H_\varphi)_s - \overline{\mathcal{H}}_\varphi\right\}u_s\,d\sigma_s. \tag{15.86}$$

Arguing as in the proof of [62, Proposition 2.7], we get

Proposition 15.4 *The following statements are equivalent:*

(a) $\psi : \Sigma^n \to M^n_\varphi \times_\rho \mathbb{R}_1$ *is a closed spacelike hypersurface with constant φ-mean curvature H_φ;*

(b) *For any weighted volume-preserving variation* $\Psi : (-\epsilon, \epsilon) \times \Sigma^n \to M^n_\varphi \times_\rho \mathbb{R}_1$ *of* $\psi : \Sigma^n \to M^n_\varphi \times_\rho \mathbb{R}_1$,

$$\frac{d}{ds}\mathcal{A}_\varphi(0) = 0;$$

(c) *For every variation* $\Psi : (-\epsilon, \epsilon) \times \Sigma^n \to M^n_\varphi \times_\rho \mathbb{R}_1$ *of* $\psi : \Sigma^n \to M^n_\varphi \times_\rho \mathbb{R}_1$,

$$\frac{d}{ds}\mathcal{J}_\varphi(0) = 0.$$

Proof We prove that $(a) \Rightarrow (c) \Rightarrow (b) \Rightarrow (a)$.

$(a) \Rightarrow (c)$ The result follows directly by (15.85) and (15.86).

$(c) \Rightarrow (b)$ We have

$$0 = \frac{d}{ds}\mathcal{J}_\varphi(0) = \frac{d}{ds}\mathcal{A}_\varphi(0) + n\overline{\mathcal{H}}_\varphi\frac{d}{ds}\mathcal{V}_\varphi(0)$$

for any variation $\Psi : (-\epsilon, \epsilon) \times \Sigma^n \to M^n_\varphi \times_\rho \mathbb{R}_1$ of $\psi : \Sigma^n \to M^n_\varphi \times_\rho \mathbb{R}_1$. Moreover, if a variation preserves the volume of $\psi : \Sigma^n \to M^n_\varphi \times_\rho \mathbb{R}_1$ then $\frac{d}{ds}\mathcal{V}_\varphi(0) = 0$. Hence,

$$\frac{d}{ds}\mathcal{A}_\varphi(0) = 0$$

for the weighted volume-preserving variations $\Psi : (-\epsilon, \epsilon) \times \Sigma^n \to M^n_\varphi \times_\rho \mathbb{R}_1$ of $\psi : \Sigma^n \to M^n_\varphi \times_\rho \mathbb{R}_1$.

$(b) \Rightarrow (a)$ Suppose there is p_0 in Σ^n such that $(H_\varphi - \overline{\mathcal{H}}_\varphi)(p_0) \neq 0$. We can assume that $(H_\varphi - \overline{\mathcal{H}}_\varphi)(p_0) > 0$. By using the definition of $\overline{\mathcal{H}}_\varphi$ in (15.84) there exists a point $q_0 \in \Sigma^n$ such that $(H_\varphi - \overline{\mathcal{H}}_\varphi)(q_0) < 0$. Indeed, by (15.85) we have

15.4 A Notion of φ-Stability in Weighted Standard Static Spacetimes

$$\int_{\Sigma^n} (H_\varphi - \overline{\mathcal{H}}_\varphi)\, d\sigma = \int_{\Sigma^n} H_\varphi\, d\sigma - \overline{\mathcal{H}}_\varphi \operatorname{Area}_\varphi(\Sigma^n)$$

$$= \int_{\Sigma^n} H_\varphi\, d\sigma - \left(\frac{1}{\operatorname{Area}_\varphi(\Sigma^n)} \int_{\Sigma^n} H_\varphi\, d\sigma\right) \operatorname{Area}_\varphi(\Sigma^n)$$

$$= 0. \tag{15.87}$$

So, if $(H_\varphi - \overline{\mathcal{H}}_\varphi)(q) > 0$ for every $q \in \Sigma^n$, since there is $p_0 \in \Sigma^n$ such that $(H_\varphi - \overline{\mathcal{H}}_\varphi)(p_0) > 0$, then

$$\int_{\Sigma^n} (H_\varphi - \overline{\mathcal{H}}_\varphi)\, d\sigma > 0,$$

against (15.87).

Thus, the sets

$$\Sigma^+ = \left\{ q \in \Sigma^n : (H_\varphi - \overline{\mathcal{H}}_\varphi)(q) > 0 \right\},$$

and

$$\Sigma^- = \left\{ q \in \Sigma^n : (H_\varphi - \overline{\mathcal{H}}_\varphi)(q) < 0 \right\}$$

are well defined.

Now, consider nonnegative smooth functions ζ_1 and ζ_2 such that $p_0 \in \operatorname{supp} \zeta_1 \subset \Sigma^+$, $\operatorname{supp} \zeta_2 \subset \Sigma^-$ and

$$\int_{\Sigma^n} (\zeta_1 + \zeta_2)(H_\varphi - \overline{\mathcal{H}}_\varphi)\, d\sigma = 0,$$

where $\operatorname{supp} \zeta_1$ and $\operatorname{supp} \zeta_2$ denote the support of ζ_1 and the support of ζ_2, respectively. If we consider the smooth function $u = (\zeta_1 + \zeta_2)(H_\varphi - \overline{\mathcal{H}}_\varphi)$ then, according to Remark 15.4, there is a weighted volume-preserving variation $\Psi : (-\epsilon, \epsilon) \times \Sigma^n \to M_\varphi^n \times_\rho \mathbb{R}_1$ of $\psi : \Sigma^n \to M_\varphi^n \times_\rho \mathbb{R}_1$ whose variational field is

$$\left.\frac{\partial \Psi}{\partial s}\right|_{s=0} = uN.$$

By hypothesis and Lemma 15.5, one has

$$0 = \frac{d}{ds} \mathcal{A}_\varphi(0) = n \int_{\Sigma^n} H_\varphi u\, d\sigma.$$

Since
$$\int_{\Sigma^n} u\, d\sigma = 0,$$
we obtain
$$0 = n \int_{\Sigma^n} H_\varphi u\, d\sigma - n\overline{\mathcal{H}}_\varphi \int_{\Sigma^n} u\, d\sigma = n \int_{\Sigma^n} (H_\varphi - \overline{\mathcal{H}}_\varphi) u\, d\sigma$$
$$= n \int_{\Sigma^n} (\zeta_1 + \zeta_2)(H_\varphi - \overline{\mathcal{H}}_\varphi)^2 d\sigma > 0,$$

which is a contradiction. Therefore, $H_\varphi = \overline{\mathcal{H}}_\varphi$ on Σ^n. □

In particular, Proposition 15.4 guarantees that a closed spacelike hypersurface $\psi : \Sigma^n \to M_\varphi^n \times_\rho \mathbb{R}_1$ is a critical point of the variational problem described above if and only if its φ-mean curvature H_φ is constant.

Definition 15.5 Let $\psi : \Sigma^n \to M_\varphi^n \times_\rho \mathbb{R}_1$ be a closed spacelike hypersurface having constant φ-mean curvature. We say that $\psi : \Sigma^n \to M_\varphi^n \times_\rho \mathbb{R}_1$ is φ-stable if

$$\frac{d^2}{ds^2} \mathcal{A}_\varphi(0) \leq 0,$$

for the weighted volume-preserving variations $\Psi : \Sigma^n \times (-\epsilon, \epsilon) \to M_\varphi^n \times_\rho \mathbb{R}_1$ of $\psi : \Sigma^n \to M_\varphi^n \times_\rho \mathbb{R}_1$.

Let $\psi : \Sigma^n \to M_\varphi^n \times_\rho \mathbb{R}_1$ be a closed spacelike hypersurface as described in Definition 15.5. Set

$$\mathcal{G} = \left\{ u \in C^\infty(\Sigma^n) : \int_{\Sigma^n} u\, d\sigma = 0 \right\}. \tag{15.88}$$

Arguing as in [62], the following φ-stability criterion can be proved.

Proposition 15.5 *With the notations considered above, $\psi : \Sigma^n \to M_\varphi^n \times_\rho \mathbb{R}_1$ is φ-stable if and only if*

$$\frac{d^2}{ds^2} \mathcal{J}_\varphi(0)(u) \leq 0$$

for every $u \in \mathcal{G}$.

15.4 A Notion of φ-Stability in Weighted Standard Static Spacetimes

Proof Suppose that $\psi : \Sigma^n \to M_\varphi^n \times_\rho \mathbb{R}_1$ is φ-stable and consider $u \in \mathcal{G}$. By Remark 15.4, there is a weighted volume-preserving variation $\Psi : (-\epsilon, \epsilon) \times \Sigma^n \to M_\varphi^n \times_\rho \mathbb{R}_1$ of $\psi : \Sigma^n \to M_\varphi^n \times_\rho \mathbb{R}_1$ whose variational field is

$$\left.\frac{\partial \Psi}{\partial s}\right|_{s=0} = uN.$$

Then,

$$\frac{d^2}{ds^2}\mathcal{V}_\varphi(0)(u) = 0.$$

Hence, by (15.82) and Definition 15.5 we obtain

$$\frac{d^2}{ds^2}\mathcal{J}_\varphi(0)(u) = \frac{d^2}{ds^2}\mathcal{A}_\varphi(0)(u) - \lambda \frac{d^2}{ds^2}\mathcal{V}_\varphi(0)(u) = \frac{d^2}{ds^2}\mathcal{A}_\varphi(0)(u) \leq 0$$

Conversely, suppose that

$$\frac{d^2}{ds^2}\mathcal{J}_\varphi(0)(u) \leq 0$$

for every $u \in \mathcal{G}$. Let $\Psi : (-\epsilon, \epsilon) \times \Sigma^n \to M_\varphi^n \times_\rho \mathbb{R}_1$ be a weighted volume-preserving variation of $\psi : \Sigma^n \to M_\varphi^n \times_\rho \mathbb{R}_1$, and let uN be the normal component of the variation vector. By Lemma 15.5,

$$\int_{\Sigma^n} u \, d\sigma = \frac{d}{ds}\mathcal{V}_\varphi(0) = 0,$$

which implies that $u \in \mathcal{G}$. Therefore, our hypotheses yield

$$0 \geq \frac{d^2}{ds^2}\mathcal{J}_\varphi(0)(u) = \frac{d^2}{ds^2}\mathcal{A}_\varphi(0)(u) - \lambda \underbrace{\frac{d^2}{ds^2}\mathcal{V}_\varphi(0)(u)}_{=0} = \frac{d^2}{ds^2}\mathcal{A}_\varphi(0)(u),$$

which, according to Definition 15.5, tells us that $\psi : \Sigma^n \to M_\varphi^n \times_\rho \mathbb{R}_1$ is φ-stable. □

The expression of the second variation of the Jacobi functional $\mathcal{J}_\varphi$ is given in the following result.

Proposition 15.6 *Let* $\psi : \Sigma^n \to M_\varphi^n \times_\rho \mathbb{R}_1$ *be a closed spacelike hypersurface having constant φ-mean curvature H_φ. If* $\Psi : (-\epsilon, \epsilon) \times \Sigma^n \to M_\varphi^n \times_\rho \mathbb{R}_1$ *is a variation of* $\psi : \Sigma^n \to M_\varphi^n \times_\rho \mathbb{R}_1$ *then the second variation*

$$\frac{d^2}{ds^2} \mathcal{J}_\varphi(0)$$

of the weighted Jacobi functional $\mathcal{J}_\varphi$ is given by

$$\frac{d^2}{ds^2} \mathcal{J}_\varphi(0)(u) = \int_{\Sigma^n} u \, \mathcal{L}_\varphi(u) \, d\sigma, \tag{15.89}$$

for any $u \in C^\infty(\partial\Omega)$, where $\mathcal{L}_\varphi : C^\infty(\Sigma^n) \to C^\infty(\Sigma^n)$ is the weighted Jacobi operator defined by

$$\mathcal{L}_\varphi = \Delta_\varphi - \left\{ \widetilde{\mathrm{Ric}}_\varphi(N^*, N^*) - \frac{1}{\rho} \widetilde{\mathrm{Hess}}\, \rho(N^*, N^*) + \Theta^2 \frac{\widetilde{\Delta}_\varphi(\rho)}{\rho^3} + |A|^2 \right\}. \tag{15.90}$$

Here, Δ_φ and $\widetilde{\Delta}_\varphi$ represent the φ-Laplacians on Σ^n and M^n, respectively, $\Theta = \langle N, K \rangle$ is the angle function, N is the future-pointing Gauss map of $\psi : \Sigma^n \to M_\varphi^n \times_\rho \mathbb{R}_1$, $\widetilde{\mathrm{Ric}}_\varphi$ and $\widetilde{\mathrm{Hess}}$ are the Bakry-Émery-Ricci tensor and the Hessian operator on M^n, $|A|^2$ represents the square of the norm of the shape operator A of $\psi : \Sigma^n \to M_\varphi^n \times_\rho \mathbb{R}_1$ and N^* is the projection of N on the tangent bundle of M^n.

Proof Since H_φ is constant, by (15.86) and (15.85) we have that

$$\frac{d^2}{ds^2} \mathcal{J}_\varphi(0)(u_0) = n \int_{\Sigma^n} \left(\frac{\partial (H_\varphi)_s}{\partial s} \bigg|_{s=0} \right) u_0 \, d\sigma$$

$$+ n \int_{\Sigma^n} \underbrace{\left(H_\varphi - \overline{\mathcal{H}}_\varphi \right)}_{=0} \frac{\partial}{\partial s} \left(u_s \, d\sigma_s \right) \bigg|_{s=0},$$

where u_s is the smooth function given in (15.81).

On the other hand, arguing as in the proof of equation (3.5) in [103], we obtain

$$n \frac{\partial (H_\varphi)_s}{\partial s} \bigg|_{s=0} = \Delta_\varphi(u_0) - \left\{ \widetilde{\mathrm{Ric}}_\varphi(N, N) + |A^2| \right\} u_0.$$

Hence, one has

$$\frac{d^2}{ds^2} \mathcal{J}_\varphi(0)(u_0) = \int_{\Sigma^n} \left\{ \Delta_\varphi(u_0) - \left\{ \widetilde{\mathrm{Ric}}_\varphi(N, N) + |A|^2 \right\} u_0 \right\} u_0 \, d\sigma. \tag{15.91}$$

Now, by (8.21) and (15.91), it follows that

15.4 A Notion of φ-Stability in Weighted Standard Static Spacetimes

$$\frac{d^2}{ds^2}\mathcal{J}_\varphi(0)(u_0) = \int_{\Sigma^n} u_0 \, \mathfrak{L}_\varphi(u_0) \, d\sigma, \tag{15.92}$$

where $\mathfrak{L}_\varphi$ is given in (15.90). To finish the proof, we observe that the expression (15.92) depends only on the hypersurface Σ^n and on the function $u_0 \in C^\infty(\Sigma^n)$. □

To show our next result, let us recall that the *eigenvalue problem* for the drift Laplacian Δ_φ on a closed Riemannian manifold Σ^n is equivalent to either the existence or not existence of nontrivial solutions (i.e. not identically zero) $u \in C^\infty(\Sigma)$ for the partial differential equation

$$\Delta_\varphi(u) + \mu u = 0,$$

on Σ^n. In this case, the corresponding function u is an *eigenfunction* associated with the *eigenvalue* μ.

By the Spectral Theorem we know that the eigenvalues of Δ_φ are determined by a sequence of eigenvalues $\{\mu_j\}_{j=0}^{+\infty}$ satisfying

$$0 = \mu_0 < \mu_1 \leq \mu_2 \leq \cdots \leq \mu_j \leq \mu_{j+1} \leq \cdots,$$

repeated according to their multiplicity, and

$$\lim_{j \to +\infty} \mu_j = +\infty;$$

see, for instance, the paper [69, Section 1]. Moreover, the variational characterization of μ_1 gives

$$\mu_1 = \min_{u \in \mathcal{G}\setminus\{0\}} \frac{-\int_{\Sigma^n} u \Delta_\varphi(u) \, d\sigma}{\int_{\Sigma^n} u^2 \, d\sigma}, \tag{15.93}$$

where $\mathcal{G}$ is defined in (15.88).

We can now present our characterization of φ-stable closed spacelike hypersurfaces in a weighted standard static spacetime.

Theorem 15.5 *Let* $\psi : \Sigma^n \to M_\varphi^n \times_\rho \mathbb{R}_1$ *be a closed spacelike hypersurface with constant φ-mean curvature. Suppose that*

$$\mu = -\widetilde{\mathrm{Ric}}_\varphi(N^*, N^*) + \frac{1}{\rho}\widetilde{\mathrm{Hess}}\,\rho(N^*, N^*) - \Theta^2 \frac{\widetilde{\Delta}_\varphi(\rho)}{\rho^3} - |A|^2$$

is a nonzero constant on Σ^n, where Δ_φ and $\widetilde{\Delta}_\varphi$ represent the φ-Laplacians on Σ^n and M^n, respectively, $\Theta = \langle N, K \rangle$ is the angle function, N is the future-pointing Gauss map of $\psi : \Sigma^n \to M^n_\varphi \times_\rho \mathbb{R}_1$, $\widetilde{\mathrm{Ric}}_\varphi$ and $\widetilde{\mathrm{Hess}}$ are the Bakry-Émery-Ricci tensor and the Hessian operator on M^n, $|A|^2$ denotes the square of the norm of the shape operator A of $\psi : \Sigma^n \to M^n_\varphi \times_\rho \mathbb{R}_1$ and N^* is the projection of N on the tangent bundle of M^n. Then $\psi : \Sigma^n \to M^n_\varphi \times_\rho \mathbb{R}_1$ is φ-stable if and only if μ is the first nonzero eigenvalue of drift Laplacian Δ_φ on Σ^n.

Proof Initially, since the φ-mean curvature of $\psi : \Sigma^n \to M^n_\varphi \times_\rho \mathbb{R}_1$ and μ are constant on Σ^n, from Proposition 8.2 we can see that μ belongs to the sequence of eigenvalues $\{\mu_j\}_{j=0}^{+\infty}$ of the drift Laplacian Δ_φ on Σ^n.

If $\mu = \mu_1$, then by (15.89), (15.90) and (15.93) we obtain

$$\frac{d^2}{ds^2} \mathcal{J}_\varphi(0)(u) = \int_{\Sigma^n} \{u \Delta_\varphi(u) + \mu u^2\} d\sigma \leq (-\mu + \mu) \int_{\Sigma^n} u^2 d\sigma = 0$$

for any $u \in \mathcal{G}$. According to Proposition 15.5, $\psi : \Sigma^n \to M^n_\varphi \times_\rho \mathbb{R}_1$ is φ-stable.

Conversely, suppose that $\psi : \Sigma^n \to M^n_\varphi \times_\rho \mathbb{R}_1$ is φ-stable, so that

$$\frac{d^2}{ds^2} \mathcal{J}_\varphi(0)(u) \leq 0,$$

for every $u \in \mathcal{G}$. Let u be an eigenfunction associated to the first nonzero eigenvalue μ_1 of Δ_φ.

Hence, by (15.89) and (15.90) we get

$$0 \geq \frac{d^2}{ds^2} \mathcal{J}_\varphi(0)(u) = (-\mu_1 + \mu) \int_{\Sigma^n} u^2 d\sigma.$$

Therefore, since $\mu_1 \leq \mu$, we must have $\mu_1 = \mu$. □

Remark 15.5 From the proof of Theorem 15.5 it is immediate to verify that any constant φ-mean curvature closed spacelike hypersurface $\psi : \Sigma^n \to M^n_\varphi \times_\rho \mathbb{R}_1$ for which

$$\mu_1 + \widetilde{\mathrm{Ric}}_\varphi(N^*, N^*) - \frac{1}{\rho} \widetilde{\mathrm{Hess}}\, \rho(N^*, N^*) + \Theta^2 \frac{\widetilde{\Delta}_\varphi(\rho)}{\rho^3} + |A|^2 \geq 0,$$

on Σ^n, is stable. Here μ_1 is the first nonzero eigenvalue of drift Laplacian Δ_φ on Σ^n. In particular, this happens when $\widetilde{\mathrm{Ric}}_\varphi \geq 0$ and ρ is constant on M^n.

L_φ-Stability of Zero φ-Mean Curvature Hypersurfaces

16.1 Introduction

It is well-known that a stable surface is area-minimizing relative to nearby surfaces with the same boundary. In variational terms, minimal surfaces are critical points of the area functional for compactly supported normal variations. In this setting, a minimal surface is stable if the second derivative of the area functional is nonnegative for such normal variations. On the other hand, when the second variation is negative for some deformation, there are nearby surfaces of smaller area, and the surface is called unstable.

In their seminal work [61], Barbosa and do Carmo proved that for a surface Σ^2 immersed in $\mathbb{R}^3$, if the area of the spherical image $N(D) \subset \mathbb{S}^2$ of a domain $D \subset \Sigma^2$ is smaller than 2π, then D must be stable. Next, do Carmo and Peng [165] proved that the planes are the only complete stable surfaces of $\mathbb{R}^3$.

Meanwhile, Shoen and Yau [274, 275] used the existence of certain area-minimizing surfaces to obtain topological obstructions to the existence of metrics with positive scalar curvature. Afterwards, Shoen [272] obtained a bound for the Gaussian curvature of stable surfaces which allows one to conclude that every complete stable minimal surface immersed a three-dimensional Riemannian manifold with nonnegative Ricci curvature is totally geodesic.

Later on, Meeks and Rosenberg [238] established a classification for the stable properly embedded orientable minimal surfaces in the Riemannian product space $\mathbb{R} \times M^2$, where M^2 is a closed orientable Riemannian surface.

More recently, Aledo and Rubio [12] obtained several stability results for minimal two-sided surfaces immersed in a wide class of three-dimensional Riemannian warped products, which includes the class of Riemannian products. As a consequence, they provided some Bernstein-type results in these ambient spaces.

Meanwhile, E.L. de Lima et al. [147, 155] established some stability results concerning hypersurfaces immersed with zero weighted mean curvature in certain semi-Riemannian warped products with density.

This chapter is dedicated to the study of the L_φ-stability of zero φ-mean curvature hypersurfaces immersed into an $(n+1)$-dimensional weighted semi-Riemannian warped product space $\overline{M}_\varphi^{n+1} = (\epsilon I \times_f M^n)_\varphi$ endowed with a weight function φ, where L_φ stands for the weighted Jacobi operator related to the second variation formula of the weighted area functional

$$\mathrm{vol}_\varphi(\Sigma^n) = \int_{\Sigma^n} e^{-\varphi} d\Sigma^n,$$

of the hypersurface Σ^n.

More precisely, here

$$L_\varphi = \Delta_\varphi + \epsilon \left(|A|^2 + \overline{\mathrm{Ric}}_\varphi(N, N) \right),$$

where Δ_φ stands the drift Laplacian, A is the Weingarten operator of Σ^n, $\overline{\mathrm{Ric}}_\varphi$ denotes the Bakry-Émery-Ricci tensor of $\overline{M}_\varphi^{n+1}$ and N is the orientation of Σ^n.

In this setting, our purpose is to present results which give a sufficient condition for these hypersurfaces to be L_φ-stable.

The results presented in this chapter are mainly based on the papers [12, 147, 155].

16.2 The Weighted Jacobi Operator L_φ

Let $(\overline{M}^{n+1}, \langle \cdot, \cdot \rangle)$ be an $(n+1)$-dimensional oriented Riemannian or Lorentzian manifold and let $\varphi : \overline{M}^{n+1} \to \mathbb{R}$ be a smooth function. We recall that the weighted manifold $\overline{M}_\varphi^{n+1}$ associated with $\overline{M}^{n+1}$ and φ is the triple $(\overline{M}^{n+1}, \langle \cdot, \cdot \rangle, e^{-\varphi} d\overline{M}^{n+1})$, where $d\overline{M}$ denotes the standard volume element of $\overline{M}^{n+1}$ induced by the metric $\langle \cdot, \cdot \rangle$. We will refer to function φ as the weight function of the weighted manifold $\overline{M}_\varphi^{n+1}$.

In this setting, for a weighted manifold $\overline{M}_\varphi^{n+1}$ we already know that an important and natural tensor is the Bakry-Émery-Ricci tensor $\overline{\mathrm{Ric}}_\varphi$, given by

$$\overline{\mathrm{Ric}}_\varphi = \overline{\mathrm{Ric}} + \overline{\mathrm{Hess}}\varphi,$$

where $\overline{\mathrm{Ric}}$ is the Ricci tensor of $\overline{M}^{n+1}$ and $\overline{\mathrm{Hess}}\varphi$ denotes the Hessian of φ on $\overline{M}^{n+1}$. In particular, if φ is constant then $\overline{\mathrm{Ric}}_\varphi$ is simply the standard Ricci tensor $\overline{\mathrm{Ric}}$ of $\overline{M}^{n+1}$.

Let $\psi : \Sigma^n \to \overline{M}_\varphi^{n+1}$ be an isometrically immersed orientable Riemannian manifold into $\overline{M}_\varphi^{n+1}$. Then Σ^n becomes automatically a weighted Riemannian manifold by the

16.2 The Weighted Jacobi Operator L_φ

weighted structure induced from $\overline{M}_\varphi^{n+1}$. In this case, according to Gromov [197], the weighted mean curvature, or simply the φ-mean curvature, H_φ of Σ^n is defined by

$$nH_\varphi = nH + \epsilon \langle \overline{\nabla}\varphi, N \rangle, \tag{16.1}$$

where H denotes the standard mean curvature of Σ^n with respect to its orientation, $\epsilon = 1$ if $\overline{M}^{n+1}$ is a Riemannian manifold, and $\epsilon = -1$ if $\overline{M}^{n+1}$ is a Lorentzian manifold. In particular, when φ is constant we have that $H_\varphi = H$ recovering the usual definition of mean curvature. When the ambient space is Riemannian and the φ-mean curvature H_φ vanishes identically on Σ^n we say that Σ^n is a φ-minimal hypersurface. In the Lorentzian setting if the φ-mean curvature H_φ vanishes identically on Σ^n, we say that Σ^n is a φ-maximal hypersurface. In both cases, by (16.1) we have that the mean curvature H of Σ^n satisfies the following equation

$$nH = -\epsilon \langle \overline{\nabla}\varphi, N \rangle. \tag{16.2}$$

It is worth to point out that the research on the geometry of hypersurfaces having constant φ-mean curvature and, in particular, the investigations on the behavior of hypersurfaces immersed into a weighted ambient space with φ-mean curvature identically zero, constitutes a recent and fruitful topic into the theory of isometric immersions. We just mention herr the papers [100, 104, 105, 114, 174] and [203, 204, 211, 233, 265, 269], as well as the references therein.

It is well-known that Σ^n has zero φ-mean curvature if and only if Σ^n is a critical point of the weighted area functional defined by

$$\mathrm{vol}_\varphi(\Sigma^n) = \int_{\Sigma^n} e^{-\varphi} d\Sigma^n,$$

for any variation of Σ^n with compact support and fixed boundary. It is natural to wonder whether these hypersurfaces has the property of to minimize (if the ambient space is Riemannian) or maximize (if the ambient space is Lorentzian) the weighted area functional; see, for instance, [100, 114, 174, 211] and the references therein.

In order to attach this problem, it is very useful to know the second variation formula of the weighted area functional. To this aim, let us recall that the φ-divergence operator on Σ^n is defined by

$$\mathrm{div}_\varphi(X) = e^\varphi \mathrm{div}(e^{-\varphi} X), \tag{16.3}$$

where X is a tangent vector field on Σ^n and div denotes the standard divergence operator on Σ^n. By (16.3) we can define the φ-Laplacian of Σ^n by

$$\Delta_\varphi u = \mathrm{div}_\varphi(\nabla u) = \Delta u - \langle \nabla\varphi, \nabla u \rangle, \tag{16.4}$$

where u is a smooth function on Σ^n, Δ denotes the Laplacian operator and ∇ stands for the Levi-Civita connection of Σ^n induced by the Levi-Civita connection $\overline{\nabla}$ by the ambient space $\overline{M}_\varphi^{n+1}$.

Now let V be a normal compactly supported variation of Σ^n and take $\zeta \in C_0^\infty(\Sigma)$ such that $V = \zeta N$, where N determines the orientation of Σ^n. If the φ-mean curvature H_φ of Σ^n vanishes identically, then it is well-known that the second variation of the weighted area functional is given by

$$\frac{d^2}{dt^2}\mathrm{vol}_\varphi(\Sigma^n)|_{t=0} = -\epsilon \int_{\Sigma^n} \zeta L_\varphi \zeta \, d\Sigma^n, \qquad (16.5)$$

where the weighted Jacobi operator L_φ is defined by

$$L_\varphi = \Delta_\varphi + \epsilon \left(|A|^2 + \overline{\mathrm{Ric}}_\varphi(N, N) \right);$$

see [114] for the Riemannian case, as well as [147] for the Lorentzian setting.

Finally, we say that Σ^n is L_φ-stable if it minimizes (resp. maximizes) the weighted are functional in the Riemannian case (resp. Lorentzian case), that is,

$$\frac{d^2}{dt^2}\mathrm{vol}_\varphi(\Sigma^n)\Big|_{t=0} \geq 0 \quad (\text{resp.}, \leq 0).$$

16.3 L_φ-Stability of φ-Minimal Hypersurfaces

Let $\overline{M}_\varphi^{n+1} = \left(I \times_f M^n \right)_\varphi$ be a weighted Riemannian warped product space and let $\psi : \Sigma^n \to \overline{M}_\varphi^{n+1}$ be a φ-minimal two-sided hypersurface. Then Eq. (16.5) says that the second variation formula of the weighted area functional is given by

$$\frac{d^2}{dt^2}\mathrm{vol}_\varphi(\Sigma^n)\Big|_{t=0} = -\int_{\Sigma^n} \zeta L_\varphi \zeta \, d\Sigma^n, \qquad (16.6)$$

where $V = \zeta N$ is a normal compactly supported variation of Σ^n and the weighted Jacobi operator L_φ is defined by

$$L_\varphi = \Delta_\varphi + |A|^2 + \overline{\mathrm{Ric}}_\varphi(N, N). \qquad (16.7)$$

In particular, (16.6) depends only on $\zeta \in C_0^\infty(\Sigma)$. In this setting, let us emphasize the following definition introduced in previous chapter:

Let Σ^n be a hypersurface as above. We say that Σ^n is L_φ-stable if, for any compactly supported smooth function $\zeta \in C_0^\infty(\Sigma)$, it holds that

16.3 L_φ-Stability of φ-Minimal Hypersurfaces

$$\frac{d^2}{dt^2}\mathrm{vol}_\varphi(\Sigma^n)\big|_{t=0} = -\int_{\Sigma^n} \zeta L_\varphi \zeta e^{-\varphi} d\Sigma^n \geq 0.$$

The following auxiliary result, which gives a sufficient condition for a φ-minimal hypersurfaces to be L_φ-stable, will be crucial in the sequel.

Lemma 16.1 *Let* $\psi : \Sigma^n \to \overline{M}_\varphi^{n+1}$ *be a f-minimal two-sided hypersurface immersed into a weighted Riemannian warped product* $\overline{M}_\varphi^{n+1} = (I \times_f M^n)_\varphi$. *If there exists a positive smooth function* $u \in C^\infty(\Sigma^n)$ *such that* $L_\varphi u \leq 0$, *then* Σ^n *is* L_φ-*stable.*

Proof Assume that there exists a function $u \in C^\infty(\Sigma^n)$ such that $L_\varphi u \leq 0$ and take $\zeta \in C_0^\infty(\Sigma)$. Then, we can choose $\eta \in C_0^\infty(\Sigma^n)$ satisfying $\zeta = \eta u$. Hence, by (16.7) we have

$$\int_{\Sigma^n} \zeta L_\varphi \zeta e^{-\varphi} d\Sigma^n = \int_\Sigma \eta u L_\varphi(\eta u) e^{-\varphi} d\Sigma^n$$

$$= \int_{\Sigma^n} \Big[\eta^2 u L_\varphi u + \eta u^2 \Delta\eta$$

$$\qquad + 2\eta u \langle \nabla u, \nabla \eta \rangle - \eta u^2 \langle \nabla \eta, \nabla \varphi \rangle \Big] e^{-\varphi} d\Sigma^n$$

$$\leq \int_{\Sigma^n} \Big[\eta u^2 \Delta\eta + 2\eta u \langle \nabla u, \nabla \eta \rangle - \eta u^2 \langle \nabla \eta, \nabla f \rangle \Big] e^{-\varphi} d\Sigma^n$$

$$= \int_{\Sigma^n} \Big[\eta u^2 \Delta\eta + \frac{1}{2}\langle \nabla u^2, \nabla \eta^2 \rangle - \eta u^2 \langle \nabla \eta, \nabla \varphi \rangle \Big] e^{-\varphi} d\Sigma^n. \quad (16.8)$$

On the other hand, we also have

$$\mathrm{div}(u^2 \nabla \eta^2) = \langle \nabla u^2, \nabla \eta^2 \rangle + u^2 \Delta\eta^2 = \langle \nabla u^2, \nabla \eta^2 \rangle + 2\eta u^2 \Delta\eta + 2u^2|\nabla \eta|^2.$$

Therefore, the weighted version of Divergence Theorem (see [100, Lemma 2.2]) ensures that

$$\int_{\Sigma^n} \zeta L_\varphi \zeta e^{-\varphi} d\Sigma^n \leq \int_{\Sigma^n} \Big[\frac{1}{2}\mathrm{div}(u^2 \nabla \eta^2) - \eta u^2 \langle \nabla \eta, \nabla f \rangle - u^2|\nabla \eta|^2\Big] e^{-\varphi} d\Sigma^n$$

$$= \int_{\Sigma^n} \Big[\frac{1}{2}\mathrm{div}_\varphi(u^2 \nabla \eta^2) - u^2|\nabla \eta|^2\Big] e^{-\varphi} d\Sigma^n$$

$$\leq -\int_{\Sigma^n} u^2|\nabla \eta|^2 e^{-\varphi} d\Sigma^n \leq 0$$

thanks to (16.8). This shows that Σ^n is L_φ-stable. □

Remark 16.1 The converse of Lemma 16.1 is also true. A direct proof can be found in [174, Lemma 2.1]; see also [211, Proposition 3].

By a splitting theorem due to Fang et al. (see [176, Theorem 1.1]) if a weighted Riemannian warped product $\overline{M}_\varphi^{n+1} = \left(I \times_f M^n\right)_\varphi$ with bounded weight function φ has $\overline{\mathrm{Ric}}_\varphi$ nonnegative, then φ must be constant along I. Motivated by this fact, in our main result we consider weighted Riemannian warped product $\overline{M}_\varphi^{n+1}$ whose weight function φ does not depend on the parameter $t \in I$, that is, $\langle \overline{\nabla}\varphi, \partial_t \rangle = 0$. For simplicity, we will denote such a manifold by $\overline{M}_\varphi^{n+1} = I \times_f M_\varphi^n$.

Now, we are ready to state the main result.

Theorem 16.1 *Let $\psi : \Sigma^n \to \overline{M}_\varphi^{n+1}$ be a f-minimal two-sided hypersurface immersed into a weighted Riemannian warped product $\overline{M}_\varphi^{n+1} = I \times_f M_\varphi^n$. Setting $\tilde{\Theta} = f\Theta$ we have*

$$L_\varphi \tilde{\Theta} = -n \frac{f''}{f} \tilde{\Theta}. \tag{16.9}$$

Moreover, the following holds:

(a) *If the angle function Θ has strict sign and the warping function satisfies $f'' \geq 0$ on Σ^n, then Σ^n is L_φ-stable;*
(b) *If Σ^n is compact, the angle function Θ has strict sign and the warping function satisfies $f'' \leq 0$ on Σ^n, then Σ^n is L_φ-stable if and only if $f'' = 0$ on Σ^n;*
(c) *If Σ^n is compact, Θ does not vanish identically and $f'' < 0$ on Σ^n, then Σ^n cannot be L_φ-stable.*

Proof Let us first prove (16.9) observing that by Proposition 1.1 we have

$$\Delta \tilde{\Theta} = -nf \partial_t^\top (H) - nf'H - nN(f') - \left(|A|^2 + \overline{\mathrm{Ric}}(N, N)\right) \tilde{\Theta}, \tag{16.10}$$

see also [98, Proposition 2.1].

Besides, since $f \partial_t$ is a conformal vector field on $\overline{M}_\varphi^{n+1}$, then $\nabla \tilde{\Theta} = -fA(\partial_t^\top)$. Hence by (16.2) on account of the main assumption on the weight function and Hessian's definition, with a straightforward computation, we obtain

$$n\partial_t^\top(H) = -\partial_t^\top \langle \overline{\nabla}\varphi, N \rangle$$
$$= -\langle \overline{\nabla}_{\partial_t} \overline{\nabla}\varphi, N\rangle + \Theta\langle \overline{\nabla}_N \overline{\nabla}\varphi, N\rangle + \langle \overline{\nabla}\varphi, A(\partial_t^\top)\rangle$$
$$= -\overline{\mathrm{Hess}}\varphi(N, \partial_t) + \Theta\overline{\mathrm{Hess}}\varphi(N, N) - f^{-1}\langle \overline{\nabla}\varphi, \nabla\tilde{\Theta}\rangle. \tag{16.11}$$

On the other hand, it is not difficult to verify that

16.3 L_φ-Stability of φ-Minimal Hypersurfaces

$$\overline{\text{Hess}}\varphi(N, \partial_t) = -f^{-1}f'\langle\overline{\nabla}\varphi, N\rangle = nf^{-1}f'H. \tag{16.12}$$

Hence, Eqs. (16.10), (16.11) and (16.12) yield

$$\Delta\tilde{\Theta} = \langle\nabla\varphi, \nabla\tilde{\Theta}\rangle - nN(f') - \left(|A|^2 + \overline{\text{Ric}}_\varphi(N, N)\right)\tilde{\Theta}.$$

Thus, by (16.4) we obtain that

$$\Delta_\varphi\tilde{\Theta} = \Delta\tilde{\Theta} - \langle\nabla\varphi, \nabla\tilde{\Theta}\rangle = -nN(f') - \left(|A|^2 + \overline{\text{Ric}}_\varphi(N, N)\right)\tilde{\Theta},$$

which implies that Eq. (16.9) holds.

To prove item (a), we observe that since the angle function has strict sign, for an appropriated choice of N, we can suppose that Θ is a positive function. Therefore, since the warping function satisfies $f'' \geq 0$ on Σ^n, it follows by Lemma 16.1 that Σ^n is L_φ-stable.

Now, let us prove item (b). We can suppose again that Θ is a positive function. Since Σ^n is L_φ-stable, we infer

$$0 \leq -\int_{\Sigma^n} \tilde{\Theta}L_\varphi\tilde{\Theta}e^{-\varphi}d\Sigma^n = \int_{\Sigma^n} n\frac{f''}{f}\tilde{\Theta}^2 e^{-\varphi}d\Sigma^n \leq 0,$$

which immediately gives $f'' = 0$ on Σ^n. The converse is trivial.

Finally, since

$$-\int_{\Sigma^n} \tilde{\Theta}L_\varphi\tilde{\Theta}e^{-\varphi}d\Sigma^n = \int_{\Sigma^n} n\frac{f''}{f}\tilde{\Theta}^2 e^{-\varphi}d\Sigma^n < 0,$$

that is, Σ^n is not L_φ-stable. Hence, item (c) is also proved. □

It is worth point out that Theorem 16.1 gives a generalization of Theorems 3, 13 and 14 due to Aledo and Rubio [12].

Remark 16.2 Let $\overline{M}_\varphi^{n+1} = I \times_f M_\varphi^n$ be a weighted Riemannian warped product and assume that its fiber M^n is compact. In particular, the slices in $\overline{M}_\varphi^{n+1}$ are compact. Assume also that $f'' \leq 0$ and

$$\Omega = \{t \in I : f''(t) = 0\},$$

is a set of isolated points. Then, for every $t \in \Omega$ such that $f'(t) = 0$, the φ-minimal slice $\Sigma_t = \{t\} \times M^n$ is L_φ-stable.

16.4 L_φ-Stability of φ-Maximal Hypersurfaces

Let $\overline{M}_\varphi^{n+1} = (-I \times_f M^n)_\varphi$ be a weighted GRW spacetime and let $\psi : \Sigma^n \to \overline{M}_\varphi^{n+1}$ be a φ-maximal spacelike hypersurface. Let $V = \zeta N$ be a normal compactly supported variation of Σ^n. Equation (16.5) yields

$$\frac{d^2}{dt^2}\mathrm{vol}_\varphi(\Sigma^n)\big|_{t=0} = \int_{\Sigma^n} \zeta L_\varphi \zeta \, d\Sigma^n,$$

where the weighted Jacobi operator L_φ is given by

$$L_\varphi = \Delta_\varphi - \left(|A|^2 + \overline{\mathrm{Ric}}_\varphi(N, N)\right). \tag{16.13}$$

Hence, the second variation of Σ^n depends only on $\zeta \in C_0^\infty(\Sigma^n)$.

Now, let Σ^n be a hypersurface as above and recall the following definition:

We say that Σ^n is L_φ-stable if, for any compactly supported smooth function $\zeta \in C_0^\infty(\Sigma)$, it holds that

$$\frac{d^2}{dt^2}\mathrm{vol}_\varphi(\Sigma^n)\big|_{t=0} = \int_{\Sigma^n} \zeta L_\varphi \zeta e^{-\varphi} d\Sigma^n \leq 0.$$

In the Lorentzian case Lemma 16.1 assumes the form given below.

Lemma 16.2 *Let $\psi : \Sigma^n \to \overline{M}_\varphi^{n+1}$ be a φ-maximal spacelike hypersurface immersed into a weighted GRW spacetime $\overline{M}_\varphi^{n+1} = -I \times_f M_\varphi^n$. If there exists a positive smooth function $u \in C^\infty(\Sigma^n)$ such that $L_\varphi u \leq 0$, then Σ^n is L_φ-stable.*

Proof Let u be such a function and take $\zeta \in C_0^\infty(\Sigma^n)$. Then, we can choose $\eta \in C_0^\infty(\Sigma^n)$ satisfying $\zeta = \eta u$. Hence, by (16.13) it follows that

$$\int_{\Sigma^n} \zeta L_\varphi \zeta e^{-\varphi} d\Sigma^n = \int_{\Sigma^n} \eta u L_\varphi(\eta u) e^{-\varphi} d\Sigma^n$$

$$= \int_{\Sigma^n} \left[\eta u \left(\Delta_\varphi \eta u - \left(|A|^2 + \overline{\mathrm{Ric}}_\varphi(N, N)\right)\eta u\right)\right] e^{-\varphi} d\Sigma^n$$

$$= \int_{\Sigma^n} \Big[\eta u \big(\eta \Delta_\varphi u + u \Delta \eta + 2\langle \nabla u, \nabla \eta\rangle - u\langle \nabla \eta, \nabla \varphi\rangle$$

$$- \left(|A|^2 + \overline{\mathrm{Ric}}_\varphi(N, N)\right)\eta u\big)\Big] e^{-\varphi} d\Sigma^n$$

$$= \int_{\Sigma^n} \left[\eta^2 u L_\varphi u + \eta u^2 \Delta \eta + 2\eta u \langle \nabla u, \nabla \eta\rangle\right.$$

16.4 L_φ-Stability of φ-Maximal Hypersurfaces

$$-\eta u^2 \langle \nabla \eta, \nabla \varphi \rangle \Big] e^{-\varphi} d\Sigma^n$$

$$\leq \int_{\Sigma^n} \Big[\eta u^2 \Delta \eta + 2\eta u \langle \nabla u, \nabla \eta \rangle - \eta u^2 \langle \nabla \eta, \nabla \varphi \rangle \Big] e^{-\varphi} d\Sigma^n$$

$$= \int_{\Sigma^n} \Big[\eta u^2 \Delta \eta + \frac{1}{2} \langle \nabla u^2, \nabla \eta^2 \rangle$$

$$-\eta u^2 \langle \nabla \eta, \nabla \varphi \rangle \Big] e^{-\varphi} d\Sigma^n. \tag{16.14}$$

On the other hand, we also have

$$\mathrm{Div}(u^2 \nabla \eta^2) = \langle \nabla u^2, \nabla \eta^2 \rangle + u^2 \Delta \eta^2$$

$$= \langle \nabla u^2, \nabla \eta^2 \rangle + 2\eta u^2 \Delta \eta + 2u^2 |\nabla \eta|^2. \tag{16.15}$$

Hence, the weighted version of Divergence Theorem, in addition to (16.14) and (16.15), yield

$$\int_{\Sigma^n} \zeta L_\varphi \zeta e^{-\varphi} d\Sigma^n \leq \int_{\Sigma^n} \Big[\frac{1}{2} \mathrm{Div}(u^2 \nabla \eta^2) - \eta u^2 \langle \nabla \eta, \nabla f \rangle - u^2 |\nabla \eta|^2 \Big] e^{-\varphi} d\Sigma^n$$

$$= \int_{\Sigma^n} \Big[\frac{1}{2} \mathrm{Div}_\varphi (u^2 \nabla \eta^2) - u^2 |\nabla \eta|^2 \Big] e^{-\varphi} d\Sigma^n$$

$$\leq - \int_{\Sigma^n} u^2 |\nabla \eta|^2 e^{-\varphi} d\Sigma^n \leq 0.$$

Therefore Σ^n is L_φ-stable. The proof is now complete. $\square$

Let $\overline{M}^{n+1} = -I \times_f M^n$ be a GRW spacetime and $\varphi : \overline{M}^{n+1} \to \mathbb{R}$ be a smooth function on $\overline{M}^{n+1}$. By a splitting theorem due to Case (see [101, Theorem 1.2]) if the weight function φ is bounded and $\overline{\mathrm{Ric}}_\varphi(T, T) \geq 0$ for all the timelike vector fields $T \in \mathfrak{X}(\Sigma)$, then φ must be constant along I. Motivated by this result, here we consider weighted GRW spacetimes $\overline{M}_\varphi^{n+1}$ whose weight function φ does not depend on the parameter $t \in I$. In other words, $\langle \overline{\nabla} \varphi, \partial_t \rangle = 0$. For simplicity, we use the notation $\overline{M}_\varphi^{n+1} = -I \times_f M_\varphi^n$.

In what follows, we take the orientation N in the same time-orientation of ∂_t, that is, $\Theta = \langle N, \partial_t \rangle \leq -1$.

Theorem 16.2 *Let $\psi : \Sigma^n \to \overline{M}_\varphi^{n+1}$ be a φ-maximal spacelike hypersurface immersed into a weighted GRW spacetime $\overline{M}_\varphi^{n+1} = -I \times_f M_\varphi^n$. Setting $\tilde{\Theta} = f \Theta$ we have*

$$L_\varphi \tilde{\Theta} = n \frac{f''}{f} \tilde{\Theta}. \tag{16.16}$$

Moreover, the following holds:

(a) *If $f'' \leq 0$ on Σ^n, then Σ^n is L_φ-stable.*
(b) *If Σ^n is compact and $f'' \geq 0$ on Σ^n, then Σ^n is L_φ-stable if and only if $f'' = 0$ on Σ^n.*
(c) *If Σ^n is compact and $f'' > 0$ on Σ^n, then Σ^n cannot be L_φ-stable.*

Proof Let us prove Eq. (16.16). By Proposition 1.1 one has

$$\Delta \tilde{\Theta} = n f \partial_t^\top(H) + n f' H - n N(p') + \left(|A|^2 + \overline{\mathrm{Ric}}(N, N)\right) \tilde{\Theta}, \qquad (16.17)$$

see also [98, Theorem 2.1].

On the o.her hand, as in the proof of Theorem 16.1, we have

$$n \partial_t^\top(H) = \overline{\mathrm{Hess}}\varphi(N, \partial_t) + \Theta \overline{\mathrm{Hess}}\varphi(N, N) + f^{-1} \langle \nabla \varphi, \nabla \tilde{\Theta} \rangle \qquad (16.18)$$

and

$$\overline{\mathrm{Hess}}\varphi(N, \partial_t) = -f^{-1} f' \langle \overline{\nabla}\varphi, N \rangle = -n f^{-1} f' H. \qquad (16.19)$$

By (16.17), (16.18) and (16.19), one has

$$\Delta \tilde{\Theta} = \langle \nabla \varphi, \nabla \tilde{\Theta} \rangle - n N(f') + \left(|A|^2 + \overline{\mathrm{Ric}}_\varphi(N, N)\right) \tilde{\Theta}.$$

Thus, by (16.4) we get that

$$\Delta_\varphi \tilde{\Theta} = \Delta \tilde{\Theta} - \langle \nabla \varphi, \nabla \tilde{\Theta} \rangle = -n N(f') + \left(|A|^2 + \overline{\mathrm{Ric}}_\varphi(N, N)\right) \tilde{\Theta}.$$

Hence, Eq. (16.16) holds.

To prove item (a), we notice that, by Eq. (16.16), we have $L_\varphi(-\tilde{\Theta}) \geq 0$. Since Θ is negative, Lemma 16.2 assures that Σ^n is L_φ-stable.

Now, let us deal with item (b). In this case, we have that $C_0^\infty(\Sigma^n) = C^\infty(\Sigma^n)$. Hence, if Σ^n is L_φ-stable, we have

$$0 \geq \int_\Sigma \tilde{\Theta} L_\varphi \tilde{\Theta} e^{-\varphi} d\Sigma^n = \int_\Sigma n \frac{f''}{f} \tilde{\Theta}^2 e^{-\varphi} d\Sigma^n \geq 0,$$

that is, $f'' = 0$ on Σ^n. The converse follows by item (a).

16.4 L_φ-Stability of φ-Maximal Hypersurfaces

Finally, we prove item (c). The definition of L_φ-stability yields

$$\int_\Sigma \tilde{\Theta} L_\varphi \tilde{\Theta} e^{-\varphi} d\Sigma = \int_\Sigma n\frac{f''}{f}\tilde{\Theta}^2 e^{-\varphi} d\Sigma^n > 0.$$

Therefore, Σ^n cannot be L_φ-stable. The proof is complete. □

Remark 16.3 Let $\overline{M}_\varphi^{n+1} = -I \times_f M_\varphi^n$ be a weighted GRW spacetime whose fiber M^n is compact. In particular, the slices in $\overline{M}_\varphi^{n+1}$ are also compact. Assume in addition that $f'' \geq 0$ and

$$\Omega = \{t \in I : f''(t) = 0\},$$

is a set of isolated points. Then, for every $t \in \Omega$ such that $f'(t) = 0$, the φ-maximal slice $\Sigma_t = \{t\} \times M^n$ is L_φ-stable.

Part V

Local Rigidity and Bifurcation of Riemannian Immersions in Semi-Riemannian Warped Products

Bifurcation of H_2-Hypersurfaces in Riemannian Warped Products

17

17.1 Introduction

According to Barbosa and do Carmo in [62], and Barbosa, do Carmo and Eschenburg in [64], any compact hypersurface Σ^n with constant mean curvature (CMC) in a Riemannian manifold $\overline{M}^{n+1}$ is a solution of the variational problem of minimizing the area functional for volume-preserving variations.

Moreover, when $\overline{M}^{n+1}$ has constant sectional curvature c (shortly, we wrote $\overline{M}_c^{n+1}$) they also established that geodesic spheres of $\overline{M}_c^{n+1}$ are the only stable critical points for this variational problem.

The set of trial maps for the variational problem should be a collection of embeddings of CMC hypersurfaces Σ^n into $\overline{M}^{n+1}$. In order to detect solutions that are not isometrically congruent, one should take into consideration the action of the diffeomorphism group of Σ^n, acting by right composition in the space of embeddings, and the action of the isometry group of $\overline{M}^{n+1}$, acting by left composition on the space of embeddings.

Note that the area and the volume functionals are invariant by the action of these two groups. The action of the diffeomorphism group of Σ^n on any set of embeddings of CMC hypersurfaces Σ^n into $\overline{M}^{n+1}$ is free, which suggests that one should consider a quotient of the space of embeddings by this action. This means that two embeddings of CMC hypersurfaces $x_1 : \Sigma^n \to \overline{M}^{n+1}$ and $x_2 : \Sigma^n \to \overline{M}^{n+1}$ will be considered equivalent if there exists a diffeomorphism $\phi : \Sigma^n \to \Sigma^n$ such that $x_2 = x_1 \circ \phi$.

As to the left action of the isometry group of $\overline{M}^{n+1}$, this is not free; nevertheless, the group is compact, and one can study a bifurcation problem for its critical orbits. Thus, the variational problem described above provides us with a framework where we can study the equivariant bifurcation in a set of equivalence classes of embeddings of CMC hypersurfaces Σ^n into $\overline{M}^{n+1}$.

For more details on Equivariant Bifurcation Theory that are related to geometric variational problems, see, for instance, the papers [23,81,83,279], as well as the references therein.

In this context, Alías and Piccione in [23] studied the bifurcation of CMC Clifford torus of the form $x_r^{n,j} : \mathbb{S}^j(r) \times \mathbb{S}^{n-j}(\sqrt{1-r^2}) \to \mathbb{S}^{n+1}$ in unit Euclidean sphere $\mathbb{S}^{n+1}$, where $j \in \{1,\ldots,n\}$ and $r \in (0,1)$. More precisely, they showed that the existence of two infinite sequences

$$x_{r_i}^{n,j} : \mathbb{S}^j(r_i) \times \mathbb{S}^{n-j}(\sqrt{1-r_i^2}) \to \mathbb{S}^{n+1}$$

and

$$x_{s_l}^{n,j} : \mathbb{S}^j(s_l) \times \mathbb{S}^{n-j}(\sqrt{1-s_l^2}) \to \mathbb{S}^{n+1}$$

that are not isometrically congruent to the CMC Clifford torus, and accumulating at some CMC Clifford torus, where $\{r_i\}_{i \geq 3}, \{s_l\}_{l \geq 3} \subset (0,1)$ are real sequences such that $\lim_{i \to \infty} r_i = 1$ and $\lim_{l \to \infty} s_l = 0$.

Furthermore, they also showed that for all other values of $r \in (0,1)$ the family of CMC Clifford torus $x_r^{n,j} : \mathbb{S}^j(r) \times \mathbb{S}^{n-j}(\sqrt{1-r^2}) \to \mathbb{S}^{n+1}$ is *locally rigid*, in the sense that any CMC embedding of $\mathbb{S}^j(r) \times \mathbb{S}^{n-j}(\sqrt{1-r^2})$ into $\mathbb{S}^{n+1}$ which is sufficiently close to $x_r^{n,j}$ must be isometrically congruent to an embedding of the CMC Clifford family.

Later on, de Lima et al. [146] adapted the methods of [23] to obtain bifurcation and local rigidity results for a family of CMC Clifford torus in 3-dimensional Berger spheres $\mathbb{S}_\tau^3$, with $\tau > 0$.

More recently, Koiso et al. [220] proved bifurcation results for (compact portions of) nodoids in the 3-dimensional Euclidian space $\mathbb{R}^3$, whose boundary consists of two fixed coaxial circles of the same radius lying in parallel planes. They showed that the degeneracy occurs at an infinite discrete sequence of instants, that are divided into four classes; and that different types of bifurcation and break of symmetry occur at each instant of three of the four classes; bifurcation does not occur at the degeneracy instants of the fourth class.

Moreover, the same authors in [221] provided criteria for the existence of bifurcation branches of fixed boundary CMC surfaces in $\mathbb{R}^3$, and they discussed stability/instability issues for the surfaces in bifurcating branches. Moreover, to illustrate the theory, they discuss an explicit example obtained from a bifurcating branch of fixed boundary unduloids in $\mathbb{R}^3$.

On the same theme, García-Martínez and Herrera [187] deduced some bifurcation and local rigidity results for a certain family of CMC hypersurfaces in a class of Riemannnian warped products of the form

$$(I \times_f M^n, d\tau^2 + f(\tau)^2 \langle \cdot, \cdot \rangle_M),$$

17.1 Introduction

namely, in product manifolds $I \times M^n$ endowed with the Riemannian metric $d\tau^2 + f(\tau)^2 \langle \cdot, \cdot \rangle_M$, where $I \subset \mathbb{R}$ is an open interval, f is a real positive function defined on I, called warped function, and $(M^n, \langle \cdot, \cdot \rangle_M)$ is a compact Riemannian manifold without boundary, called Riemannian fiber.

These results were obtained considering some appropriate hypotheses that depend of the behavior of the eigenvalues of the Laplacian operator on $(M^n, \langle \cdot, \cdot \rangle_M)$. In particular, they were able to establish sufficient conditions to ensure the existence of one-parameter families locally rigid on the 3-dimensional Riemannian fiber

$$(I \times \mathbb{S}^2, \psi_{K,E}(\tau)^{-2} d\tau^2 + \tau^2 (d\theta^2 + \sin^2 \theta d\phi^2))$$

of Anti-de Sitter Schwarzschild spacetime provided with the Lorentzian metric

$$ds^2 = -\psi_{K,E}(\tau)^{-2} dt^2 + \psi_{K,E}(\tau)^{-2} d\tau^2 + \tau^2 (d\theta^2 + \sin^2 \theta d\phi^2),$$

with constants $E, K > 0$, and one-parameter families with bifurcation instants on the 3-dimensional Riemannian fiber $(I \times \mathbb{S}^2, \psi_{K,E}(\tau)^{-2} d\tau^2 + \tau^2 (d\theta^2 + \sin^2 \theta d\phi^2))$ of de Sitter Schwarzschild spacetime endowed with the Lorentzian metric ds^2, with constants $E < 0$ and $K > 0$, where in both cases

$$\psi_{K,E}(\tau) = \sqrt{1 - 2K/\tau + E/\tau^2},$$

and I is the maximal connected and open interval where $\psi_{K,E}$ is well defined.

On the other hand, an extension of the hypersurfaces in a Riemannian manifold $\overline{M}^{n+1}$ described above are those that admit constant *second mean curvature* H_2, namely, those whose mean of the sum of the products two by two of its principal curvatures is constant. For simplicity, we call such hypersurfaces only H_2-hypersurfaces. We have that H_2 defines an intrinsic geometric quantity of a hypersurface Σ^n of $\overline{M}^{n+1}$, which is directly related to its normalized scalar curvature S.

For instance, if $\overline{M}_c^{n+1}$ has constant sectional curvature c, it follows from Gauss equation of Σ^n that $H_2 = S - c$, so that, in this case, H_2 is constant on Σ^n if and only if S is constant on Σ^n. In this sense, Alencar et al. [16] extended the results of [62] and [64] to the context of compact H_2-hypersurfaces immersed into $\overline{M}_c^{n+1}$. More specifically, they showed that compact H_2-hypersurfaces of $\overline{M}_c^{n+1}$ are the critical points of the so-called 1-area functional $\mathcal{A}_1(t)$ (see Eq. (17.11)) for volume-preserving variations and, for the case $c \geq 0$, they also proved a compact H_2-hypersurface is stable if and only if it is a geodesic sphere.

After, the variational problem of minimizing 1-area functional $\mathcal{A}_1(t)$ for volume-preserving variations was addressed by Elbert [172] when the ambient space $\overline{M}^{n+1}$ is Einstein, obtaining again that compact H_2-hypersurfaces are the corresponding critical points. Moreover, assuming that the Morse index of an H_2-hypersurface Σ^n is finite (with positive constant second mean curvature H_2), and considering a polynomial or exponential growth of the 1-area of Σ^n, an upper estimate for H_2 is established in [172].

In particular, as an application of this estimate, she obtains a result due to Chern [116], which states that there are no complete graphs in the n-dimensional Euclidean space $\mathbb{R}^{n+1}$ with positive constant second mean curvature H_2.

In this chapter, motivated by all the works described above, in a warped product

$$(I \times_f M^n, d\tau^2 + f(\tau)^2 \langle \cdot, \cdot \rangle_M)$$

with compact (without boundary) Riemannian manifold fiber $(M^n, \langle \cdot, \cdot \rangle_M)$, our purpose is to investigate the existence of bifurcation instants or the local rigidity of a certain family $\{\Omega_\tau\}_{\tau \in (\tau_1, \tau_2]}$ of open sets whose boundaries are H_2-hypersurfaces.

The results that will be present are based mainly on the paper [290].

17.2 Some Preliminaries

Let $\overline{M}^{n+1}$ be a $(n+1)$-dimensional orientable Riemannian manifold ($n \geq 2$) with metric tensor $\langle \cdot, \cdot \rangle$, Levi-Civita connection $\overline{\nabla}$ and curvature tensor $\overline{R}$. We denote by $\mathfrak{X}(\overline{M}^{n+1})$ the set of vector fields of class C^∞ on $\overline{M}^{n+1}$ and by $C^\infty(\overline{M}^{n+1})$ the ring of real functions of class C^∞ on $\overline{M}^{n+1}$. In this context, we consider hypersurfaces $\psi : \Sigma^n \to \overline{M}^{n+1}$, namely, isometric immersions from a connected, n-dimensional orientable Riemannian manifold Σ^n into $\overline{M}^{n+1}$.

Since Σ^n is orientable, one can choose a globally defined unit normal vector field N on Σ^n. The shape operator of $\psi : \Sigma^n \to \overline{M}^{n+1}$ with respect to N is given by

$$A : \mathfrak{X}(\Sigma^n) \to \mathfrak{X}(\Sigma^n)$$
$$Y \mapsto A(Y) = -\overline{\nabla}_Y N.$$

For each fixed $p \in \Sigma^n$, $A_p : T_p\Sigma^n \to T_p\Sigma^n$ is a self-adjoint linear map, the spectral theorem allows us to choose on $T_p\Sigma^n$ an orthonormal basis $\{e_1, \ldots, e_n\}$ of eigenvectors of A_p, with corresponding eigenvalues $\kappa_1(p), \ldots, \kappa_n(p)$, respectively.

The functions $\kappa_1, \ldots, \kappa_n$ on Σ^n thus defined are called principal curvatures of $\psi : \Sigma^n \to \overline{M}^{n+1}$. Moreover, it is well-known that the curvature tensor R of $\psi : \Sigma^n \to \overline{M}^{n+1}$ is described in terms of A and $\overline{R}$ by the so called Gauss equation, which can be written as

$$R(U, V)W = (\overline{R}(U, V)W)^\top + \langle A(U), W \rangle A(V) - \langle A(V), W \rangle A(U) \qquad (17.1)$$

for all $U, V, W \in \mathfrak{X}(\Sigma^n)$, where $(\cdot)^\top$ stands for tangential components on Σ^n.

Along this chapter, we will deal with the first three mean curvatures of the hypersurface $\psi : \Sigma^n \to \overline{M}^{n+1}$, namely

17.2 Some Preliminaries

$$\begin{cases} H_1 = \dfrac{1}{n}\sum_{i=1}^{n}\kappa_i, \\ H_2 = \dfrac{2}{n(n-1)}\sum_{i<j}\kappa_i\kappa_j, \\ H_3 = \dfrac{6}{n(n-1)(n-2)}\sum_{i<j<k}\kappa_i\kappa_j\kappa_k. \end{cases} \quad (17.2)$$

We have that H_1 is the mean curvature of $\psi: \Sigma^n \to \overline{M}^{n+1}$, which is the main extrinsic curvature of the hypersurface. On the other hand, the second mean curvature H_2 defines a geometric quantity which is related to the scalar curvature S of $\psi: \Sigma^n \to \overline{M}^{n+1}$. Indeed, it follows by the Gauss equation (17.1) that the (non-normalized) Ricci curvature Ric_Σ of $\psi: \Sigma^n \to \overline{M}^{n+1}$ is given by

$$\mathrm{Ric}_\Sigma(U,V) = \mathrm{Ric}_{\overline{M}}(U,V) - \langle \overline{R}(U,N)V, N\rangle + nH_1\langle A(U), V\rangle - \langle A(U), A(V)\rangle,$$

for any $U, V \in \mathfrak{X}(\Sigma^n)$, where $\mathrm{Ric}_{\overline{M}}$ stands for the Ricci curvature of $\overline{M}^{n+1}$.

Therefore, S obeys the relation

$$S = \overline{S} - 2\mathrm{Ric}_{\overline{M}}(N, N) + n(n-1)H_2, \quad (17.3)$$

where $\overline{S}$ stands for the scalar curvature of $\overline{M}^{n+1}$. For instance, if there is one $\overline{\varrho} \in \mathbb{R}$ such that the Ricci curvature of $\overline{M}^{n+1}$ verifies the condition

$$\mathrm{Ric}_{\overline{M}}(N, N) = \overline{\varrho} = \mathrm{const.} \quad \text{on} \quad \Sigma^n, \quad (17.4)$$

by (17.3) and (17.4) we get that S and H_2 are related by

$$S = (n-1)(\overline{\varrho} + nH_2). \quad (17.5)$$

If $\overline{M}^{n+1}$ is Einstein, condition (17.4) is naturally valid for any hypersurface. Moreover, there is a larger class of manifolds that verify this condition for a considerable set of hypersurfaces, that will be evidenced at the beginning of Sect. 17.4.

If $\psi: \Sigma^n \to \overline{M}^{n+1}$ has constant second mean curvature H_2, we will simply say that $\psi: \Sigma^n \to \overline{M}^{n+1}$ is an H_2-hypersurface.

Moreover, let us consider the Newton transformation $T: \mathfrak{X}(\Sigma^n) \to \mathfrak{X}(\Sigma^n)$ associated with $\psi: \Sigma^n \to \overline{M}^{n+1}$ given by

$$T = nH_1\mathrm{Id} - A, \quad (17.6)$$

where $\mathrm{Id}: \mathfrak{X}(\Sigma^n) \to \mathfrak{X}(\Sigma^n)$ denotes the identity map.

Finally, associated to the Newton transformation T one has the well-known Cheng-Yau's square operator [113]

$$\square : C^\infty(\Sigma^n) \to C^\infty(\Sigma^n)$$
$$u \mapsto \square(u) = \text{tr}\,(T \circ \text{Hess}_\Sigma\, u)\,, \tag{17.7}$$

where Hess_Σ stands for the Hessian operator on Σ^n.

17.3 The Variational Problem and the Notion of Bifurcation Instants

Let $\overline{M}^{n+1}$ be a Riemannian manifold as in the previous section and let $\mathcal{M}$ be the space of open subsets $\Omega \subset \overline{M}^{n+1}$ with compact closure $\overline{\Omega}$ and whose smooth compact boundary $\partial\Omega$ is a closed, connected and orientable hypersurface. Here, when we refer to a closed hypersurface we mean that the considered hypersurface is compact without boundary. We denote by $d\overline{M}^{n+1}$ and dV the volume elements of $\overline{M}^{n+1}$ and $\partial\Omega$, respectively. If $\Omega \in \mathcal{M}$, the unit normal vector field globally defined on $\partial\Omega$ will be denoted by N. Moreover, according to [60], we write the volume and area element of Ω as follows

$$\text{Vol}(\Omega) = \int_\Omega d\overline{M}^{n+1}, \quad \text{and} \quad \text{Area}(\partial\Omega) = \int_{\partial\Omega} dV$$

as well as

$$1 - \text{Area}(\partial\Omega) = n\int_{\partial\Omega} H_1 dV,$$

where H_1 is the mean curvature of $\partial\Omega$ with respect to N.

17.3.1 Description of the Variational Problem

Let $\Omega \in \mathcal{M}$. A variation of $\partial\Omega$ is a smooth map

$$X : (-\epsilon, \epsilon) \times \partial\Omega \to \overline{M}^{n+1}$$
$$(t, p) \mapsto X(t, p)$$

such that the following conditions hold:

1. For all $t \in (-\epsilon, \epsilon)$, the map

$$X_t : \partial\Omega \to \overline{M}^{n+1}$$
$$p \mapsto X_t(p) = X(t, p) \tag{17.8}$$

17.3 The Variational Problem and the Notion of Bifurcation Instants

is an immersion;
2. $X(0, p) = \iota(p)$ for all $p \in \partial\Omega$, where $\iota : \partial\Omega \hookrightarrow \overline{\Omega}$ is the inclusion map.

In this context, given $\Omega \in \mathcal{M}$ and a variation $X : (-\epsilon, \epsilon) \times \partial\Omega \to \overline{M}^{n+1}$ of $\partial\Omega$ we adopted the notation $\partial\Omega_t = X_t(\partial\Omega)$. For values of t small enough, $\partial\Omega_t$ is also a connected and oriented smooth submanifold. Moreover, it bounds an open subset Ω_t whose closure is also compact. Thus, $X : (-\epsilon, \epsilon) \times \partial\Omega \to \overline{M}^{n+1}$ induces us naturally a variation of the open subset Ω denoted by Ω_t, which is also an element of $\mathcal{M}$.

In the sequel, dV_t denotes the volume element of the metric induced on $\partial\Omega$ by (17.8) and by N_t the unit normal vector field of (17.8). When $t = 0$, these objects coincide with those already defined on $\partial\Omega$.

The variational field associated to $X : (-\epsilon, \epsilon) \times \partial\Omega \to \overline{M}^{n+1}$ is the vector field given by

$$\left.\frac{\partial X}{\partial t}\right|_{t=0}.$$

Letting

$$u_t = \left\langle \frac{\partial X}{\partial t}, N_t \right\rangle, \tag{17.9}$$

we get

$$\left.\frac{\partial X}{\partial t}\right|_{t=0} = u_0 N + \left(\left.\frac{\partial X}{\partial t}\right|_{t=0}\right)^\top.$$

The balance of volume of $X : (-\epsilon, \epsilon) \times \partial\Omega \to \overline{M}^{n+1}$ is the functional

$$\begin{aligned} \mathcal{V} : (-\epsilon, \epsilon) &\to \mathbb{R} \\ t &\mapsto \mathcal{V}(t) = \mathrm{Vol}(\Omega_t) \end{aligned} \tag{17.10}$$

and we say that the variation $X : (-\epsilon, \epsilon) \times \partial\Omega \to \overline{M}^{n+1}$ is volume-preserving of Ω if $\mathcal{V}(t) = \mathcal{V}(0)$, for all $t \in (-\epsilon, \epsilon)$.

The formula of the first variation of $\mathcal{V}(t)$ is given in the following result. See [296] for a detailed proof.

Lemma 17.1 *If $\Omega \in \mathcal{M}$ and $X : (-\epsilon, \epsilon) \times \partial\Omega \to \overline{M}^{n+1}$ is a variation of $\partial\Omega$, then*

$$\frac{d}{dt}\mathcal{V}(t) = \int_{\partial\Omega_t} u_t \, dV_t,$$

for each $t \in (-\epsilon, \epsilon)$, where u_t is the smooth function defined in (17.9). In particular, $X : (-\epsilon, \epsilon) \times \partial\Omega \to \overline{M}^{n+1}$ is volume-preserving of Ω if and only if

$$\int_{\partial\Omega_t} u_t \, dV_t = 0$$

for all $t \in (-\epsilon, \epsilon)$.

Remark 17.1 By Barbosa et al. [64, Lemma 2.2], we have that if $u_0 : \partial\Omega \to \mathbb{R}$ is a smooth function such that

$$\int_{\partial\Omega} u_0 \, dV = 0,$$

then there exists a volume-preserving variation $X : (-\epsilon, \epsilon) \times \partial\Omega \to \overline{M}^{n+1}$ of $\partial\Omega$ whose variational field is

$$\left. \frac{\partial X}{\partial t} \right|_{t=0} = u_0 N.$$

Taking into account [60], the 1-area functional associated to a variation $X : (-\epsilon, \epsilon) \times \partial\Omega \to \overline{M}^{n+1}$ is given by

$$\begin{aligned} \mathcal{A}_1 : (-\epsilon, \epsilon) &\to \mathbb{R} \\ t &\mapsto \mathcal{A}_1(t) = 1 - \operatorname{Area}(\partial\Omega_t) = n \int_{\partial\Omega_t} H_1^t \, dV_t, \end{aligned} \quad (17.11)$$

where $H_1^t = H_1(t, \cdot)$ denotes the mean curvature of $\partial\Omega_t$ with respect to the metric induced by the immersion X_t defined in (17.8).

The next result follows by Elbert [172, Proposition 3.2].

Lemma 17.2 If $\Omega \in \mathcal{M}$ and $X : (-\epsilon, \epsilon) \times \partial\Omega \to \overline{M}^{n+1}$ is a variation of $\partial\Omega$, then

$$\frac{\partial}{\partial t} H_2^t = \frac{2}{n(n-1)} \square_t(u_t)$$

$$+ \left\{ n H_1^t H_2^t - (n-2) H_3^t + \frac{2}{n(n-1)} \operatorname{tr}(T_t \circ \overline{R}_t) \right\} u_t$$

$$+ \left\langle \left(\frac{\partial X}{\partial t} \right)^\top, \nabla(H_2^t) \right\rangle,$$

where $\square_t$ is the Cheng-Yau's square operator on $\partial\Omega_t$, $H_2^t = H_2(t, \cdot)$ and $H_3^t = H_3(t, \cdot)$ are the second and third mean curvatures of $\partial\Omega_t$, respectively, u_t is the function defined

17.3 The Variational Problem and the Notion of Bifurcation Instants

in (17.9), T_t is the Newton transformation on $\partial\Omega_t$ and $\overline{R}_t$ is the linear operator on $\partial\Omega_t$ given by $\overline{R}_t(Y) = \overline{R}(N_t, Y)N_t$ for any $Y \in \mathfrak{X}(\partial\Omega_t)$.

The previous lemma allows us to compute the first variation of the 1-area functional $\mathcal{A}_1(t)$ (cf. of [172, Proposition 3.4]).

Lemma 17.3 *If $\Omega \in \mathcal{M}$ and $X : (-\epsilon, \epsilon) \times \partial\Omega \to \overline{M}^{n+1}$ is a variation of $\partial\Omega$, then*

$$\frac{d}{dt}\mathcal{A}_1(t) = \int_{\partial\Omega_t} \left\{-n(n-1)H_2^t + \mathrm{Ric}_{\overline{M}}(N_t, N_t)\right\} u_t \, dV_t,$$

for all $t \in (-\epsilon, \epsilon)$, where u_t is the function defined in (17.9) and $\mathrm{Ric}_{\overline{M}}$ is the Ricci curvature of $\overline{M}^{n+1}$.

In order to characterize the open subsets Ω of $\overline{M}^{n+1}$ whose boundary $\partial\Omega$ is a closed hypersurface with constant second mean curvature, we consider the variational problem

(VP-1) *Minimizing the 1-area functional $\mathcal{A}_1(t)$ given in (17.11) for any variation of $\partial\Omega$ that preserves the volume of Ω.*

The Lagrange multiplier method leads us then to the Jacobi functional

$$\begin{aligned}\mathcal{F}^\lambda : (-\epsilon, \epsilon) &\to \mathbb{R} \\ t &\mapsto \mathcal{F}^\lambda(t) = \mathcal{A}_1(t) + \lambda \mathcal{V}(t),\end{aligned} \quad (17.12)$$

where $\lambda \in \mathbb{R}$ is a constant to be determined. As an immediate consequence of Lemmas 17.1 and 17.3 we get that the first variation of $\mathcal{F}^\lambda(t)$ takes the following form

$$\frac{d}{dt}\mathcal{F}^\lambda(t) = \int_{\partial\Omega_t} \left\{-n(n-1)H_2^t + \mathrm{Ric}_{\overline{M}}(N_t, N_t) + \lambda\right\} u_t \, dV_t, \quad (17.13)$$

for each $t \in (-\epsilon, \epsilon)$.

To make an appropriate choice of λ, we assume from now on that there is $\overline{\varrho} \in \mathbb{R}$ such that the Ricci curvature $\mathrm{Ric}_{\overline{M}}$ of $\overline{M}^{n+1}$ satisfies the condition

$$\mathrm{Ric}_{\overline{M}}(N_t, N_t) = \overline{\varrho} = \mathrm{const.} \quad \text{on} \quad \partial\Omega_t, \quad \text{for every } t \in (-\epsilon, \epsilon). \quad (17.14)$$

If $\overline{M}^{n+1}$ is Einstein, (17.14) is naturally valid, but there is a larger class of manifolds that verify this condition, which will be described in Sect. 17.4.

In addition, let

$$\Lambda = \frac{1}{\text{Area}(\partial\Omega)} \int_{\partial\Omega} H_2 \, dV \qquad (17.15)$$

be the integral mean of the second mean curvature H_2 on $\partial\Omega$.
If H_2 is constant, we have that

$$\Lambda = H_2. \qquad (17.16)$$

Hence, if we choose

$$\lambda = n(n-1)\Lambda - \overline{\varrho}, \qquad (17.17)$$

by (17.13) we have

$$\frac{d}{dt}\mathcal{F}^\lambda(t) = -n(n-1)\int_{\partial\Omega_t}\{H_2^t - \Lambda\} u_t \, dV_t, \qquad (17.18)$$

for every $t \in (-\epsilon, \epsilon)$.
In particular,

$$\frac{d}{dt}\mathcal{F}^\lambda(0) = -n(n-1)\int_{\partial\Omega}\{H_2 - \Lambda\} u_0 \, dV. \qquad (17.19)$$

Now, arguing as in [62, Proposition 2.7], the following result holds.

Proposition 17.1 *Let $\Omega \in \mathcal{M}$. Assume that the Ricci curvature $\text{Ric}_{\overline{M}}$ of $\overline{M}^{n+1}$ satisfies (17.14). The following statements are equivalent:*

(a) $\partial\Omega$ is a H_2-hypersurface with constant second mean curvature

$$H_2 = \frac{\lambda + \overline{\varrho}}{n(n-1)};$$

(b) For any variation $X : (-\epsilon, \epsilon) \times \partial\Omega \to \overline{M}^{n+1}$ of $\partial\Omega$ which preserves the balance of volume of Ω, one has

$$\frac{d}{dt}\mathcal{A}_1(0) = 0;$$

(c) For any variation $X : (-\epsilon, \epsilon) \times \partial\Omega \to \overline{M}^{n+1}$ of $\partial\Omega$, we have that

17.3 The Variational Problem and the Notion of Bifurcation Instants

$$\frac{d}{dt}\mathcal{F}^\lambda(0) = 0.$$

Proof We prove that $(a) \Rightarrow (c) \Rightarrow (b) \Rightarrow (a)$.
$(a) \Rightarrow (c)$ The result directly follows by (17.16), (17.17) and (17.19).
$(c) \Rightarrow (b)$ By (17.12), one has

$$0 = \frac{d}{dt}\mathcal{F}^\lambda(0) = \frac{d}{dt}\mathcal{A}_1(0) + \lambda \frac{d}{dt}\mathcal{V}(0)$$

for any variation $X : (-\epsilon, \epsilon) \times \partial\Omega \to \overline{M}^{n+1}$ of $\partial\Omega$.

Now, if the variation preserves the volume of Ω, then

$$\frac{d}{dt}\mathcal{V}(0) = 0.$$

Hence, it follows that

$$\frac{d}{dt}\mathcal{A}_1(0) = 0$$

for volume-preserving variations $X : (-\epsilon, \epsilon) \times \partial\Omega \to \overline{M}^{n+1}$ of $\partial\Omega$.
$(b) \Rightarrow (a)$ Arguing by contradiction, suppose that there exists p_0 in $\partial\Omega$ such that

$$\left(H_2 - \frac{\lambda + \overline{\varrho}}{n(n-1)}\right)(p_0) \neq 0.$$

We can assume that

$$\left(H_2 - \frac{\lambda + \overline{\varrho}}{n(n-1)}\right)(p_0) > 0. \tag{17.20}$$

The definition of Λ in (14.5) yields the existence of a point $q_0 \in \Sigma^n$ such that

$$\left(H_2 - \frac{\lambda + \overline{\varrho}}{n(n-1)}\right)(q_0) < 0.$$

Indeed, by (17.15) and (17.17), we have that

$$\int_{\partial\Omega}\left(H_2 - \frac{\lambda + \overline{\varrho}}{n(n-1)}\right)dV = \int_{\partial\Omega} H_2\,dV - \Lambda\,\text{Area}(\partial\Omega)$$

$$= \int_{\partial\Omega} H_2\,dV - \frac{1}{\text{Area}(\partial\Omega)}\left(\int_{\partial\Omega} H_2\,dV\right)\text{Area}(\partial\Omega)$$

$$= 0. \tag{17.21}$$

Thus, if $\left(H_2 - \frac{\lambda + \overline{\varrho}}{n(n-1)}\right)(q) > 0$ for every $q \in \partial\Omega$, since there is $p_0 \in \partial\Omega$ such that (17.20) is valid, one has

$$\int_{\partial\Omega} \left(H_2 - \frac{\lambda + \overline{\varrho}}{n(n-1)}\right) dV > 0,$$

which contradicts (17.21).

So, we have the sets

$$\partial\Omega^+ = \left\{ q \in \partial\Omega : \left(H_2 - \frac{\lambda + \overline{\varrho}}{n(n-1)}\right)(q) > 0 \right\},$$

and

$$\partial\Omega^- = \left\{ q \in \partial\Omega : \left(H_2 - \frac{\lambda + \overline{\varrho}}{n(n-1)}\right)(q) < 0 \right\}$$

are well defined.

Now, let us consider nonnegative smooth functions φ and ψ on $\partial\Omega$ such that $p_0 \in \operatorname{supp}\varphi \subset \partial\Omega^+$, $\operatorname{supp}\psi \subset \partial\Omega^-$ and

$$\int_{\partial\Omega} (\varphi + \psi)\left(-n(n-1)H_2 + \overline{\varrho} + \frac{\lambda + \overline{\varrho}}{n(n-1)}\right) dV = 0,$$

where $\operatorname{supp}\varphi$ and $\operatorname{supp}\psi$ denote the support of φ and the support of ψ, respectively.

If we consider the smooth function

$$u_0 = (\varphi + \psi)\left(-n(n-1)H_2 + \overline{\varrho} + \frac{\lambda + \overline{\varrho}}{n(n-1)}\right),$$

then, according to Remark 17.1, there is a volume-preserving variation $X : (-\epsilon, \epsilon) \times \partial\Omega \to \overline{M}^{n+1}$ of $\partial\Omega$ whose variational field is

$$\left.\frac{\partial X}{\partial t}\right|_{t=0} = u_0 N.$$

Next, by our hypothesis, Lemma 17.3 and (17.14) we get

$$0 = \frac{d}{dt}\mathcal{A}_1(0) = \int_{\partial\Omega} (-n(n-1)H_2 + \overline{\varrho}) u_0 \, dV.$$

Furthermore, since $\int_{\partial\Omega} u_0 \, dV = 0$, then

17.3 The Variational Problem and the Notion of Bifurcation Instants

$$
\begin{aligned}
0 &= \int_{\partial\Omega} (-n(n-1)H_2 + \overline{\varrho}) u_0 \, dV \\
&= \int_{\partial\Omega} (-n(n-1)H_2 + \overline{\varrho}) u_0 \, dV + \frac{\lambda + \overline{\varrho}}{n(n-1)} \int_{\partial\Omega} u_0 \, dV \\
&= \int_{\partial\Omega} \left(-n(n-1)H_2 + \overline{\varrho} + \frac{\lambda + \overline{\varrho}}{n(n-1)} \right) u_0 \, dV \\
&= \int_{\partial\Omega} (\varphi + \psi) \left(-n(n-1)H_2 + \overline{\varrho} + \frac{\lambda + \overline{\varrho}}{n(n-1)} \right)^2 dV > 0,
\end{aligned}
$$

which constitutes an absurd.

Therefore, we must have $H_2 = \dfrac{\lambda + \overline{\varrho}}{n(n-1)}$ on $\partial\Omega$. □

Hence, if the Riemannian manifold $\overline{M}^{n+1}$ verifies (17.14), by Proposition 17.1 we have that the solutions of (VP-1) are open subsets Ω of $\overline{M}^{n+1}$ whose boundary $\partial\Omega$ is a closed H_2-hypersurface with constant second mean curvature H_2 equal to

$$H_2 = \frac{\lambda + \overline{\varrho}}{n(n-1)}, \tag{17.22}$$

with $\lambda, \overline{\varrho} \in \mathbb{R}$. Now, let us consider the following variational problem:

(VP-2) *Minimizing the 1-area functional $\mathcal{A}_1(t)$ given in (17.11) for any variation of $\partial\Omega$, not necessarily a volume-preserving variation of Ω,*

By Proposition 17.1 we obtain that the solutions of (VP-2) are exactly the solutions of the initial variational problem (VP-1).

Remark 17.2 As observed in [187], our approach is valid in a more general setting. Assume that $\mathcal{M}$ is the set of open subsets $\Omega \subset \overline{M}^{n+1}$ whose boundary $\partial\Omega$ is the union of two disjoint sets

$$\partial\Omega = \Sigma_1^n \cup \Sigma_2^n.$$

Assume that one of them, for instance Σ_1^n, is fixed. Consequently, the variations considered of $\partial\Omega$ only affects Σ_2^n. Under this assumption, the solutions of (VP-1) or (VP-2) will be open subsets Ω such that their boundaries are union of a (fixed) set Σ_1^n and of a closed hypersurface Σ_2^n with constant second mean curvature.

The formula for the second variation of the functional $\mathcal{F}^\lambda$ is given below.

Proposition 17.2 *Let Ω be an open subset of an $(n+1)$-dimensional Riemannian manifold $\overline{M}^{n+1}$ ($n \geq 2$) whose boundary $\partial\Omega$ is a closed H_2-hypersurface, with constant second mean curvature H_2 given in (17.22), and let $X : (-\epsilon, \epsilon) \times \partial\Omega \to \overline{M}^{n+1}$ be a variation of $\partial\Omega$. Assume that the Ricci curvature $\mathrm{Ric}_{\overline{M}}$ of $\overline{M}^{n+1}$ satisfies (17.14). Then*

$$\frac{d^2}{dt^2} \mathcal{F}^\lambda(0)(u) = -2 \int_{\partial\Omega} u \, \mathcal{J}(u) \, dV, \qquad (17.23)$$

for any $u \in C^\infty(\partial\Omega)$, where $\mathcal{J} : C^\infty(\partial\Omega) \to C^\infty(\partial\Omega)$ is the Jacobi operator associated with the variational problems (VP-1) and (VP-2) defined by

$$\mathcal{J} = \Box + \left\{ \frac{n(n-1)}{2} (nH_1 H_2 - (n-2)H_3) + \mathrm{tr}\left(T \circ \overline{R}_0\right) \right\}. \qquad (17.24)$$

Proof For any variation $X : (-\epsilon, \epsilon) \times \partial\Omega \to \overline{M}^{n+1}$ of $\partial\Omega$ let us consider the function $u_0 \in C^\infty(\partial\Omega)$ defined in (17.9). Since H_2 is constant on $\partial\Omega$, by (17.16), (17.18) and Lemma 17.2, we get

$$\begin{aligned}
\frac{d^2}{dt^2} \mathcal{F}^\lambda(0)(u_0) &= -n(n-1) \int_{\partial\Omega} \left(\frac{\partial}{\partial t} H_2^t \Big|_{t=0} \right) u_0 \, dV \\
&\quad - n(n-1) \int_{\partial\Omega} \underbrace{(H_2 - \Lambda)}_{=0} \frac{\partial}{\partial t} (u_t \, dV_t) \Big|_{t=0} \qquad (17.25) \\
&= -\int_{\partial\Omega} \{2\Box(u_0) + \{n(n-1)(nH_1 H_2 \\
&\quad - (n-2)H_3) + 2\,\mathrm{tr}\left(T \circ \overline{R}_0\right)\}u_0\} u_0 \, dV.
\end{aligned}$$

Now, for any $u \in C^\infty(\partial\Omega)$, considering the variations $X : (-\epsilon, \epsilon) \times \partial\Omega \to \overline{M}^{n+1}$ of $\partial\Omega$ whose variational field is

$$\frac{\partial X}{\partial t}\bigg|_{t=0} = uN,$$

we obtain that (17.25) is also valid for every $u \in C^\infty(\partial\Omega)$. This shows the formula of the second variation of a solution of (VP-2).

Furthermore, if $X : (-\epsilon, \epsilon) \times \partial\Omega \to \overline{M}^{n+1}$ is a variation of $\partial\Omega$ which preserves the balance of volume of Ω, then, for $u_0 \in C^\infty(\partial\Omega)$ defined in (17.9), by Lemma 17.1 we have that

$$\int_{\partial\Omega} u_0 \, dV = 0,$$

17.3 The Variational Problem and the Notion of Bifurcation Instants

and, in addition, the expression (17.25) is valid for such u_0. Finally, for any function $u \in C^\infty(\partial\Omega)$ such that

$$\int_{\partial\Omega} u\, dV = 0,$$

by Remark 17.1 we have a variation $X : (-\epsilon, \epsilon) \times \partial\Omega \to \overline{M}^{n+1}$ of $\partial\Omega$ which preserves the balance of volume of Ω such that the variational field is

$$\left.\frac{\partial X}{\partial t}\right|_{t=0} = uN.$$

The conclusion easily follows. □

Remark 17.3 In Eq. (17.24) of Proposition 17.2, □ is the Cheng-Yau's square operator on $\partial\Omega$. Furthermore H_1 and H_3 are the first and third mean curvatures of $\partial\Omega$ respectively. Finally, T is the Newton transformation on $\partial\Omega$ and $\overline{R}_0$ is the linear operator on $\partial\Omega$ given by $\overline{R}_0(Y) = \overline{R}(N, Y)N$ for any $Y \in \mathfrak{X}(\partial\Omega)$. With respect to the functions on $\partial\Omega$ to be evaluated in the second variation (17.23) of a solution of (VP-1), they have to be considered according to Remark 17.1, that is, smooth functions on $\partial\Omega$ whose integral mean is zero. On the other hand, any smooth function on $\partial\Omega$ can be evaluated on the second variation (17.23) of a solution of (VP-2).

17.3.2 The Notion of Bifurcation Instants for H_2-Hypersurfaces

Let us consider the one-parameter family $\{\Omega_\tau\}_\tau \subset M$ of open subsets in $\overline{M}^{n+1}$ such that the boundary of each Ω_τ, denoted by $\partial\Omega_\tau$, is a closed H_2^τ-hypersurface with constant second mean curvature H_2^τ, where τ varies on a prescribed interval $I \subset \mathbb{R}$. For every $\tau \in I$, let N_τ be the unit normal vector field globally defined on $\partial\Omega_\tau$. We assume that there is $\overline{\rho} \in \mathbb{R}$ such that the Ricci curvature $\mathrm{Ric}_{\overline{M}}$ of $\overline{M}^{n+1}$ satisfies

$$\mathrm{Ric}_{\overline{M}}(N_\tau, N_\tau) = \overline{\varrho} = \mathrm{const.} \quad \text{on} \quad \partial\Omega_\tau \quad \text{for every } \tau \in I. \tag{17.26}$$

In this setting, taking into account Sect. 17.3.1, we have that each Ω_τ is a solution of a certain variational problem (VP-2). More precisely, each Ω_τ is a critical point for the Jacobi functional

$$I \ni \tau \longmapsto \mathcal{F}^{\lambda(\tau)} = \mathcal{A}_1 + \lambda(\tau)\mathcal{V}$$

defined in (17.12), where

$$\lambda(\tau) = n(n-1)H_2^\tau - \overline{\varrho}$$

Moreover, associated with each closed H_2^{τ}-hypersurface $\partial\Omega_\tau$, by Proposition 17.2 we have that the second variation of $\mathcal{F}^{\lambda(\tau)}$ is given by

$$\frac{d^2}{dt^2}\mathcal{F}^{\lambda(\tau)}(0)(u) = -2\int_{\partial\Omega_\tau} u\,\mathcal{J}_\tau(u)\,dV_\tau, \tag{17.27}$$

for any $u \in C^\infty(\partial\Omega_\tau)$, where dV_τ is the volume element on $\partial\Omega_\tau$ and

$$\mathcal{J}_\tau = \Box_\tau + \left\{\frac{n(n-1)}{2}\left(nH_1^\tau H_2^\tau - (n-2)H_3^\tau\right) + \mathrm{tr}\left(T_\tau \circ \overline{R}_\tau\right)\right\} \tag{17.28}$$

is the Jacobi operator on $\partial\Omega_\tau$. Here, $\Box_\tau$ is the Cheng-Yau's square operator on $\partial\Omega_\tau$, H_1^τ, H_2^τ and H_3^τ are the first three mean curvatures of $\partial\Omega_\tau$ with respect to unit normal vector field N_τ. Finally, $T_\tau : \mathfrak{X}(\partial\Omega_\tau) \to \mathfrak{X}(\partial\Omega_\tau)$ is the Newton transformation on $\partial\Omega_\tau$ and $\overline{R}_\tau : \mathfrak{X}(\partial\Omega_\tau) \to \mathfrak{X}(\partial\Omega_\tau)$ is the linear operator given by $\overline{R}_\tau(Y) = \overline{R}(N_\tau, Y)N_\tau$ for every $Y \in \mathfrak{X}(\partial\Omega_\tau)$.

With respect to our family $\{\Omega_\tau\}_{\tau\in I}$ of solutions of (VP-2), we need to adopt some notions and results that correspond to equivariant bifurcation theory for geometric variational problems; see [23, 81, 83, 279] for more comments and remarks.

Let us first recall that two elements Ω_{τ_1} and Ω_{τ_1} of $\{\Omega_\tau\}_{\tau\in I}$ are said to be isometrically congruent when there is an isometry ψ of $\overline{M}^{n+1}$ that carries the image of $x_1 : \partial\Omega_{\tau_1} \to \overline{M}^{n+1}$ onto the image of $x_2 : \partial\Omega_{\tau_2} \to \overline{M}^{n+1}$ (cf. [23, Section 1.2]), where x_1 and x_2 are the immersions of $\partial\Omega_{\tau_1}$ and $\partial\Omega_{\tau_2}$ into $\overline{M}^{n+1}$, respectively, namely, if there exists a diffeomorphism $\phi : \partial\Omega_{\tau_1} \to \partial\Omega_{\tau_2}$ and an isometry ψ of $\overline{M}^{n+1}$ such that the following diagram commutes

$$\begin{array}{ccc} \partial\Omega_{\tau_1} & \xrightarrow{x_1} & \overline{M}^{n+1} \\ \phi \downarrow & & \downarrow \psi \\ \partial\Omega_{\tau_2} & \xrightarrow{x_2} & \overline{M}^{n+1} \end{array}$$

According to [81], $\widetilde{\tau} \in I$ is said to be a bifurcation instant of $\{\Omega_\tau\}_{\tau\in I}$ if there exists a sequence $\{\tau_n\}_{n\in\mathbb{N}} \subset I$ and a sequence $\{\Omega_{\tau_n}\}_{n\in\mathbb{N}} \subset \{\Omega_\tau\}_{\tau\in I}$ such that

(a) $\lim_{n\to\infty} \tau_n = \widetilde{\tau}$;

(b) $\lim_{n\to\infty} x_n = \widetilde{x}$, where $x_n : \Omega_{\tau_n} \to \overline{M}^{n+1}$ and $\widetilde{x} : \Omega_{\widetilde{\tau}} \to \overline{M}^{n+1}$ are, respectively, the immersions of Ω_{τ_n} and $\Omega_{\widetilde{\tau}}$ into $\overline{M}^{n+1}$;

(c) For every $n \in \mathbb{N}$, x_n is not isometrically congruent to $\widetilde{x}$.

17.3 The Variational Problem and the Notion of Bifurcation Instants

Furthermore, by following [83], if $\tilde{\tau} \in I$ is not a bifurcation instant, the family $\{\Omega_\tau\}_{\tau \in I}$ is said to be *locally rigid* at $\tilde{\tau}$.

One of the classical criterion to determine when a instant $\tilde{\tau} \in I$ is of bifurcation is related with the so-called Morse index (cf. [23,81]). We recall that the Morse index of Ω_τ, which will be denoted by

$$\mathrm{Ind}\left(\mathcal{F}^{\lambda(\tau)}, \Omega_\tau\right),$$

is equal to the dimension of the maximal subspace where the second variation of the Jacobi functional $\mathcal{F}^{\lambda(\tau)}$ is negative definite.

Equivalently, $\mathrm{Ind}\left(\mathcal{F}^{\lambda(\tau)}, \Omega_\tau\right)$ is the number of negative eigenvalues (counted with multiplicity) of the Jacobi operator $\mathcal{J}_\tau$. With our notations, a real number $\widehat{\mu}(\tau)$ is an eigenvalue of $\mathcal{J}_\tau$ if and only if

$$\mathcal{J}_\tau(u) + \widehat{\mu}(\tau) u = 0$$

for some function $u \in C^\infty(\partial \Omega_\tau)$. By Alías and Piccione [23, Proposition 2.7] we obtain that $\mathrm{Ind}\left(\mathcal{F}^{\lambda(\tau)}, \Omega_\tau\right)$ is finite in $I \subset \mathbb{R}$. Intuitively, $\mathrm{Ind}\left(\mathcal{F}^{\lambda(\tau)}, \Omega_\tau\right)$ measures the number of independent directions in which the hypersurface $\partial \Omega_\tau$ fails to minimize the 1-area functional $\mathcal{A}_1(t)$ defined in (17.11).

Essentially, a variation of the Morse index $\mathrm{Ind}\left(\mathcal{F}^{\lambda(\tau)}, \Omega_\tau\right)$ along the interval $I \subset \mathbb{R}$ will indicate the existence of a bifurcation instant. More precisely, under suitable Fredholmness assumptions (cf. [23] and [81]), we have that if there are $\tau_1, \tau_2 \in I$, with $\tau_1 < \tau_2$, such that the second variation of the Jacobi functional $\mathcal{F}^{\lambda(\tau)}$ is nonsingular (namely, the eigenvalues of the Jacobi operator $\mathcal{J}_{\tau_j}$ are nonzero) for $j \in \{1, 2\}$ and

$$\mathrm{Ind}(\mathcal{F}^{\lambda(\tau_1)}, \Omega_{\tau_1}) \neq \mathrm{Ind}(\mathcal{F}^{\lambda(\tau_2)}, \Omega_{\tau_2}),$$

then $\{\Omega_\tau\}_{\tau \in I}$ admits a bifurcation instant at some $\tau_* \in (\tau_1, \tau_2)$. On the other hand, according to [83], using the Implicit Function Theorem, we obtain that if

$$\frac{d^2}{dt^2} \mathcal{F}^{\lambda(\tilde{\tau})}(0)$$

is nonsingular for same $\tilde{\tau} \in I$, then the family $\{\Omega_\tau\}_{\tau \in I}$ is locally rigid at $\tilde{\tau}$. In particular, if $\mathrm{Ind}\left(\mathcal{F}^{\lambda(\tau)}, \Omega_\tau\right) = 0$ for every $\tau \in I$, $\{\Omega_\tau\}_{\tau \in I}$ does not have bifurcation instants.

In this chapter, we will study the local rigidity and the bifurcation instants of $\{\Omega_\tau\}_{\tau \in I}$ by analyzing the spectrum of $\mathcal{J}_\tau$ for all $\tau \in I$. Essentially, we will determine the number of negative eigenvalues for each τ (counted with their multiplicity) and we will study the variation of such a number.

17.4 Bifurcation and Locally Rigidity Results for H_2-Hypersurfaces

For an open interval $I \subset \mathbb{R}$ and a given n-dimensional Riemannian manifold M^n ($n \geq 2$) with metric tensor $\langle \cdot, \cdot \rangle_M$, consider the product $I \times_f M^m$ endowed with warped metric tensor

$$\langle \cdot, \cdot \rangle = d\tau^2 + f(\tau)^2 \langle \cdot, \cdot \rangle_M,$$

where $f : I \to \mathbb{R}$ is a positive smooth function on I.

In other words, $I \times_f M^n$ is a Riemannian warped product with Riemannian base $(I, d\tau^2)$, Riemannian fiber $(M^n, \langle \cdot, \cdot \rangle_M)$ and warping function f. In this context, for every $\tau \in I$ we have that the slice

$$\Sigma_\tau^n = \{\tau\} \times M^n \subset I \times_f M^n \tag{17.29}$$

is a totally umbilical hypersurface in $I \times_f M^n$ (see for instance [242]), oriented by the unit normal vector field $N_\tau = -\partial_\tau$, and whose shape operator is given by

$$\begin{aligned} A_\tau : \mathfrak{X}(\Sigma_\tau^n) &\to \mathfrak{X}(\Sigma_\tau^n) \\ Y &\mapsto A_\tau(Y) = -\overline{\nabla}_Y(-\partial_\tau) = \frac{f'(\tau)}{f(\tau)} Y. \end{aligned} \tag{17.30}$$

Actually, the induced metric on Σ_τ^n is given by $f(\tau)^2 \langle \cdot, \cdot \rangle_M$, which means that Σ_τ^n is homotetic to M^n with scale factor $f(\tau)$. Therefore, the correspondence

$$I \ni \tau \mapsto \Sigma_\tau^n = \{\tau\} \times M^n$$

determines a foliation of $I \times_f M^n$ by totally umbilical hypersurfaces, whose first three (constant) mean curvatures (see equations in (17.2)) are given respectively by

$$H_1^\tau = \frac{f'(\tau)}{f(\tau)}, \quad H_2^\tau = \left(\frac{f'(\tau)}{f(\tau)}\right)^2, \quad H_3^\tau = \left(\frac{f'(\tau)}{f(\tau)}\right)^3. \tag{17.31}$$

Moreover, the Ricci curvature $\mathrm{Ric}_{I \times_f M^n}(\cdot, \cdot)$ of $I \times_f M^n$ obeys the condition

$$\mathrm{Ric}_{I \times_f M^n}(N_\tau, N_\tau) = -n \frac{f''(\tau)}{f(\tau)} = \mathrm{const.} \quad \text{on } \Sigma_\tau^n, \tag{17.32}$$

that is, the Riemanniann warped product $I \times_f M^m$ satisfies (17.4).

By (17.32) we observe that the slices (17.29) of the Riemannian warped product $I \times_f M^n$ verify the conditions (17.14) and (17.26) when the warped function $f : I \to \mathbb{R}$ verifies the ordinary differential equation

$$nf''(\tau) + \overline{\varrho} f(\tau) = 0, \quad \tau \in I, \tag{17.33}$$

17.4 Bifurcation and Locally Rigidity Results for H_2-Hypersurfaces

whose solutions are given by

$$f(\tau) = \begin{cases} c_1 \cosh\left(\sqrt{\frac{-\overline{\varrho}}{n}}\,\tau\right) + c_2 \sinh\left(\sqrt{\frac{-\overline{\varrho}}{n}}\,\tau\right) & \text{if } \overline{\varrho} < 0 \\ a > 0 \quad \text{or} \quad c_1\tau + c_2 & \text{if } \overline{\varrho} = 0 \\ c_1 \cos\left(\sqrt{\frac{\overline{\varrho}}{n}}\,\tau\right) + c_2 \sin\left(\sqrt{\frac{\overline{\varrho}}{n}}\,\tau\right) & \text{if } \overline{\varrho} > 0, \end{cases}$$

where $c_1, c_2 \in \mathbb{R}$ are constants and, in each case, the interval of definition $I \subset \mathbb{R}$ of f is the maximal one where f is positive. In Table 17.1 we collect the Riemannian warped products arising from Eq. (17.33).

For all the warped functions described in Table 17.1, when the Riemannian fiber M^n is closed, we have that the Riemannian warped product $I \times_f M^n$ support a family of open subsets which can be realized as solutions of the variational problem that was described in Sect. 17.3.1. To see this, let τ_1 and τ_2 be arbitrary numbers in $I \subset \mathbb{R}$ and we consider the family

$$\{\Omega_\tau\}_{\tau \in (\tau_1, \tau_2]}$$

of open subsets of $I \times_f M^n$ defined by

$$\Omega_\tau = (\tau_1, \tau) \times M^n, \quad \tau \in (\tau_1, \tau_2]. \tag{17.34}$$

Thus, assuming M^n closed, we have that the boundary $\partial\Omega_\tau$ of each Ω_τ is the disjoint union $\partial\Omega_\tau = \Sigma^n_{\tau_1} \cup \Sigma^n_\tau$ of two closed hypersurfaces $\Sigma^n_{\tau_1} = \{\tau_1\} \times M^n$ (fixed) and $\Sigma^n_\tau = \{\tau\} \times M^n$. Since the variations of $\partial\Omega_\tau$ only affects Σ^n_τ and taking into account that Σ^n_τ is a closed H^τ_2-hypersurface, Remark 17.2 assures us that each element of $\{\Omega_\tau\}_{\tau \in (\tau_1, \tau_2]}$ is a

Table 17.1 Riemannian warped products satisfying the ordinary differential equation (17.33)

Riemannian warped product	$\overline{\varrho}$	c_1	c_2
$(-\infty, +\infty) \times_{e^\tau} M^n$	$-n$	1	1
$(-\infty, +\infty) \times_{\cosh \tau} M^n$	$-n$	1	0
$(0, +\infty) \times_{\sinh \tau} M^n$	$-n$	0	1
$(-\infty, +\infty) \times M^n$	0	...	...
$(0, +\infty) \times_\tau M^n$	0	1	0
$(-\pi/2, \pi/2) \times_{\cos \tau} M^n$	n	1	0
$(0, \pi) \times_{\sin \tau} M^n$	n	0	1
$(0, \pi/2) \times_{\sin \tau + \cos \tau} M^n$	n	1	1

solution of the variational problem (VP-2). Moreover, by (17.24), the differential operator $\mathcal{J}_\tau : C^\infty(\Sigma_\tau^n) \to C^\infty(\Sigma_\tau^n)$ given by

$$\mathcal{J}_\tau(u) = \Box_\tau(u) + \left\{\frac{n(n-1)}{2}\left(nH_1^\tau H_2^\tau - (n-2)H_3^\tau\right) + \mathrm{tr}\left(T_\tau \circ \overline{R}_\tau\right)\right\} u \qquad (17.35)$$

is the Jacobi operator associated with our variational problem, where $\Box_\tau$ is the Cheng-Yau's square operator on Σ_τ^n, H_1^τ, H_2^τ and H_3^τ are the first three mean curvatures of Σ_τ^n given in (17.31), $T_\tau : \mathfrak{X}(\Sigma_\tau^n) \to \mathfrak{X}(\Sigma_\tau^n)$ is the Newton transformation on Σ_τ^n and $\overline{R}_\tau : \mathfrak{X}(\Sigma_\tau^n) \to \mathfrak{X}(\Sigma_\tau^n)$ is the linear operator given by

$$\overline{R}_\tau(Y) = \overline{R}(\partial_\tau, Y)\partial_\tau \qquad (17.36)$$

for every $Y \in \mathfrak{X}(\Sigma_\tau^n)$.

Now, let $I \times_f M^n$ be a Riemannian warped product with closed Riemannian fiber M^n and whose warped function satisfies the ordinary differential equation (17.33). The main idea is to study suitable conditions for which either the local rigidity or the existence of bifurcation instants of the family of open subsets $\{\Omega_\tau\}_{\tau \in (\tau_1, \tau_2]}$, defined in (17.34), hold.

In the next result we give some expressions that will allow us to write the Jacobi operator $\mathcal{J}_\tau$ in terms of the warped function f, of the Laplacian Δ on M^n, as well as of the constant $\overline{\varrho}$.

Proposition 17.3 *With the above notations, the following facts hold:*

(a) $T_\tau = (n-1)\dfrac{f'(\tau)}{f(\tau)} \mathrm{Id}_\tau$, *where* Id_τ *denotes the identity map on* $\mathfrak{X}(\Sigma_\tau^n)$;

(b) $\Box_\tau = (n-1)\dfrac{f'(\tau)}{f(\tau)} \Delta_\tau$, *where* Δ_τ *is the Laplacian operator on* Σ_τ^n;

(c) $\Box_\tau = (n-1)\dfrac{f'(\tau)}{f(\tau)^3} \Delta$, *where* Δ *is the Laplacian operator on* M^n;

(d) *The i-th eigenvalue $\mu_i(\tau)$ of the Cheng-Yau's square operator $\Box_\tau$ on Σ_τ^n is*

$$\mu_i(\tau) = (n-1)\frac{f'(\tau)}{f(\tau)^3}\mu_i,$$

where μ_i is the i-th eigenvalue of the Laplacian operator Δ on M^n;

(e) $\mathrm{tr}\left(T_\tau \circ \overline{R}_\tau\right) = (n-1)\dfrac{f'(\tau)}{f(\tau)} \overline{\varrho}$;

(f) $\mathcal{J}_\tau = (n-1)\dfrac{f'(\tau)}{f(\tau)^3} (\Delta + Q_0)$, *where*

$$Q_0 = n\left(f'(\tau)\right)^2 + f(\tau)^2 \overline{\varrho} \qquad (17.37)$$

17.4 Bifurcation and Locally Rigidity Results for H_2-Hypersurfaces

is a constant on (τ_1, τ_2);

(g) The i-th eigenvalue $\widehat{\mu}_i(\tau)$ of the Jacobi operator $\mathcal{J}_\tau$ on Σ_τ^n is

$$\widehat{\mu}_i(\tau) = (n-1)\frac{f'(\tau)}{f(\tau)^3}(\mu_i - Q_0),$$

where μ_i is the i-th eigenvalue of the Laplacian operator Δ on M^n.

Proof Item (a) is obtained immediately by (17.6) and (17.30). In order to prove item (b), let us observe that by (17.7) and item (a) one has

$$\Box_\tau(u) = \mathrm{tr}\left(T_\tau\left(\mathrm{Hess}_{\Sigma_\tau^n}(u)\right)\right) = (n-1)\frac{f'(\tau)}{f(\tau)}\,\mathrm{tr}\left(\mathrm{Hess}_{\Sigma_\tau^n}(u)\right) = \frac{f'(\tau)}{f(\tau)}\Delta_\tau(u)$$

for every $u \in C^\infty(\Sigma_\tau^n)$.

Now, through the natural identification of $C^\infty(\Sigma_\tau^n)$ with $C^\infty(M^n)$, item (c) follows by item (b) noting that the induced metric on Σ_τ^n is given by $f(\tau)^2\langle\cdot,\cdot\rangle_M$. Moreover, item (d) directly follows by (c).

Let us prove now item (e). To this aim, let $\{E_1, \ldots, E_n\}$ be an orthonormal frame defined in a neighborhood of some point of Σ_τ^n and let $K_{\overline{M}}(\partial_\tau, E_j)$ be the sectional curvature of $\overline{M}^{n+1}$ along the plane generated by ∂_τ and E_j, $j \in \{1, \ldots, n\}$.

Then, by (17.36), item (a) and (17.14), we get

$$\frac{1}{n-1}\mathrm{tr}\left(T_\tau \circ \overline{R}_\tau\right) = \frac{f'(\tau)}{f(\tau)}\sum_{j=1}^n \langle \overline{R}_\tau(E_j), E_j\rangle = \frac{f'(\tau)}{f(\tau)}\sum_{j=1}^n \langle \overline{R}(\partial_\tau, E_j)\partial_\tau, E_j\rangle$$

$$= \frac{f'(\tau)}{f(\tau)}\sum_{j=1}^n K_{\overline{M}}(\partial_\tau, E_j) = \frac{f'(\tau)}{f(\tau)}\mathrm{Ric}_{\overline{M}}(\partial_\tau, \partial_\tau) = \frac{f'(\tau)}{f(\tau)}\overline{\varrho}.$$

Hence, (e) is proved.

Now, by (17.31), (17.35), as well as items (c) and (e), we have

$$\mathcal{J}_\tau = (n-1)\frac{f'(\tau)}{f(\tau)^3}\Delta + \left\{\frac{n(n-1)}{2}\left(n\frac{f'(\tau)}{f(\tau)}\left(\frac{f'(\tau)}{f(\tau)}\right)^2\right.\right.$$

$$\left.\left. -(n-2)\left(\frac{f'(\tau)}{f(\tau)}\right)^3 + (n-1)\frac{f'(\tau)}{f(\tau)}\overline{\varrho}\right)\right\}$$

$$= (n-1)\frac{f'(\tau)}{f(\tau)^3}\Delta + \left\{n(n-1)\left(\frac{f'(\tau)}{f(\tau)}\right)^3 + (n-1)\frac{f'(\tau)}{f(\tau)}\overline{\varrho}\right\}$$

$$= (n-1)\frac{f'(\tau)}{f(\tau)^3}\Delta + \frac{(n-1)f'(\tau)}{f(\tau)^3}\left(n\left(f'(\tau)\right)^2 + f(\tau)^2\overline{\varrho}\right).$$

To end the proof of item (f), it remains to show that

$$Q : (\tau_1, \tau_2) \to \mathbb{R}$$
$$\tau \mapsto Q(\tau) = n\left(f'(\tau)\right)^2 + f(\tau)^2 \overline{\varrho}$$

is a constant function. To this scope, by (17.33) we observe that

$$Q'(\tau) = 2f'(\tau)\underbrace{(nf''(\tau) + f(\tau)\overline{\varrho})}_{0} = 0$$

for every $\tau \in (\tau_1, \tau_2)$. Consequently, there exists $Q_0 \in \mathbb{R}$ such that $Q(\tau) = Q_0$ for every $\tau \in (\tau_1, \tau_2)$.

Finally, item (g) directly follows by (f). $\square$

For a better understanding of the statements in the following results, let us remember that the *spectrum* of the Laplacian operator Δ on a closed Riemannian manifold M^n (cf. [107, Section 1.3]) is determined by a sequence of eigenvalues $\{\mu_i\}_{i=0}^{+\infty}$ satisfying

$$0 = \mu_0 < \mu_1 \leq \mu_2 \leq \cdots \leq \mu_i \leq \mu_{i+1} \leq \cdots,$$

repeated according to their multiplicity, and

$$\lim_{i \to +\infty} \mu_i = +\infty.$$

Our first result provides some simple sufficient conditions to get the local rigidity of the family $\{\Omega_\tau\}_{\tau \in (\tau_1, \tau_2]}$.

Theorem 17.1 *Let $I \subset \mathbb{R}$ be an open interval, let M^n be a closed n-dimensional Riemannian manifold ($n \geq 2$) and let $I \times_f M^n$ be a Riemannian warped product, whose warped function $f : I \to \mathbb{R}$ satisfies the ordinary differential equation (17.33). Let $\{\Omega_\tau\}_{\tau \in (\tau_1, \tau_2]}$ be a family of open subsets of $I \times_f M^n$ of the form*

$$\Omega_\tau = (\tau_1, \tau) \times M^n,$$

where τ_1 and τ_2 are fixed numbers in $I \subset \mathbb{R}$ with $\tau_1 < \tau_2$. If

(a) *$Q_0 \neq \mu_i$ for every $i \in \mathbb{N}$, where Q_0 is the constant defined in (17.37) and μ_i is the i-th eigenvalue of the Laplacian operator Δ on M^n, and*
(b) *$f'(\tau) \neq 0$ for every $\tau \in (\tau_1, \tau_2)$,*

then $\{\Omega_\tau\}_{\tau \in (\tau_1, \tau_2]}$ is locally rigid at each $\tau \in (\tau_1, \tau_2)$.

17.4 Bifurcation and Locally Rigidity Results for H_2-Hypersurfaces

Table 17.2 Families of open sets that are locally rigid according to Theorem 17.1

Riemannian warped product	Family of open sets	Q_0
$(-\infty, +\infty) \times_{\cosh \tau} \mathbb{S}^n(r)$ with $r > 0$	$\{\Omega_\tau\}_{\tau \in (\tau_1, \tau_2]}$ with τ_1, τ_2 in $(-\infty, 0)$ and $\tau_1 < \tau_2$, or with τ_1, τ_2 in $(0, +\infty)$ and $\tau_1 < \tau_2$	$-n$
$(0, +\infty) \times_{\sinh \tau} \mathbb{S}^n(r)$ with $r > 1$ or $0 < r < 1$	$\{\Omega_\tau\}_{\tau \in (\tau_1, \tau_2]}$ with τ_1, τ_2 in $(0, +\infty)$ and $\tau_1 < \tau_2$	n
$(0, +\infty) \times_\tau \mathbb{S}^n(r)$ with $r > 1$ or $0 < r < 1$	$\{\Omega_\tau\}_{\tau \in (\tau_1, \tau_2]}$ with τ_1, τ_2 in $(0, +\infty)$ and $\tau_1 < \tau_2$	n
$(-\pi/2, \pi/2) \times_{\cos \tau} \mathbb{S}^n(r)$ with $r > 1$ or $0 < r < 1$	$\{\Omega_\tau\}_{\tau \in (\tau_1, \tau_2]}$ with τ_1, τ_2 in $(-\pi/2, 0)$ and $\tau_1 < \tau_2$, or with τ_1, τ_2 in $(0, \pi/2)$ and $\tau_1 < \tau_2$	n
$(0, \pi) \times_{\sin \tau} \mathbb{S}^n(r)$ with $r > 1$ or $0 < r < 1$	$\{\Omega_\tau\}_{\tau \in (\tau_1, \tau_2]}$ with τ_1, τ_2 in $(0, \pi/2)$ and $\tau_1 < \tau_2$, or with τ_1, τ_2 in $(\pi/2, \pi)$ and $\tau_1 < \tau_2$	n
$(0, \pi/2) \times_{\sin \tau + \cos \tau} \mathbb{S}^n(r)$ with $r > 1$ or $0 < r < 1$	$\{\Omega_\tau\}_{\tau \in (\tau_1, \tau_2]}$ with τ_1, τ_2 in $(0, \pi/4)$ and $\tau_1 < \tau_2$, or with τ_1, τ_2 in $(\pi/4, \pi/2)$ and $\tau_1 < \tau_2$	n

Proof Taking into account our assumptions, item (g) of Proposition 17.3 yields that the i-th eigenvalue $\widehat{\mu}_i(\tau)$ of the Jacobi operator $\mathcal{J}_\tau$ on Σ_τ^n is such that

$$\widehat{\mu}_i(\tau) = (n-1) \frac{f'(\tau)}{f(\tau)^3} (\mu_i - Q_0) \neq 0.$$

Hence, the second variation

$$\frac{d^2}{dt^2} \mathcal{F}^{\lambda(\tau)}(0),$$

given in (17.27) is nonsingular for every $\tau \in (\tau_1, \tau_2)$ and, therefore, the family $\{\Omega_\tau\}_{\tau \in (\tau_1, \tau_2]}$ is locally rigid at each $\tau \in (\tau_1, \tau_2)$. □

Let $\mathbb{S}^n(r)$ be the n-dimensional Euclidean sphere of radius $r > 0$. We know that the eigenvalues μ_i of the Laplacian operator Δ on $\mathbb{S}^n(r)$ are given by

$$\mu_i = \frac{i(i+n-1)}{r^2}, \quad i \in \mathbb{N}; \qquad (17.38)$$

see [107, Section 2.4].

Then, from Table 17.1 we can investigate the families $\{\Omega_\tau\}_{\tau \in (\tau_1, \tau_2]}$ of open sets in the warped products $I \times_f \mathbb{S}^n(r)$ of the form $\Omega_\tau = (\tau_1, \tau) \times \mathbb{S}^n(r)$, with $\tau_1, \tau_2 \in I \subset \mathbb{R}$ and $\tau_1 < \tau_2$, that verify the conditions of Theorem 17.1. In Table 17.2 we collect the results of this analysis.

From the Table 17.2 we observe that the first case can be extended to a broader warped product class, exchanging the Euclidean sphere $\mathbb{S}^n(r)$ by any closed Riemannian manifold M^n.

Corollary 17.1 *Let $(-\infty, +\infty) \times_{\cosh \tau} M^n$ be a Riemannian warped product, with closed Riemannian fiber M^n ($n \geq 2$), and let $\{\Omega_\tau\}_{\tau \in (\tau_1, \tau_2]}$ be a family of open subsets of $(-\infty, +\infty) \times_{\cosh \tau} M^n$ of the form $\Omega_\tau = (\tau_1, \tau) \times M^n$, where τ_1 and τ_2 are fixed numbers either in $(-\infty, 0)$ or in $(0, +\infty)$, in both cases with $\tau_1 < \tau_2$. Then $\{\Omega_\tau\}_{\tau \in (\tau_1, \tau_2]}$ is locally rigid at each $\tau \in (\tau_1, \tau_2)$.*

In the next result we have established a criterion that guarantees the existence of bifurcation instants of the family $\{\Omega_\tau\}_{\tau \in (\tau_1, \tau_2]}$.

Theorem 17.2 *Let $I \subset \mathbb{R}$ be an open interval, let M^n be a closed n-dimensional Riemannian manifold ($n \geq 2$) and let $I \times_f M^n$ be a Riemannian warped product, whose warped function $f : I \to \mathbb{R}$ satisfies the ordinary differential equation (17.33). Moreover, let $\{\Omega_\tau\}_{\tau \in (\tau_1, \tau_2]}$ be a family of open subsets of $I \times_f M^n$ of the form*

$$\Omega_\tau = (\tau_1, \tau) \times M^n,$$

where τ_1 and τ_2 are fixed numbers in $I \subset \mathbb{R}$ with $\tau_1 < \tau_2$. Suppose that

(a) *$Q_0 \neq \mu_i$ for every $i \in \mathbb{N}$, where Q_0 is the constant defined in (17.37) and μ_i is the i-th eigenvalue of the Laplacian operator Δ on M^n, and*
(b) *There exist numbers $\delta_0, \eta_0 \in (\tau_1, \tau_2)$ with $\delta_0 < \eta_0$ such that either $f'(\delta_0) > 0$ and $f'(\eta_0) < 0$, or $f'(\delta_0) < 0$ and $f'(\eta_0) > 0$.*

Then $\{\Omega_\tau\}_{\tau \in (\tau_1, \tau_2]}$ admits at least a bifurcation instant at some $\tau_* \in (\delta_0, \eta_0)$.

Proof Since $f > 0$ on $I \subset \mathbb{R}$, by item (g) of Proposition 17.3 and from our hypotheses involving Q_0 and f' we obtain that the eigenvalue $\widehat{\mu}_i(\delta_0)$ and $\widehat{\mu}_i(\eta_0)$ of the Jacobi operators $\mathcal{J}_{\delta_0}$ and $\mathcal{J}_{\eta_0}$ are such that

$$\widehat{\mu}_i(\delta_0) = (n-1)\frac{f'(\delta_0)}{f(\delta_0)^3}(\mu_i - Q_0) \neq 0 \qquad (17.39)$$

and

$$\widehat{\mu}_i(\eta_0) = (n-1)\frac{f'(\eta_0)}{f(\eta_0)^3}(\mu_i - Q_0) \neq 0 \qquad (17.40)$$

for all $i \in \mathbb{N}$, respectively. Furthermore, for some $i_0 \in \mathbb{N}$, by (17.39) and (17.40),

$$\widehat{\mu}_{i_0}(\delta_0)\widehat{\mu}_{i_0}(\eta_0) = (n-1)^2 \frac{f'(\delta_0)f'(\eta_0)}{f(\delta_0)^3 f(\eta_0)^3}(\mu_{i_0} - Q_0)^2 < 0, \qquad (17.41)$$

since the hypothesis (b) guarantees that $f'(\delta_0)f'(\eta_0) < 0$.

17.4 Bifurcation and Locally Rigidity Results for H_2-Hypersurfaces

Now, by (17.27), (17.39) and (17.40) we get that the second variations

$$\frac{d^2}{dt^2}\mathcal{F}^{\lambda(\delta_0)}(0) \quad \text{and} \quad \frac{d^2}{dt^2}\mathcal{F}^{\lambda(\eta_0)}(0)$$

are nonsingular. Furthermore, by (17.41) we obtain that the eigenvalue $\widehat{\mu}_i(\tau)$ of the Jacobi operator $\mathcal{J}_\tau = (n-1)(f'(\tau)/f(\tau)^3)(\Delta + Q_0)$ which corresponds to $i = i_0$ change sign between τ_1 and τ_2. Consequently, since the eigenvalues of the Jacobi operator $\mathcal{J}_\tau$ are ordered, the number of negative eigenvalues between τ_1 and τ_2 changes. Therefore, one has

$$\text{Ind}(\mathcal{F}^{\lambda(\tau_1)}, \Omega_{\tau_1}) \neq \text{Ind}(\mathcal{F}^{\lambda(\tau_2)}, \Omega_{\tau_2})$$

and the result follows. □

Taking into account once again the eigenvalues of the Laplacian operator Δ of the Euclidean spheres $\mathbb{S}^n(r)$, giving in (17.38), we can list in Table 17.3 some examples of families $\{\Omega_\tau\}_{\tau \in (\tau_1, \tau_2]}$ of open sets in warped products $I \times_f \mathbb{S}^n(r)$ of the form $\Omega_\tau = (\tau_1, \tau) \times \mathbb{S}^n(r)$, with $\tau_1, \tau_2 \in I \subset \mathbb{R}$ and $\tau_1 < \tau_2$, that verify the conditions of Theorem 17.2.

We remark that the first case of Table 17.3 above can be extended to a larger class of warped products, exchanging the Euclidean sphere $\mathbb{S}^n(r)$ by any closed Riemannian manifold M^n.

Corollary 17.2 *Let* $(-\infty, +\infty) \times_{\cosh \tau} M^n$ *be a Riemannian warped product, with closed Riemannian fiber* M^n ($n \geq 2$), *and let* $\{\Omega_\tau\}_{\tau \in (\tau_1, \tau_2]}$ *be a family of open subsets of* $(-\infty, +\infty) \times_{\cosh \tau} M^n$ *of the form* $\Omega_\tau = (\tau_1, \tau) \times M^n$, *where* τ_1 *and* τ_2 *are fixed numbers such that* $\tau_1 \in (-\infty, 0)$ *and* $\tau_2 \in (0, +\infty)$. *If* δ_0 *and* η_0 *are two real numbers such that* $\tau_1 < \delta_0 < 0 < \eta_0 < \tau_2$, *then* $\Omega_\tau = (\tau_1, \tau) \times M^n$ *admits at least a bifurcation instant at some* $\tau_* \in (\delta_0, \eta_0)$.

Table 17.3 Families of open sets that admit a bifurcation instant according to Theorem 17.2

Riemannian warped product	Family of open sets	Q_0
$(-\infty, +\infty) \times_{\cosh \tau} \mathbb{S}^n(r)$ with $r > 0$	$\{\Omega_\tau\}_{\tau \in (\tau_1, \tau_2]}$ with τ_1, τ_2, δ_0 and η_0 such that $-\infty < t_1 < \delta_0 < 0 < \eta_0 < \tau_2 < +\infty$	$-n$
$(-\pi/2, \pi/2) \times_{\cos \tau} \mathbb{S}^n(r)$ with $r > 1$ or $0 < r < 1$	$\{\Omega_\tau\}_{\tau \in (\tau_1, \tau_2]}$ with τ_1, τ_2, δ_0 and η_0 such that $-\pi/2 < t_1 < \delta_0 < 0 < \eta_0 < \tau_2 < \pi/2$	n
$(0, \pi) \times_{\sin \tau} \mathbb{S}^n(r)$ with $r > 1$ or $0 < r < 1$	$\{\Omega_\tau\}_{\tau \in (\tau_1, \tau_2]}$ with τ_1, τ_2, δ_0 and η_0 such that $0 < \tau_1 < \delta_0 < \pi/2 < \eta_0 < \tau_2 < \pi$	n
$(0, \pi/2) \times_{\sin \tau + \cos \tau} \mathbb{S}^n(r)$ with $r > 1$ or $0 < r < 1$	$\{\Omega_\tau\}_{\tau \in (\tau_1, \tau_2]}$ with τ_1, τ_2, δ_0 and η_0 such that $0 < \tau_1 < \delta_0 < \pi/4 < \eta_0 < \tau_2 < \pi/2$	n

The existence of bifurcation instants of the family $\{\Omega_t\}_{t \in (t_1, t_2]}$ can be obtained as follows.

Theorem 17.3 *Let $I \subset \mathbb{R}$ be an open interval, let M^n be a closed n-dimensional Riemannian manifold ($n \geq 2$) and let $I \times_f M^n$ be a Riemannian warped product, whose warped function $f : I \to \mathbb{R}$ satisfies the ordinary differential equation (17.33) for some nonzero constant real $\overline{\varrho}$. Moreover, let $\{\Omega_\tau\}_{\tau \in (\tau_1, \tau_2]}$ be a family of open subsets of $I \times_f M^n$ of the form $\Omega_\tau = (\tau_1, \tau) \times M^n$, where τ_1 and τ_2 are fixed numbers in $I \subset \mathbb{R}$ with $\tau_1 < \tau_2$. Suppose that*

(a) $Q_0 \neq \mu_i$ *for all $i \in \mathbb{N}$, where Q_0 is the constant defined in (17.37) and μ_i is the i-th eigenvalue of the Laplacian operator Δ on M^n, and*
(b) *There exists $\tau_* \in (\tau_1, \tau_2)$ such that $f'(\tau_*) = 0$.*

Then $\{\Omega_\tau\}_{\tau \in (\tau_1, \tau_2]}$ admits a bifurcation instant in τ_*.

Proof By item (g) of Proposition 17.3, for every $\delta_0, \eta_0 \in (\tau_1, \tau_2)$ with

$$\delta_0 < \tau_* < \eta_0$$

and for $i_0 \in \mathbb{N}$ we have

$$\widehat{\mu}_{i_0}(\delta_0)\widehat{\mu}_{i_0}(\eta_0) = (n-1)^2 \frac{f'(\delta_0) f'(\eta_0)}{f(\delta_0)^3 f(\eta_0)^3} \left(\mu_{i_0} - Q_0\right)^2. \qquad (17.42)$$

Since $f > 0$ on I, $\overline{\varrho} \neq 0$ and $-nf''(\tau) = \overline{\varrho} f(\tau)$ on I (see Eq. (17.33)) then $f''(\tau) \neq 0$ on I, which asserts that f' is strictly increasing or strictly decreasing on I. So, by (b), since $\delta_0 < t_* < \eta_0$ one has

$$f'(\delta_0) < 0 < f'(\eta_0) \quad \text{or} \quad f'(\eta_0) < 0 < f'(\delta_0).$$

In both cases, $f'(\delta_0) f'(\eta_0) < 0$. Hence, thanks to (17.42) and (a), we have that

$$\widehat{\mu}_{i_0}(\delta_0)\widehat{\mu}_{i_0}(\eta_0) < 0.$$

In addition, using again item (g) of Proposition 17.3, by (a) we get that (17.39) and (17.40) are satisfied.

Arguing as in the proof of Theorem 17.2 the conclusion follows. □

With slight modifications, it is immediate to observe that the families of open sets described in Table 17.3 can fit under the conditions of Theorem 17.3. We recorded this new configuration in Table 17.4.

By Theorem 17.3 and the first case of Table 17.4, we get the following result.

17.4 Bifurcation and Locally Rigidity Results for H_2-Hypersurfaces

Table 17.4 Families that admit a bifurcation instant at τ_* in (τ_1, τ_2) according to Theorem 17.3

Riemannian Warped product	Family of open sets	Q_0
$(-\infty, +\infty) \times_{\cosh \tau} \mathbb{S}^n(r)$ with $r > 0$	$\{\Omega_\tau\}_{\tau \in (\tau_1, \tau_2]}$ with τ_1, τ_2, and τ_* such that $-\infty < \tau_1 < \tau_* = 0 < \tau_2 < +\infty$	$-n$
$(-\pi/2, \pi/2) \times_{\cos \tau} \mathbb{S}^n(r)$ with $r > 1$ or $0 < r < 1$	$\{\Omega_\tau\}_{\tau \in (\tau_1, \tau_2]}$ with τ_1, τ_2, and τ_* such that $-\pi/2 < \tau_1 < \tau_* = 0 < \tau_2 < \pi/2$	n
$(0, \pi) \times_{\sin \tau} \mathbb{S}^n(r)$ with $r > 1$ or $0 < r < 1$	$\{\Omega_\tau\}_{\tau \in (\tau_1, \tau_2]}$ with τ_1, τ_2, and τ_* such that $0 < \tau_1 < \tau_* = \pi/2 < \tau_2 < \pi$	n
$(0, \pi/2) \times_{\sin \tau + \cos \tau} \mathbb{S}^n(r)$ with $r > 1$ or $0 < r < 1$	$\{\Omega_\tau\}_{\tau \in (\tau_1, \tau_2]}$ with τ_1, τ_2, and τ_* such that $0 < \tau_1 < \tau_* = \pi/4 < \tau_2 < \pi/2$	n

Corollary 17.3 *Let $(-\infty, +\infty) \times_{\cosh \tau} M^n$ be a Riemannian warped product, with closed Riemannian fiber M^n ($n \geq 2$), and let $\{\Omega_\tau\}_{\tau \in (\tau_1, \tau_2]}$ be a family of open subsets of $(-\infty, +\infty) \times_{\cosh \tau} M^n$ of the form $\Omega_\tau = (\tau_1, \tau) \times M^n$, where τ_1 and τ_2 are fixed numbers such that $\tau_1 \in (-\infty, 0)$ and $\tau_2 \in (0, +\infty)$. Then $\{\Omega_\tau\}_{\tau \in (\tau_1, \tau_2]}$ admits a bifurcation point in $\tau_* = 0$.*

Remark 17.4 The requirement on the constant Q_0 that appears in the hypotheses of Theorems 17.1, 17.2 and 17.3, can be interpreted as a geometric condition on the Riemannian fiber M^n of the warped product $I \times_f M^n$. Indeed, let us first observe by (17.37) that the constant $\bar{\varrho}$ admits the expression

$$\bar{\varrho} = \frac{Q_0}{f(\tau)^2} - n H_2^\tau.$$

Consequently, by (17.5) we obtain that the scalar curvature S^τ of Σ_τ^n is given by $S^\tau = Q_0/f(\tau)^2$. Since the induced metric on Σ_τ^n is $f(\tau)^2 \langle \cdot, \cdot \rangle_M$, we have that the scalar curvature S^M of M^n and Q_0 are related by

$$S^M = (n-1)Q_0.$$

Therefore, what is requested in item (a) of Theorems 17.1, 17.2 and 17.3 can be interpreted as the requirement that the constant scalar curvature S^M of M^n does not belong to the spectrum of the Laplacian operator Δ of M^n.

18 Bifurcation of Spacelike Hypersurfaces with Constant Mean Curvature in Spacetimes

18.1 Introduction

Stability questions concerning hypersurfaces with constant mean curvature H (shortly, H-hypersurfaces) in Riemannian manifolds $\overline{M}^{n+1}$ ($n \geq 2$) began with Barbosa and do Carmo in [62], and Barbosa, do Carmo and Eschenburg in [64].

In these papers, they introduced the notion of stability and proved that any closed H-hypersurface immersed into $\overline{M}^{n+1}$ is a solution of the variational problem of minimizing the area functional for volume-preserving variations. Furthermore, when $\overline{M}^{n+1}$ has constant sectional curvature c, they also established that geodesic spheres are the only stable solutions for this variational problem.

In the Lorentz context, Barbosa and Oliker [63] obtained an analogous result to those obtained in [62] and [64], proving that spacelike H-hypersurfaces in Lorentz manifolds are also critical points of the area functional for any variation that keep the volume constant. They also computed the second variation formula and showed, for the de Sitter space $\mathbb{S}_1^{n+1}$, that spheres maximize the area functional for volume-preserving variations.

Meanwhile, Barros et al. [67] have studied the problem of *strong* stability (that is, stability with respect to not necessarily volume-preserving variations) for closed spacelike H-hypersurfaces in a *generalized Robertson-Walker* (GRW) spacetime, giving a characterization of maximal spacelike hypersurfaces and spacelike slices. By a GRW spacetime, we mean a Lorentz warped product

$$(-I \times_f M^n, -dt^2 + f^2 \langle \cdot, \cdot \rangle_M),$$

with 1-dimensional negative definite base $(I, -dt^2)$, Riemannian fiber $(M^n, \langle \cdot, \cdot \rangle_M)$ and warping function f. We also recall that a GRW spacetime $-I \times_f M^n$ is said to be spatially closed when M^n is closed.

In the current chapter, we use Equivariant Bifurcation Theory in order to establish sufficient conditions that allow us to guarantee the existence of bifurcation instants or the local rigidity of family $\{\Omega_\tau\}_{\tau \in I}$ of open sets in a time-oriented Lorentzian manifold $\overline{M}^{n+1}$, whose boundaries $\partial\Omega_\tau$ are closed (compact without boundary) spacelike H-hypersurfaces, that is, closed spacelike hypersurfaces with constant mean curvature H.

In particular, we do this study for a certain family of open sets in a spatially closed generalized Robertson-Walker spacetime, namely, Lorentz warped products

$$(-I \times_f M^n, -dt^2 + f^2 \langle \cdot, \cdot \rangle_M),$$

with 1-dimensional negative definite base $(I, -dt^2)$, closed Riemannian fiber $(M^n, \langle \cdot, \cdot \rangle_M)$ and warping function $f : I \to (0, +\infty)$.

In this case, the results are obtained considering some appropriate hypotheses that depend of f and of the behavior of the eigenvalues of Laplacian operator of M^n. Applications to such families in some spacetimes, like the de Sitter space, are also given.

The results that will be present are based mainly on the paper [286].

18.2 The Variational Problems (VP-1) and (VP-2)

Let $\overline{M}^{n+1}$ be a $(n+1)$-dimensional time-oriented Lorentzian manifold ($n \geq 2$) with metric tensor $\langle \cdot, \cdot \rangle$ and let $\mathcal{M}$ be the open of open subsets $\Omega \subset \overline{M}^{n+1}$ with compact closure $\overline{\Omega}$ and whose boundary $\partial\Omega$ is a closed spacelike hypersurface.

Here, $\Omega \in \mathcal{M}$ if and only if $\partial\Omega$ is spacelike and Riemannian with respect to the induced metric on $\partial\Omega$, via the natural immersion. Moreover, $\partial\Omega$ is *closed* means that $\partial\Omega$ is compact without boundary. We denote by $\mathfrak{X}(\overline{M}^{n+1})$ the set of vector fields of class C^∞ on $\overline{M}^{n+1}$ and by $C^\infty(\overline{M}^{n+1})$ the ring of real functions of class C^∞ on $\overline{M}^{n+1}$. Similar notations will be used on $\partial\Omega$.

We recall that a vector field $X \in \mathfrak{X}(\overline{M}^{n+1})$ is said to be timelike if $\langle X, X \rangle < 0$ on $\overline{M}^{n+1}$; spacelike if $\langle X, X \rangle > 0$ on $\overline{M}^{n+1}$; a unit vector field if $\langle X, X \rangle = \pm 1$ on $\overline{M}^{n+1}$; see [251, Chapter 3].

Furthermore, as $\overline{M}^{n+1}$ is time-orientable by a timelike vector field (cf. [251, Lemma 5.32]), say a certain $K \in \mathfrak{X}(\overline{M}^{n+1})$, and $\partial\Omega$ is a spacelike hypersurface, then $\partial\Omega$ is orientable and one can choose a globally defined unit normal vector field N on $\partial\Omega$ having the same time-orientation of $\overline{M}^{n+1}$, that is, $\langle K, N \rangle < 0$.

Such N is said the future-pointing Gauss map of $\partial\Omega$. Now, let

$$A : \mathfrak{X}(\partial\Omega) \to \mathfrak{X}(\partial\Omega) \qquad (18.1)$$
$$V \mapsto A(V) = -\overline{\nabla}_V N$$

the shape operator of $\partial\Omega$ with respect to N. The *mean curvature* H of $\partial\Omega$ is defined by

18.2 The Variational Problems (VP-1) and (VP-2)

$$H : \partial\Omega \to \mathbb{R}$$
$$p \mapsto H(p) = -\frac{1}{n} \operatorname{tr}(A_p). \tag{18.2}$$

We denote by $d\overline{M}^{n+1}$ and dV the volume elements of $\overline{M}^{n+1}$ and $\partial\Omega$, respectively. Furthermore, if $\Omega \in \mathcal{M}$, one has that

$$\operatorname{Vol}(\Omega) = \int_\Omega d\overline{M}^{n+1} \quad \text{and} \quad \operatorname{Area}(\partial\Omega) = \int_{\partial\Omega} dV$$

denote the volume of Ω and the area of $\partial\Omega$, respectively.

Let $\Omega \in \mathcal{M}$. A variation of $\partial\Omega$ is a smooth map

$$X : (-\epsilon, \epsilon) \times \partial\Omega \to \overline{M}^{n+1},$$

satisfying the following two conditions:

1. For every $t \in (-\epsilon, \epsilon)$, the map

$$X_t : \partial\Omega \to \overline{M}^{n+1}$$
$$p \mapsto X(p) = X(t, p) \tag{18.3}$$

is a spacelike hypersurface;
2. $X(0, p) = \iota(p)$ for every $p \in \partial\Omega$, where $\iota : \partial\Omega \hookrightarrow \overline{\Omega}$ is the inclusion map.

In this context, given $\Omega \in \mathcal{M}$ and a variation $X : (-\epsilon, \epsilon) \times \partial\Omega \to \overline{M}^{n+1}$ of $\partial\Omega$, we adopted the notation $\partial\Omega_t = X_t(\partial\Omega)$. We notice that, for values of t small enough, $\partial\Omega_t$ is a closed spacelike hypersurface. Furthermore, $\partial\Omega_t$ bounds an open subset Ω_t whose closure is also compact. Thus, $X : (-\epsilon, \epsilon) \times \partial\Omega \to \overline{M}^{n+1}$ induces naturally a variation of the open subset denoted by Ω_t, which is also an element of the set $\mathcal{M}$.

Now, let us denote by dV_t and N_t be the volume element and the future-pointing Gauss map on $\partial\Omega_t$ of the immersion (18.3), respectively. When $t = 0$, these objects coincide with those already defined on $\partial\Omega$.

The variational field associated to $X : (-\epsilon, \epsilon) \times \partial\Omega \to \overline{M}^{n+1}$ is the vector field

$$\left.\frac{\partial X}{\partial t}\right|_{t=0} \in \mathfrak{X}(X((-\epsilon, \epsilon) \times \partial\Omega)).$$

Letting

$$u_t = -\left\langle \frac{\partial X}{\partial t}, N_t \right\rangle \in C^\infty(\partial\Omega_t), \tag{18.4}$$

we get

$$\left.\frac{\partial X}{\partial t}\right|_{t=0} = u_0 N + \left(\left.\frac{\partial X}{\partial t}\right|_{t=0}\right)^\top,$$

where $(\cdot)^\top$ stands for tangential component.

We recall that the balance of volume and the area functionals associated with $X : (-\epsilon, \epsilon) \times \partial\Omega \to \overline{M}^{n+1}$ are defined by

$$\begin{aligned} \mathcal{V} : (-\epsilon, \epsilon) &\to \mathbb{R} \\ t &\mapsto \mathcal{V}(t) = \text{Vol}(\Omega_t). \end{aligned}$$

and

$$\begin{aligned} \mathcal{A} : (-\epsilon, \epsilon) &\to \mathbb{R} \\ t &\mapsto \mathcal{A}(t) = \text{Area}(\partial\Omega_t) = \int_{\partial\Omega_t} dV_t, \end{aligned} \tag{18.5}$$

respectively; see [63, Section 3].

Furthermore, we say that $X : (-\epsilon, \epsilon) \times \partial\Omega \to \overline{M}^{n+1}$ is volume-preserving of Ω if $\mathcal{V}(t) = \mathcal{V}(0)$ for every $t \in (-\epsilon, \epsilon)$.

The formula of the first variation of $\mathcal{V}(t)$ and $\mathcal{A}(t)$ are given in the next lemma below; see [63, Section 3] for a detailed proof.

Lemma 18.1 *If $\Omega \in \mathcal{M}$ and $X : (-\epsilon, \epsilon) \times \partial\Omega \to \overline{M}^{n+1}$ is a variation of $\partial\Omega$, then*

(a) $\dfrac{d}{dt}\mathcal{V}(t) = \displaystyle\int_{\partial\Omega_t} u_t dV_t,\ t \in (-\epsilon, \epsilon)$, *where u_t is the smooth function defined in (18.4).*
In particular, $X : (-\epsilon, \epsilon) \times \partial\Omega \to \overline{M}^{n+1}$ *is volume-preserving of Ω if and only if*

$$\int_{\partial\Omega_t} u_t dV_t = 0, \quad \text{for any } t \in (-\epsilon, \epsilon).$$

(b) $\dfrac{d}{dt}\mathcal{A}(t) = \displaystyle\int_{\partial\Omega_t} n H_t u_t dV_t,\ \text{for every } t \in (-\epsilon, \epsilon),$ *where $H_t = H(t, \cdot)$ denotes the mean curvature of $\partial\Omega_t$.*

Remark 18.1 Let $\Omega \in \mathcal{M}$. According to [63, Lemma 3.2], if $u_0 \in C^\infty(\partial\Omega)$ is such that

$$\int_{\partial\Omega} u_0 dV = 0,$$

18.2 The Variational Problems (VP-1) and (VP-2)

then there exists a volume-preserving variation $X : (-\epsilon, \epsilon) \times \partial\Omega \to \overline{M}^{n+1}$ of $\partial\Omega$ whose variational field is

$$\left.\frac{\partial X}{\partial t}\right|_{t=0} = u_0 N.$$

In order to characterize open subsets Ω of the Lorentzian manifold $\overline{M}^{n+1}$ whose boundary $\partial\Omega$ is a *spacelike H-hypersurface*, namely, a spacelike hypersurface with constant mean curvature H, we consider the following variational problem

(VP-1) *Maximizing the area functional $\mathcal{A}(t)$ given in (18.5) for any variation of $\partial\Omega$ that preserves the volume of Ω;*

and

(VP-2) *Maximizing the area functional $\mathcal{A}(t)$ given in (18.5) for any variation of $\partial\Omega$, not necessarily a volume-preserving variation of Ω.*

The Lagrange multiplier method leads us then to the *Jacobi functional*

$$\begin{aligned}\mathcal{F}^\lambda : (-\epsilon, \epsilon) &\to \mathbb{R} \\ t &\mapsto \mathcal{F}^\lambda(t) = \mathcal{A}(t) + \lambda \mathcal{V}(t),\end{aligned} \quad (18.6)$$

where $\lambda \in \mathbb{R}$ is a constant to be determined.

As an immediate consequence of Lemma 18.1 we get that the first variation of $\mathcal{F}^\lambda(t)$ takes the following form

$$\frac{d}{dt}\mathcal{F}^\lambda(t) = \int_{\partial\Omega_t} (nH_t - \lambda)\, u_t\, dV_t, \quad t \in (-\epsilon, \epsilon), \quad (18.7)$$

where u_t is the smooth function defined in (18.4). Let

$$\overline{H} = \frac{1}{\text{Area}(\partial\Omega)} \int_{\partial\Omega} H\, dV$$

be the integral mean of the mean curvature H on $\partial\Omega$.

If H constant, we have $\overline{H} = H$. So, if we choose $\lambda = n\overline{H}$, by (18.7) we have that

$$\frac{d}{dt}\mathcal{F}^\lambda(t) = n \int_{\partial\Omega_t} (H_t - \overline{H}) u_t\, dV_t, \quad \text{for any } t \in (-\epsilon, \epsilon). \quad (18.8)$$

By (18.8) and Remark 18.1 we get the following result; see [63, Proposition 3.3] for additional comments and remarks.

Lemma 18.2 *Let $\Omega \in \mathcal{M}$. The following statements are equivalent:*

(a) $\partial\Omega$ is a closed spacelike H-hypersurface with constant mean curvature

$$H = \frac{\lambda}{n};$$

(b) For any variation $X : (-\epsilon, \epsilon) \times \partial\Omega \to \overline{M}^{n+1}$ of $\partial\Omega$ which preserves the balance of volume of Ω,

$$\frac{d}{dt}\mathcal{A}(0) = 0;$$

(c) $\dfrac{d}{dt}\mathcal{F}^\lambda(0) = 0$ for any variation $X : (-\epsilon, \epsilon) \times \partial\Omega \to \overline{M}^{n+1}$ of $\partial\Omega$ we have

$$\frac{d}{dt}\mathcal{F}^\lambda(0) = 0.$$

Hence, by Lemma 18.2 the set of solutions of the variational problem (VP-1) coincides with the set of solutions of (VP-2). More precisely, this set of solutions is formed by all the open subsets Ω of the time-oriented Lorentz manifold $\overline{M}^{n+1}$ whose boundary $\partial\Omega$ is a closed spacelike H-hypersurface with constant mean curvature H equal to

$$H = \frac{\lambda}{n}, \quad \text{with } \lambda \in \mathbb{R}. \tag{18.9}$$

Remark 18.2 As observed in [187], our approach is valid in a more general framework. More precisely, assume that $\mathcal{M}$ is the set of the open subsets $\Omega \subset \overline{M}^{n+1}$ whose boundary $\partial\Omega$ is the union of two disjoint sets $\partial\Omega = \Sigma_1^n \cup \Sigma_2^n$. We will assume that one of them, for instance Σ_1^n, is fixed. Consequently, the variations considered of $\partial\Omega$ only affects Σ_2^n. Under this assumption, the solutions of (VP-1), or (VP-2), will be the open subsets Ω of $\overline{M}^{n+1}$ such that their boundaries $\partial\Omega$ are union of a (fixed) set Σ_1^n and a closed spacelike H-hypersurface Σ_2^n.

By Barros et al. [67, Proposition 2.3], in the following result, we get the formula for the second variation of $\mathcal{F}^\lambda$.

Lemma 18.3 *Let $\overline{M}^{n+1}$ ($n \geq 2$) be a time-oriented Lorentz manifold and let $\Omega \subset \overline{M}^{n+1}$ be an open subset whose boundary $\partial\Omega$ is a closed spacelike H-hypersurface with constant mean curvature H given in (18.9). The second variation of the Jacobi functional $\mathcal{F}^\lambda$ is given by*

18.2 The Variational Problems (VP-1) and (VP-2)

$$\frac{d^2}{dt^2}\mathcal{F}^\lambda(0) : C^\infty(\partial\Omega) \to \mathbb{R}$$
$$u \mapsto \frac{d^2}{dt^2}\mathcal{F}^\lambda(0)(u) = \int_{\partial\Omega} u\,\mathcal{J}(u)\,dV, \tag{18.10}$$

where

$$\mathcal{J} : C^\infty(\partial\Omega) \to C^\infty(\partial\Omega)$$
$$u \mapsto \mathcal{J}(u) = \Delta u - (\overline{\mathrm{Ric}}(N,N) + |A|^2)u \tag{18.11}$$

is the Jacobi operator on $\partial\Omega$ associated with the variational problems (VP-1) and (VP-2) described above. Here, $\overline{\mathrm{Ric}}$ is the Ricci tensor of $\overline{M}^{n+1}$, A is the shape operator of $\partial\Omega$ with respect to N and Δ is the Laplacian on $\partial\Omega$.

In Lemma 18.3, with respect to the set of functions on $\partial\Omega$ to be evaluated in the second variation (18.10) of a solution of (VP-1), they have to be considered according to Remark 18.1, that is, they are smooth functions on $\partial\Omega$ whose integral mean is zero. On the other hand, any smooth function on $\partial\Omega$ can be evaluated on the second variation (18.10) of a solution of (VP-2).

We observed that, since $\partial\Omega$ is closed, Green's Identity for Δ ensures that

$$\int_{\partial\Omega} u\,\mathcal{J}(v)\,dV = \int_{\partial\Omega} v\,\mathcal{J}(u)\,dV$$

for any $u, v \in C^\infty(\partial\Omega)$. The following definition now makes sense.

Definition 18.1 The index form Q of a solution $\Omega \subset \overline{M}^{n+1}$ of the variational problem (VP-1) or (VP-2) is the quadratic form associated to the symmetric bilinear form

$$\mathcal{B} : C^\infty(\partial\Omega) \times C^\infty(\partial\Omega) \to \mathbb{R}$$
$$(u, v) \mapsto \mathcal{B}(u, v) = \int_{\partial\Omega} u\,\mathcal{J}(v)\,dV,$$

where $\mathcal{J}$ is the Jacobi operator on $\partial\Omega$ defined in (18.11).

Thus, one has that

$$Q : C^\infty(\partial\Omega) \to \mathbb{R}$$
$$u \mapsto Q(u) = \int_{\partial\Omega} u\,\mathcal{J}(u)\,dV \tag{18.12}$$

is the index form associated with each solution Ω of (VP-1) or (VP-2).

18.3 Notions of Bifurcation Instants for Spacelike H-Hypersurfaces

In the sequel, we consider the one-parameter family $\{\Omega_\tau\}_\tau \subset M$ of open subsets in the time-oriented Lorentz manifold $\overline{M}^{n+1}$ ($n \geq 2$) such that the boundary of each Ω_τ, denoted by $\partial\Omega_\tau$, is a closed spacelike $H(\tau)$-hypersurface with constant mean curvature $H(\tau)$, where τ varies on a prescribed interval $I \subset \mathbb{R}$.

For every $\tau \in I$, let N_τ be the future-pointing Gauss map of $\partial\Omega_\tau$. In this context, according to Sect. 18.2, we have that each Ω_τ is a solution of the variational problems (VP-1) and (VP-2). More specifically, each Ω_τ is a critical point for the Jacobi functional

$$I \ni \tau \longmapsto \mathcal{F}^{\lambda(\tau)} = \mathcal{A} + \lambda(\tau)\mathcal{V}$$

defined in (15.82), where the constant $\lambda(\tau)$ is chosen in such a way that $\lambda(\tau) = nH(\tau)$; see Eq. (18.9). Furthermore, for each closed spacelike $H(\tau)$-hypersurface $\partial\Omega_\tau$, as a consequence of Lemma 18.3, we have the index form (see Eq. (18.12))

$$\begin{aligned} Q_\tau : C^\infty(\partial\Omega_\tau) &\to \mathbb{R} \\ u &\mapsto Q_\tau(u) = \int_{\partial\Omega_\tau} u\,\mathcal{J}_\tau(u)\,dV_\tau, \end{aligned} \quad (18.13)$$

where dV_τ is the volume element on $\partial\Omega_\tau$ and

$$\begin{aligned} \mathcal{J}_\tau : C^\infty(\partial\Omega_\tau) &\to C^\infty(\partial\Omega_\tau) \\ u &\mapsto \mathcal{J}_\tau(u) = \Delta_\tau(u) - (\overline{\mathrm{Ric}}(N_\tau, N_\tau) + |A_\tau|^2)\,u \end{aligned} \quad (18.14)$$

is the Jacobi operator defined on $\partial\Omega_\tau$.

Here, Δ_τ is the Laplacian operator on $\partial\Omega_\tau$, $\overline{\mathrm{Ric}}$ is the Ricci tensor of $\overline{M}^{n+1}$ and A_τ is the shape operator of $\partial\Omega_\tau$ with respect to N_τ.

For these closed spacelike $H(\tau)$-hypersurfaces one can the following eigenvalue problem

(D-1) The usual *Dirichlet eigenvalue problem* associated with the index form Q_τ acting on the whole space of functions $C^\infty(\partial\Omega_\tau)$;

and

(D-2) The so-called *twisted Dirichlet eigenvalue problem* associated with the same quadratic form Q_τ, but restricted to the subspace of functions

$$\mathcal{G}(\partial\Omega_\tau) = \left\{ u \in C^\infty(\partial\Omega_\tau) : \int_{\partial\Omega_\tau} u\,dV_\tau = 0 \right\}. \quad (18.15)$$

18.3 Notions of Bifurcation Instants for Spacelike H-Hypersurfaces

According to the main notations, given $\widehat{\mu}(\tau) \in \mathbb{R}$, the Dirichlet eigenvalue problem (D-1) consists of determining whether or not nontrivial solutions exist (i.e., not identically zero) for the following elliptic equation

$$\mathcal{J}_\tau(u) + \widehat{\mu}(\tau)u = 0, \quad u \in C^\infty(\partial\Omega_\tau),$$

while that the twisted Dirichlet eigenvalue problem (D-2) studies the existence or not of nontrivial solutions for the following elliptic equation

$$\mathcal{J}_\tau(u) + \widehat{\mu}(\tau)u = 0, \quad u \in \mathcal{G}(\partial\Omega_\tau).$$

The numbers $\widehat{\mu}(\tau) \in \mathbb{R}$ for which there is a nontrivial solution to any of the above problems are the eigenvalues of Jacobi operator $\mathcal{J}_\tau$. In this case, the corresponding function u is an eigenfunction associated with the eigenvalue $\widehat{\mu}(\tau)$. Similarly, there are two different notions of index. The strong index, denoted by

$$i^s_{\text{Morse}}(\mathcal{F}^{\lambda(\tau)}, \Omega_\tau)$$

associated with (D-1), and the weak index, denoted by

$$i^w_{\text{Morse}}(\mathcal{F}^{\lambda(\tau)}, \Omega_\tau)$$

and associated with (D-2). Specifically, we have the following

Definition 18.2 The strong index $i^s_{\text{Morse}}(\mathcal{F}^{\lambda(\tau)}, \Omega_\tau)$ is the maximum dimension of any subspace of functions of $C^\infty(\partial\Omega_\tau)$ on which the index form Q_τ is negative definite. On the other hand, the weak index $i^w_{\text{Morse}}(\mathcal{F}^{\lambda(\tau)}, \Omega_\tau)$ is the maximum dimension of any subspace of functions of $\mathcal{G}(\partial\Omega_\tau)$ on which the index form Q_τ is negative definite.

Obviously, from a geometric point of view the weak index is more natural than the strong index. However, from an analytical point of view, the strong index is more natural and easier to use.

With respect to the family $\{\Omega_\tau\}_{\tau \in I}$ of solutions of (VP-1) and (VP-2), we need to adopt some notions and results that correspond to equivariant bifurcation theory for geometric variational problems. We refer to [23, 82, 83, 279], as well as the references therein, for additional comments and remarks.

Let us first remember (cf. [23, Subsection 1.2]) that two elements Ω_{τ_1} and Ω_{τ_2} of $\{\Omega_\tau\}_{\tau \in I}$ are said to be isometrically congruent when there is an isometry ψ of $\overline{M}^{n+1}$ that carries the image of $x_1 : \partial\Omega_{\tau_1} \to \overline{M}^{n+1}$ onto the image of $x_2 : \partial\Omega_{\tau_2} \to \overline{M}^{n+1}$, where x_1 and x_2 are the immersions of $\partial\Omega_{\tau_1}$ and $\partial\Omega_{\tau_2}$ into $\overline{M}^{n+1}$, respectively.

According to [82, Section 4] and [83, Section 3] we introduce the following notion of bifurcation instants that will be crucial in the sequel.

Definition 18.3 With the above notations:

(a) We say that $\tilde{\tau} \in I$ is a bifurcation instant of $\{\Omega_\tau\}_{\tau \in I}$ if there exists a sequence $\{\tau_n\}_{n \in \mathbb{N}} \subset I$ and a sequence $\{\Omega_{\tau_n}\}_{n \in \mathbb{N}} \subset \{\Omega_\tau\}_{\tau \in I}$ such that
   (1) $\lim_{n \to \infty} \tau_n = \tilde{\tau}$,
   (2) $\lim_{n \to \infty} x_n = \tilde{x}$, where $x_n : \Omega_{\tau_n} \to \overline{M}^{n+1}$ and $\tilde{x} : \Omega_{\tilde{\tau}} \to \overline{M}^{n+1}$ are the immersions of Ω_{τ_n} and $\Omega_{\tilde{\tau}}$ into $\overline{M}^{n+1}$, respectively,
   (3) For every $n \in \mathbb{N}$, x_n is not isometrically congruent to $\tilde{x}$;
(b) If $\tilde{\tau} \in I$ is not a bifurcation instant, we say that the family $\{\Omega_\tau\}_{\tau \in I}$ is locally rigid at $\tilde{\tau}$.

One of the classical criterion to determine when $\tilde{\tau} \in I$ is a bifurcation instant with respect to the variational problem (VP-1) (resp., (VP-2)) is related with the weak index (resp., with the strong index); cf. [23, 82] and [279]. Essentially, the variations of $i^w_{\text{Morse}}(\mathcal{F}^{\lambda(\tau)}, \Omega_\tau)$ and $i^s_{\text{Morse}}(\mathcal{F}^{\lambda(\tau)}, \Omega_\tau)$ along the interval $I \subset \mathbb{R}$ will indicate the existence or not of a bifurcation instant for (VP-1) and (VP-2), respectively.

More precisely:

For the Variational Problem (VP-1): under suitable Fredholmness assumptions (cf. [23, Section 2], [82, Subsection 5.3] and [279, Section 3]), we get that if there are $\tau_1, \tau_2 \in I$, with $\tau_1 < \tau_2$, such that the index form Q_{τ_j} is nonsingular on the subspace $\mathcal{G}(\partial \Omega_{\tau_j})$, i.e., any eigenvalue of the Jacobi operator $\mathcal{J}_{\tau_j}$ associated with nonconstant functions is nonzero, for $j \in \{1, 2\}$ and

$$i^w_{\text{Morse}}(\mathcal{F}^{\lambda(\tau_1)}, \Omega_{\tau_1}) \neq i^w_{\text{Morse}}(\mathcal{F}^{\lambda(\tau_2)}, \Omega_{\tau_2}),$$

then the family $\{\Omega_\tau\}_{\tau \in I}$ admits a bifurcation instant at some $\tau_* \in (\tau_1, \tau_2)$. According to [83], by using the Implicit Function Theorem we obtain that if $Q_{\tilde{\tau}}$ is nonsingular on $\mathcal{G}(\partial \Omega_{\tilde{\tau}})$, for some $\tilde{\tau} \in I$, then the family $\{\Omega_\tau\}_{\tau \in I}$ is locally rigid at $\tilde{\tau}$.

For the Variational Problem (VP-2): under suitable Fredholmness assumptions (cf. [23, Section 2], [82, Subsection 5.3] and [279, Section 2]), we have that if there are two instants $\tau_1, \tau_2 \in I$, with $\tau_1 < \tau_2$, such that the index form Q_{τ_j} is nonsingular on the space $C^\infty(\partial \Omega_{\tau_j})$, i.e., any eigenvalue of the Jacobi operator $\mathcal{J}_{\tau_j}$ is nonzero, for $j \in \{1, 2\}$ and

$$i^s_{\text{Morse}}(\mathcal{F}^{\lambda(\tau_1)}, \Omega_{\tau_1}) \neq i^s_{\text{Morse}}(\mathcal{F}^{\lambda(\tau_2)}, \Omega_{\tau_2})$$

then $\{\Omega_\tau\}_{\tau \in I}$ admits a bifurcation instant at some $\tau_* \in (\tau_1, \tau_2)$. Moreover, by using the Implicit Function Theorem again we obtain that if $Q_{\tilde{\tau}}$ is nonsingular on $C^\infty(\partial \Omega_{\tilde{\tau}})$ for some $\tilde{\tau} \in I$, then $\{\Omega_\tau\}_{\tau \in I}$ is locally rigid at $\tilde{\tau}$.

In particular, if $i^W_{\text{Morse}}(\mathcal{F}^{\lambda(\tau)}, \Omega_\tau) = 0$ (resp., $i^s_{\text{Morse}}(\mathcal{F}^{\lambda(\tau)}, \Omega_\tau) = 0$) for every $\tau \in I$, then the family $\{\Omega_\tau\}_{\tau \in I}$ does not have bifurcation instants for the variational problem (VP-1) (resp., (VP-2)).

We will study the local rigidity and the bifurcation instants of the family $\{\Omega_\tau\}_{\tau \in I}$ of solutions of (VP-1) or (VP-2) by analyzing the spectrum of the Jacobi operator $\mathcal{J}_\tau$ for every $\tau \in I$. Essentially, we will determine the number of negative eigenvalues (counted according to their multiplicity) for each τ and we will study the variation of this parameter.

18.4 Bifurcation and Locally Rigidity Results for (VP-1)

Our first result provides some simple sufficient conditions to get the local rigidity of $\{\Omega_\gamma\}_{\gamma \in I}$ for the variational problem (VP-1).

Theorem 18.1 *Let $\overline{M}^{n+1}$ be a time-oriented Lorentz manifold ($n \geq 2$) and let $\{\Omega_\tau\}_{\tau \in I}$ be a family of open subsets of $\overline{M}^{n+1}$ such that each boundary $\partial \Omega_\tau$ is a closed spacelike $H(\tau)$-hypersurface. If for every $\tau \in I$ the function*

$$\Phi(\tau) = \overline{\text{Ric}}(N_\tau, N_\tau) + |A_\tau|^2$$

is a nonnegative constant on $\partial \Omega_\tau$, then $\{\Omega_\tau\}_{\tau \in I}$ is locally rigid at each $\tau \in I$.

Proof Since $\Phi(\tau)$ is constant, by (18.14) we notice that the eigenfunctions of the Jacobi operator $\mathcal{J}_\tau$ will coincide with the eigenfunctions of the Laplacian operator Δ_τ on $\partial \Omega_\tau$. More specifically, if u is an eigenfunction of Δ_τ associated with an eigenvalue $\mu(\gamma)$ then u is eigenfunction of $\mathcal{J}_\tau$ with eigenvalue

$$\widehat{\mu}(\tau) = \mu(\tau) + \Phi(\tau).$$

Moreover, we know (cf. for instance [107, Section 1.3]) that the eigenvalues of Δ_τ are determined by a sequence of eigenvalues $\{\mu_j(\gamma)\}_{j=0}^{+\infty}$ satisfying

$$0 = \mu_0(\tau) < \mu_1(\tau) \leq \cdots \leq \mu_j(\tau) \leq \mu_{j+1}(\tau) \leq \cdots,$$

repeated according to their multiplicity, and $\lim_{j \to +\infty} \mu_j(\tau) = +\infty$.

Then, the eigenvalues $\widehat{\mu}_j(\tau)$ of the Jacobi operator $\mathcal{J}_\tau$ have the following form

$$\widehat{\mu}_j(\tau) = \mu_j(\tau) + \Phi(\tau) \quad \text{for every } j \in \mathbb{N}. \tag{18.16}$$

So, since $\Phi(\tau) \geq 0$ on $\partial \Omega_\tau$, by (18.16),

$$\widehat{\mu}_j(\tau) = \mu_j(\tau) + \Phi(\tau) > \mu_1(\tau) + \Phi(\tau) > 0 \quad \text{for any } j \in \mathbb{N} \setminus \{0\}. \tag{18.17}$$

Therefore, the eigenvalues of $\mathcal{J}_\tau$ associated with nonconstant functions are positive and, by (18.13), (18.14) and (18.17) we get that the index form Q_τ is nonsingular on the subspace $\mathcal{G}(\partial \Omega_\tau)$, given in (18.15), for every $\tau \in I$. The proof is now complete. □

A meaningful consequence of Theorem 18.1 is the following result.

Corollary 18.1 *Let $\overline{M}^{n+1}$ ($n \geq 2$) be a time-oriented Lorentz manifold and let $\{\Omega_\tau\}_{\tau \in I}$ be a family of open subsets of $\overline{M}^{n+1}$ such that each boundary $\partial \Omega_\tau$ is a closed spacelike $H(\tau)$-hypersurface. We have that $\{\Omega_\tau\}_{\tau \in I}$ is locally rigid at each $\tau \in I$ if*

$$\Phi(\tau) = \overline{\text{Ric}}(N_\tau, N_\tau) + |A_\tau|^2$$

is constant on each $\partial \Omega_\tau$ and either one of the following statements holds:

(a) $\mu_1(\tau) + \overline{\text{Ric}}(N_\tau, N_\tau) > 0$, or
(b) $\mu_1(\tau) + \overline{\text{Ric}}(N_\tau, N_\tau) + |A_\tau|^2 > 0$,

where $\mu_1(\tau)$ represents the nonzero first eigenvalue of the Laplacian operator Δ_τ on $\partial \Omega_\tau$.

We recall that a Lorentzian manifold $\overline{M}^{n+1}$ obeys the *Timelike Convergence Condition* (TCC) if its Ricci tensor $\overline{\text{Ric}}$ satisfies

$$\overline{\text{Ric}}(Z, Z) \geq 0 \quad \text{for any timelike vector filed } Z \in \mathfrak{X}(\overline{M}^{n+1}); \tag{18.18}$$

see [251, Section 12.3].

The above condition formally express the attraction property of the gravity. We say the condition holds *strictly* if the inequality $\geq$ is replaced by the strict inequality $>$ in (18.18) and, in this case, we will say that $\overline{M}^{n+1}$ obeys the *Timelike Convergence Strict Condition* (TCSC).

Arguing as in the proof of Theorem 18.1 the following result holds.

Corollary 18.2 *Let $\overline{M}^{n+1}$ ($n \geq 2$) be a time-oriented Lorentz manifold satisfying the TCC or TCSC. Any family $\{\Omega_\tau\}_{\tau \in I}$ of open subsets of $\overline{M}^{n+1}$ such that each boundary $\partial \Omega_\tau$ is a closed spacelike $H(\tau)$-hypersurface and the function*

$$\Phi(\tau) = \overline{\text{Ric}}(N_\tau, N_\tau) + |A_\tau|^2$$

is constant on each $\partial \Omega_\tau$, is locally rigid at $\tau \in I$.

The existence of bifurcation points of the family $\{\Omega_\gamma\}_{\gamma \in I}$ is studied below.

Theorem 18.2 Let $\overline{M}^{n+1}$ be a time-oriented Lorentz manifold ($n \geq 2$) and let $\{\Omega_\tau\}_{\tau \in I}$ be a family of open subsets of $\overline{M}^{n+1}$ such that each boundary $\partial \Omega_\tau$ is a closed spacelike $H(\tau)$-hypersurface. For all $\tau \in I$, assume that the function

$$\Phi(\tau) = \overline{\mathrm{Ric}}(N_\tau, N_\tau) + |A_\tau|^2$$

is constant on $\partial \Omega_\tau$. If there are two values $\tau_1, \tau_2 \in I$, with $\tau_1 < \tau_2$, such that the eigenvalues $\widehat{\mu}_j(\tau_1)$ and $\widehat{\mu}_j(\tau_2)$ of the Jacobi operators $\mathcal{J}_{\tau_1}$ and $\mathcal{J}_{\tau_2}$ (respectively) satisfy

(a) $\widehat{\mu}_j(\tau_1) \neq 0$ and $\widehat{\mu}_j(\tau_2) \neq 0$ for every $j \in \mathbb{N} \setminus \{0\}$,
(b) There exists $j_0 \in \mathbb{N} \setminus \{0\}$ such that $\left(\widehat{\mu}_{j_0}(\tau_1)\right)\left(\widehat{\mu}_{j_0}(\tau_2)\right) < 0$,

then there exists a bifurcation instant $\tau_* \in (\tau_1, \tau_2)$.

Proof Let us first observe that our hypotheses about Φ allow us to say that the eigenvalues of $\mathcal{J}_\tau$ admit the expression (18.16). So, by (18.13) and (18.14) we note that hypothesis (a) ensures that the index form Q_{τ_j} is nonsingular on the subspace $\mathcal{G}(\partial \Omega_{\tau_j})$ (see (18.15)) for $j \in \{1, 2\}$. On the order hand, we observe that assumption (b) ensures that the eigenvalue of the Jacobi operator which corresponds to $j = j_0$ changes sign between τ_1 and τ_2. Moreover, as the eigenvalues of the Jacobi operator are ordered, the number of negative eigenvalues between τ_1 and τ_2 changes. Therefore, one has

$$i^w_{\mathrm{Morse}}(\mathcal{F}^{\lambda(\tau_1)}, \Omega_{\tau_1}) \neq i^w_{\mathrm{Morse}}(\mathcal{F}^{\lambda(\tau_2)}, \Omega_{\tau_2}),$$

and the result follows. □

18.5 Bifurcation and Locally Rigidity Results for (VP-2)

According to the terminology introduced in [24], a particular class of time-oriented Lorentzian manifolds is given by the generalized Robertson-Walker (GRW) spacetimes denoted by $-I \times_f M^n$ ($n \geq 2$), namely, product manifolds $I \times M^n$ endowed with *warped metric tensor*

$$\langle \cdot, \cdot \rangle = -d\tau^2 + f(\tau)^2 \langle \cdot, \cdot \rangle_M,$$

where $I \subset \mathbb{R}$ is an open interval with the metric tensor $-d\tau^2$, M^n is an n-dimensional Riemannian manifold with metric tensor $\langle \cdot, \cdot \rangle_M$ and $f : I \to \mathbb{R}$ is a positive smooth function on I. In other words, $-I \times_f M^n$ is nothing but a warped product with Lorentzian base $(I, -d\tau^2)$, Riemannian fiber $(M^n, \langle \cdot, \cdot \rangle_M)$ and warping function f.

In particular, if M^n has constant sectional curvature, then $-I \times_f M^n$ is classically called a Robertson-Walker (RW) spacetime. Moreover, we recall that a GRW spacetime $-I \times_f M^n$ is said to be spatially closed when M^n is closed.

We observe that the vector field given by

$$K(\tau, p) = f(\tau) \partial_\tau|_{(\tau,p)}, \quad (\tau, p) \in -I \times_f M^n, \tag{18.19}$$

determines a (non-vanishing) future-pointing conformal timelike vector field on $-I \times_f M^n$ which is also closed, in the sense that its metrically equivalent 1-form is closed; see [243, Section 3]. Indeed, we have that

$$\overline{\nabla}_V K = f'(\tau) V, \tag{18.20}$$

for every smooth vector field $V \in \mathfrak{X}(-I \times_f M^n)$, where $\overline{\nabla}$ denotes the Levi-Civita connection of $-I \times_f M^n$.

For every fixed $\tau \in I$, the slice

$$\Sigma_\tau^n = \{\tau\} \times M^n$$

is a spacelike hypersurface of $-I \times_f M^n$. Actually, the induced metric on Σ_τ^n is given by $f(\tau)^2 \langle \cdot, \cdot \rangle_M$, which means that Σ_τ^n is homotetic to M^n with scale factor $f(\tau)$. The restriction of ∂_τ to Σ_τ^n gives a future-pointing Gauss map for it.

By (18.19) and (18.20) it follows that

$$\overline{\nabla}_V \partial_\tau = \overline{\nabla}_V \left(\frac{1}{f(\tau)} K\right) = -\frac{1}{(f(\tau))^2} \langle V, \overline{\nabla}\overline{f}\rangle K + \frac{1}{f(\tau)} f'(\tau) V \tag{18.21}$$

for any $V \in \mathfrak{X}(-I \times_f M^n)$, where $\overline{\nabla}\overline{f}$ denotes the gradient on $-I \times_f M^n$ of the function $\overline{f}(\tau, y) = f(\tau)$.

We observe that the gradient of the projection $\pi_I(\tau, p) = \tau$ on $-I \times_f M^n$ is given by $\overline{\nabla}\pi_I = -\langle \overline{\nabla}\pi_I, \partial_\tau\rangle \partial_\tau = -\partial_\tau$. Hence, if $\overline{f} = f \circ \pi_I$, then $\overline{\nabla}\overline{f} = f'(\tau)\overline{\nabla}\pi_I = -f'(\tau)\partial_\tau$ and (18.21) assume the form

$$\overline{\nabla}_V \partial_\tau = \frac{f'(\tau)}{f(\tau)} (V + \langle V, \partial_\tau\rangle \partial_\tau), \quad V \in \mathfrak{X}(-I \times_f M^{n+1}).$$

In particular, one has

$$\overline{\nabla}_V \partial_t = \frac{f'(\tau)}{f(\tau)} V \quad \text{for any } V \in \mathfrak{X}(\Sigma_\tau^n).$$

This means that Σ_τ^n is a totally umbilical spacelike hypersurface in $-I \times_f M^n$ with shape operator (see (18.1)) with respect to the future-pointing Gauss map ∂_τ given by

18.5 Bifurcation and Locally Rigidity Results for (VP-2)

$$A_\tau : \mathfrak{X}(\Sigma_\tau^n) \to \mathfrak{X}(\Sigma_\tau^n)$$
$$V \mapsto A_\tau(V) = -\overline{\nabla}_V(-\partial_t) = \frac{f'(\tau)}{f(\tau)} V. \tag{18.22}$$

Therefore, the correspondence

$$I \ni \tau \mapsto \Sigma_\tau^n \subset -I \times_f M^n$$

determines a foliation of $-I \times_f M^n$ by totally umbilical spacelike $H(\tau)$-hypersurface with constant mean curvature (see (18.2)) given by

$$H(\tau) = -\frac{1}{n} \mathrm{tr}(A_\tau) = -\frac{f'(\tau)}{f(\tau)}.$$

When the GRW spacetime $-I \times_f M^n$ is spatially closed, we have that $-I \times_f M^n$ support a family of open subsets which can be realized as solutions of the variational problem (VP-2). In order to prove this claim, let $\tau_1, \tau_2 \in I \subset \mathbb{R}$ and consider the family

$$\{\Omega_\tau\}_{\tau \in (\tau_1, \tau_2]}.$$

of open subsets of $-I \times_f M^n$ defined by

$$\Omega_\tau = (\tau_1, \tau) \times M^n, \quad \tau \in (\tau_1, \tau_2]$$

Thus, assuming that the Riemannian fiber M^n of $-I \times_f M^n$ is closed, we have that the boundary $\partial \Omega_\tau$ of each Ω_τ is the disjoint union $\partial \Omega_\tau = \Sigma_{\tau_1}^n \cup \Sigma_\tau^n$ of two closed spacelike hypersurfaces $\Sigma_{\tau_1}^n = \{\tau_1\} \times M^n$ (fixed) and $\Sigma_\tau^n = \{\tau\} \times M^n$. Since the variations of $\partial \Omega_\tau$ only affects Σ_τ^n, taking into account that Σ_τ^n is a closed spacelike H_τ-hypersurface, Remark 18.2 ensures that each element of the family $\{\Omega_\tau\}_{\tau \in (\tau_1, \tau_2]}$ is a solution of the variational problem (VP-2). In addition, according to Sect. 18.3, associated with each $\Omega_\tau = (\tau_1, \tau) \times M^n$, we have the index form

$$Q_\tau : C^\infty(\Sigma_\tau^n) \to \mathbb{R}$$
$$u \mapsto Q_\tau(u) = \int_{\Sigma_\tau^n} u \, \mathcal{J}_\tau(u) \, dV_\tau, \tag{18.23}$$

where

$$\mathcal{J}_\tau : C^\infty(\Sigma_\tau^n) \to C^\infty(\Sigma_\tau^n)$$
$$u \mapsto \mathcal{J}_\tau(u) = \Delta_\tau(u) - (\overline{\mathrm{Ric}}(\partial_\tau, \partial_\tau) + |A_\tau|^2) u, \tag{18.24}$$

is the corresponding Jacobi operator.

Here, Δ_τ is the Laplacian operator on Σ_τ^n, $\overline{\mathrm{Ric}}$ is the Ricci tensor of the spatially closed GRW spacetime $-I \times_f M^n$ and A_τ is the shape operator of Σ_τ^n with respect to ∂_τ.

We notice that:

(i) The induced metric on Σ_τ^n is $f(\tau)^2 \langle \cdot, \cdot \rangle_M$, through the natural identification of $C^\infty(\Sigma_\tau^n)$ with $C^\infty(M^n)$. Moreover, the Laplacian operator Δ on M^n is related to Δ_τ by $\Delta = f(\tau)^2 \Delta_\tau$;

(ii) By (18.22), one has $|A_\tau|^2 = n \dfrac{(f'(\tau))^2}{(f(\tau))^2}$;

(iii) By O'Neill [251, Corollary 7.43], one has $\overline{\mathrm{Ric}}(\partial_\tau, \partial_\tau) = -n \dfrac{f''(\tau)}{f(\tau)}$.

Thus, the Jacobi operator $\mathcal{J}_\tau$ given in (18.24) can be written as

$$\mathcal{J}_\tau = \frac{1}{(f(\tau))^2} (\Delta - n\Psi(\tau)), \qquad (18.25)$$

where

$$\Psi(\tau) = (f'(\tau))^2 - f''(\tau) f(\tau), \quad \tau \in I. \qquad (18.26)$$

Now, let us recall that the eigenvalues of the Laplacian operator Δ on a closed Riemannian manifold M^n are determined by a sequence $\{\mu_i\}_{i=0}^{+\infty}$ satisfying

$$0 = \mu_0 < \mu_1 \leq \mu_2 \leq \cdots \leq \mu_i \leq \mu_{i+1} \leq \cdots, \qquad (18.27)$$

repeated according to their multiplicity, and

$$\lim_{i \to +\infty} \mu_i = +\infty; \qquad (18.28)$$

see [107, Section 1.3].

Moreover, since the function Ψ only depends of the warped function f, which is constant in each Σ_τ^n, by (18.25) we get that the i-th eigenvalue $\widehat{\mu}_i(\tau)$ of the Jacobi operator $\mathcal{J}_\tau$ on Σ_τ^n admits the following expression

$$\widehat{\mu}_i(\tau) = \frac{1}{(f(\tau))^2} \{\mu_i + n\Psi(\tau)\}. \qquad (18.29)$$

The next result provides some simple sufficient conditions to get the local rigidity of the family $\{\Omega_\tau\}_{\tau \in (\tau_1, \tau_2]}$, in terms of the eigenvalues of the Laplacian operator Δ of M^n and the warped function f.

18.5 Bifurcation and Locally Rigidity Results for (VP-2)

Table 18.1 Some families that are locally rigid according to Corollary 18.3

Spatially closed GRW spacetime	Family of open sets	Ψ
$-(0, +\infty) \times_\tau M^n$	$\{\Omega_\tau\}_{\tau \in (\tau_1, \tau_2]}$ with $\tau_1, \tau_2 \in (0, +\infty)$ and $\tau_1 < \tau_2$	1
$-(0, +\infty) \times_{\sinh \tau} M^n$	$\{\Omega_\tau\}_{\tau \in (\tau_1, \tau_2]}$ with $\tau_1, \tau_2 \in (0, +\infty)$ and $\tau_1 < \tau_2$	1
$-(0, \pi) \times_{\sin \tau} M^n$	$\{\Omega_\tau\}_{\tau \in (\tau_1, \tau_2]}$ with $\tau_1, \tau_2 \in (0, \pi)$ and $\tau_1 < \tau_2$	1
$-(-\pi/2, \pi/2) \times_{\cos \tau} M^n$	$\{\Omega_\tau\}_{\tau \in (\tau_1, \tau_2]}$ with $\tau_1, \tau_2 \in (-\pi/2, \pi/2)$ and $\tau_1 < \tau_2$	1
$-(1, +\infty) \times_{\log \tau} M^n$	$\{\Omega_\tau\}_{\tau \in (\tau_1, \tau_2]}$ with $\tau_1, \tau_2 \in (1, +\infty)$ and $\tau_1 < \tau_2$	$\frac{1}{\tau^2}(1 + \log \tau)$

Theorem 18.3 *Let* $-I \times_f M^n$ *($n \geq 2$) be a spatially closed GRW spacetime and let* $\{\Omega_\tau\}_{\tau \in (\tau_1, \tau_2]}$ *be a family of open subsets of* $-I \times_f M^n$ *given by* $\Omega_\tau = (\tau_1, \tau) \times M^n$, *with* $\tau \in (\tau_1, \tau_2]$, *where* τ_1 *and* τ_2 *are fixed numbers in* $I \subset \mathbb{R}$, *with* $\tau_1 < \tau_2$. *If*

$$\Psi(\tau) \neq -\frac{\mu_i}{n}, \tag{18.30}$$

for every $i \in \mathbb{N}$ *and for every* $\tau \in (\tau_1, \tau_2]$, *where* Ψ *is the function defined in* (18.26) *and* μ_i *is the i-th eigenvalue of the Laplacian operator* Δ *on* M^n, *then* $\{\Omega_\tau\}_{\tau \in (\tau_1, \tau_2]}$ *is locally rigid at each* $\tau \in (\tau_1, \tau_2)$.

Proof By (18.29) and (18.30) we have that the eigenvalues $\widehat{\mu}_i(\tau)$ of the Jacobi operator $\mathcal{J}_\tau$ are nonzero for every $\tau \in (\tau_1, \tau_2]$ and, therefore, the index form $\mathcal{Q}_\tau$ given in (18.23) is nonsingular on the space of smooth functions $C^\infty(\partial \Omega_\tau)$ for every $\tau \in (\tau_1, \tau_2)$. The conclusion follows. □

A special case of Theorem 18.3 is the following locally rigidity result.

Corollary 18.3 *Let* $-I \times_f M^n$ *($n \geq 2$) be a spatially closed GRW spacetime and let* $\{\Omega_\tau\}_{\tau \in (\tau_1, \tau_2]}$ *be a family of open subsets of* $-I \times_f M^n$ *given by* $\Omega_\tau = (\tau_1, \tau) \times M^n$, *with* $\tau \in (\tau_1, \tau_2]$, *where* τ_1 *and* τ_2 *are fixed numbers in* $I \subset \mathbb{R}$, *with* $\tau_1 < \tau_2$. *If the function* Ψ *defined in* (18.26) *is positive then* $\{\Omega_\tau\}_{\tau \in (\tau_1, \tau_2]}$ *is locally rigid at each* $\tau \in (\tau_1, \tau_2)$.

In Table 18.1 some spatially closed GRW spacetimes, for which the assumptions of Corollary 18.3 hold, are presented.

The above result can be applied to the de Sitter space of positive constant sectional curvature $1/r^2$, with $r \in (0, +\infty)$.

More precisely, let us denote by $\mathbb{L}^{n+2}$ the $(n+2)$-dimensional Lorentz-Minkowski space ($n \geq 2$), that is, the real vector space $\mathbb{R}^{n+2}$, endowed with the Lorentz metric

$$\langle v, w \rangle = \sum_{i=1}^{n+1} v_i w_i - v_{n+2} w_{n+2},$$

for all $v, w \in \mathbb{R}^{n+2}$. We define the $(n+1)$-dimensional de Sitter space $\mathbb{S}_1^{n+1}(r)$ of radius $r \in (0, +\infty)$ as the following hyperquadric of $\mathbb{L}^{n+2}$

$$\mathbb{S}_1^{n+1}(r) = \left\{ p \in \mathbb{L}^{n+2} : \langle p, p \rangle = r^2 \right\}.$$

From the above definition it is easy to show that the metric induced from $\langle \cdot, \cdot \rangle$ turns $\mathbb{S}_1^{n+1}(r)$ into a Lorentz manifold with constant sectional curvature $1/r^2$. In order to simplify the notation, when $r=1$ we denote $\mathbb{S}_1^{n+1}(1)$ simply by $\mathbb{S}_1^{n+1}$.

In General Relativity, $\mathbb{S}_1^{n+1}$ is the maximally symmetric vacuum solution of the Einstein's field equations with positive cosmological constant (corresponding to a positive vacuum energy density and negative pressure).

Choose a unit timelike vector $a \in \mathbb{L}^{n+2}$. Then $V(p) = a - \langle p, a \rangle p$, $p \in \mathbb{S}_1^{n+1}(r)$, is a conformal and closed timelike vector field that foliates the de Sitter space by means of umbilical round spheres

$$M_\tau^n = \{ p \in \mathbb{S}_1^{n+1}(r) : \langle p, a \rangle = \tau \}, \quad \tau \in \mathbb{R}.$$

In the context of RW spacetimes, the de Sitter space of radius $r \in (0, +\infty)$ can be identified as the warped product

$$\mathbb{S}_1^{n+1}(r) = -(-\infty, +\infty) \times_{\cosh \tau} \mathbb{S}^n(r),$$

where $\mathbb{S}^n(r)$ means n-dimensional Euclidean sphere of radius $r \in (0, +\infty)$. We observe (cf. [243], Section 4) that there is a lot of possible choices for the unit timelike vector $a \in \mathbb{L}^{n+2}$ and, hence, a lot of ways to describe $\mathbb{S}_1^{n+1}(r)$ as such a RW.

Furthermore, each slice

$$\{\tau\} \times \mathbb{S}^n(r) \subset -(-\infty, +\infty) \times_{\cosh \tau} \mathbb{S}^n(r)$$

is isometric to the umbilical round spheres

$$M_\tau^n \subset \mathbb{S}_1^{n+1}(r).$$

By (18.26) we have the function Ψ associated to $-(-\infty, +\infty) \times_{\cosh \tau} \mathbb{S}^n(r)$ is $\Psi = -1$. Moreover, we know (cf. [107], Section 2.4]) that the eigenvalues μ_i of the Laplacian operator Δ on $\mathbb{S}^n(r)$ are given by

18.5 Bifurcation and Locally Rigidity Results for (VP-2)

$$\mu_i = \frac{i(i+n-1)}{r^2}, \quad i \in \mathbb{N}. \tag{18.31}$$

Now, let us consider a family $\{\Omega_\tau\}_{\tau \in (\tau_1, \tau_2]}$ of open subsets $\Omega_\tau = (\tau_1, \tau) \times \mathbb{S}^n(r)$, with $\tau_1, \tau_2 \in (-\infty, +\infty)$ and $\tau_1 < \tau_2$. Now, by choosing $0 < r < 1$ we have that condition (18.30) of Theorem 18.3 holds. Therefore, we can state the next result.

Corollary 18.4 *In de Sitter space $\mathbb{S}_1^{n+1}(r) = -(-\infty, +\infty) \times_{\cosh \tau} \mathbb{S}^n(r)$ $(n \geq 2)$ of radius $r \in (0, 1)$, the family $\{\Omega_\tau\}_{\tau \in (\tau_1, \tau_2]}$ of open subsets $\Omega_\tau = (\tau_1, \tau) \times \mathbb{S}^n(r)$, with $\tau_1, \tau_2 \in (-\infty, +\infty)$ and $\tau_1 < \tau_2$, is locally rigid at each $\tau \in (\tau_1, \tau_2)$.*

In the next result, requesting this time that the relation (18.30) is not valid, we establish a characterization of the bifurcation instants for the family $\{\Omega_\tau\}_{\tau \in (\tau_1, \tau_2]}$.

Theorem 18.4 *Let $-I \times_f M^n$ $(n \geq 2)$ be a spatially closed GRW spacetime and let $\{\Omega_\tau\}_{\tau \in (\tau_1, \tau_2]}$ be a family of open subsets of $-I \times_f M^n$ given by $\Omega_\tau = (\tau_1, \tau) \times M^n$, with $\tau \in (\tau_1, \tau_2]$, where τ_1 and τ_2 are fixed numbers in $I \subset \mathbb{R}$, with $\tau_1 < \tau_2$. If $\tau_* \in (\tau_1, \tau_2)$ and there is $i_0 \in \mathbb{N}$ such that*

$$\Psi(\tau_*) = -\frac{\mu_{i_0}}{n}, \tag{18.32}$$

and

$$\Psi'(\tau_*) \neq 0, \tag{18.33}$$

where Ψ is the function defined in (18.26) and μ_{i_0} is the i_0-th eigenvalue of the Laplacian operator Δ on M^n, then τ_* is a bifurcation instant of $\{\Omega_\tau\}_{\tau \in (\tau_1, \tau_2]}$.

Proof Taking into account hypothesis (18.33), by continuity we can choose $\tau_*', \overline{\tau}_*' \in (\tau_1, \tau_2)$ sufficiently close to τ_*, with $\tau_*' < \tau_* < \overline{\tau}_*'$, such that

$$-\frac{\mu_{i_0-1}}{n} > \Psi(\tau_*') > -\frac{\mu_{i_0}}{n} > \Psi(\overline{\tau}_*') > -\frac{\mu_{i_0+1}}{n}. \tag{18.34}$$

or

$$-\frac{\mu_{i_0-1}}{n} > \Psi(\overline{\tau}_*') > -\frac{\mu_{i_0}}{n} < \Psi(\tau_*') > -\frac{\mu_{i_0+1}}{n}. \tag{18.35}$$

So, by (18.27) and either (18.34) or (18.35), we get

$$\cdots > -\frac{\mu_{i_0-k}}{n} > \cdots > \Psi(\tau_*') > -\frac{\mu_{i_0}}{n} > \Psi(\overline{\tau}_*') > \cdots > -\frac{\mu_{i_0+k}}{n} > \cdots \tag{18.36}$$

or

$$\cdots > -\frac{\mu_{i_0-k}}{n} > \cdots > \Psi(\overline{\tau}'_*) > -\frac{\mu_{i_0}}{n} > \Psi(\tau'_*) > \cdots > -\frac{\mu_{i_0+k}}{n} > \cdots \quad (18.37)$$

for every $k \in \{1, 2, \ldots\}$. Then, by (18.29) and either (18.36) or (18.37), one has

$$\widehat{\mu}_i(\tau'_*) = \frac{1}{f(\tau'_*)^2} \{\mu_i + n\Psi(\tau'_*)\} \neq 0$$

and

$$\widehat{\mu}_i(\overline{\tau}'_*) = \frac{1}{f(\overline{\tau}'_*)^2} \{\mu_i + n\Psi(\overline{\tau}'_*)\} \neq 0$$

for any $i \in \mathbb{N}$. That is, the eigenvalues $\widehat{\mu}_i(\tau'_*)$ and $\widehat{\mu}_i(\overline{\tau}'_*)$ of the Jacobi operators $\mathcal{J}_{\tau'_*}$ and $\mathcal{J}_{\overline{\tau}'_*}$ are nonzero and, therefore, the index forms $Q_{\tau'_*}$ and $Q_{\overline{\tau}'_*}$ are nonsingular on space of smooth functions $C^\infty(\partial\Omega_{\tau'_*})$ and $C^\infty(\partial\Omega_{\overline{\tau}'_*})$, respectively.

Furthermore, by using (18.29) and either (18.34) or (18.35) we get

$$\widehat{\mu}_{i_0}(\tau'_*)\widehat{\mu}_{i_0}(\overline{\tau}'_*) = \frac{1}{(f(\tau'_*))^2(f(\overline{\tau}'_*))^2} \{\mu_{i_0} + n\Psi(\tau'_*)\} \{\mu_{i_0} + n\Psi(\overline{\tau}'_*)\} < 0.$$

Hence, the eigenvalue $\widehat{\mu}_{i_0}$ changes sign between τ'_* and $\overline{\tau}'_*$. As the eigenvalues $\widehat{\mu}_i$ are ordered, we conclude that the number of negative eigenvalues between $\widehat{\mu}(\tau'_*)$ and $\widehat{\mu}(\overline{\tau}'_*)$ changes. This ensures that

$$\mathrm{i}^{\mathrm{s}}_{\mathrm{Morse}}(\mathcal{F}^{\lambda(\tau'_*)}, \Omega_{\tau'_*}) \neq \mathrm{i}^{\mathrm{s}}_{\mathrm{Morse}}(\mathcal{F}^{\lambda(\overline{\tau}'_*)}, \Omega_{\overline{\tau}'_*}).$$

Therefore, we can conclude that the family $\{\Omega_\tau\}_{\tau \in (\tau_1, \tau_2]}$ admits a bifurcation instant in the interval $(\tau'_*, \overline{\tau}'_*) \subset (\tau_1, \tau_2)$. Since $\tau'_*, \overline{\tau}'_* \in (\tau_1, \tau_2)$ were considered arbitrarily, with $\tau'_* < \tau_* < \overline{\tau}'_*$, we can conclude that τ_* is a bifurcation instant for the family $\{\Omega_\tau\}_{\tau \in (\tau_1, \tau_2]}$.
□

The reasoning above allow us to state a criterion that guarantees the existence of bifurcation instants for the family $\{\Omega_\tau\}_{\tau \in (\tau_1, \tau_2]}$ of $-I \times_f M^n$.

Corollary 18.5 *Let* $-I \times_f M^n$ *($n \geq 2$) be a spatially closed GRW spacetime and let* $\{\Omega_\tau\}_{\tau \in (\tau_1, \tau_2]}$ *be a family of open subsets of* $-I \times_f M^n$ *given by* $\Omega_\tau = (\tau_1, \tau) \times M^n$, *with* $\tau \in (\tau_1, \tau_2]$, *where* τ_1 *and* τ_2 *are fixed numbers in* $I \subset \mathbb{R}$, $\tau_1 < \tau_2$. *If there exist instants* $\tau_*, \overline{\tau}_* \in (\tau_1, \tau)$ *with* $\tau_* < \overline{\tau}_*$ *such that either*

$$\Psi(\tau_*) < -\frac{\mu_{i_0}}{n} < \Psi(\overline{\tau}_*) \quad or \quad \Psi(\overline{\tau}_*) < -\frac{\mu_{i_0}}{n} < \Psi(\tau_*) \quad (18.38)$$

18.5 Bifurcation and Locally Rigidity Results for (VP-2)

for some $i_0 \in \mathbb{N}$, where Ψ is the function defined in (18.26) and μ_{i_0} is the i_0-th eigenvalue of the Laplacian operator Δ on M^n, then $\{\Omega_\tau\}_{\tau \in (\tau_1, \tau_2]}$ admits at least a bifurcation instant at some $\tau_* \in (\tau_0, \overline{\tau}_0)$.

Proof Taking into account the continuity of Ψ, by (18.38) there exists some $\tau_* \in (\tau_*, \overline{\tau}_*)$ such that $\Psi(\tau_*) = -\mu_{i_0}/n$. The result follows directly by Theorem 18.4. □

In the following result we will show that if the function Ψ admits a certain asymptotic behavior at the extremes of the interval (τ_1, τ_2), then the family $\{\Omega_\tau\}_{\tau \in (\tau_1, \tau_2]}$ has infinitely many bifurcation instants.

Corollary 18.6 *Let* $-I \times_f M^n$ ($n \geq 2$) *be a spatially closed GRW spacetime and let* $\{\Omega_\tau\}_{\tau \in (\tau_1, \tau_2]}$ *be a family of open subsets of* $-I \times_f M^n$ *given by* $\Omega_\tau = (\tau_1, \tau) \times M^n$, *with* $\tau \in (\tau_1, \tau_2]$, *where* τ_1 *and* τ_2 *are fixed numbers in* $I \subset \mathbb{R}$, *with* $\tau_1 < \tau_2$. *If*

$$\lim_{\tau \to \tau_1} \Psi(\tau) = -\infty \quad \text{or} \quad \lim_{\tau \to \tau_2} \Psi(\tau) = -\infty,$$

where Ψ is the function defined in (18.26), then the family $\{\Omega_\tau\}_{\tau \in (\tau_1, \tau_2]}$ *admits infinitely many bifurcation instants.*

Proof Let us assume that

$$\lim_{\tau \to \tau_1} \Psi(\tau) = -\infty. \tag{18.39}$$

Taking $\tau_0 \in (\tau_1, \tau_2)$, since

$$\lim_{i \to \infty} -\frac{\mu_i}{n} = -\infty,$$

(see (18.28)), there exists i_0 such that

$$\Psi(\tau_0) > -\frac{\mu_{i_0}}{n}.$$

On the other hand, by (18.39)

$$-\frac{\mu_{i_0}}{n} > \Psi(\overline{\tau}_0),$$

for some $\overline{\tau}_0 \in (\tau_1, \tau_2)$. Then, by Corollary 18.5, there exists a bifurcation instant in $(\tau_0, \overline{\tau}_0)$ or $(\overline{\tau}_0, \tau_0)$.

Now, it easily seen that the above arguments ensures the existence of another bifurcation instant in the interval $(\overline{\tau}_0, \tau_1)$ or $(\overline{\tau}_1, \overline{\tau}_0)$. So, by induction, the existence of infinitely many instants of bifurcation is proved.

The case $\lim_{\tau \to \tau_2} \Psi(\tau) = -\infty$ is similar. □

A careful analysis of the proof of Corollary 18.6 ensures that if $\lim_{\tau \to \tau_1} \Psi(\tau) = -\infty$, for every $\tau \in (\tau_1, \tau_2)$, then the family $\{\Omega_\tau\}_{\tau \in (\tau_1, \tau]}$ admits infinitely many bifurcation instants. Moreover, in the case

$$\lim_{\tau \to \tau_2} \Psi(\tau) = -\infty,$$

the family $\{\Omega_\tau\}_{\tau \in (\tau_1, \tau]}$ has just a finite many bifurcation instants and as τ yields τ_2 the bifurcation instants are being captured.

Remark 18.3 We finish this chapter with some applications of Corollaries 18.5 and 18.6. To this aim, let us consider the spatially closed GRW spacetime $-(0, 1) \times_{-\ln t} M^n$. By (18.26) we have

$$\Psi(\tau) = \frac{1 + \ln t}{t^2}, \quad \tau \in (0, 1).$$

Then, by Corollary 18.5, the family $\{\Omega_\tau\}_{\tau \in (0, 1]}$ of open sets of the form $\Omega_\tau = (0, \tau) \times M^n$, admits a bifurcation instant in $\tau = e^{-1}$. Furthermore, one has

$$\lim_{\tau \to 0} \Psi(\tau) = -\infty.$$

So, by Corollary 18.6, there exist infinitely many bifurcation instants on the interval $(0, \tau)$, for every $\tau \in (0, 1]$.

Finally, let us consider the spatially closed GRW spacetime $-(-\infty, +\infty) \times_{\exp(\tau^m)} M^n$, where $m \in \mathbb{N}$ and $m \geq 2$. By (18.26), one has

$$\Psi(\tau) = -m(m-1) \exp(2\tau^m) \tau^{m-2}, \quad \tau \in (-\infty, +\infty).$$

Hence, it follows that

$$\lim_{\tau \to \infty} \Psi(\tau) = -\infty.$$

By Corollary 18.6, we obtain that the family $\{\Omega_\tau\}_{\tau \in (0, +\infty)}$ of open sets given by $\Omega_\tau = (0, \tau) \times M^n$ has infinitely many bifurcation instants.

On the other hand, if m is even, then

$$\lim_{\tau \to -\infty} \Psi(\tau) = -\infty.$$

In this case, the family $\{\Omega_\tau\}_{\tau \in (-\infty, 0)}$ of open sets given by $\Omega_\tau = (-\infty, \tau) \times M^n$ also have infinitely many bifurcation instants.

Bifurcation of φ-Minimal Hypersurfaces in a Weighted Killing Warped Product

19.1 Introduction

Throughout this chapter, we consider an $(n+1)$-dimensional Riemannian manifold $\overline{M}^{n+1}$ ($n \geq 2$) endowed with a Killing vector field Y. Suppose that the distribution of all vector fields of $\overline{M}^{n+1}$ that are orthogonal to Y is of constant rank and integrable.

Given an integral leaf M^n of that distribution, let $\Psi : \mathbb{I} \times M^n \to \overline{M}^{n+1}$ be the flow generated by Y with initial values in M^n, where $\mathbb{I}$ is a maximal interval of definition. Without loss of generality, in the sequel we consider $\mathbb{I} = \mathbb{R}$. In this setting, the space $\overline{M}^{n+1}$ can be viewed as the Killing warped product $M^n \times_\rho \mathbb{R}$, that is the product manifold $M^n \times \mathbb{R}$ endowed with the warping metric

$$\langle \cdot, \cdot \rangle = \pi_M^*(\langle \cdot, \cdot \rangle_M) + (\rho \circ \pi_P)^2 \pi_\mathbb{R}^*(dt^2), \tag{19.1}$$

where π_M and $\pi_\mathbb{R}$ denote the canonical projections from $M^n \times \mathbb{R}$ onto each factor, $\langle \cdot, \cdot \rangle_M$ is the induced Riemannian metric on the base M^n, dt^2 denotes the usual Riemannian metric in $\mathbb{R}$ and $\rho = |Y| > 0$ is the warping function.

As usual, we denote by $C^\infty(M^n \times_\rho \mathbb{R})$ the ring of real functions of class C^∞ on $M^n \times_\rho \mathbb{R}$ and by $\mathfrak{X}(M^n \times_\rho \mathbb{R})$ the $C^\infty(M^n \times_\rho \mathbb{R})$-module of vector fields of class C^∞ on $M^n \times_\rho \mathbb{R}$. Finally, let $\overline{\nabla}$ and $\widetilde{\nabla}$ be the Levi-Civita connections of $M^n \times_\rho \mathbb{R}$ and M^n, respectively.

Now, let $(M^n \times_\rho \mathbb{R})_\varphi$ be a weighted Killing warped product, namely, a Killing warped product $M^n \times_\rho \mathbb{R}$ endowed with a weighted volume form $d\overline{\sigma} = e^{-\varphi} d\overline{v}$, where $\varphi \in C^\infty(M^n \times_\rho \mathbb{R})$ is a real-valued function, called weighted function (or density function), and $d\overline{v}$ is the volume element induced by the warping metric $\langle \cdot, \cdot \rangle$ defined in (19.1).

For the space $(M^n \times_\rho \mathbb{R})_\varphi$, it is worth to recall that the Bakry-Émery-Ricci tensor $\overline{\mathrm{Ric}}_\varphi$ is defined by

© The Author(s), under exclusive license to Springer Nature Switzerland AG 2025
H. Fernandes de Lima et al., *Immersions in Warped Product Spaces*, Frontiers in Elliptic and Parabolic Problems, https://doi.org/10.1007/978-3-031-78042-4_19

$$\overline{\mathrm{Ric}}_\varphi = \overline{\mathrm{Ric}} + \overline{\mathrm{Hess}}\varphi, \tag{19.2}$$

where $\overline{\mathrm{Ric}}$ and $\overline{\mathrm{Hess}}$ are the Ricci tensor and the Hessian operator in $M^n \times_\rho \mathbb{R}$, respectively.

In this context, we deal here with hypersurfaces Σ^n immersed into weighted Killing warped products $\left(M^n \times_\rho \mathbb{R}\right)_\varphi$ which are two-sided. This condition means that there is a globally defined unit normal vector field N on the space.

Moreover, let us denote by ∇ the Levi-Civita connection of Σ^n and let A be the shape operator of Σ^n with respect to N, so that at each $p \in \Sigma^n$, A restricts to a self-adjoint linear map

$$A_p : T_p\Sigma^n \to T_p\Sigma^n$$
$$v \mapsto A_p v = -\overline{\nabla}_v N.$$

According to Gromov [197], we also recall that the weighted mean curvature, or simply the φ-mean curvature, H_φ of Σ^n is given by

$$nH_\varphi = nH + \langle \overline{\nabla}\varphi, N \rangle, \tag{19.3}$$

where H denotes the standard mean curvature of Σ^n with respect to its orientation N. In particular, when φ is constant we have that $H_\varphi = H$ and the usual definition of mean curvature is recovered. Finally, a two-sided hypersurface Σ^n is called a φ-minimal hypersurface if its φ-mean curvature H_φ is identically zero.

The study of the geometry of hypersurfaces immersed in weighted Riemannian manifolds having constant φ-mean curvature and, in particular, the investigations on the behavior of φ-minimal hypersurfaces immersed in a weighted ambient space constitute a recent and fruitful topic into the theory of isometric immersions. It has been already approached by many authors and we may cite, for instance, the papers [100, 105, 114, 203, 204], as well as [211, 233, 265, 269].

Motivated by a splitting theorem due to Fang, Li and Zhang (see [176, Theorem 1.1]), in a weighted Killing warped product $M_\varphi^n \times_\rho \mathbb{R}$ (see (19.4)) with closed Riemannian manifold base M^n, in this chapter we study the existence of bifurcation instants of a certain family $\{\Omega_\gamma\}_{\gamma \in (t_1, t_2]}$ (see (19.17)) of open subsets Ω_γ of $M_\varphi^n \times_\rho \mathbb{R}$ whose boundaries $\partial\Omega_\gamma$ are φ-minimal hypersurfaces.

To achieve this goal, we use some results of Equivariant Bifurcation Theory (cf. [23, 81, 83, 279]) to establish our notions of bifurcation instants and local rigidity associated with the family $\{\Omega_\gamma\}_{\gamma \in (t_1, t_2]}$.

The results presented in this chapter are mainly based on the paper [291].

19.2 The Setup of Ambient Space

Let Σ^n be a two-sided hypersurface immersed into $\left(M^n \times_\rho \mathbb{R}\right)_\varphi$ as described in Sect. 19.1. As in Chap. 5, the φ-divergence on Σ^n is defined by

$$\operatorname{div}_\varphi : \mathfrak{X}(\Sigma^n) \to C^\infty(\Sigma^n)$$
$$X \mapsto \operatorname{div}_\varphi X = \operatorname{div}(X) - \langle \nabla\varphi, X \rangle,$$

where $\operatorname{div}(\cdot)$ denotes the standard divergence on Σ^n. So, the drift Laplacian (or φ-Laplacian) of Σ^n is given by

$$\Delta_\varphi : C^\infty(\Sigma^n) \to C^\infty(\Sigma^n)$$
$$u \mapsto \Delta_\varphi(u) = \operatorname{div}_\varphi \nabla u = \Delta u - \langle \nabla\varphi, \nabla u \rangle$$

where Δ is the standard Laplacian on Σ^n.

We have that the Killing vector field Y determines in $M^n \times_\rho \mathbb{R}$ a codimension one foliation by totally geodesic slices

$$M^n \times \{t\}, \quad (\forall t \in \mathbb{R})$$

with respect to the orientation determined by Y. Moreover, assuming that the weighted function $\varphi \in C^\infty(M^n \times_\rho \mathbb{R})$ is invariant along the flow determinate by Y, by (19.3) we get that each slice $M^n \times \{t\}$ is φ-minimal.

We observe that the following result is a consequence of a Cheeger-Gromoll type splitting theorem due to Wei and Wylie (cf. [294, Theorem 6.1]; see also [176, Theorem 1.1]):

Let $\overline{M}_\varphi^{n+1}$ be a weighted Riemannian manifold that contains a line. If the Bakry-Émery-Ricci tensor of $\overline{M}_\varphi^{n+1}$ is nonnegative and the weighted function φ is bounded, then φ must be constant along the line.

Consequently, in any weighted Killing warped product $\left(M^n \times_\rho \mathbb{R}\right)_\varphi$ having nonnegative Bakry-Émery-Ricci tensor and with bounded weighted function φ, we have that φ does not depend on the parameter of the flow associated to the Killing vector field Y.

Motivated by the configuration described in the last two paragraphs above, along this chapter we will consider Killing warped products $M^n \times_\rho \mathbb{R}$ endowed with a weighted function φ which does not depends on the parameter $t \in \mathbb{R}$, that is, $\langle \overline{\nabla}\varphi, Y \rangle = 0$. We denote by

$$M_\varphi^n \times_\rho \mathbb{R}, \tag{19.4}$$

this ambient space.

19.3 The Second Variation of the Weighted Area Functional

Let $\mathcal{M}$ be the set of open subsets Ω of a weighted Killing warped product of the form given in (19.4) with compact closure $\overline{\Omega}$ and whose smooth compact boundary $\partial\Omega$ is a closed, connected and orientable hypersurface. Furthermore, for any $\Omega \in \mathcal{M}$, the symbol $\mathrm{Area}_\varphi(\partial\Omega)$ denotes the φ-area of $\partial\Omega$.

If $\Omega \in \mathcal{M}$, the globally unit normal vector field defined on $\partial\Omega$ will be denoted by N. We recall that a variation of $\partial\Omega$ is a smooth mapping

$$X : (-\epsilon, \epsilon) \times \partial\Omega \to M_\varphi^n \times_\rho \mathbb{R} \tag{19.5}$$

such that

1. For every $s \in (-\epsilon, \epsilon)$, the map

$$\begin{aligned} X_s : \partial\Omega &\to M_\varphi^n \times_\rho \mathbb{R} \\ p &\mapsto X_s(p) = X(s, p) \end{aligned} \tag{19.6}$$

is an immersion;
2. $X(0, p) = \iota(p)$ for every $p \in \partial\Omega$, where $\iota : \partial\Omega \hookrightarrow \overline{\Omega}$ is the inclusion map.

Moreover, the variational field associated to (19.5) is given by

$$\left.\frac{\partial X}{\partial s}\right|_{s=0}.$$

In this context, given $\Omega \in \mathcal{M}$ and a variation $X : (-\epsilon, \epsilon) \times \partial\Omega \to M_\varphi^n \times_\rho \mathbb{R}$ of $\partial\Omega$ we adopt the notation $\partial\Omega_s = X_s(\partial\Omega)$, where X_s is the immersion given by (19.6). For values of s small enough, $\partial\Omega_s$ is also a connected and oriented n-dimensional smooth submanifold. Moreover, $\partial\Omega_s$ bounds an open subset Ω_s whose closure is also compact. Thus, the variation $X : (-\epsilon, \epsilon) \times \partial\Omega \to M_\varphi^n \times_\rho \mathbb{R}$ described above induces a variation of the open subset Ω denoted by Ω_s, which is also an element of the set $\mathcal{M}$.

In the sequel, by $d(\partial\Omega_s)$ we denote the volume element of the metric induced on $\partial\Omega_s$ by X_s and by N_s, the unit normal vector field along X_s. Moreover, we also consider in $\partial\Omega_s$ the weighted volume form given by $d\sigma_s = e^{-\varphi} d(\partial\Omega_s)$. When $s = 0$ these objects coincide with ones defined in $\partial\Omega$, respectively.

According to [103, 211], every function $u \in C^\infty(\partial\Omega)$ with

$$\int_{\Sigma^n} u \, d\sigma = 0$$

induces a variation of $\partial\Omega$ of the type (19.5), with variational normal field

19.3 The Second Variation of the Weighted Area Functional

$$\frac{\partial X}{\partial s}\bigg|_{s=0} = uN,$$

and whose first variation of the weighted area functional

$$\begin{aligned}\mathcal{A}_\varphi : (-\epsilon, \epsilon) &\to \mathbb{R} \\ s &\mapsto \mathcal{A}_\varphi(s) = \text{Area}_\varphi(\partial\Omega_s) = \int_{\partial\Omega_s} d\sigma_s\end{aligned} \quad (19.7)$$

is given by

$$\frac{d}{ds}\mathcal{A}_\varphi(s) = -n \int_{\partial\Omega_s} (H_\varphi)_s \, u \, d\sigma_s, \quad (19.8)$$

for every $s \in (-\epsilon, \epsilon)$, where $(H_\varphi)_s = H_\varphi(s, \cdot)$ denotes the φ-mean curvature of $\partial\Omega_s$ with respect to the metric induced by the immersion X_s defined in (19.6).

In particular, one has

$$\frac{d}{ds}\mathcal{A}_\varphi(0) = -n \int_{\partial\Omega} H_\varphi \, u \, d\sigma. \quad (19.9)$$

As a consequence of (19.9), any open subset Ω of $M_\varphi^n \times_\rho \mathbb{R}$ whose boundary $\partial\Omega$ is a closed φ-minimal hypersurface is characterized as a critical point of the weighted area functional $\mathcal{A}_\varphi$.

The formula for the second variation of $\mathcal{A}_\varphi$ is given in the following result.

Proposition 19.1 *Let $\Omega \in M$ be open subset of $M_\varphi^n \times_\rho \mathbb{R}$ whose boundary $\partial\Omega$ is a closed φ-minimal hypersurface. Then*

$$\frac{d^2}{ds^2}\mathcal{A}_\varphi(0)(u) = -\int_{\partial\Omega} u \, \mathcal{J}_\varphi(u) \, d\sigma, \quad (19.10)$$

for any $u \in C^\infty(\partial\Omega)$, where $\mathcal{J}_\varphi : C^\infty(\partial\Omega) \to C^\infty(\partial\Omega)$ is the weighted Jacobi operator given by

$$\mathcal{J}_\varphi = \Delta_\varphi + \widetilde{\text{Ric}}_\varphi(N^*, N^*) - \frac{1}{\rho}\widetilde{\text{Hess}}\,\rho(N^*, N^*) - \langle N, Y\rangle^2 \frac{\widetilde{\Delta}_\varphi(\rho)}{\rho^3} + |A|^2. \quad (19.11)$$

Here, Y is the Killing vector field on $M_\varphi^n \times_\rho \mathbb{R}$, $\rho = |Y| > 0$, N is the unit normal vector field on $\partial\Omega$, Δ_φ and $\widetilde{\Delta}_\varphi$ represent the φ-Laplacians on $\partial\Omega$ and M_φ^n, respectively, $\widetilde{\text{Ric}}_\varphi$ and $\widetilde{\text{Hess}}$ are the Bakry-Émery-Ricci tensor and the Hessian operator on M_φ^n, $|A|^2$ represents the square of the norm of the shape operator A of $\partial\Omega$ with respect to N and N^* is the orthogonal projection of N on the tangent bundle of M^n.

Proof By (19.8), it follows that

$$\frac{d^2}{ds^2}\mathcal{A}_\varphi(0)(u) = -n\int_{\partial\Omega}\left(\frac{\partial (H_\varphi)_s}{\partial s}\bigg|_{s=0}\right)u d\sigma$$

$$-n\int_{\partial\Omega}\underbrace{H_\varphi}_{=0}\frac{\partial}{\partial s}(u d\sigma_s)\bigg|_{s=0}.$$

Arguing as in the proof of equation (3.5) of [103], we obtain

$$n\frac{\partial (H_\varphi)_s}{\partial s}\bigg|_{s=0} = \Delta_\varphi(u) + \left\{\overline{\mathrm{Ric}}_\varphi(N, N) + |A|^2\right\}u. \tag{19.12}$$

Hence, one has

$$\frac{d^2}{ds^2}\mathcal{A}_\varphi(0)(u) = -\int_{\partial\Omega}\left\{\Delta_\varphi(u) + \left\{\overline{\mathrm{Ric}}_\varphi(N, N) + |A|^2\right\}u\right\} u d\sigma. \tag{19.13}$$

On the other hand, denoting by N^* and $N^\perp$ the orthogonal projections of N over the tangent and normal bundles of M^n, respectively, and taking into account that φ is invariant along the flow determinate by Y, by O'Neill [251, Proposition 7.35] we obtain

$$\overline{\mathrm{Hess}}\varphi(N, N) = \langle\overline{\nabla}_N\overline{\nabla}\varphi, N\rangle$$

$$= \langle\overline{\nabla}_N\overline{\nabla}\varphi, N^* + N^\perp\rangle$$

$$= \widetilde{\mathrm{Hess}}\varphi(N^*, N^*) + \frac{1}{\rho}\langle\widetilde{\nabla}\varphi, \widetilde{\nabla}\rho\rangle|N^\perp|^2$$

$$= \widetilde{\mathrm{Hess}}\varphi(N^*, N^*) + \frac{1}{\rho^3}\langle\widetilde{\nabla}\varphi, \widetilde{\nabla}\rho\rangle\langle N, Y\rangle^2. \tag{19.14}$$

Moreover, by O'Neill [251, Corollary 7.43] we get that

$$\overline{\mathrm{Ric}}(N, N) = \widetilde{\mathrm{Ric}}(N^*, N^*) - \frac{1}{\rho}\widetilde{\mathrm{Hess}}\rho(N^*, N^*) - \langle N, Y\rangle^2\frac{\widetilde{\Delta}(\rho)}{\rho^3} \tag{19.15}$$

Now, by (19.2), (19.14) and (19.15), we have that

$$\overline{\mathrm{Ric}}_\varphi(N, N) = \widetilde{\mathrm{Ric}}_\varphi(N^*, N^*) - \frac{1}{\rho}\widetilde{\mathrm{Hess}}\rho(N^*, N^*) - \langle N, Y\rangle^2\frac{\widetilde{\Delta}_\varphi(\rho)}{\rho^3} \tag{19.16}$$

Therefore, by (19.13) and (19.16) we get (19.10) and (19.11), as claimed. □

As observed in a similar case in [187], our approach in this section is valid in a more general context. Assume that $\mathcal{M}$ is the set of open subsets $\Omega \subset M_\varphi^n \times_\rho \mathbb{R}$ whose boundary $\partial \Omega$ is union of two disjoint sets Σ_1^n and Σ_2^n, i.e. $\partial \Omega = \Sigma_1^n \cup \Sigma_2^n$. We will assume that one of them, for instance Σ_1^n, is a fixed set.

Consequently the variations considered of $\partial \Omega$ only affects Σ_2^n. Under this assumption, the critical points of weighted area functional $\mathcal{A}_\varphi$ will be open subsets Ω such that their boundaries are union of a (fixed) set Σ_1^n and a closed φ-minimal hypersurface Σ_2^n.

19.4 Bifurcation Instants for φ-Minimal Hypersurfaces in $M_\varphi^n \times_\rho \mathbb{R}$

When M^n is closed, the weighted Killing warped product $M_\varphi^n \times_\rho \mathbb{R}$ naturally admits a family of open subsets that can be realized as critical points of the weighted area functional defined in (19.7).

Indeed, for $t_1, t_2 \in \mathbb{R}$ with $t_1 < t_2$, we consider the family of open sets $\{\Omega_\gamma\}_{\gamma \in (t_1, t_2]}$ given by

$$\Omega_\gamma = M^n \times (t_1, \gamma), \quad \gamma \in (t_1, t_2], \tag{19.17}$$

whose boundary $\partial \Omega_\gamma$ of each Ω_γ is a disjoint union

$$\partial \Omega_\gamma = \Sigma_1^n \cup \Sigma_2^n(\gamma),$$

of a fixed set $\Sigma_1^n = M^n \times \{t_1\}$ and other set $\Sigma_2^n(\gamma) = M^n \times \{\gamma\}$.

We have that each $\Sigma_2^n(\gamma)$, $\gamma \in (t_1, t_2]$, is an φ-minimal totally geodesic closed hypersurface. So, as noted at the end of Sect. 19.3, since the variations of $\partial \Omega_\tau$ only affects $\Sigma_2^n(\gamma)$, we conclude that each element of $\Omega_{\gamma \in (t_1, t_2]}$ is a critical point of the weighted area functional $\mathcal{A}_{\varphi;\gamma}$, defined as in (19.7).

For these critical points, noting that ∂_t is the vector field on $M_\varphi^n \times_\rho \mathbb{R}$ that determines the orientation of each $\Sigma_2^n(\gamma)$, $\gamma \in (t_1, t_2]$, from Proposition 19.1 we have that the second variation of $\mathcal{A}_{\varphi;\gamma}$ and the weighted Jacobi operator $\mathcal{J}_{\varphi;\gamma}$ on each $\partial \Omega_\gamma$ are given by

$$\frac{d^2}{ds^2} \mathcal{A}_{\varphi;\gamma}(0)(u) = -\int_{\Sigma_2(\gamma)} u \, \mathcal{J}_{\varphi;\gamma}(u) \, d\sigma, \tag{19.18}$$

and

$$\mathcal{J}_{\varphi;\gamma}(u) = \Delta_{\varphi;\gamma}(u) - \frac{1}{\rho} \widetilde{\Delta}_\varphi(\rho) u \tag{19.19}$$

for any $u \in C^\infty(\Sigma_2^n(\gamma))$, respectively, where $\Delta_{\varphi;\gamma}$ represents the φ-Laplacian on $\Sigma_2^n(\gamma)$, $\widetilde{\Delta}_\varphi$ is the φ-Laplacian on M_φ^n, $\rho = |Y| > 0$ and Y is the Killing vector field that determines on $M_\varphi^n \times_\rho \mathbb{R}$ the foliation by totally geodesic closed slices $M^n \times \{t\}$ for $t \in \mathbb{R}$.

With respect to the family of open sets $\{\Omega_\gamma\}_{\gamma \in (t_1, t_2]}$ given by (19.17), we need to adopt some notions and results that correspond to equivariant bifurcation theory for geometric variational problems. We refer to [23, 81, 83, 279] for additional comments and remarks.

Let us first recall that two elements Ω_{γ_1} and Ω_{γ_2} of $\{\Omega_\gamma\}_{\gamma \in (t_1, t_2]}$ are said to be isometrically congruent when there is an isometry ψ of $M_\varphi^n \times_\rho \mathbb{R}$ that carries the image of $x_1 : \partial\Omega_{\gamma_1} \to M_\varphi^n \times_\rho \mathbb{R}$ onto the image of $x_2 : \partial\Omega_{\gamma_2} \to M_\varphi^n \times_\rho \mathbb{R}$ (cf. [23, Section 1.2]), where x_1 and x_2 are the immersions of $\partial\Omega_{\gamma_1}$ and $\partial\Omega_{\gamma_2}$ into $M_\varphi^n \times_\rho \mathbb{R}$, respectively. In other words, two elements Ω_{γ_1} and Ω_{γ_2} of $\{\Omega_\gamma\}_{\gamma \in (t_1, t_2]}$ are said to be isometrically congruent if there exists a diffeomorphism $\phi : \partial\Omega_{\gamma_1} \to \partial\Omega_{\gamma_2}$ and an isometry ψ of $M_\varphi^n \times_\rho \mathbb{R}$ such that the following diagram commutes

$$\begin{array}{ccc} \partial\Omega_{\gamma_1} & \xrightarrow{x_1} & M_\varphi^n \times_\rho \mathbb{R} \\ \phi \downarrow & & \downarrow \psi \\ \partial\Omega_{\gamma_2} & \xrightarrow{x_2} & M_\varphi^n \times_\rho \mathbb{R} \end{array}$$

Taking into account the results in [81], $\widetilde{\gamma} \in (t_1, t_2]$ is said to be a bifurcation instant for the family $\{\Omega_\gamma\}_{\gamma \in (t_1, t_2]}$ if there exists a sequence $\{\gamma_n\}_{n \in \mathbb{N}} \subset (t_1, t_2]$ and a sequence $\{\Omega_{\gamma_n}\}_{n \in \mathbb{N}} \subset \{\Omega_\gamma\}_{\gamma \in (t_1, t_2]}$ such that

(a) $\lim_{n \to \infty} \gamma_n = \widetilde{\gamma}$,
(b) $\lim_{n \to \infty} x_n = \widetilde{x}$, where $x_n : \Omega_{\gamma_n} \to M_\varphi^n \times_\rho \mathbb{R}$ and $\widetilde{x} : \Omega_{\widetilde{\gamma}} \to M_\varphi^n \times_\rho \mathbb{R}$ are the immersions of Ω_{γ_n} and $\Omega_{\widetilde{\gamma}}$ into $M_\varphi^n \times_\rho \mathbb{R}$, respectively,
(c) For all $n \in \mathbb{N}$, x_n is not isometrically congruent to $\widetilde{x}$.

Furthermore, according to [83], if $\widetilde{\gamma} \in (t_1, t_2]$ is not a bifurcation instant, the family $\{\Omega_\gamma\}_{\gamma \in (t_1, t_2]}$ is said to be locally rigid at $\widetilde{\gamma}$.

One of the classical criterion to determine when a instant $\widetilde{\gamma} \in (t_1, t_2]$ is of bifurcation is related with the so-called Morse index associated with the variational problem of minimizing the weighted area functional defined in (19.7); see, for instance, [23, 81]. Following this approach, we define the Morse index of Ω_γ, which will be denoted by

$$\mathrm{Ind}_\varphi\left(\mathcal{A}_{\varphi;\gamma}, \Omega_\gamma\right),$$

as the dimension of the maximal subspace where the second variation of $\mathcal{A}_{\varphi;\gamma}$ (see (19.18)) is negative definite.

Equivalently, $\text{Ind}_\varphi\left(\mathcal{A}_{\varphi;\gamma}, \Omega_\gamma\right)$ is the number of negative eigenvalues (counted with their multiplicity) of the weighted Jacobi operator $\mathcal{J}_{\varphi;\gamma}$ given in (19.19). With our notations, a real number $\widehat{\mu}(\gamma)$ is an eigenvalue of $\mathcal{J}_{\varphi;\gamma}$ if and only if

$$\mathcal{J}_{\varphi;\gamma}(u) + \widehat{\mu}(\gamma)u = 0$$

for some function $u \in C^\infty(\partial\Omega_\gamma)$.

Moreover, by using the same arguments of [23, Proposition 2.7] we obtain that $\text{Ind}_\varphi\left(\mathcal{A}_{\varphi;\gamma}, \Omega_\gamma\right)$ is finite on $(t_1, t_2] \subset \mathbb{R}$. Intuitively, $\text{Ind}_\varphi\left(\mathcal{A}_{\varphi;\gamma}, \Omega_\gamma\right)$ measures the number of independent directions in which the φ-minimal hypersurface $\partial\Omega_\gamma$ fails to minimize the weighted area functional $\mathcal{A}_{\varphi;\gamma}$.

Essentially, a variation of $\text{Ind}_\varphi\left(\mathcal{A}_{\varphi;\gamma}, \Omega_\gamma\right)$ along the interval $(t_1, t_2] \subset \mathbb{R}$ indicates the existence of a bifurcation instant. More precisely, under suitable Fredholmness assumptions (cf. [23, 81]), we have that if there are $\gamma_1, \gamma_2 \in (t_1, t_2]$, with $\gamma_1 < \gamma_2$, such that the second variation of $\mathcal{A}_{\varphi;\gamma_j}$ is nonsingular (namely, the eigenvalues of the weighted Jacobi operator $\mathcal{J}_{\varphi;\gamma_j}$ are nonzero) for $j \in \{1,2\}$ and

$$\text{Ind}_\varphi\left(\mathcal{A}_{\varphi;\gamma_1}, \Omega_{\gamma_1}\right) \neq \text{Ind}_\varphi\left(\mathcal{A}_{\varphi;\gamma_2}, \Omega_{\gamma_2}\right),$$

then $\{\Omega_\gamma\}_{\gamma \in (t_1, t_2]}$ admits a bifurcation instant at some $\gamma_* \in (\gamma_1, \gamma_2)$.

On the other hand, according to [83], using the Implicit Function Theorem, we obtain that if the second variation of $\mathcal{A}_{\varphi;\widetilde{\gamma}}(0)$ is nonsingular for same $\widetilde{\gamma} \in (t_1, t_2]$, then the family $\{\Omega_\gamma\}_{\gamma \in (t_1, t_2]}$ is locally rigid at $\widetilde{\gamma}$. In particular, when $\text{Ind}_\varphi\left(\mathcal{A}_{\varphi;\gamma}, \Omega_\gamma\right) = 0$ for every $\gamma \in (t_1, t_2]$, the family $\{\Omega_\gamma\}_{\gamma \in (t_1, t_2]}$ does not have bifurcation instants.

Here, we study the local rigidity and the bifurcation instants of $\{\Omega_\gamma\}_{\gamma \in (t_1, t_2]}$ by analyzing the spectrum of $\mathcal{J}_{\varphi;\gamma}$ for every $\gamma \in (t_1, t_2]$. Essentially, we will determine the number of negative eigenvalues for each γ (counted with their multiplicity) and we will study how this number changes.

19.5 Main Results of Bifurcation for φ-Minimal Hypersurfaces

In the scenario described in Sect. 19.4, our first result provides sufficient conditions to obtain the local rigidity of $\{\Omega_\gamma\}_{\gamma \in (t_1, t_2]}$ in terms of the behavior of the spectrum of the drift Laplacian $\widetilde{\Delta}_\varphi$ of closed Riemannian manifold M^n_φ.

Theorem 19.1 *Let M^n be an n-dimensional closed Riemannian manifold and, for $t_1, t_2 \in \mathbb{R}$ with $t_1 < t_2$, let $\Omega_{\gamma \in (t_1, t_2]}$ be the family of open subsets of the weighted Killing warped product $M^n_\varphi \times_\rho \mathbb{R}$ given by (19.17). Let $\widetilde{\Delta}_\varphi$ be the φ-Laplacian on M^n_φ. If ρ is an eigenfunction of $\widetilde{\Delta}_\varphi$ (with associated eigenvalue c) and the first nonzero eigenvalue*

$\mu_\varphi^1(\gamma)$ of the φ-Laplacian $\Delta_{\varphi;\gamma}$ on $\Sigma_2(\gamma) = M^n \times \{\gamma\}$, $\gamma \in (t_1, t_2]$, satisfies

$$\mu_\varphi^1(\gamma) > c, \qquad (19.20)$$

then $\{\Omega_\gamma\}_{\gamma \in (t_1, t_2]}$ is locally rigid at each $\gamma \in (t_1, t_2]$.

Proof It is known that the eigenvalues of the φ-Laplacian $\Delta_{\varphi;\gamma}$ on $\Sigma_2^n(\gamma)$ are given by a sequence

$$\left\{\mu_\varphi^j(\gamma)\right\}_{j=0}^{+\infty}$$

satisfying

$$0 = \mu_\varphi^0(\gamma) < \mu_\varphi^1(\gamma) \leq \cdots \leq \mu_\varphi^j(\gamma) \leq \mu_\varphi^{j+1}(\gamma) \leq \cdots,$$

repeated according to their multiplicity, and

$$\lim_{j \to +\infty} \mu_\varphi^j(\gamma) = +\infty;$$

see, for instance, [299, Section 1].

Now, since ρ is an eigenfunction of $\widetilde{\Delta}_\varphi$ with associated eigenvalue c, the weighted Jacobi operator $\mathcal{J}_{\varphi;\gamma}$ given in (19.19) can be written as

$$\mathcal{J}_{\varphi;\gamma} = \Delta_{\varphi;\gamma} + c.$$

Next, since c is a nonnegative constant, we have that if u is an eigenfunction of $\Delta_{\varphi;\gamma}$ associated with an eigenvalue $\mu_\varphi(\gamma)$ then u is eigenfunction of $\mathcal{J}_{\varphi;\gamma}$ with eigenvalue

$$\widehat{\mu}_\varphi(\gamma) = \mu_\varphi(\gamma) - c. \qquad (19.21)$$

So, by (19.20) and (19.21), the eigenvalues $\widehat{\mu}_\varphi^j(\gamma)$ of $\mathcal{J}_{\varphi;\gamma}$ satisfy the condition

$$\widehat{\mu}_\varphi^j(\gamma) = \mu_\varphi^j(\gamma) - c \geq \mu_\varphi^1(\gamma) - c > 0$$

for all $j \in \mathbb{N}$.

Hence, the second variation of $\mathcal{A}_{\varphi;\gamma}$ is nonsingular for all $\gamma \in (t_1, t_2]$ and, therefore, the family $\{\Omega_\gamma\}_{\gamma \in (t_1, t_2]}$ is locally rigid at each $\gamma \in (t_1, t_2]$. □

Remark 19.1 Considering once more the behavior of the eigenvalues of the φ-Laplacian on an arbitrary closed weighted manifold M_φ^n, by Theorem 19.1 we obtain the following consequence:

19.5 Main Results of Bifurcation for φ-Minimal Hypersurfaces

The family of open subsets $\{\Omega_\gamma\}_{\gamma \in (t_1, t_2]}$ of the weighted product $M_\varphi^n \times \mathbb{R}$ given by (19.17) is always locally rigid at each $\gamma \in (t_1, t_2]$.

In the last result of this chapter, we get a criterion that guarantees the existence of bifurcation instants of $\{\Omega_\gamma\}_{\gamma \in (t_1, t_2]}$.

Theorem 19.2 *Let M^n be an n-dimensional closed Riemannian manifold and, for $t_1, t_2 \in \mathbb{R}$ with $t_1 < t_2$, let $\Omega_{\gamma \in (t_1, t_2]}$ be the family of open subsets of the weighted Killing warped product $M_\varphi^n \times_\rho \mathbb{R}$ given by (19.17). Let $\widetilde{\Delta}_\varphi$ be the φ-Laplacian on M_φ^n. If ρ is an eigenfunction of $\widetilde{\Delta}_\varphi$ (with associated eigenvalue c) and if there are two values $\gamma_1, \gamma_2 \in (t_1, t_2]$, with $\gamma_1 < \gamma_2$, such that the eigenvalues $\widehat{\mu}_\varphi^j(\gamma_1)$ and $\widehat{\mu}_\varphi^j(\gamma_2)$ of the weighted Jacobi operators $\mathcal{J}_{\varphi;\gamma_1}$ and $\mathcal{J}_{\varphi;\gamma_2}$ (respectively) satisfy*

(a) $\widehat{\mu}_\varphi^j(\gamma_1) \neq 0$ and $\widehat{\mu}_\varphi^j(\gamma_2) \neq 0$ for every $j \in \mathbb{N}$,
(b) there exists $j_0 \in \mathbb{N}$ such that

$$\left(\widehat{\mu}_\varphi^{j_0}(\gamma_1)\right)\left(\widehat{\mu}_\varphi^{j_0}(\gamma_2)\right) < 0,$$

then $\Omega_{\gamma \in (t_1, t_2]}$ admits a bifurcation instant in $\gamma_* \in (\gamma_1, \gamma_2)$.

Proof By (19.21) we notice that the hypotheses contained in item (*a*) assures us that the second variation of the weighted Jacobi operator $\mathcal{J}_{\varphi;\gamma_j}$ is nonsingular for $j \in \{1, 2\}$.

On the order hand, we observe that hypothesis (*b*) assures us that the eigenvalue of the weighted Jacobi operator which corresponds to $j = j_0$ changes sign between γ_1 and γ_2.

Moreover, since the eigenvalues of the one-parameter family of weighted Jacobi operators are ordered, the number of negative eigenvalues between γ_1 and γ_2 changes. Therefore, one has

$$\mathrm{Ind}_\varphi\left(\mathcal{A}_{\varphi;\gamma_1}, \Omega_{\gamma_1}\right) \neq \mathrm{Ind}_\varphi\left(\mathcal{A}_{\varphi;\gamma_2}, \Omega_{\gamma_2}\right).$$

The conclusion is achieved. ◻

Bifurcation of Hypersurfaces with Constant φ-Mean Curvature in $M_\varphi^n \times_\rho \mathbb{R}$

20.1 Introduction

It is well-known that Killing vector fields are important mathematical objects which have been widely used in order to understand the geometry of submanifolds and, more particularly, of hypersurfaces immersed in Riemannian spaces.

Into this branch, Alías et al. [28] extended the classical Bernstein's theorem [77] to the context of complete minimal surfaces in Riemannian spaces of nonnegative Ricci curvature carrying a Killing vector field. This was done under the assumption that the sign of the angle function between a global Gauss mapping and the Killing vector field remains unchanged along the surface.

Afterwards, Dajczer et al. [134] defined a notion of Killing graph in a class of Riemannian manifolds endowed with a Killing vector field and solved the corresponding Dirichlet problem, for prescribed mean curvature, under suitable assumptions involving the domain geometry data and the Ricci curvature of the ambient space.

Later on, Dajczer and de Lira [131] showed that an entire Killing graph of constant mean curvature contained in a slab must be a totally geodesic slice, under certain restrictions on the curvature of the ambient space. More recently, in [132] the same authors revisited this thematic treating the case when the entire Killing graph of constant mean curvature lies inside a possible unbounded region.

Furthermore, Cunha et al. [127] applied suitable maximum principles in order to obtain Bernstein type properties concerning CMC hypersurfaces Σ^n immersed in a Killing warped product $(M^n \times_\rho \mathbb{R}, \langle \cdot, \cdot \rangle_M + \rho^2 dt)$, namely, in product manifolds $M^n \times \mathbb{R}$ endowed with the warping metric $\langle \cdot, \cdot \rangle_M + \rho^2 dt$, where M^n is a Riemannian manifold with Riemannian tensor $\langle \cdot, \cdot \rangle_M$, called Riemannian base, and ρ is a real positive function defined on M^n, called warping function.

To obtain these results, the authors assumed that M^n satisfies certain constraints and that ρ is concave on M^n. Afterwards, in [157], de Lima, Medeiros, Lima Jr. and Santos obtained Liouville type results concerning hypersurfaces Σ^n immersed in a weighted Killing warped product $M_\varphi^n \times_\rho \mathbb{R}$, where the weighted function φ does not depend on the parameter $t \in \mathbb{R}$; see Sect. 19.2. To this purpose, they assumed suitable boundedness on the Bakry-Émery-Ricci tensor of the base M^n. Furthermore, some rigidity results via constraints on the height function of the hypersurface have been proved.

Along this direction, let us consider a weighted Killing warped product $M_\varphi^n \times_\rho \mathbb{R}$ endowed with a weighted function φ which does not depends on the parameter $t \in \mathbb{R}$. Our purpose in this chapter is to study the notions of local rigidity and bifurcation instants for a family of open sets $\{\Omega_\gamma\}_\gamma$ of $M_\varphi^n \times_\rho \mathbb{R}$ whose boundaries $\partial \Omega_\gamma$ are closed hypersurfaces with constant weighted mean curvature $H_\varphi(\gamma)$, where γ varies on a prescribed real interval $I \subset \mathbb{R}$.

In terms of the weighted mean curvature, this study is a continuation of that performed in Chap. 19 and it is also mainly based on reference [291].

20.2 Description of the Variational Problem

In the sequel, we consider the weighted Killing warped product $M_\varphi^n \times_\rho \mathbb{R}$ defined in Sect. 19.2. If a hypersurface Σ^n immersed in $M_\varphi^n \times_\rho \mathbb{R}$ has constant φ-mean curvature H_φ, we will shortly that Σ^n is a H_φ-hypersurface.

Let M be the set of open subsets Ω of $M_\varphi^n \times_\rho \mathbb{R}$ with compact closure $\overline{\Omega}$ and whose smooth compact boundary $\partial \Omega$ is a closed, connected and orientable hypersurface. For any $\Omega \in M$, as usual

$$\mathrm{Vol}_\varphi(\Omega) \quad \text{and} \quad \mathrm{Area}_\varphi(\partial \Omega),$$

will denote the φ-volume and em φ-area of Ω and $\partial \Omega$, respectively.

If $\Omega \in M$, the globally unit normal vector field defined on $\partial \Omega$ will be denoted by N. Moreover, for any $\Omega \in M$, we define a variation of $\partial \Omega$ as a smooth mapping

$$\begin{aligned} X : (-\epsilon, \epsilon) \times \partial \Omega &\to M_\varphi^n \times_\rho \mathbb{R} \\ (s, p) &\mapsto X(s, p), \end{aligned}$$

satisfying the following two conditions:

1. For every $s \in (-\epsilon, \epsilon)$, the map

$$\begin{aligned} X_s : \partial \Omega &\to M_\varphi^n \times_\rho \mathbb{R} \\ p &\mapsto X_s(p) = X(s, p) \end{aligned} \tag{20.1}$$

20.2 Description of the Variational Problem

is an immersion;
2. $X(0, p) = \iota(p)$ for every $p \in \partial\Omega$, where $\iota : \partial\Omega \hookrightarrow \overline{\Omega}$ is the inclusion map.

In this framewor, given $\Omega \in \mathcal{M}$ and a variation $X : (-\epsilon, \epsilon) \times \partial\Omega \to M_\varphi^n \times_\rho \mathbb{R}$ of $\partial\Omega$ we adopt the notation $\partial\Omega_s = X_s(\partial\Omega)$, where X_s is the immersion (20.1).

For values of s small enough, $\partial\Omega_s$ is also a connected and oriented n-dimensional smooth submanifold. Moreover, $\partial\Omega_s$ bounds an open subset Ω_s whose closure is also compact. Thus, the variation $X : (-\epsilon, \epsilon) \times \partial\Omega \to M_\varphi^n \times_\rho \mathbb{R}$ described above induces a variation of the open subset Ω denoted by Ω_s, which is also an element of $\mathcal{M}$. Now, let us denote by $d(\partial\Omega_s)$ the volume element of the metric induced on $\partial\Omega_s$ by X_s and by N_s be the unit normal vector field along X_s.

Moreover, we also consider in $\partial\Omega_s$ the weighted volume form given by $d\sigma_s = e^{-\varphi} d(\partial\Omega_s)$. When $s = 0$ these objects coincide with ones defined in $\partial\Omega$, respectively. Finally, the field associated to the variation $X : (-\epsilon, \epsilon) \times \partial\Omega \to M_\varphi^n \times_\rho \mathbb{R}$ is the vector field given by

$$\frac{\partial X}{\partial s}\bigg|_{s=0}.$$

Letting

$$u_s = \left\langle \frac{\partial X}{\partial s}, N_s \right\rangle, \tag{20.2}$$

we get

$$\frac{\partial X}{\partial s}\bigg|_{s=0} = u_0 N + \left(\frac{\partial X}{\partial s}\bigg|_{s=0}\right)^\top,$$

where $(\cdot)^\top$ stands for tangential component.

The weighted volume functional associated to $X : (-\epsilon, \epsilon) \times \partial\Omega \to M_\varphi^n \times_\rho \mathbb{R}$ is given by

$$\begin{aligned} \mathcal{V}_\varphi : (-\epsilon, \epsilon) &\to \mathbb{R} \\ s &\mapsto \mathcal{V}_\varphi(s) = \mathrm{Vol}_\varphi(\Omega_s) = \int_{\Omega_s} d\overline{\sigma}, \end{aligned}$$

and we say that $X : (-\epsilon, \epsilon) \times \partial\Omega \to M_\varphi^n \times_\rho \mathbb{R}$ is weighted volume-preserving of Ω if $\mathcal{V}_f(s) = \mathcal{V}_f(0)$, for every $s \in (-\epsilon, \epsilon)$.

The following result is well-known and, in the context of weighted manifolds, it can be found in [103].

Lemma 20.1 *If $\Omega \in \mathcal{M}$ and $X : (-\epsilon, \epsilon) \times \partial\Omega \to M_\varphi^n \times_\rho \mathbb{R}$ is a variation of $\partial\Omega$ then*

$$\frac{d}{dt}\mathcal{V}_\varphi(s) = \int_{\partial\Omega_s} u_s \, d\sigma_s$$

for every $s \in (-\epsilon, \epsilon)$, where u_s is the function defined in (20.2). In particular, $X : (-\epsilon, \epsilon) \times \partial\Omega \to M_\varphi^n \times_\rho \mathbb{R}$ is weighted volume-preserving of Ω if and only if

$$\int_{\partial\Omega_s} u_s \, d\sigma_s = 0$$

for any $s \in (-\epsilon, \epsilon)$.

Remark 20.1 Is not difficult to verify that [64, Lemma 2.2] remains valid in the context of weighted Riemannian manifolds. More precisely, if $u \in C^\infty(\partial\Omega)$ is such that

$$\int_{\partial\Omega} u \, d\sigma = 0,$$

then there exists a weighted volume-preserving variation $X : (-\epsilon, \epsilon) \times \partial\Omega \to M_\varphi^n \times_\rho \mathbb{R}$ of $\partial\Omega$ whose variational field is given by

$$\left.\frac{\partial X}{\partial s}\right|_{s=0} = uN.$$

Now, the weighted area functional associated to $X : (-\epsilon, \epsilon) \times \partial\Omega \to M_\varphi^n \times_\rho \mathbb{R}$ is defined by

$$\begin{aligned}\mathcal{A}_\varphi : (-\epsilon, \epsilon) &\to \mathbb{R} \\ s &\mapsto \mathcal{A}_\varphi(s) = \mathrm{Area}_\varphi(\partial\Omega_s) = \int_{\partial\Omega_s} d\sigma_s.\end{aligned} \qquad (20.3)$$

Arguing as in the proof of [103, Lemma 3.2], the following result holds.

Lemma 20.2 *If $\Omega \in \mathcal{M}$ and $X : (-\epsilon, \epsilon) \times \partial\Omega \to M_\varphi^n \times_\rho \mathbb{R}$ is a variation of $\partial\Omega$, then*

$$\frac{d}{ds}\mathcal{A}_\varphi(s) = -n \int_{\partial\Omega_s} (H_\varphi)_s \, u_s \, d\sigma_s$$

for every $s \in (-\epsilon, \epsilon)$, where u_s is the function given in (20.2) and $(H_\varphi)_s = H_\varphi(s, \cdot)$ denotes the φ-mean curvature of $\partial\Omega_s$ with respect to the metric induced by the immersion X_s defined in (20.1).

20.2 Description of the Variational Problem

In order to characterize open subsets Ω of $M_\varphi^n \times_\rho \mathbb{R}$ whose boundary are closed hypersurfaces with constant φ-mean curvature, we consider the following variational problem:

(VP-1) *Minimizing the weighted area functional $\mathcal{A}_\varphi$ defined in (20.3) for any variation of $\partial\Omega$ that preserves the weighted volume of Ω.*

The Lagrange multiplier method leads us then to the associated weighted Jacobi functional

$$\mathcal{F}_\varphi^\lambda : (-\epsilon, \epsilon) \to \mathbb{R}$$
$$s \mapsto \mathcal{F}_\varphi^\lambda(s) = \text{Area}_\varphi(\partial\Omega_s) + \lambda \text{Vol}_\varphi(\Omega_s), \tag{20.4}$$

where λ is a constant to be determined.

As an immediate consequence of Lemmas 20.2 and 20.1 we get that the first variation of $\mathcal{F}_\varphi^\lambda$ takes the following form

$$\frac{d}{ds} \mathcal{F}_\varphi^\lambda(s) = \frac{d}{ds} \mathcal{A}_\varphi(s) + \lambda \frac{d}{ds} \mathcal{V}_\varphi(s)$$
$$= \int_{\partial\Omega_s} \{-n(H_f)_s + \lambda\} u_s \, d\sigma_s. \tag{20.5}$$

In order to determine the best possible choice of the parameter λ, let

$$\overline{\mathcal{H}} = \frac{1}{\text{Area}_\varphi(\partial\Omega)} \int_{\partial\Omega} H_\varphi \, d\sigma$$

be an integral mean of the φ-mean curvature H_f on $\partial\Omega$. We emphasize that if H_φ is constant, we have

$$\overline{\mathcal{H}} = H_\varphi. \tag{20.6}$$

Therefore, taking $\lambda = n\overline{\mathcal{H}}$, by (20.5) we have

$$\frac{d}{ds} \mathcal{F}_\varphi^\lambda(s) = -n \int_{\partial\Omega_s} \left\{ (H_\varphi)_s - \overline{\mathcal{H}} \right\} u_s \, d\sigma_s. \tag{20.7}$$

In particular, one has

$$\frac{d}{ds} \mathcal{F}_\varphi^\lambda(0) = -n \int_{\partial\Omega} \left\{ H_\varphi - \overline{\mathcal{H}} \right\} u_0 \, d\sigma. \tag{20.8}$$

Now, by (20.8), arguing as in [62, Proposition 2.7] the following result holds.

Proposition 20.1 *Let $\Omega \in \mathcal{M}$. The following statements are equivalent:*

(a) $\partial\Omega$ is a closed H_φ-hypersurface with constant φ-mean curvature H_φ equal to

$$H_\varphi = \frac{\lambda}{n};$$

(b) For the weighted volume-preserving variations $X : (-\epsilon, \epsilon) \times \partial\Omega \to M_\varphi^n \times_\rho \mathbb{R}$ of $\partial\Omega$, we have

$$\frac{d}{ds}\mathcal{A}_\varphi(0) = 0;$$

(c) For any variation $X : (-\epsilon, \epsilon) \times \partial\Omega \to M_\varphi^n \times_\rho \mathbb{R}$ of $\partial\Omega$, we have

$$\frac{d}{ds}\mathcal{F}_\varphi^\lambda(0) = 0.$$

Hence, by Proposition 20.1 we have that the solutions of (VP-1) are open subsets Ω of $M_\varphi^n \times_\rho \mathbb{R}$ whose boundary $\partial\Omega$ is a closed H_φ-hypersurface with constant second mean curvature H_φ equal to

$$H_\varphi = \frac{\lambda}{n}, \qquad (20.9)$$

with $\lambda \in \mathbb{R}$. Now, let us consider the variational problem

(VP-2) *Minimizing the weighted area functional $\mathcal{A}_\varphi$ defined in (20.3) for any variation of $\partial\Omega$, not necessarily a weighted volume-preserving variation of Ω.*

By Proposition 20.1 the solutions of (VP-2) coincide with the solutions of the variational problem (VP-1).

By (20.4) and (20.9) we observe that, if $\lambda = 0$, the set of solutions of the variable problems (VP-1) and (VP-2) is formed by the open subsets Ω of $M_\varphi^n \times_\rho \mathbb{R}$ whose boundary $\partial\Omega$ are closed φ-minimal hypersurfaces; see Chap. 19 in which a similar approach has been used in order to study a specific family of open subsets of $M_\varphi^n \times_\rho \mathbb{R}$.

Remark 20.2 As observed in [187], our approach is valid in a more general setting. More precisely, assume that $\mathcal{M}$ is the set of open subsets $\Omega \subset M_\varphi^n \times_\rho \mathbb{R}$ whose boundary $\partial\Omega$ is union of two disjoint sets Σ_1^n and Σ_2^n, ie. $\partial\Omega = \Sigma_1^n \cup \Sigma_2^n$. We will assume that one of them, for instance Σ_1^n, is a fixed set. Consequently, the variations considered of $\partial\Omega$ only affects Σ_2^n. Under this assumption, the solutions of (VP-1) or (VP-2) will be open subsets Ω such that their boundaries are union of a (fixed) set Σ_1^n and a closed H_φ-hypersurface Σ_2^n with constant φ-mean curvature H_φ given by (20.9).

In the following result we get the formula for the second variation of $\mathcal{F}_\varphi^\lambda$.

20.2 Description of the Variational Problem

Proposition 20.2 *Let $\Omega \in \mathcal{M}$ be open subset of $M_\varphi^n \times_\rho \mathbb{R}$ whose boundary $\partial \Omega$ is a compact H_φ-hypersurface, with constant φ-mean curvature H_φ given by (20.9). Then the second variation of the weighted Jacobi functional $\mathcal{F}_\varphi^\lambda$ is given by*

$$\frac{d^2}{ds^2}\mathcal{F}_\varphi^\lambda(0)(u) = -\int_{\partial\Omega} u\,\mathcal{J}_\varphi(u)\,d\sigma \tag{20.10}$$

for any $u \in C^\infty(\partial\Omega)$, where $\mathcal{J}_\varphi : C^\infty(\partial\Omega) \to C^\infty(\partial\Omega)$ is the weighted Jacobi operator given by

$$\mathcal{J}_\varphi = \Delta_\varphi + \widetilde{\mathrm{Ric}}_\varphi(N^*, N^*) - \frac{1}{\rho}\widetilde{\mathrm{Hess}}\,\rho(N^*, N^*) - \langle N, Y\rangle^2 \frac{\widetilde{\Delta}_\varphi(\rho)}{\rho^3} + |A|^2. \tag{20.11}$$

Here, Y is the Killing vector field on $M_\varphi^n \times_\rho \mathbb{R}$, $\rho = |Y| > 0$, N is the unit normal vector field on $\partial\Omega$, Δ_φ and $\widetilde{\Delta}_\varphi$ represent the φ-Laplacians on $\partial\Omega$ and M_φ^n, respectively, $\widetilde{\mathrm{Ric}}_\varphi$ and $\widetilde{\mathrm{Hess}}$ are the Bakry-Émery-Ricci tensor and the Hessian operator on M_φ^n, $|A|^2$ represents the square of the norm of the shape operator A of $\partial\Omega$ with respect to the orientation given by N and N^* is the orthogonal projection of N on the tangent bundle of M^n.

With respect to the functions on $\partial\Omega$ to be evaluated in (20.10) for a solution of (VP-1), they have to be considered according to Remark 20.1, that is, smooth functions on $\partial\Omega$ whose integral mean is zero; and, on the other hand, any smooth function on $\partial\Omega$ can be evaluated in (20.10) for a solution of (VP-2).

Proof For any variation $X : (-\epsilon, \epsilon) \times \partial\Omega \to M_\varphi^n \times_\rho \mathbb{R}$ of $\partial\Omega$ let us consider the function $u_0 \in C^\infty(\partial\Omega)$ defined in (20.2). Since H_φ is constant, by (20.7), (20.6), (19.12) and (19.16) we obtain, respectively, the following expression for the second variation of $\mathcal{F}_\varphi^\lambda$:

$$\frac{d^2}{ds^2}\mathcal{F}_\varphi^\lambda(0)(u_0) = -n\int_{\partial\Omega}\left(\frac{\partial(H_\varphi)_s}{\partial s}\bigg|_{s=0}\right)u_0\,d\sigma$$

$$-n\int_{\partial\Omega}\underbrace{(H_\varphi - \mathcal{H})}_{=0}\frac{\partial}{\partial s}(u_s\,d\sigma_s)\bigg|_{s=0}$$

$$= -\int_{\partial\Omega}\left\{\Delta_\varphi(u_0) + \left\{\widetilde{\mathrm{Ric}}_\varphi(N,N) + |A|^2\right\}u_0\right\}u_0\,d\sigma$$

$$= -\int_{\partial\Omega}\left\{\Delta_\varphi(u_0) + \left\{\widetilde{\mathrm{Ric}}_f(N^*, N^*) - \frac{1}{\rho}\widetilde{\mathrm{Hess}}\rho(N^*, N^*)\right.\right.$$

$$\left.\left. -\langle N, Y\rangle^2\frac{\widetilde{\Delta}_f(\rho)}{\rho^3} + |A|^2\right\}u_0\right\}u_0\,d\sigma. \tag{20.12}$$

Now, for any $u \in C^\infty(\partial\Omega)$, considering variations $X : (-\epsilon, \epsilon) \times \partial\Omega \to M_\varphi^n \times_\rho \mathbb{R}$ of $\partial\Omega$ whose variational field is

$$\frac{\partial X}{\partial t}\bigg|_{t=0} = uN,$$

one has that the last expression (20.12) is also valid for every $u \in C^\infty(\partial\Omega)$.

Taking into account the set of functions on $\partial\Omega$ that are admissible for a solution of (VP-2), we conclude that the arguments stated above are valid to provide the formula of the second variation of $\mathcal{F}_\varphi^\lambda$ for solutions of (VP-2).

For those solutions of (VP-1), if $X : (-\epsilon, \epsilon) \times \partial\Omega \to M_\varphi^n \times_\rho \mathbb{R}$ is a variation of $\partial\Omega$ which preserve the weighted volume of Ω then, for $u_0 \in C^\infty(\partial\Omega)$ defined in (20.2), by Lemma 20.1 we have that

$$\int_{\partial\Omega} u_0 \, dV = 0.$$

In addition, the expression (20.12) is valid for such u_0. Finally, for any function $u \in C^\infty(\partial\Omega)$ such that

$$\int_{\partial\Omega} u \, dV = 0,$$

by Remark 20.1 we get a variation $X : (-\epsilon, \epsilon) \times \partial\Omega \to M_\varphi^n \times_\rho \mathbb{R}$ of $\partial\Omega$ which preserve the weighted volume of Ω and such that

$$\frac{\partial X}{\partial t}\bigg|_{t=0} = uN.$$

The conclusion is achieved. □

We conclude this section observing that the weighted Jacobi operator $\mathcal{J}_\varphi$, given in (20.11), belongs into the class of differential operators which are usually referred to as Schrödinger operators, that is, operators of the form $\Delta + q$, where Δ is the standard Laplacian on $\partial\Omega$ and q is a continuous function on $\partial\Omega$; see, for instance, [174]. In particular, we notice that the behavior of the eigenvalues of $\mathcal{J}_\varphi$ is well-known. This fact will play an important role in order to obtain the main results of this chapter.

20.3 The Notion of Bifurcation for H_φ-Hypersurfaces in $M_\varphi^n \times_\rho \mathbb{R}$

Let us consider the one-parameter family $\{\Omega_\gamma\}_\gamma$ of open subsets in a weighted Killing warped product $M_\varphi^n \times_\rho \mathbb{R}$ such that the boundary of each Ω_γ, denoted by $\partial\Omega_\gamma$, is a

20.3 The Notion of Bifurcation for H_φ-Hypersurfaces in $M_\varphi^n \times_\rho \mathbb{R}$

closed $H_\varphi(\gamma)$-hypersurface with constant φ-mean curvature $H_\varphi(\gamma)$, where γ varies on a prescribed interval $I \subset \mathbb{R}$.

In this context, as a consequence of the results in Sect. 20.2, we have that each Ω_γ is a solution of a certain variational problem of type (VP-2). More precisely, each Ω_τ is a critical point for the one-parameter family of the weighted Jacobi functionals

$$I \ni \gamma \mapsto \mathcal{F}_\varphi^{\lambda(\gamma)} = \mathcal{A}_\varphi + \lambda(\gamma)\mathcal{V}_\varphi$$

defined in (20.4), where

$$\lambda(\gamma) = nH_\varphi(\gamma).$$

Moreover, by Proposition 20.2, associated with each closed $H_\varphi(\gamma)$-hypersurface $\partial\Omega_\gamma$, we have that the second variation of $\mathcal{F}_\varphi^{\lambda(\gamma)}$ is given by

$$\frac{d^2}{ds^2}\mathcal{F}_\varphi^{\lambda(\tau)}(0)(u) = -\int_{\partial\Omega} u\,\mathcal{J}_{\varphi;\gamma}(u)\,d\sigma, \tag{20.13}$$

for any $u \in C^\infty(\partial\Omega_\gamma)$, where

$$\mathcal{J}_{\varphi;\gamma} = \Delta_{\varphi;\gamma} + \widetilde{\mathrm{Ric}}_\varphi(N_\gamma^*, N_\gamma^*) - \frac{1}{\rho}\widetilde{\mathrm{Hess}}\rho(N_\gamma^*, N_\gamma^*) - \langle N_\gamma, Y\rangle^2\frac{\widetilde{\Delta}_\varphi(\rho)}{\rho^3} + |A_\gamma|^2 \tag{20.14}$$

is the weighted Jacobi operator on $\partial\Omega_\gamma$.

Here, $\Delta_{\varphi;\gamma}$ and $\widetilde{\Delta}_\varphi$ are the φ-Laplacians on $\partial\Omega_\gamma$ and M_φ^n, respectively, $\widetilde{\mathrm{Ric}}_\varphi$ and $\widetilde{\mathrm{Hess}}$ are the Bakry-Émery-Ricci tensor and the Hessian operator in M_φ^n, A_γ is the shape operator of $\partial\Omega_\gamma$ with respect to normal vector field N_γ and N_γ^* is the orthogonal projection of N_γ on the tangent bundle of M^n.

With respect to the family $\{\Omega_\gamma\}_{\gamma\in I}$ of solutions of (VP-2), some notions and results, that correspond to Equivariant Bifurcation Theory for geometric variational problems, are necessary; see [23, 81, 83, 279].

Let us recall that two elements Ω_{γ_1} and Ω_{γ_2} of $\{\Omega_\gamma\}_{\gamma\in I}$ are said to be isometrically congruent if there is an isometry ψ of $M_\varphi^n \times_\rho \mathbb{R}$ that carries the image of $x_1 : \partial\Omega_{\gamma_1} \to M_\varphi^n \times_\rho \mathbb{R}$ onto the image of $x_2 : \partial\Omega_{\gamma_2} \to M_\varphi^n \times_\rho \mathbb{R}$ (cf. [23, Section 1.2]), where x_1 and x_2 are the immersions of $\partial\Omega_{\gamma_1}$ and $\partial\Omega_{\gamma_2}$ into $M_\varphi^n \times_\rho \mathbb{R}$, respectively.

In other words, two elements Ω_{γ_1} and Ω_{γ_2} of $\{\Omega_\gamma\}_{\gamma\in I}$ are said to be isometrically congruent if there exists a diffeomorphism $\phi : \partial\Omega_{\gamma_1} \to \partial\Omega_{\gamma_2}$ and an isometry ψ of $M_\varphi^n \times_\rho \mathbb{R}$ such that the following diagram commutes

$$\begin{array}{ccc} \partial\Omega_{\gamma_1} & \xrightarrow{x_1} & M_\varphi^n \times_\rho \mathbb{R} \\ \phi \downarrow & & \downarrow \psi \\ \partial\Omega_{\gamma_2} & \xrightarrow{x_2} & M_\varphi^n \times_\rho \mathbb{R} \end{array}$$

According to [81], $\tilde{\gamma} \in I$ is said to be a bifurcation instant for the family $\{\Omega_\gamma\}_{\gamma \in I}$ if there exists a sequence $\{\gamma_n\}_{n \in \mathbb{N}} \subset I$ and a sequence $\{\Omega_{\gamma_n}\}_{n \in \mathbb{N}} \subset \{\Omega_\gamma\}_{\gamma \in I}$ such that

(a) $\lim_{n \to \infty} \gamma_n = \tilde{\gamma}$,
(b) $\lim_{n \to \infty} x_n = \tilde{x}$, where $x_n : \Omega_{\gamma_n} \to M_\varphi^n \times_\rho \mathbb{R}$ and $\tilde{x} : \Omega_{\tilde{\gamma}} \to M_\varphi^n \times_\rho \mathbb{R}$ are the immersions of Ω_{γ_n} and $\Omega_{\tilde{\gamma}}$ into $M_\varphi^n \times_\rho \mathbb{R}$, respectively,
(c) For all $n \in \mathbb{N}$, x_n is not isometrically congruent to $\tilde{x}$.

Furthermore, as in [83], if $\tilde{\gamma} \in I$ is not a bifurcation instant, the family $\{\Omega_\gamma\}_{\gamma \in I}$ is said to be locally rigid at $\tilde{\gamma}$.

One of the classical tools to study bifurcation is related with the notion of Morse index; see, for instance, [23,81]. Thus, let us define the Morse index of Ω_γ, which will be denoted by

$$\mathrm{Ind}_\varphi(\mathcal{F}_\varphi^{\lambda(\gamma)}, \Omega_\gamma),$$

as the dimension of the maximal subspace where the second variation of the weighted Jacobi functional $\mathcal{F}_\varphi^{\lambda(\gamma)}$ is negative definite.

Equivalently, $\mathrm{Ind}_\varphi(\mathcal{F}_\varphi^{\lambda(\gamma)}, \Omega_\gamma)$ is the number of negative eigenvalues (counted with their multiplicity) of the weighted Jacobi operator $\mathcal{J}_{\varphi;\gamma}$ given in (20.14). With our notations, a real number $\hat{\mu}(\gamma)$ is an eigenvalue of $\mathcal{J}_{\varphi;\gamma}$ if and only if

$$\mathcal{J}_{\varphi;\gamma}(u) + \hat{\mu}(\gamma)u = 0$$

for some function $u \in C^\infty(\partial\Omega_\gamma)$.

Furthermore, by using the same arguments of [23, Proposition 2.7] we obtain that $\mathrm{Ind}_\varphi(\mathcal{F}_\varphi^{\lambda(\gamma)}, \Omega_\gamma)$ is finite on $I \subset \mathbb{R}$. Intuitively, $\mathrm{Ind}_\varphi(\mathcal{F}_\varphi^{\lambda(\gamma)}, \Omega_\gamma)$ measures the number of independent directions in which the $H_\varphi(\gamma)$-hypersurface $\partial\Omega_\gamma$ fails to minimize the weighted area functional.

Essentially, a variation of $\mathrm{Ind}_\varphi(\mathcal{F}_\varphi^{\lambda(\gamma)}, \Omega_\gamma)$ along the interval $I \subset \mathbb{R}$ indicates the existence of a bifurcation instant. More precisely, under suitable Fredholmness assumptions (cf. [23, 81]), we have that if there are $\gamma_1, \gamma_2 \in I$, with $\gamma_1 < \gamma_2$, such that the second variation of the weighted Jacobi functional $\mathcal{F}_\varphi^{\lambda(\gamma_j)}$ is nonsingular (namely, the eigenvalues of the weighted Jacobi operator $\mathcal{J}_{\varphi;\gamma_j}$ are nonzero) for $j \in \{1, 2\}$ and

$$\mathrm{Ind}_\varphi(\mathcal{F}_\varphi^{\lambda(\gamma_1)}, \Omega_{\gamma_1}) \neq \mathrm{Ind}_\varphi(\mathcal{F}_\varphi^{\lambda(\gamma_2)}, \Omega_{\gamma_2}),$$

then $\{\Omega_\gamma\}_{\gamma \in I}$ admits a bifurcation instant at some $\gamma_* \in (\gamma_1, \gamma_2)$.

On the other hand, according to [83], by using the Implicit Function Theorem, we obtain that if the second variation of $\mathcal{F}_\varphi^{\lambda(\widetilde{\gamma})}(0)$ is nonsingular for same $\widetilde{\gamma} \in I$, then the family $\{\Omega_\gamma\}_{\gamma \in I}$ is locally rigid at $\widetilde{\gamma}$. In particular, when $\mathrm{Ind}_\varphi(\mathcal{F}_\varphi^{\lambda(\gamma)}, \Omega_\gamma) = 0$ for all $\gamma \in I$, the family $\{\Omega_\gamma\}_{\gamma \in I}$ does not have bifurcation instants.

Here, we will study the local rigidity and the bifurcation instants of $\{\Omega_\gamma\}_{\gamma \in I}$ by analyzing the spectrum of $\mathcal{J}_{\varphi;\gamma}$ for all $\gamma \in I$. Essentially, we will determine the number of negative eigenvalues for each γ (counted with their multiplicity) and we study how this number changes.

20.4 Main Results of Bifurcation for H_φ-Hypersurfaces

The main result below provides some simple sufficient conditions to get the local rigidity of the family $\{\Omega_\gamma\}_{\gamma \in I}$ of solutions of the variational problem (VP-2) described in Sect. 20.3.

Theorem 20.1 *Let $\{\Omega_\gamma\}_{\gamma \in I}$ be a family of open subsets of the weighted Killing warped product $M_\varphi^n \times_\rho \mathbb{R}$ whose boundaries $\partial\Omega_\gamma$ are closed $H_\varphi(\gamma)$-hypersurfaces. If, for every $\gamma \in I$, the function*

$$Q_\varphi(\gamma) = \widetilde{\mathrm{Ric}}_\varphi\left(N_\gamma^*, N_\gamma^*\right) - \frac{1}{\rho}\widetilde{\mathrm{Hess}}\,\rho\left(N_\gamma^*, N_\gamma^*\right) - \langle N_\gamma, Y\rangle^2 \frac{\widetilde{\Delta}_\varphi(\rho)}{\rho^3} + |A_\gamma|^2$$

is constant on $\partial\Omega_\gamma$ and the first nonzero eigenvalue $\mu_\varphi^1(\gamma)$ of the φ-Laplacian $\Delta_{\varphi;\gamma}$ on $\partial\Omega_\gamma$ satisfies

$$\mu_\varphi^1(\gamma) - Q_\varphi(\gamma) > 0, \tag{20.15}$$

then $\{\Omega_\gamma\}_{\gamma \in I}$ is locally rigid at each γ. In particular, such a family is locally rigid if one of the following conditions holds:

(a) $\widetilde{\mathrm{Ric}}_\varphi(N_\gamma^*, N_\gamma^*) - \frac{1}{\rho}\widetilde{\mathrm{Hess}}\,\rho(N_\gamma^*, N_\gamma^*) - \langle N_\gamma, Y\rangle^2 \frac{\widetilde{\Delta}_\varphi(\rho)}{\rho^3} \leq -|A_\gamma|^2;$

(b) *Either*

$$\widetilde{\mathrm{Ric}}_\varphi(N_\gamma^*, N_\gamma^*) - \frac{1}{\rho}\widetilde{\mathrm{Hess}}\,\rho(N_\gamma^*, N_\gamma^*) - \langle N_\gamma, Y\rangle^2 \frac{\widetilde{\Delta}_\varphi(\rho)}{\rho^3} < 0 \quad \text{and} \quad \mu_\varphi^1(\gamma) \geq |A_\gamma|^2,$$

or

$$\widetilde{\mathrm{Ric}}_\varphi(N_\gamma^*, N_\gamma^*) - \frac{1}{\rho}\widetilde{\mathrm{Hess}}\rho(N_\gamma^*, N_\gamma^*) - \langle N_\gamma, Y\rangle^2 \frac{\widetilde{\Delta}\varphi(\rho)}{\rho^3} \le 0 \quad \text{and} \quad \mu_\varphi^1(\gamma) > |A_\gamma|^2.$$

Proof Since $Q_\varphi(\gamma)$ is constant, by (20.14) we have that the eigenfunctions of the weighted Jacobi operator $\mathcal{J}_{\varphi;\gamma}$ coincide with the eigenfunctions of φ-Laplacian $\Delta_{\varphi;\gamma}$.

More precisely, if u is an eigenfunction of $\Delta_{\varphi;\gamma}$ associated with an eigenvalue $\mu_\varphi(\gamma)$ then u is eigenfunction of $\mathcal{J}_{f;\gamma}$ with eigenvalue

$$\widehat{\mu}_\varphi(\gamma) = \mu_\varphi(\gamma) - Q_\varphi(\gamma).$$

Furthermore, by the Spectral Theorem we know that the eigenvalues of $\Delta_{\varphi;\gamma}$ are given by a sequence $\{\mu_\varphi^j(\gamma)\}_{j=0}^{+\infty}$ satisfying

$$0 = \mu_\varphi^0(\gamma) < \mu_\varphi^1(\gamma) \le \cdots \le \mu_\varphi^j(\gamma) \le \mu_\varphi^{j+1}(\gamma) \le \cdots,$$

repeated according to their multiplicity, and

$$\lim_{j\to+\infty} \mu_\varphi^j(\gamma) = +\infty;$$

see, for instance, [299, Section 1]. So, the eigenvalues $\widehat{\mu}_\varphi^j(\gamma)$ of $\mathcal{J}_{\varphi;\gamma}$ have the following form

$$\widehat{\mu}_f^j(\gamma) = \mu_f^j(\gamma) - Q_f(\gamma), \qquad (20.16)$$

for every $j \in \mathbb{N}$.

Consequently, by (20.15) and (20.16) we obtain

$$\widehat{\mu}_\varphi^j(\gamma) = \mu_\varphi^j(\gamma) - Q_\varphi(\gamma) \ge \mu_\varphi^1(\gamma) - Q_\varphi(\gamma) > 0$$

for every $j \in \mathbb{N}$. Hence, the second variation of the weighted Jacobi functional $\mathcal{F}_f^{\lambda(\gamma)}(0)$ given in (20.13) is nonsingular for every $\gamma \in I$ and, therefore, the family $\{\Omega_\gamma\}_{\gamma \in I}$ is locally rigid at each $\gamma \in I$. □

We close this chapter giving a sufficient condition concerning the existence of bifurcation instants of the family $\{\Omega_\gamma\}_{\gamma \in I}$.

Theorem 20.2 *Let $\{\Omega_\gamma\}_{\gamma \in I}$ be a family of open subsets of the weighted Killing warped product $M_\varphi^n \times_\rho \mathbb{R}$ whose boundaries $\partial\Omega_\gamma$ are closed $H_\varphi(\gamma)$-hypersurfaces. Suppose that, for every $\gamma \in I$, the function*

20.4 Main Results of Bifurcation for H_φ-Hypersurfaces

$$Q_\varphi(\gamma) = \widetilde{\mathrm{Ric}}_\varphi\left(N_\gamma^*, N_\gamma^*\right) - \frac{1}{\rho}\widetilde{\mathrm{Hess}}\,\rho\left(N_\gamma^*, N_\gamma^*\right) - \langle N_\gamma, Y\rangle^2 \frac{\widetilde{\Delta}_\varphi(\rho)}{\rho^3} + |A_\gamma|^2$$

is constant on $\partial\Omega_\gamma$. If there are two values γ_1 and γ_2, with $\gamma_1 < \gamma_2$, such that the eigenvalues $\widehat{\mu}_\varphi^j(\gamma_1)$ and $\widehat{\mu}_\varphi^j(\gamma_2)$ of the weighted Jacobi operators $\mathcal{J}_{\varphi;\gamma_1}$ and $\mathcal{J}_{\varphi;\gamma_2}$ (respectively) satisfy

(a) $\widehat{\mu}_\varphi^j(\gamma_1) \neq 0$ and $\widehat{\mu}_\varphi^j(\gamma_2) \neq 0$ for every $j \in \mathbb{N}$,
(b) there exists $j_0 \in \mathbb{N}$ such that

$$\left(\widehat{\mu}_\varphi^{j_0}(\gamma_1)\right)\left(\widehat{\mu}_\varphi^{j_0}(\gamma_2)\right) < 0,$$

then there exists a bifurcation instant $\gamma_* \in (\gamma_1, \gamma_2)$.

Proof By (20.13) and (20.14) we notice that the condition on the function $Q_\varphi(\gamma)$ and hypothesis (a) assures that the second variation of the weighted Jacobi functional $\mathcal{F}_\varphi^{\lambda(\gamma_j)}$ is nonsingular for $j \in \{1, 2\}$.

On the order hand, we observe that assumption (b) ensures that the eigenvalue of the weighted Jacobi operator which corresponds to $j = j_0$ changes sign between γ_1 and γ_2.

Moreover, as the eigenvalues of the one-parameter family of weighted Jacobi functionals are ordered, we have that the number of negative eigenvalues between γ_1 and γ_2 changes. Therefore, one has

$$\mathrm{Ind}_\varphi(\mathcal{F}_\varphi^{\lambda(\gamma_1)}, \Omega_{\gamma_1}) \neq \mathrm{Ind}_\varphi(\mathcal{F}_\varphi^{\lambda(\gamma_2)}, \Omega_{\gamma_2}),$$

and the conclusion follows. □

References

1. M. Aarons, Mean curvature flow with a forcing term in Minkowski space. Calc. Var. PDE **25**, 205–246 (2005)
2. L.V. Ahlfors, Sur le type dune surface de Riemann. C.R. Acad. Sc. Paris **201**, 30–32 (1935)
3. R. Aiyama, On the Gauss map of complete space-like hypersurfaces of constant mean curvature in Minkowski space. Tsukuba J. Math. **16**, 353–361 (1992)
4. A.L. Albujer, New examples of entire maximal graphs in $\mathbb{H}^2 \times \mathbb{R}_1$. Differ. Geom. Appl. **26**, 456–462 (2008)
5. A.L. Albujer, L.J. Alías, Calabi-Bernstein results for maximal surfaces in Lorentzian product spaces. J. Geom. Phys. **59**, 620–631 (2009)
6. A.L. Albujer, L.J. Alías, Spacelike hypersurfaces with constant mean curvature in the steady state space. Proc. Am. Math. Soc. **137**, 711–721 (2009)
7. A.L. Albujer, J.A. Aledo, L.J. Alías, On the scalar curvature of in hypersurfaces in spaces with Killing field. Adv. Geom. **10**, 487–503 (2010)
8. A.L. Albujer, F. Camargo, H.F. de Lima, Complete spacelike hypersurfaces with constant mean curvature in $-\mathbb{R} \times \mathbb{H}^n$. J. Math. Anal. Appl. **368**, 650–657 (2010)
9. A.L. Albujer, F.E.C. Camargo, H.F. de Lima, Complete spacelike hypersurfaces in a Robertson-Walker spacetime. Math. Proc. Cambridge Philos. Soc. **151**, 271–282 (2011)
10. A.L. Albujer, H.F. de Lima, A.M. Oliveira, M.A.L. Velásquez, Rigidity of spacelike hypersurfaces in spatially weighted generalized Robertson-Walker spacetimes. Differ. Geom. Appl. **50**, 140–154 (2017)
11. A.L. Albujer, H.F. de Lima, A.M. Oliveira, M.A.L. Velásquez, ϕ-parabolicity and the uniqueness of spacelike hypersurfaces immersed in a spatially weighted GRW spacetime. Mediterr. J. Math. **15**, 84 (2018)
12. J.A. Aledo, R.M. Rubio, Stable minimal surfaces in Riemannian warped products. J. Geom. Anal. **27**, 65–78 (2017)
13. J.A. Aledo, L.J. Alías, A. Romero, Integral formulas for compact space-like hypersurfaces in de Sitter space: applications to the case of constant higher order mean curvature. J. Geom. Phys. **31**, 195–208 (1999)
14. J.A. Aledo, J.M. Espinar, J.A. Gálvez, Height estimates for surfaces with positive constant mean curvature in $\mathbb{M}^2 \times \mathbb{R}$. Illinois J. Math. **52**, 203–211 (2008)
15. J.A. Aledo, R.M. Rubio, J.J. Salamanca, Space-like hypersurfaces with functionally bounded mean curvature in Lorentzian warped products and generalized Calabi-Bernstein-type problems. Proc. Royal Soc. Edinburgh Sect. A **149**, 849–868 (2019)
16. H. Alencar, M. do Carmo, A.G. Colares, Stable hypersurfaces with constant scalar curvature. Math. Z. **213**, 117–131 (1993)

17. H. Alencar, M. do Carmo, H. Rosenberg, On the first eigenvalue of the linearized operator of the RTH mean curvature of a hypersurface. Ann. Global Anal. Geom. **11**, 387–395 (1993)
18. A. Alexandrov, A characteristic property of spheres. Ann. Mat. Pura Appl. **58**, 303–315 (1962)
19. L.J. Alías, A.G. Colares, Uniqueness of spacelike hypersurfaces with constant higher order mean curvature in Generalized Robertson-Walker spacetimes. Math. Proc. Cambridge Philos. Soc. **143**, 703–729 (2007)
20. L.J. Alías, M. Dajczer, Uniqueness of constant mean curvature surfaces properly immersed in a slab. Comment. Math. Helv. **81**, 653–663 (2006)
21. L.J. Alías, M. Dajczer, Constant mean curvature hypersurfaces in warped product spaces. Proc. Edinb. Math. Soc. **50**, 511–526 (2007)
22. L.J. Alías, J.M. Malacarne, Spacelike hypersurfaces with constant higher order mean curvature in Minkowski space-time. J. Geom. Phys. **41**, 359–375 (2002)
23. L.J. Alías, P. Piccione, Bifurcation of constant mean curvature tori in Euclidean spheres. J. Geom. Anal. **23**(2), 677–708 (2013)
24. L.J. Alías, A. Romero, M. Sánchez, Uniqueness of complete spacelike hypersurfaces with constant mean curvature in Generalized Robertson-Walker spacetimes. Gen. Relat. Grav. **27**, 71–84 (1995)
25. L.J. Alías, A. Romero, M. Sánchez, Spacelike hypersurfaces of constant mean curvature and Calabi-Bernstein type problems. Tôhoku Math. J. **49**, 337–345 (1997)
26. L.J. Alías, A. Brasil, Jr., A.G. Colares, Integral formulae for spacelike hypersurfaces in conformally stationary spacetimes and applications. Proc. Edinburgh Math. Soc. **46**, 465–488 (2003)
27. L.J. Alías, T. Kurose, G. Solanes, Hadamard-type theorems for hypersurfaces in hyperbolic spaces. Differ. Geom. Appl. **24**, 492–502 (2006)
28. L.J. Alías, M. Dajczer, J.B. Ripoll, A Bernstein-type theorem for Riemannian manifolds with a Killing field. Ann. Glob. Anal. Geom. **31**, 363–373 (2007)
29. L.J. Alías, D. Impera, M. Rigoli, Spacelike hypersurfaces of constant higher order mean curvature in generalized Robertson-Walker spacetimes. Math. Proc. Cambridge Philos. Soc. **152**, 365–383 (2012)
30. L.J. Alías, A.G. Colares, H.F. de Lima, On the rigidity of complete spacelike hypersurfaces immersed in a generalized Robertson-Walker spacetime. Bull. Braz. Math. Soc. **44**, 195–217 (2013)
31. L.J. Alías, D. Impera, M. Rigoli, Hypersurfaces of constant higher order mean curvature in warped products. Trans. Am. Math. Soc. **365**, 591–621 (2013)
32. L.J. Alías, A.G. Colares, H.F. de Lima, Uniqueness of entire graphs in warped products. J. Math. Anal. Appl. **430**, 60–75 (2015)
33. L.J. Alías, P. Mastrolia, M. Rigoli, *Maximum Principles and Geometric Applications*. Springer Monographs in Mathematics (Springer, Cham, 2016), xvii+570pp.
34. L.J. Alías, V. Cánovas, A.G. Colares, Marginally trapped submanifolds in generalized Robertson-Walker Spacetimes. Gen. Relativ. Gravit. **49**, 1–23 (2017)
35. L.J. Alías, A. Caminha, F.Y. do Nascimento, A maximum principle at infinity with applications to geometric vector fields. J. Math. Anal. Appl. **474**, 242–247 (2019)
36. L.J. Alías, J.H. de Lira, M. Rigoli, Mean curvature flow solitons in the presence of conformal vector fields. J. Geom. Anal. **30**, 1466–1529 (2020)
37. L.J. Alías, A. Caminha, F.Y. do Nascimento, A maximum principle related to volume growth and applications. Ann. Mat. Pura Appl. **200**, 1637–1650 (2021)
38. H.V.Q. An, D.V. Cuong, N.T.M. Duyenb, D.T. Hieub, T.L. Nam, On entire f-maximal graphs in the Lorentzian product $\mathbb{G}^n \times \mathbb{R}_1$. J. Geom. Phys. **114**, 587–592 (2017)

39. H. Anciaux, Marginally trapped submanifolds in space forms with arbitrary signature. Pac. J. Math. **272**, 257–274 (2014)
40. H. Anciaux, N. Cipriani, Codimension two marginally trapped submanifolds in Robertson-Walker spacetimes. J. Geom. Phys. **88**, 105–112 (2015)
41. H. Anciaux, Y. Godoy, Marginally trapped submanifolds in Lorentzian space forms and in the Lorentzian product of a space form by the real line. J. Math. Phys. **56**, 023502, 12pp. (2015)
42. L. Andersson, M. Mars, W. Simon, Local existence of dynamical and trapping horizons. Phys. Rev. Lett. **95**, 111102 (2005)
43. L. Andersson, M. Eichmair, J. Metzger, Jang's equation and its applications to marginally trapped surfaces, in *Complex Analysis and Dynamical Systems IV: Part 2. General Relativity, Geometry, and PDE*, Contemporary Mathematics, vol. 554 (AMS and Bar-Ilan, Providence, 2011)
44. R. Antonia, G. Molica Bisci, H.F. de Lima, M.S. Santos, Rigidity and nonexistence of complete hypersurfaces via Liouville type results and other maximum principles, with applications to entire graphs. Asymp. Anal. **135**, 363–398 (2023)
45. C.P. Aquino, H.F. de Lima, On the rigidity of constant mean curvature complete vertical graphs in warped products. Differ. Geom. Appl. **29**, 590–506 (2011)
46. C.P. Aquino, H.F. de Lima, On the Gauss map of complete CMC hypersurfaces in the hyperbolic space. J. Math. Anal. Appl. **386**, 862–869 (2012)
47. C.P. Aquino, H.F. de Lima, On the unicity of complete hypersurfaces immersed in a semi-riemannian warped product. J. Geom. Anal. **24**, 1126–1143 (2014)
48. C.P. Aquino, H.F. de Lima, E.A. Lima, Jr., Complete CMC spacelike hypersurfaces immersed in a Lorentzian product space. Arch. Math. **104**, 577–587 (2015)
49. C.P. Aquino, J.G. Araújo, H.F. de Lima, Rigidity of complete hypersurfaces in warped product spaces via higher order mean curvatures. Beitr. Algebra Geom. **57**, 391–405 (2016)
50. C.P. Aquino, H.I. Baltazar, H.F. de Lima, A new Calabi-Bernstein type result in spatially closed generalized Robertson-Walker spacetimes. Milan J. Math. **85**, 235–245 (2017)
51. C.P. Aquino, H.I. Baltazar, H.F. de Lima, New Calabi-Bernstein type results in Lorentzian product spaces with density. Nonl. Anal. **197**, 111855 (2020)
52. J.G. Araújo, H.F. de Lima, M.A.L. Velásquez, Submanifolds immersed in a warped product: rigidity and nonexistence. Proc. Am. Math. Soc. **147**, 811–821 (2019)
53. J.G. Araújo, H.F. de Lima, W.F. Gomes, M.A.L. Velásquez, Submanifolds immersed in a warped product with density. Bull. Belg. Math. Soc. Simon Stevin **27**, 683–696 (2020)
54. J.G. Araújo, H.F. de Lima, E.A. Lima, Jr., M.S. Santos, On the geometry of submanifolds in certain warped products. Int. J. Math. **32**, 2150044 (2021)
55. J.G. Araújo, H.F. de Lima, W.F. Gomes, Rigidity of hypersurfaces and Moser-Bernstein type results in certain warped products, with applications to pseudo-hyperbolic spaces. Aequat. Math. **96**, 1159–1177 (2022)
56. J.G. Araújo, H.F. de Lima, W.F. Gomes, On the rigidity of mean curvature flow solitons in certain semi-Riemannian warped products. Kodai Math. J. **46**, 62–74 (2023)
57. J.G. Araújo, H.F. de Lima, W.F. Gomes, On the mean curvature flow solitons in Riemannian spaces endowed with a Killing vector field. Eur. J. Math. **10**, 11 (2024)
58. A. Ashtekar, B. Krishnan, Dynamical horizons and their properties. Phys. Rev. D **68**, 104030 (2003)
59. D. Bakry, M. Émery, Diffusions hypercontractives, in *Seminaire de Probabilites, XIX, 1983/84*. Lecture Notes in Mathematics, vol. 1123 (Springer, Berlin, 1985), pp. 177–206
60. J.L.M. Barbosa, A.G. Colares, Stability of hypersurfaces with constant r-mean curvature. Ann. Global Anal. Geom. **15**, 277–297 (1997)

61. J.L.M. Barbosa, M. do Carmo, On the size of a stable minimal surface in $\mathbb{R}^3$. Am. J. Math. **98**, 515–528 (1976)
62. J.L.M. Barbosa, M. do Carmo, Stability of hypersurfaces with constant mean curvature. Math. Z. **185**, 339–353 (1984)
63. J.L.M. Barbosa, V. Oliker, Spacelike hypersurfaces with constant mean curvature in Lorentz spaces. Mat. Contemp. **4**, 27–44 (1993)
64. J. Barbosa, M. do Carmo, J. Eschenburg, Stability of hypersurfaces with constant mean curvature. Math. Z. **197**, 123–138 (1988)
65. A.P. Barreto, F. Fontenele, Some remarks on the Pigola-Rigoli-Setti version of the Omori-Yau maximum principle. Bull. Australian Math. Soc. **89**, 337–342 (2014)
66. A. Barros, P. Sousa, Compact graphs over a sphere of constant second order mean curvature. Proc. Am. Math. Soc. **137**, 3105–3114 (2009)
67. A. Barros, A. Brasil, A. Caminha, Stability of spacelike hypersurfaces in foliated spaces. Differ. Geom. Appl. **26**, 357–365 (2008)
68. R. Bartnik, Existence of maximal surfaces in asymptotically flat spacetimes. Commun. Math. Phys. **94**, 155–175 (1984)
69. M. Batista, M.P. Cavalcante, J. Pyo, Some isoperimetric inequalities and eigenvalue estimates in weighted manifolds. J. Math. Anal. Appl. **419**, 617–626 (2014)
70. M. Batista, H.F. de Lima, F.R. dos Santos, On the classification of MOTS in the de Sitter space. Manuscript. Math. **162**, 159–169 (2020)
71. M. Batista, G. Molica Bisci, H.F. de Lima, Entire translating graphs in weighted product spaces: rigidity and nonexistence results. Differ. Geom. Appl. **83**, 101899 (2022)
72. M. Batista, G. Molica Bisci, H.F. de Lima, Spacelike translating solitons of the mean curvature flow in Lorentzian product spaces with density. Math. Eng. **5**, 1–18 (2023)
73. M. Batista, G. Molica Bisci, H.F. de Lima, W.F. Gomes, Solitons of the spacelike mean curvature flow in a generalized Robertson-Walker spacetime. New York J. Math. **29**, 554–579 (2023)
74. M. Batista, H.F. de Lima, W.F. Gomes, Rigidity of mean curvature flow solitons and uniqueness of solutions of the mean curvature flow soliton equation in certain warped products. Mediter. J. Math. **20**, 199 (2023)
75. M. Batista, G. Molica Bisci, H.F. de Lima, W.F. Gomes, Nonexistence of mean curvature flow solitons with polynomial volume growth immersed in certain semi-Riemannian warped products. Adv. Nonl. Anal. **13**, 20240034 (2024)
76. J.K. Beem, P.E. Ehrlich, K.L. Easley, *Global Lorentzian Geometry* (Marcel Dekker, New York, 1996)
77. S. Bernstein, Sur les surfaces d'efinies au moyen de leur courboure moyenne ou totale. Ann. Ec. Norm. Sup. **27**, 233–256 (1910)
78. S. Bernstein, Sur une théorème de géometrie et ses applications aux équations dérivées partielles du type elliptique. Commun. Soc. Math. Kharkov **15**, 38–45 (1915)
79. G.P. Bessa, S. Pigola, A.G. Setti, On the L^1-Liouville property of stochastically incomplete manifolds. Potential Anal. **39**, 313–324 (2013)
80. A.L. Besse, *Einstein Manifolds* (Springer, Berlin, 1987)
81. R.G. Bettiol, P. Piccione, G. Siciliano, *Equivariant Bifurcation in Geometric Variational Problems*. Progress in Nonlinear Differential Equations, Birkhäuser (Proceedings of the Workshop on Nonlinear Differential Equations) (João Pessoa, Brazil, 2012)
82. R.G. Bettiol, P. Piccione, G. Siciliano, Equivariant bifurcation in geometric variational problems. Progress Nonlinear Differ. Equ. Appl. **85**, 103–133 (2014)
83. R.G. Bettiol, P. Piccione, G. Siciliano, On the equivariant implicit function theorem with low regularity and applications to geometric variational problems. Proc. Edinb. Math. Soc. **58**, 53–80 (2015)

84. G.M. Bisci, J.H.H. de Lacerda, H.F. de Lima, M.A.L. Velásquez, On the higher order mean curvatures of spacelike hypersurfaces in pp-wave spacetimes. Discrete Cont. Dynam. Syst. Ser. S **16**, 3212–3257 (2023)
85. G.M. Bisci, H.F. de Lima, A.V.F. Leite, M.A.L. Velásquez, Uniqueness and nonexistence of spacelike translating solitons in GRW spacetimes. Discrete Cont. Dynam. Syst. Ser. S (2024). https://doi.org/10.3934/dcdss.2024173
86. R.L. Bishop, B. O'Neill, Manifolds of negative curvature. Trans. Am. Math. Soc. **145**, 1–49 (1969)
87. S. Bochner, Vector fields and Ricci curvature. Bull. Am. Math. Soc. **52**, 776–797 (1946)
88. E. Bombieri, E. de Giorgi, E. Giusti, Minimal cones and the Bernstein problem. Invent. Math. **7**, 243–268 (1969)
89. E. Bombieri, E. de Giorgi, M. Miranda, Una maggiorazione a priori relativa alle ipersuperfici minimali non parametriche. Arch. Ration. Mech. Anal. **32**, 255–267 (1969)
90. A. Borbély, A remark on the Omori-Yau maximum principle. Kuwait J. Sci. Eng. **39**, 45–56 (2012)
91. K. Brighton, A liouville-type theorem for smooth metric measure spaces. J. Geom. Anal. **23**, 562–570 (2013)
92. E. Calabi, Examples of Bernstein problems for some nonlinear equations. Proc. Symp. Pure Math. **15**, 223–230 (1970)
93. F. Camargo, A. Caminha, M. da Silva, H.F. de Lima, On the r-stability of spacelike hypersurfaces. J. Geom. Phys. **60**, 1402–1410 (2010)
94. F. Camargo, A. Caminha, H.F. de Lima, M. Velásquez, r-stable spacelike hypersurfaces in conformally stationary spacetimes. Kodai Math. J. **34**, 339–351 (2011)
95. F.E.C. Camargo, A. Caminha, H.F. de Lima, U. Parente, Generalized maximum principles and the rigidity of complete spacelike hypersurfaces. Math. Proc. Cambridge Philos. Soc. **153**, 541–556 (2012)
96. A. Caminha, A rigidity theorem for complete CMC hypersurfaces in Lorentz manifolds. Differ. Geom. Appl. **24**, 652–659 (2006)
97. A. Caminha, The geometry of closed conformal vector fields on Riemannian spaces. Bull. Braz. Math. Soc. **42**, 277–300 (2011)
98. A. Caminha, H.F de Lima, Complete vertical graphs with constant mean curvature in semi-Riemannian warped products. Bull. Belg. Math. Soc. Simon Stevin **16**, 91–105 (2009)
99. A. Caminha, H.F. de Lima, Complete spacelike hypersurfaces in conformally stationary Lorentz Manifolds. Gen. Relat. Grav. **41**, 173–189 (2009)
100. A. Cañnete, C. Rosales, Compact stable hypersurfaces with free boundary in convex solid cones with homogeneous densities. Cal. Var. Partial Differ. Equ. **51**, 887–913 (2014)
101. J.S. Case, Singularity theorems and the Lorentzian splitting theorem for the Bakry-Émery-Ricci tensor. J. Geom. Phys. **60**, 477–490 (2010)
102. J. Case, Y.J. Shu, G. Wei, Rigidity of quasi-Einstein metrics. Differ. Geom. Appl. **29**, 93–100 (2011)
103. K. Castro, C. Rosales, Free boundary stable hypersurfaces in manifolds with density and rigidity results. J. Geom. Phys. **79**, 14–28 (2014)
104. M.P. Cavalcante, H.F. de Lima, M.S. Santos, New Calabi-Bernstein type results in weighted generalized Robertson-Walker spacetimes. Acta Math. Hungar. **145**, 440–454 (2015)
105. M.P. Cavalcante, H.F. de Lima, M.S. Santos, On Bernstein-type properties of complete hypersurfaces in weighted warped products. Ann. Mat. **195**, 309–322 (2016)

106. N. Charalambous, Z. Lu, The L1 Liouville property on weighted manifolds. Contemp. Math. **653**, 65–80 (2015). *Complex Analysis and Dynamical Systems VI: Part 1: PDE, Differential Geometry, Radon Transform*, ed. by M. L. Agranovsky, M. Ben-Artzi, G. Galloway, L. Karp, D. Khavinson, S. Reich, G. Weinstein, L. Zalcman
107. I. Chavel, *Eigenvalues in Riemannian Geometry* (Academic Press, Cambridge, 1984)
108. B.Y. Chen, Classification of marginally trapped Lorentzian flat surfaces in $\mathbb{E}_2^4$ and its application to biharmonic surfaces. J. Math. Anal. Appl. **340**, 861–875 (2008)
109. B.Y. Chen, *Differential Geometry of Warped Product Manifolds and Submanifolds* (World Scientific, New Jersey, 2017)
110. B.Y. Chen, J. Van der Veken, Marginally trapped surfaces in Lorentzian space forms with positive relative nullity. Class. Quant. Grav. **24**, 551–563 (2007)
111. X. Cheng, H. Rosenberg, Embedded positive constant r-mean curvature hypersurfaces in $M^m \times \mathbb{R}$. An. Acad. Bras. Cienc. **77**, 183–199 (2005)
112. S.Y. Cheng, S.T. Yau, Maximal spacelike hypersurfaces in the Lorentz-Minkowski space. Ann. Math. **104**, 407–419 (1976)
113. S.Y. Cheng, S.T. Yau, Hypersurfaces with constant scalar curvature. Math. Ann. **225**, 195–204 (1977)
114. X. Cheng, T. Mejia, D. Zhou, Stability and compactness for complete f-minimal surfaces. Trans. Am. Math. Soc. **367**, 4041–4059 (2015)
115. S.S. Chern, On the curvatures of a piece of hypersurfaces in Euclidean space. Abh. Math. Semin. Univ. Hamb. **29**, 77–91 (1965)
116. S.S. Chern, On the differential geometry of a piece of a submanifold in Euclidean space, in *Proceedings of US-Japan Seminar in Differential Geometry (Kyoto, 1965)* (Nippon Hyoronsha, Tokyo, 1966), pp. 17–21
117. S.S. Chern, Simple proofs of two theorems on minimal surfaces. Enseign. Math. **15**, 53–61 (1969)
118. Y. Choquet-Bruhat, J. York, The cauchy problem, in *General Relativity and Gravitation*, ed. by A. Held (Plenum Press, New York, 1980)
119. P.T. Chruściel, G.J. Galloway, D. Solis, Topological censorship for Kaluza-Klein space-times. Ann. Henri Poincaré **10**, 893–912 (2009)
120. A.G. Colares, H.F. de Lima, Space-like hypersurfaces with positive constant r-mean curvature in Lorentzian product spaces. Gen. Relativ. Gravit. **40**, 2131–2147 (2008)
121. A.G. Colares, H.F. de Lima, On the rigidity of spacelike hypersurfaces immersed in the steady state space $\mathcal{H}^{n+1}$. Publ. Math. Debrecen **81**, 103–119 (2012)
122. A.G. Colares, E.L. de Lima, H.F. de Lima, Generalized linear weingarten spacelike hypersurfaces in GRW spacetimes: height estimates and half-space theorems. Mediterr. J. Math. **18**, 264 (2021)
123. G. Colombo, L. Mari, M. Rigoli, Remarks on mean curvature flow solitons in warped products. Disc. Cont. Dynam. Syst. Ser. S **13**, 1957–1991 (2020)
124. J. Corvino, Scalar curvature deformation and a gluing construction for the Einstein constraint equations. Commun. Math. Phys. **214**, 137–189 (2000)
125. J. Corvino, D. Pollack, Scalar curvature and the einstein constraint equations, in *Surveys in Geometric Analysis and Relativity*. Advanced Lectures in Mathematics, ed. by H.L. Bray, W.P. Minicozzi II, vol. XX (International Press, Vienna, 2011), pp. 145–188
126. J. Corvino, R.M. Schoen, On the asymptotics for the vacuum Einstein constraint equations. J. Differ. Geom. **73**, 185–217 (2006)
127. A.W. Cunha, E.L. de Lima, H.F. de Lima, E.A. Lima, Jr., A.A. Medeiros, Bernstein type properties of two-sided hypersurfaces immersed in a Killing warped product. Stud. Math. **233**, 183–196 (2016)

128. A.W. Cunha, H.F. de Lima, E.A. Lima, Jr., M.S. Santos, Weakly trapped submanifolds immersed in generalized Robertson-Walker spacetimes. J. Math. Anal. Appl. **484**, 123734 (2020)
129. J.F. da Silva, H.F. de Lima, M.A. Velásquez, The stability of hypersurfaces revisited. Monatsh. Math. **179**, 293–303 (2016)
130. M. Dajczer, J.H. de Lira, Conformal Killing graphs with prescribed mean curvature. J. Geom. Anal. **22**, 780–799 (2012)
131. M. Dajczer, J.H. de Lira, Entire bounded constant mean curvature Killing graphs. J. Math. Pures Appl. **103**, 219–227 (2015)
132. M. Dajczer, J.H. de Lira, Entire unbounded constant mean curvature Killing graphs. Bull. Braz. Math. Soc. **48**, 187–198 (2017)
133. M. Dajczer, R. Tojeiro, *Submanifolds Theory, Beyond an Introduction*. Universitext (Springer, New York, 2019)
134. M. Dajczer, P. Hinojosa, J.H. de Lira, Killing graphs with prescribed mean curvature. Calc. Var. Partial Differ. Equ. **33**, 231–248 (2008)
135. E. de Giorgi, Una estensione del teorema di Bernstein. Ann. Sc. Norm. Super. Pisa Cl. Sci. **19**, 79–85 (1965)
136. H.F. de Lima, A sharp estimate for compact spacelike hypersurfaces with constant r-mean curvature in the Lorentz-Minkowski space and application. Differ. Geom. Appl. **26**, 445–455 (2008)
137. H.F. de Lima, On Bernstein-type properties of complete spacelike hypersurfaces immersed in a generalized Robertson-Walker spacetime. J. Geom. **103**, 219–229 (2012)
138. H.F. de Lima, E.A. Lima, Jr., Generalized maximum principles and the unicity of complete spacelike hypersurfaces immersed in a Lorentzian product space. Beitr. Algebra Geom. **55**, 59–75 (2014)
139. E.L. de Lima, H.F. de Lima, Height estimates and topology at infinity of hypersurfaces immersed in a certain class of warped products. Aequat. Math. **92**, 737–761 (2018)
140. E.L. de Lima, H.F. de Lima, Height estimates and half-space theorems for hypersurfaces in product spaces of type $\mathbb{R} \times M^n$. Ann. Acad. Bras. Cienc. **93**, e20190329 (2021)
141. H.F. de Lima, U.L. Parente, A Bernstein type theorem in $\mathbb{R} \times \mathbb{H}^n$. Bull. Brazilian Math. Soc. **43**, 17–26 (2012)
142. H.F. de Lima, U.L. Parente, On the geometry of maximal spacelike hypersurfaces in generalized Robertson-Walker spacetimes. Ann. Mat. Pura Appl. **192**, 649–663 (2013)
143. H.F. de Lima, M.S. Santos, Height estimates and half-space type theorems in weighted product spaces with nonnegative Bakry-Émery-Ricci curvature. Ann. Univ. Ferrara **63**, 323–332 (2017)
144. H.F. de Lima, M.A.L. Velásquez, Uniqueness of complete spacelike hypersurfaces via their higher order mean curvatures in a conformally stationary spacetime. Math. Nachr. **287**, 1223–1240 (2014)
145. H.F. de Lima, E.A. Lima, Jr., U. Parente, Hypersurfaces with prescribed angle function. Pac. J. Math. **269**, 393–406 (2014)
146. L.L. de Lima, J.H. de Lira, P. Piccione, Bifurcation of clifford tori in Berger 3-spheres. Quart. J. Math. **65**, 1345–1362 (2014)
147. E.L. de Lima, H.F. de Lima, F.R. dos Santos, On the stability of f-maximal spacelike hypersurfaces in weighted generalized Robertson-Walker spacetimes. Bull. Pol. Acad. Sci. Math. **64**, 199–208 (2016)
148. H.F. de Lima, A.F. de Sousa, M.A.L. Velásquez, Strongly stable linear Weingarten hypersurfaces immersed in the hyperbolic space. Mediterr. J. Math. **13**, 2147–2160 (2016)
149. H.F. de Lima, A.M.S. Oliveira, M.S. Santos, Rigidity of complete spacelike hypersurfaces with constant weighted mean curvature. Beiträge Algebra Geom. **57**, 623–635 (2016)

150. E.L. de Lima, H.F. de Lima, E.A. Lima, Jr., A.A. Medeiros, Parabolicity and rigidity of spacelike hypersurfaces immersed in a Lorentzian Killing warped product. Comment. Math. Univ. Carolinae **58**, 183–196 (2017)
151. H.F. de Lima, A.M.S. Oliveira, M.S. Santos, Rigidity of entire graphs in weighted product spaces with nonnegative Bakry-Émery-Ricci tensor. Adv. Geom. **17**, 53–59 (2017)
152. H.F. de Lima, A.M. Oliveira, M.S. Santos, M.A.L. Velásquez, f-stability of spacelike hypersurfaces in weighted spacetimes. Acta Math. Hungar. **153**, 334–349 (2017)
153. H.F. de Lima, A.M. Oliveira, M.A.L. Velásquez, On the uniqueness of complete two-sided hypersurfaces immersed in a class of weighted warped products. J. Geom. Anal. **27**, 2278–2301 (2017)
154. E.L. de Lima, H.F. de Lima, C.P. Aquino, Sharp height estimate in Lorentz-Minkowski space revisited. Bull. Belg. Math. Soc. Simon Stevin **25**, 29–38 (2018)
155. E.L. de Lima, H.F. de Lima, F.R. dos Santos, On the stability and parabolicity of complete f-minimal hypersurfaces in weighted warped products. Results Math. **73**, 14 (2018)
156. E.L. de Lima, H.F. de Lima, E.A. Lima, Jr., A.A. Medeiros, Parabolicity and rigidity of two-sided hypersurfaces in Killing warped products. Beitr. Algebra Geom. **59**, 453–463 (2018)
157. H.F. de Lima, E.A. Lima, Jr., A.A. Medeiros, M.S. Santos, Liouville type results for two-sided hypersurfaces in weighted Killing warped products. Bull. Braz. Math. Soc. **49**, 43–55 (2018)
158. E.L. de Lima, H.F. de Lima, E.A. Lima, Jr., A.A. Medeiros, Constant mean curvature spacelike hypersurfaces in standard static spaces: rigidity and parabolicity. Hokkaido Math. J. **49**, 297–323 (2020)
159. E.L. de Lima, H.F. de Lima, A.F. Ramalho, M.A.L. Velásquez, Spacelike hypersurfaces immersed in weighted standard static spacetimes: uniqueness, nonexistence and stability. Result. Math. **75**, 76 (2020)
160. H.F. de Lima, A.F. Ramalho, M.A.L. Velásquez, Solutions to mean curvature equations in weighted standard static spacetimes. Electron. J. Differ. Equ. **2020**(83), 1–19 (2020)
161. H.F. de Lima, W.F. Gomes, M.S. Santos, M.A.L. Velásquez, On the geometry of spacelike mean curvature flow solitons immersed in a GRW spacetime. J. Australian Math. Soc. **116**, 221–256 (2024)
162. J.H.S. de Lira, F. Martín, Translating solitons in Riemannian products. J. Differ. Equ. **266**, 7780–7812 (2019)
163. M.P. do Carmo, *Riemannian Geometry* (Birkhäuser, Basel, 1992)
164. M.P. do Carmo, H.R. Lawson, Jr., On Alexandrov-Bernstein theorems in hyperbolic space. Duke Math. J. **50**, 995–1003 (1983)
165. M.P. do Carmo, C.K. Peng, Stable complete minimal surfaces in $\mathbb{R}^3$ are planes. Bull. Am. Math. Soc. **1**, 903–906 (1979)
166. D. Eardley, L. Smarr, Time functions in numerical relativity: marginally bound dust collapse. Phys. Rev. D **19**, 2239–2259 (1979)
167. R.S. Earp, Parabolic and hyperbolic screw motion in $\mathbb{H}^2 \times \mathbb{R}$. J. Aust. Math. Soc. **85**, 113–143 (2008)
168. K. Ecker, On mean curvature flow of spacelike hypersurfaces in asymptotically flat spacetime. J. Austral. Math. Soc. Ser. A **55**, 41–59 (1993)
169. K. Ecker, Interior estimates and longtime solutions for mean curvature flow of noncompact spacelike hypersurfaces in Minkowski space. J. Differ. Geom. **46**, 481–498 (1997)
170. K. Ecker, Mean curvature flow of spacelike hypersurfaces near null initial data. Commun. Anal. Geom. **11**, 181–205 (2003)
171. K. Ecker, G. Huisken, Parabolic methods for the construction of spacelike slices of prescribed mean curvature in cosmological spacetimes. Commun. Math. Phys. **135**, 595–613 (1991)

172. M.F. Elbert, Constant positive 2-mean curvature hypersurfaces. Illinois J. Math. **46**, 247–267 (2002)
173. M. Émery, *Stochastic Calculus on Manifolds* (Springer, Berlin, 1989)
174. J.M. Espinar, Gradient Schrödinger operators, manifolds with density and applications. J. Math. Anal. Appl. **455**, 1505–1528 (2017)
175. J.M. Espinar, J.A. Gálvez, H. Rosenberg, Complete surfaces with positive extrinsic curvature in product spaces. Comment. Math. Helv. **84**, 351–386 (2009)
176. F. Fang, X.D. Li, Z. Zhang, Two generalizations of Cheeger-Gromoll splitting theorem via Bakry-Émery Ricci curvature. Ann. Inst. Fourier **59**, 563–573 (2009)
177. A. Fischer, J. Marsden, V. Moncrief, The structure of the space of solutions of Einstein's equations. I. One Killing field. Ann. Inst. H. Poincare **33**, 147–194 (1980)
178. H. Flanders, Remark on mean curvature. J. Lond. Math. Soc. **41**, 364–366 (1966)
179. W.H. Fleming, On the oriented Plateau problem. Rend. Circ. Mat. Palermo **11**, 69–90 (1962)
180. J.L. Flores, S. Haesen, M. Ortega, New examples of marginally trapped surfaces and tubes in warped spacetimes. Class. Quant. Gravity **27**, 145021, 18pp. (2010)
181. F. Fontenele, S.L. Silva, A tangency principle and applications. Illinois J. Math. **45**, 213–228 (2001)
182. A.G. Freitas, H.F. de Lima, E.A. Lima, Jr., M.S. Santos, Weakly trapped submanifolds in standard static spacetimes. Ark. Mat. **57**, 317–332 (2019)
183. A.G. Freitas, H.F. de Lima, E.A. Lima, Jr., M.S. Santos, Submanifolds immersed in Riemannian spaces endowed with a Killing vector field: nonexistence and rigidity. Differ. Geom. Appl. **75**, 101714 (2021)
184. A.G. Freitas, H.F. de Lima, M.S. Santos, J.S. Sindeaux, Nonexistence and rigidity of spacelike mean curvature flow solitons immersed in a GRW spacetime. Ann. Glob. Anal. Geom. **63**, 2 (2023)
185. G.J. Galloway, E. Woolgar, Cosmological singularities in Bakry-Émery spacetimes. J. Geom. Phys. **86**, 359–369 (2014)
186. S. Gao, G. Li, C. Wu, Translating spacelike graphs by mean curvature flow with prescribed contact angle. Arch. Math. **103**, 499–508 (2014)
187. S.C. García-Martínez, J. Herrera, Rigidity and bifurcation results for CMC hypersurfaces in warped product spaces. J. Geom. Anal. **26**, 1186–1201 (2016)
188. S.C. García-Martínez, D. Impera, Height estimates and half-space theorems for spacelike hypersurfaces in generalized Robertson-Walker spacetimes. Differ. Geom. Appl. **32**, 46–67 (2014)
189. S.C. García-Martínez, D. Impera, M. Rigoli, A sharp height estimate for compact hypersurfaces with constant k-mean curvature in warped product spaces. Proc. Edinb. Math. Soc. **58**, 403–419 (2015)
190. L. Gärding, An inequality for hyperbolic polynomials. J. Math. Mech. **8**, 957–965 (1959)
191. R.E. Greene, H. Wu, *Function Theory on Manifolds which Possess a Pole*. Lecture Notes in Mathematics, vol. 699 (Springer, New York, 1979)
192. A. Grigor'yan, On the existence of positive fundamental solution of the Laplace equation on Riemannian manifolds (in Russian). Matem. Sbornik **128**, 354–363 (1985). Engl. transl. Math. USSR Sb. **56** (1987), 349–358
193. A.A. Grigor'yan, Stochastically complete manifolds and summable harmonic functions. Izv. Akad. Nauk SSSR Ser. Mat. **52**, 1102–1108 (1988); translation in Math. USSR-Izv. **33**, 425–432 (1989)
194. A. Grigor'yan, Analytic and geometric background of recurrence and non-explosion of the Brownian motion on Riemannian manifolds. Bull. Am. Math. Soc. **36**, 135–249 (1999)

195. A. Grigor'yan, Escape rate of Brownian motion on Riemannian manifolds. Appl. Anal. **71**(1–4), 63–89 (1999)
196. A. Grigor'yan, L. Saloff-Coste, Dirichlet heat-kernel in the exterior of a compact set. Commun. Pure Appl. Math. **55**, 93–133 (2002)
197. M. Gromov, Isoperimetry of waists and concentration of maps. Geom. Funct. Anal. **13**, 178–215 (2003)
198. G. Hardy, J.E. Littlewood, G. Pólya, *Inequalities*, 2nd edn. (Cambridge Mathematical Library, Cambridge, 1989)
199. A. Hatcher, *Algebraic Topology* (Cambridge University Press, Cambridge, 2002)
200. S.W. Hawking, G. Ellis, *The Large Scale Structure of Space-Time* (Cambridge University Press, Cambridge, 1973)
201. S.W. Hawking, R. Penrose, The singularities of gravitational collapse and cosmology. Proc. R. Soc. Lond. A **314**, 529–548 (1970)
202. E. Heinz, On the nonexistence of a surface of constant mean curvature with finite area and prescribed rectifiable boundary. Arch. Rational Mech. Anal. **35**, 249–252 (1969)
203. D.T. Hieu, T.L. Nam, Bernstein type theorem for entire weighted minimal graphs in $\mathbb{G}^n \times \mathbb{R}$. J. Geom. Phys. **81**, 87–91 (2014)
204. P.T. Ho, The structure of ϕ-stable minimal hypersurfaces in manifolds of nonnegative p-scalar curvature. Math. Ann. **348**, 319–332 (2010)
205. D. Hoffman, J. Lira, H. Rosenberg, Constant mean curvature surfaces in $M^2 \times \mathbb{R}$. Trans. Am. Math. Soc. **358**, 491–507 (2006)
206. E. Hopf, *Elementare Bemerkungen über die Lösungen partieller Differentialgleichungen zweiter Ordnung vom elliptischen Typus*. Sitzungsberichte Preussiche Akade mie Wissenschaften, Berlin (1927), pp. 147–152
207. E. Hopf, On S. Bernstein's theorem on surfaces $z(x, y)$ of nonpositive curvature. Proc. Am. Math. Soc. **1**, 80–85 (1950)
208. A. Huber, On subharmonic functions and differential geometry in the large. Comment. Math. Helv. **32**, 13–72 (1957)
209. G. Huisken, Flow by mean curvature convex surfaces into spheres. J. Differ. Geom. **20**, 237–266 (1984)
210. G. Huisken, S.T. Yau, Definition of center of mass for isolated physical system and unique foliations by stable spheres with constant curvature. Invent. Math. **124**, 281–311 (1996)
211. D. Impera, M. Rimoldi, Stability properties and topology at infinity of f-minimal hypersurfaces. Geom. Ded. **178**, 21–47 (2015)
212. D. Impera, S. Pigola, A.G. Setti, Potential theory for manifolds with boundary and applications to controlled mean curvature graphs. J. Reine Angew. Math. **733**, 121–159 (2017)
213. H. Jian, Translating solitons of mean curvature flow of noncompact spacelike hypersurfaces in Minkowski space. J. Differ. Equ. **220**, 147–162 (2006)
214. H. Ju, J. Lu, H. Jian, Translating solutions to mean curvature flow with a forcing term in Minkowski space. Commun. Pure Appl. Anal. **9**, 963–973 (2010)
215. M. Kanai, Rough isometries and combinatorial approximations of geometries of noncompact Riemannian manifolds. J. Math. Soc. Japan **37**, 391–413 (1985)
216. M. Kanai, Rough isometries and the parabolicity of Riemannian manifolds. J. Math. Soc. Japan **38**, 227–238 (1986)
217. D.S. Kim, Y.H. Kim, Compact Einstein warped product spaces with nonpositive scalar curvature. Proc. Am. Math. Soc. **131**, 2573–2576 (2003)
218. S. Kobayashi, K. Nomizu, *Foundations of Differential Geometry, Vol. II* (Interscience, New York, 1969)

219. N. Koiso, Hypersurfaces of Einstein manifolds. Ann. Sci. l'École. Normale Supérieure **14**, 433–443 (1981)
220. M. Koiso, B. Palmer, P. Piccione, Bifurcation and symmetry breaking of nodoids with fixed boundary. Adv. Calc. Var. **8**, 337–370 (2015)
221. M. Koiso, B. Palmer, P. Piccione, Stability and bifurcation for surfaces with constant mean curvature. J. Math. Soc. Japan **69**, 1519–1554 (2017)
222. N. Korevaar, R. Kusner, W.H. Meeks, III, B. Solomon, Constant mean curvature surfaces in hyperbolic space. Am. J. Math. **114**, 1–43 (1992)
223. M. Kriele, *Spacetime* (Springer, Berlin, 1999)
224. W. Kühnel, Conformal transformations between Einstein spaces, in *Conformal Geometry*, ed. by R.S. Kulkarni, U. Pinkall. Aspects of Mathematics, vol. E12 (Vieweg-Verlag, Braunschweig, 1988), pp. 105–146
225. B. Lambert, A note on the oblique derivative problem for graphical mean curvature flow in Minkowski space. Abh. Math. Semin. Univ. Hambg. **82**, 115–120 (2012)
226. B. Lambert, The perpendicular Neumann problem for mean curvature flow with a timelike cone boundary condition. Trans. Am. Math. Soc. **366**, 3373–3388 (2014)
227. B. Lambert, J.D. Lotay, Spacelike mean curvature flow. J. Geom. Anal. **31**, 1291–1359 (2021)
228. J.M. Latorre, A. Romero, Uniqueness of noncompact spacelike hypersurfaces of constant mean curvature in generalized robertson-walker spacetimes. Geom. Ded. **93**, 1–10 (2002)
229. P. Li, *Curvature and Function Theory on Riemannian Manifolds*. Surveys in Differential Geometry, vol. 2 (International Press, Hong Kong, 2000), pp. 375–432
230. G. Li, I. Salavessa, Mean curvature flow of spacelike graphs. Math. Z. **269**, 697–719 (2011)
231. A. Lichnerowicz, Variétés Riemanniennes à tenseur C non négatif. C. R. Acad. Sci. Paris Sér. A-B **271**, A650–A653 (1970)
232. A. Lichnerowicz, Variétés Kählériennes à première classe de Chern non negative et variétés Riemanniennes à courbure de Ricci généralisée non negative. J. Differ. Geom. **6**, 47–94 (1971)
233. G. Liu, Stable weighted minimal surfaces in manifolds with nonnegative Bakry-Émery Ricci tensor. Commun. Anal. Geom. **21**, 1061–1079 (2013)
234. R. López, Area Monotonicity for spacelike surfaces with constant mean curvature. J. Geom. Phys. **52**, 353–363 (2004)
235. M. Mars, J.M.M. Senovilla, Trapped surfaces and symmetries. Class. Quant. Gravity **20**, 293–300 (2003)
236. J.E. Marsden, F.J. Tipler, Maximal hypersurfaces and foliations of constant mean curvature in general relativity. Phys. Rep. **66**, 109–139 (1980)
237. M. McGonagle, J. Ross, The hyperplane is the only stable, smooth solution to the isoperimetric problem in Gaussian space. Geom. Dedicata **178**, 277–296 (2015)
238. W.H. Meeks, H. Rosenberg, Stable minimal surfaces in $M \times \mathbb{R}$. J. Differ. Geom. **68**, 515–534 (2004)
239. P. Miao, L.F. Tam, On the volume functional of compact manifolds with boundary with constant scalar curvature. Calc. Var. PDE **36**, 141–171 (2009)
240. J. Milnor, On deciding whether a surface is parabolic or hyperbolic. Am. Math. Month. **84**, 43–46 (1977)
241. S. Montiel, An integral inequality for compact spacelike hypersurfaces in the de Sitter space and applications to the case of constant mean curvature. Ind. Univ. Math. J. **37**, 909–917 (1988)
242. S. Montiel, Unicity of constant mean curvature hypersurfaces in some riemannian manifolds. Ind. Univ. Math. J. **48**, 711–748 (1999)
243. S. Montiel, Uniqueness of spacelike hypersurfaces of constant mean curvature in foliated spacetimes. Math. Ann. **314**, 529–553 (1999)

244. S. Montiel, Complete non-compact spacelike hypersurfaces of constant mean curvature in de Sitter spaces. J. Math. Soc. Jpn. **55 (4)**, 915–938 (2003)
245. F. Morgan, Manifolds with density. Not. Am. Math. Soc. **52**, 853–858 (2005)
246. J. Moser, On Harnack's theorem for elliptic differential equations. Commun. Pure Appl. Math. **14**, 577–591 (1961)
247. O. Munteanu, J. Wang, Smooth metric measure spaces with non-negative curvature. Commun. Anal. Geom. **19**, 451–486 (2011)
248. S. Nishikawa, On maximal spacetime hypersurfaces in a Lorentzian manifold. Nagoya Math. J. **95**, 117–124 (1984)
249. A.M.S. Oliveira, H.F. de Lima, On the existence of f-maximal spacelike hypersurfaces in certain weighted manifolds. Bull. Aust. Math. Soc. **96**, 317–325 (2017)
250. H. Omori, Isometric immersions of Riemannian manifolds. J. Math. Soc. Jpn. **19**, 205–214 (1967)
251. B. O'Neill, *Semi-Riemannian Geometry with Applications to Relativity* (Academic Press, London, 1983)
252. J.A.S. Pelegrín, M. Rigoli, Constant mean curvature spacelike hypersurfaces in spatially open GRW spacetimes. J. Geom. Anal. **29**, 3293–3307 (2018)
253. R. Penrose, Gravitational collapse and space-time singularities. Phys. Rev. Lett. **14**, 57 (1965)
254. S. Pigola, M. Rigoli, A.G. Setti, A remark on the maximum principle and stochastic completeness. Proc. Am. Math. Soc. **131**, 1283–1288 (2003)
255. S. Pigola, M. Rigoli, A.G. Setti, Vanishing theorems on Riemannian manifolds, and geometric applications. J. Funct. Anal. **229**, 424–461 (2005)
256. S. Pigola, M. Rigoli, A.G. Setti, Maximum principles on Riemannian manifolds and applications. Mem. Am. Math. Soc. **174**, 822 (2005)
257. T. Piran, Problems and solutions in numerical relativity. Ann. NY Acad. **375**, 1–14 (1981)
258. P. Pucci, J. Serrin, *The Strong Maximum Principle*. Progress in Nonlinear Differential Equations and their Applications, vol. 73 (Birkhauser, Basel, 2007), x+235pp.
259. M. Rimoldi, Rigidity results for Lichnerowicz Bakry-Émery Ricci tensors. Ph.D. Thesis, Università degli Studi di Milano, Milano (2011)
260. A. Romero, R.M. Rubio, J.J. Salamanca, Uniqueness of complete maximal hypersurfaces in spatially parabolic generalized Robertson-Walker spacetimes. Class. Quant. Gravity **30**, 1–13 (2013)
261. A. Romero, R.M. Rubio, J.J. Salamanca, Parabolicity of spacelike hypersurfaces in Generalized Robertson-Walker spacetimes. Applications to uniqueness results. Int. J. Geom. Methods Mod. Phys. **10**, 1360014, 8pp. (2013)
262. A. Romero, R.M. Rubio, J.J. Salamanca, A new approach for uniqueness of complete maximal hypersurfaces in spatially parabolic GRW spacetimes. J. Math. Anal. Appl. **419**, 355–372 (2014)
263. A. Romero, R.M. Rubio, J.J. Salamanca, Complete maximal hypersurfaces in certain spatially open generalized Robertson-Walker spacetimes. RACSAM **109**, 451–460 (2015)
264. A. Romero, R.M. Rubio, J.J. Salamanca, New examples of Moser-Bernstein problems for some nonlinear elliptic partial differential equations arising in geometry. Ann. Fennici Math. **46**, 781–794 (2021)
265. C. Rosales, A. Cañete, V. Bayle, F. Morgan, On the isoperimetric problem in Euclidean space with density. Calc. Var. PDE **31**, 27–46 (2008)
266. H. Rosenberg, Hypersurfaces of constant curvature in space forms. Bull. Sci. Math. **117**, 217–239 (1993)
267. H. Rosenberg, F. Schulze, J. Spruck, The half-space property and entire positive minimal graphs in $M \times \mathbb{R}$. J. Differ. Geom. **95**, 321–336 (2013)

268. R.K. Sachs, H. Wu, *General Relativity for Mathematicians*. Graduate Texts in Mathematics, vol. 48 (Springer, New York, 1977)
269. J.J. Salamanca, I.M.C. Salavessa, Uniqueness of ϕ-minimal hypersurfaces in warped product manifolds. J. Math. Anal. Appl. **422**, 1376–1389 (2015)
270. L. Saloff-Coste, *Aspects of Sobolev-Type Inequalities*. London Mathematical Society Lecture Note Series, vol. 289 (Cambridge University Press, Cambridge, 2002)
271. J.M.M. Senovilla, Trapped surfaces, horizons and exact solutions in higher dimensions. Class. Quant. Gravity **19**, 113 (2002)
272. R. Schoen, *Estimates for Stable Minimal Surfaces in Three Dimensional Manifolds*. Annals of Mathematics Studies, vol. 103 (Princeton University Press, Princeton, 1983)
273. R. Schoen, S.T. Yau, Proof of the positive mass theorem. I. Commun. Math. Phys. **65**, 45–76 (1979)
274. R. Schoen, S.T. Yau, Existence of incompressible minimal surfaces and the topology of three-dimensional manifolds with nonnegative scalar curvature. Ann. Math. **110**, 127–142 (1979)
275. R. Schoen, S.T. Yau, On the structure of manifolds with positive scalar curvature. Manuscript. Math. **28**, 159–183 (1979)
276. R. Schoen, S.T. Yau, Proof of the positive mass theorem. II. Commun. Math. Phys. **79**, 231–260 (1981)
277. J.A. Schouten, *Der Ricci-Kalkül* (Springer, Berlin, 1924)
278. J. Simons, Minimal varieties in Riemannian manifolds. Ann. Math. **88**, 62–105 (1968)
279. J. Smoller, G. Wasserman, Bifurcation and symmetry breaking. Invent. Math. **100**, 63–95 (1990)
280. D. Stroock, *An Introduction to the Analysis of Paths on a Riemannian Manifold*. Mathematical Surveys and Monographs, vol. 4 (American Matheamtical Society, Providence, 2000)
281. S.M. Stumbles, Hypersurfaces of constant mean curvature. Ann. Phys. **133**, 28–56 (1981)
282. Y. Tashiro, Complete Riemannian manifolds and some vector fields. Trans. Am. Math. Soc. **117**, 251–275 (1965)
283. A.E. Treibergs, Entire spacelike hypersurfaces of constant mean curvature in minkowski space. Invent. Math. **66**, 39–56 (1982)
284. M. Troyanov, Parabolicity of manifolds. Siberian Adv. Math. **9**, 125–150 (1999)
285. C. Udriște, *Convex Functions and Optimization Methods on Riemannian Manifolds* (Springer, Dordrecht, 1994)
286. M.A.L. Velásquez, A.F.A. Ramalho, On the bifurcation of spacelike hypersurfaces with constant mean curvature in spacetimes. Differ. Geom. Appl. **88**, 102000 (2023)
287. M.A. Velásquez, A.F. de Sousa, H.F. de Lima, On the stability of hypersurfaces in space forms. J. Math. Anal. Appl. **406**, 134–146 (2013)
288. M.A. Velásquez, H.F. de Lima, J.F. da Silva, A.M. Oliveira, A new stability notion of closed hypersurfaces in the hyperbolic space. J. Math. Soc. Jpn. **71**, 413–428 (2019)
289. M.A.L. Velásquez, H.F. de Lima, J.F. da Silva, A.M. Oliveira, Stable compact spacelike hypersurfaces in the de Sitter space as maxima of a linear combination of area and volume. Manuscript. Math. **159**, 229–245 (2019)
290. M.A.L. Velásquez, A.F.A. Ramalho, H.F. de Lima, J.F. da Silva, J.Q. Oliveira, Bifurcation and local rigidity of constant second mean curvature hypersurfaces in Riemannian warped products. Nonlinear. Anal. **197**, 111865 (2020)
291. M.A.L. Velásquez, H.F. de Lima, A.F.A. Ramalho, Local rigidity, bifurcation, and stability of H_f-hypersurfaces in weighted Killing warped products. Publ. Mat. **65**, 363–388 (2021)
292. M.A.L. Velásquez, A.F.A. Ramalho, H.F. de Lima, M.S. Santos, A.M.S. Oliveira, Conformal killing graphs in foliated Riemannian spaces with density: rigidity and stability. Comment. Math. Univ. Carolin. **62**, 175–200 (2021)

293. M. Vieira, Harmonic forms on manifolds with non-negative Bakry-Émery-Ricci curvature. Arch. Math. **101**, 581–590 (2013)
294. G. Wei, W. Wyllie, Comparison geometry for the Bakry-Émery Ricci tensor. J. Differ. Geom. **83**, 377–405 (2009)
295. Y.L. Xin, On the Gauss image of a spacelike hypersurface with constant mean curvature in Minkowski space. Comment. Math. Helv. **66**, 590–598 (1991)
296. Y.L. Xin, *Minimal Submanifolds and Related Topics* (World Scientific, Singapore, 2003)
297. S.T. Yau, Harmonic functions on complete Riemannian manifolds. Commun. Pure Appl. Math. **28**, 201–228 (1975)
298. S.T. Yau, Some function-theoretic properties of complete Riemannian manifolds and their applications to geometry. Ind. Univ. Math. J. **25**, 659–670 (1976)
299. L. Zeng, Eigenvalues of the drifting Laplacian on complete noncompact Riemannian manifolds. Nonlinear Anal. **141**, 1–15 (2016)
300. S. Zhang, Stability of spacelike hypersurfaces in de Sitter space. Proc. Am. Math. Soc. **143**, 851–857 (2015)

The manufacturer's authorised representative in the EU is Springer Nature Customer Service Centre GmbH, Europaplatz 3, 69115 Heidelberg, Germany. If you have any concerns regarding our products, please contact ProductSafety@springernature.com

Printed and bound by CPI Group (UK) Ltd, Croydon, CR0 4YY